Animal Welfare

Volume 20

The Animal Welfare series has been designed to help contribute towards a culture of respect for animals and their welfare by producing academic texts addressing how best to provide for the welfare of the animal species that are managed and cared for by humans. Books in the series do not provide a detailed blue-print for the management of each species, but they do describe and discuss the major welfare concerns, often in relation to the wild progenitors of the managed animals. Welfare has been considered in relation to animals' needs, concentrating on nutrition, behaviour, reproduction and the physical and social environment. Economic effects of animal welfare provision were also considered where relevant, as were key areas where further research is required.

More information about this series at http://www.springer.com/series/5675

Tore S. Kristiansen • Anders Fernö •
Michail A. Pavlidis • Hans van de Vis
Editors

The Welfare of Fish

Co-ordinating Editor: Marieke Cassia Gartner

Editors
Tore S. Kristiansen
Institute of Marine Research
Bergen, Norway

Anders Fernö
Department of Biology
University of Bergen
Bergen, Norway

Michail A. Pavlidis
Department of Biology
University of Crete
Heraklion, Greece

Hans van de Vis
IMARES
Wageningen University & Research Centre
Yerseke, The Netherlands

ISSN 1572-7408
Animal Welfare
ISBN 978-3-030-41677-5 ISBN 978-3-030-41675-1 (eBook)
https://doi.org/10.1007/978-3-030-41675-1

This Springer imprint is published by the registered company Springer Nature Switzerland AG.
The registered company address is: Gewerbestrasse 11, 6330 Cham, Switzerland

Animal Welfare Series Preface

Animal welfare is attracting increasing interest worldwide, especially in developed countries where the knowledge and resources are available to, at least potentially, provide better management systems for farm animals, as well as companion, zoo, laboratory, and performance animals. The key requirements for adequate food, water, a suitable environment, companionship, and health are important for animals kept for all of these purposes.

There has been increased attention given to animal welfare in the West in recent years. This derives largely from the fact that the relentless pursuit of financial reward and efficiency, to satisfy market demands, has led to the development of intensive animal management systems that challenge the conscience of many consumers in this part of the world, particularly in the farm and laboratory animal sectors. Livestock are the world's biggest land users, and the farmed animal population is increasing rapidly to meet the needs of an expanding human population. This results in a tendency to allocate fewer resources to each animal and to value individual animals less, for example in the case of farmed poultry where flocks of over twenty thousand birds are not uncommon. In these circumstances, the importance of each individual's welfare is diminished.

In developing countries, human survival is still a daily uncertainty, so that provision for animal welfare has to be balanced against human welfare. Animal welfare is usually a priority only if it supports the output of the animal, be it food, work, clothing, sport, or companionship. However, in many situations the welfare of animals is synonymous with the welfare of the humans who look after them, because happy, healthy animals will be able to assist humans best in their struggle for survival. In principle the welfare needs of both humans and animals can be provided for, in both developing and developed countries, if resources are properly husbanded. In reality, the inequitable division of the world's riches creates physical and psychological poverty for humans and animals alike in many parts of the world.

Increased attention to welfare issues is just as evident for zoo, companion, laboratory, sport, and wild animals. Of growing importance is the ethical management of breeding programmes, since genetic manipulation is now technically advanced, but there is less public tolerance of the breeding of extreme animals if it comes at the expense of animal welfare. The quest for producing novel genotypes has fascinated breeders for centuries. Dog and cat breeders have produced a variety of deformities that have adverse effects on their welfare, but nowadays the breeders are just as active in the laboratory, where the mouse is genetically manipulated with equally profound effects.

The intimate connection between animals and humans that was once so essential for good animal welfare is rare nowadays, having been superseded by technologically efficient production systems where animals on farms and in laboratories are tended by increasingly few humans in the drive to enhance labour efficiency. With today's busy lifestyles, companion animals too may suffer from reduced contact with humans, although their value in providing companionship, particularly for certain groups such as the elderly, is beginning to be recognised. Animal consumers also rarely have any contact with the animals that are kept for their benefit.

In this estranged, efficient world, people struggle to find the moral imperatives to determine the level of welfare that they should afford to animals within their charge. A few people, and in particular many companion animal owners, strive for what they believe to be the highest levels of welfare provision, while others, deliberately or through ignorance, keep animals in impoverished conditions in which their health and wellbeing can be extremely poor. Today's multiple moral codes for animal care and use are derived from a broad range of cultural influences, including media reports of animal abuse, guidelines on ethical consumption, and campaigning and lobbying groups.

This series has been designed to contribute towards a culture of respect for animals and their welfare by producing learned treatises about the provision for the welfare of the animal species that are managed and cared for by humans. The early species-focused books were not detailed management blueprints; rather they described and considered the major welfare concerns, often with reference to the behaviour of the wild progenitors of the managed animals. Welfare was specifically focused on animals' needs, concentrating on nutrition, behaviour, reproduction, and the physical and social environment. Economic effects of animal welfare provision were also considered where relevant, as were key areas where further research is required.

In this volume, the series again departs from a single species focus to address the welfare of fish. Editors Tore S. Kristiansen, Anders Fernö, Michail A. Pavlidis, and Hans van de Vis have drawn from their extensive research in this field to gather a large group of authors who consider the topic from a variety of angles. This is an emerging science, which hitherto has attracted little attention, so it is necessary to start with the basics, from defining the neurology of fish and the anatomy of the fish brain to asking such questions as "Do fish experience pain?" It is evident from the contributed chapters that there are serious welfare issues that should be a focus of immediate attention. Prominent among these are the following concerns: the effects on welfare of catch and release fishing and intensive fish farming and of anthropogenic environmental changes. The issue of how pet fish behave and how similar their behaviour is to their wild counterparts is addressed, and, similarly, how laboratory fish are treated. Considering that there are around 34,000 species of fish, and that fish are considered by a few not to be sentient, the book also describes in detail the learning capacity of fish, as well as other aspects of their cognition. Also included are details of the remarkable discoveries that fish have different personalities and a considerable memory. The book will undoubtedly become a standard reference work in this emerging area of animal welfare science, and it is hoped that it will stimulate a new determination to address the risks to welfare that are the focus of the book.

Atlanta, USA
Brisbane, Australia

Marieke Gartner
Clive Phillips

Preface

On a beautiful cold day in March I (Tore) am sailing along the coast of western Norway watching the endless ocean as it disappears into the west. Beneath the blue surface lies the world of fishes—a world completely unknown to most of us. When I am snorkelling or diving I always get the feeling that this is a completely new and different place, a world where I can float weightlessly over deep abysses surrounded by quite different kinds of organisms and landscapes. The multitude of fishes evolved to live in this environment while I obviously did not. While the mackerel seem to swim effortlessly at high speed, I must struggle not to be swept away by the water currents—and I have air for just a few more seconds!

Ever since my childhood, I have loved to lie on a landing stage in the harbour watching wrasses and other fish swimming below. Who are here today? How do they behave and relate to each other? Do they know who is who? For hours I could simply lie there lost in the underwater world. Something new could happen anytime! A big pollack might enter the scene, and suddenly the behaviour of the "locals" would change. Not to panic, but clearly paying more attention to the potential threat, while the pollack was obviously aware that it had been detected and that a sudden attack was useless when its prey were alert.

Since then, the lives of fishes have become an important part of the professional life of all of the editors, as professors and researchers, and "fish welfare", which was unknown when I was young, has become a widely used concept, at least in Europe. We have studied the welfare of farmed and wild fish, and we have carried out a lot of fundamental research on fish behaviour, cognition and stress physiology, and other welfare-related topics. We have learned a lot about fish and have discovered the enormous diversity and richness of the world of fishes, but we have also realised that so far we have visited only a few islands in our wide ocean of ignorance.

But now, we are near the end of our journey to create *The Welfare of Fish*—a journey that has led us to some unexpected places, both geographically and scientifically. The editors first met in the small village of Mochlos on Crete to discuss the contents and how we should approach this challenge issued by Springer. This ought to be an easy task, we thought, since together we know quite a lot about fish welfare, but when we revisited our "knowledge", we soon realised that the challenge was a greater challenge than we had expected. In the first place, there are around 34,000 fish species, which are probably much more different from each other than, for example, frogs and humans. Of these, we had some knowledge about only a few, while most of them were totally unknown to us. Then, even though fish welfare is a multidimensional topic, the most important aspect for us as scientists is to understand the subjective experience of fishes: How do their qualitative experiences enable them to cope with their world? We soon understood we needed help from scientists who knew more than us about the different topics involved, and we began to contact experts we knew from previous cooperation and from the literature—and almost all of them were willing to contribute.

We have not tried to tell a story in which all the pieces finally come together; each chapter can be read independently. The authors have different backgrounds, in addition to their own view of what "fish welfare" means, and we have not tried to arrive at a consensus. The book is not only about fish welfare and how we should behave towards the fish we interact with, but also an attempt to introduce the reader to the world they inhabit. We have chosen to tell some stories about who fish are and how and where they live, how they function, their cognitive capacities, and how they behave in and cope with their social and physical environment. We have also tried to get a glimpse of their brains and to speculate about how new knowledge and theories in neurobiology can offer us new insights and hypotheses about what it is like to be a fish (Fig. 1). So here comes a brief introduction to the following 21 chapters.

The history of the concept "fish welfare" is quite short, going back only a few decades in time. In Chap. 1, Tore S. Kristiansen and Marc Bracke briefly revisit the origins of the animal welfare movement, the various definitions of welfare, how

Fig. 1 In the Thalassokosmos Aquarium at Heraklion in Crete, I met these two rockfish face-to-face. We were staring at each other through the thin glass surface that separated our worlds, only a few inches away. They were obviously aware of each other and of me, and I felt that "someone" was watching me. They were obviously socially aware beings, not mere somethings. (Photo: Tore S. Kristiansen)

concern for the welfare of domestic animals also gradually began to include fish, and some of the controversies this concept has led to. At least in Europe, fish welfare now has become an important part of animal welfare legislation, and the relationship between the expanding aquaculture sector and the science of fish welfare has become increasingly important. Even though European legislation has brought fish within the moral circle, they are still treated differently and with less concern than land animals. Fishes' suffering can be studied by scientific methods, but just how much suffering is still a matter of ethics. In Chap. 2, Bernice Bovenkerk and Franck Meijboom introduce you to the different theories of animal ethics and discuss the ethical and moral concerns and controversies about how we treat fish.

In order to improve the welfare of fish in aquaculture, public and private aquaria, and experimental research, we need to know how fish live in their natural environments, and Chaps. 3–5 provide a background to the ecology and behaviour of fish and how they are shaped by their environment. In Chap. 3 Anders Fernö, Otte Bjelland, and Tore S. Kristiansen offer a glimpse of the extremely diverse world of fishes and describe how they have adapted to a variety of habitats with regard to life history traits, spatial dynamics, and social structures and examine how physiological, behavioural, and ecological traits affect the probability that an individual fish will survive and reproduce. The "personality" of a species may determine how well it adapts to farms, aquaria, or experimental tanks. In order to adapt, fish need to possess the relevant mechanisms and tools, but anthropogenic environmental changes may affect physiological and developmental processes and bring fish close to their levels of tolerance and thereby impair their welfare. In Chap. 4, Felicity Huntingford explores the fascinating behaviour of fish in the wild, focusing on their use of space, feeding, predator avoidance, aggression, and courtship. Her chapter discusses the mechanisms that control the expression of a behaviour, how it develops, the functions it serves, and its phylogenetic history. Cultured fish retain the natural behaviour and instincts of their wild counterparts and this has implications for their welfare; Huntingford suggests behaviourally informed ways of mitigating adverse effects. Chapter 5, by Victoria Braithwaite and Ida Ahlbeck Bergdahl, examines the effects of early life experience on behavioural development in captive fish species, focusing on zebrafish in research facilities and fish in hatcheries that are later released for restocking purposes. The rearing environments influence the kinds of adult phenotypes that are expressed, and environmental variability through physical and social enrichment could improve cognitive skills and the ability to respond to challenging situations. For fish reared in hatchery environments and later released, an appropriate behavioural repertoire improves post-release survival rates.

In Chaps. 6–9, we go to the fish brain and investigate what it enables a fish to experience and to do. In Chap. 6, Alexander and Kurt Kotrschal give an overview of the anatomy, functionality, and evolution of the fish nervous system. The brain does not at all look the same in all species, and fish has the largest range of brain anatomy of all vertebrates. The authors discuss the evolution of brain anatomy and brain plasticity. The high metabolic costs of brain tissue limit brain evolution, and relative brain size is positively associated with social and environmental complexity, feeding ecology, and type of parental care. Finally, they report on recent experiments on the evolutionary costs and benefits of brain size in guppies artificially selected for large and small brains. In Chap. 7, Anders Fernö, Ole Folkedal, Jonatan Nilsson, and Tore S. Kristiansen explore what takes place inside the fish brain, focusing on learning, cognition, and consciousness. All fish species studied so far can learn and remember, but not everything is learned equally well, and the potential costs of making the wrong decision seem to influence the number of related events needed to form an association. Although the brains of fish and mammals differ in many ways, some fish species have an impressive ability to do complex tasks, which suggests advanced cognitive abilities. However, cognitive capacity depends on the environmental and social complexity that a given species encounters in its natural environment and also differs between populations, coping styles, and sexes. We ought therefore to adopt an ecological perspective on the mental abilities of fish. How well fish cope with the farming situation depends on their behavioural flexibility and cognitive capacity. In Chap. 8, Ruud van den Bos discusses awareness in fish and other animals, including humans. This is one aspect of the moral question of how to treat fish. Awareness may be perceived as a limited mental working space consisting of mental states, such as feelings or cognitions. Different strategies have been adopted to determine whether fish have awareness and what they are aware of. Chap. 9, by Tore S. Kristiansen and Anders Fernö, presents a novel way to look at how the brain works. Recent significant advances in neuroscience turn the traditional picture of perception upside down and suggest that the brain is not reactive but predictive and that its main goal is to perform allostatic regulation of the body budget and activities. The brain continuously attempts to predict sensory inputs from the external world as well as proprioceptive signals of body movements and interoceptive signals from other internal processes, based on earlier experiences from similar contexts. The predictive brain paradigm could help us to better understand behaviour and physiology and improve fish welfare.

Whether fish feel pain has been one of the most hotly debated questions regarding fish welfare of recent years, since this has important implications for how we should treat fishes. In Chap. 10, Lynne Sneddon discusses the scientific knowledge and indices for pain perception in fish and attempts to answer the question: Can fish experience pain?

It has long been known that the physiological stress responses in fish are very similar to those in mammals. In Chap. 11, Angelico Madaro, Tore S. Kristiansen, and Michail A. Pavlidis take a new look at stress in fish and use the concept of allostasis as the basis of an understanding how well fish have adapted to changes in their environment. Their chapter reviews the concepts, neurobiology, physiology, and molecular biology that underlie the processes of allostasis and acute and chronic stress and take a look at the ontogeny of the stress response in some fish species. Fishes within the same group or species also display a wide range of individual variation in behaviour and responses to stress. In Chap. 12, Ida Beitnes Johansen, Erik Höglund, and Øyvind Øverli review how key components of a stress coping style (i.e. behaviour, physiology, neuroendocrinology, neuronal plasticity, and immunity) are subject to great individual and heritable variation, and how such specific trait characteristics can influence the welfare of fish.

In order to find out how well the fish cope in their farms, fish farmers need to monitor and assess the welfare and performance of their fish and the quality of the rearing environment and procedures. Moreover, in many countries fish are protected by the same animal welfare legislation as land animals, and laws and regulations require fish farmers to have sufficient competence, technology, and equipment to ensure their animals' welfare. However, since there are no well-established methods for assessing or documenting fish welfare, it is difficult for the fish farmer to know how to comply with the regulations and for animal welfare and for the Food Authorities to control or enforce them. In Chap. 13, Lars H. Stien, Marc Bracke, and Tore S. Kristiansen discuss the theoretical foundations of welfare assessment, how animal welfare is related to the fundamental needs of animals, what welfare indicators can be used, and challenges related to fish welfare assessment.

Global aquaculture has rapidly expanded over the last few decades. In 1996, the reported production of farmed fish was only 17 million tonnes but this had tripled to more than 54 million tonnes 20 years later in 2016. In Chap. 14, Hans van de Vis, Jelena Kolarevic, Lars H. Stien, Tore S. Kristiansen, Marien Gerritzen, Karin van de Braak, Wout Abbink, Bjørn-Steinar Sæther, and Chris Noble address some of the differing welfare challenges farmed fish face when reared under different production systems and being subject to differing handling operations. They also give an overview of system- and operation-specific threats, with examples from commonly farmed species in each rearing system and operation, and summarise some potential operational mitigation strategies.

Welfare problems are not limited to farmed fish. Billions of fish in public and private aquaria and experimental tanks are exposed to challenges that can impair their welfare. In Chap. 15, Thomas Torgersen investigates welfare problems in ornamental fish kept as display or hobby animals. Welfare problems can be divided into those that arise from stressful and damaging transient processes before the fish end up in the tank and those that arise from discrepancies between the requirements of the fish and the environment provided by their keeper. The death of some fish is a proxy for poor welfare among survivors, but even fish that are surviving and growing may experience poor welfare based on their subjective experience of the quality of life. A fish can adapt its physiology to a suboptimal environment by a process of physiological acclimation, by moving to another part of the tank, or by accepting its current situation, with each of these three ways of coping having positive and negative consequences for different species. Chapter 16, by Anne Christine Utne-Palm and Adrian Smith, covers fish used as laboratory animals. The huge diversity of fishes makes it difficult to adequately address species-specific needs. The authors provide an overview of the species of fish most often used in research and the legislation governing their use and describe efforts to improve welfare and provide guidelines for planning and carrying out experiments.

Chapters 17–19 deal with whether and how fish that are captured by fishing gear suffer from impaired welfare and what we can do about to mitigate their suffering. Chapter 17, by Mike Breen, Neil Anders, Odd-Børre Humborstad, Jonatan Nilsson,

Maria Tenningen, and Aud Vold, deals with commercial fisheries. There has been little research that specifically addresses catch welfare in commercial fisheries. A number of stressors have the potential to impair welfare during the catching process. These include confinement and crowding, removal from the water, and physical trauma during handling. Welfare can be improved for example by limiting the duration of fishing operations and reducing catch volumes, and ethical harvesting practices could be encouraged and incentivised. Chapter 18, by Odd-Børre Humborstad, Chris Noble, Bjørn-Steinar Sæther, Kjell Øivind Midling, and Michael Breen, investigates capture-based aquaculture that combines capture fisheries with aquaculture practices that keep the catch alive for storage or feeding. An intense focus on welfare is already present in capture-based aquaculture of cod. The authors discuss welfare challenges in relation to capture, transport, and live storage. Sorting out fish is important to reduce the risk of poor welfare during the aquaculture phase. In commercial settings on board vessels, rapid, robust, and user-friendly operational welfare indicators are needed, and behaviour in the sorting bin and reflex impairment are usually used as proxies for vitality. Chapter 19, by Keno Ferter, Steven Cooke, Odd-Børre Humborstad, Jonatan Nilsson, and Robert Arlinghaus, investigates welfare problems in recreational fishing. Catching fish that are not utilised for food raises special ethical problems. Angling and other recreational fishing practices inevitably have some negative impacts on fish welfare, and the authors focus on how to minimise these impacts by altering fishers' behaviour and practices.

Anthropogenic disturbances are a major consideration for the welfare of wild fish. Disturbances include any external forces that alter ecosystem structure, such as toxic chemical pollutants that may cause direct and/or indirect mortality or habitat alteration that compromises living spaces and resource availability. In Chap. 20, Kathryn Hassell, Luke Barrett, and Tim Dempster review the effects of anthropogenic pollution in aquatic environments on the welfare of fish, first by summarising known effects for specific classes of pollutants, and then through case studies that highlight such effects within an ecosystem through time.

This book clearly demonstrates that fully understanding the cognitive abilities of fish is not a straightforward matter, nor is the degree to which fish are conscious and can experience emotions. Even so, we cannot avoid making decisions that have significant impacts on fish welfare. In Chap. 21, the editors draw their own conclusions regarding what they have learnt from this book, express their views on fish welfare, and suggest what we should do even if we lack complete knowledge. Finally, we suggest new lines of research.

Bergen, Norway — Tore S. Kristiansen
Bergen, Norway — Anders Fernö
Heraklion, Greece — Michail A. Pavlidis
Yerseke, The Netherlands — Hans van de Vis

Contents

Chapter 1
A Brief Look into the Origins of Fish Welfare Science

Tore S. Kristiansen and Marc B. M. Bracke

Very little of the great cruelty shown by men can really be attributed to cruel instinct. Most of it comes from thoughtlessness or inherited habit. The roots of cruelty, therefore, are not so much strong as widespread. But the time must come when inhumanity protected by custom and thoughtlessness will succumb before humanity championed by thought. Let us work that this time may come.

Albert Schweitzer, Reverence for Life

Abstract Every year, humans kill or injure trillions of fishes in fisheries, recreational fishing, aquaculture, and through the destruction or contamination of their habitats. However, until recently fish welfare has been paid little attention. The recent, at least partial, inclusion of fish within the moral circle can be seen as a natural/logical consequence of the increased attention paid to animal welfare in general, and in particular to the welfare of farmed fish in the rapidly growing intensive fish farming industry. The concern for fish welfare was first raised by animal protection groups in the early 1990s, and by the end of that decade, fish welfare had started to receive attention from scientists, food authorities, politicians, and the aquaculture industry. After the turn of the millennium, fish welfare blossomed into a research topic and became a prioritized and integrated part of animal welfare legislation in Europe. This chapter tells the story about the rise of animal welfare as a topic of concern, and especially fish welfare science, including the controversy concerning pain and consciousness in fish.

Keywords Animal welfare · Five freedoms · Welfare definitions · Moral circle · Pain · Welfare research · Laws and regulations

---

T. S. Kristiansen (✉)
Institute of Marine Research, Bergen, Norway
e-mail: torek@hi.no

M. B. M. Bracke
Wageningen Livestock Research, Wageningen, The Netherlands
e-mail: Marc.Bracke@wur.nl

T. S. Kristiansen et al. (eds.), *The Welfare of Fish*, Animal Welfare 20,
https://doi.org/10.1007/978-3-030-41675-1_1

1.1 Introduction

Fishing is an ancient practice dating back at least 42,000 years (O'Connor et al. 2011), and the domestication of fish species began more than 4000 years ago (Beveridge and Little 2002). Humans have probably had a strong relationship to fish for several thousand years, but we know little about people's attitudes towards fish through time. A quote from the seventh century Chinese poet Du Fu (杜甫) expresses his compassion for fish and that others should be compassionate too:

> I see shining fish struggling within tight nets, while I hear orioles singing carefree tunes. Even trivial creatures know the difference between freedom and bondage. Sympathy and compassion should be but natural to the human heart
>
> Du Fu (712–770 AD)

Another example can be found in the earliest biography of Saint Francis of Assisi, commissioned by Pope Gregory IX and completed in 1230, which tells us that:

> He was inspired by the same fatherly affection towards fish. When they were caught, and he had the opportunity, he threw them back into the water alive, instructing them to take care not to be caught again.
>
> Thomas de Celano, First life of St. Francis

Maybe these men were exceptions to the rule, since until recently fishes have mostly been kept outside of the moral circle of animals that we consider to have interests or sentience (Lund et al. 2007). The recent, at least partial, inclusion of fish within the moral circle can be seen as a logical consequence of the increased attention being paid to animal welfare in general, and in particular to the welfare of farmed fish in the rapidly growing intensive fish farming industry.

Fish and fisheries are of great importance to human trade and welfare, and cultures, cities, and nations have been based on them (Hart and Reynolds 2002). Millions of people are directly or indirectly involved in catching and processing fish, which are a very important source of proteins, minerals, and unsaturated fatty acids for humans and their domestic pets and livestock (FAO 2016). Every year, humans kill or injure trillions of fishes in fisheries, recreational fishing, aquaculture, and through the destruction or contamination of their habitats (Lymbery 2002). However, until recently fish welfare has been paid little attention. Concern for fish welfare was first raised by animal protection groups in the early 1990s (Lymbery 1992) and during that decade the topic of "fish welfare" started to appear in a few scientific journals, at aquaculture conferences, and in calls for research funding. By the end of the 1990s, fish welfare had started to receive attention from scientists, food authorities, politicians, and the aquaculture industry. Both EU and national research funding institutions began prioritizing fish welfare, and various European authorities e.g. the European Commission and the European Food Safety Authority (EFSA), put fish welfare on the map. Thus, at least in Europe, farmed fish have started to enter the moral and legal circles, but in most parts of the world, fish are still lacking legal protection.

But even if wild fish theoretically enjoy the same legal protection as farmed fish and other vertebrates, in practice this is ignored also in Europe. Recreational fishing

Fig. 1.1 Young hunter. Fish are still not killed and treated like mammals

is a very popular activity performed by millions of people (Arlinghaus et al. 2007a, b; Cooke and Sneddon 2007). It is seen as a relaxing recreational activity, associated with positive experiences of nature, including excitement and the satisfaction of catching your own food. We even let small children catch fish and some play with live fish on land (Fig. 1.1). "Catch and release" fishing, where the aim is to experience the joy of catching the fish but not use it for food, has become increasingly popular (Arlinghaus et al. 2007b). In contrast, catching and killing most other vertebrate animals by hand is something most people find aversive and have never done. Fishing is the exception! In our experience, almost all of us become excited and happy when they catch a fish on a hook and line—perhaps releasing our old predatory instincts? When it comes to removing the hook and killing the struggling fish, we are reluctant to cut its throat the first time, but soon most of us get convinced by more experienced fishers that this is an acceptable and normal thing to do. In traditional fishing cultures there is usually little or no consideration for the suffering or welfare of fish, and, if anything, the primary management goal has been to optimize sustainable harvests. Accordingly, fish are "harvested" from the sea and the catches are quantified in terms of weight and not numbers. In this sense, fish tend to be treated more like vegetables than animals. However, the quote above from Albert Schweizer reminds us that long-lasting traditions do not necessarily mean that our actions are ethical and should continue.

During the past few decades aquaculture has been the fastest growing animal production industry in the world; in Norway, for example the value of farmed fish now surpasses by far the value of landings of wild fish (Torrissen et al. 2011). Whilst fisheries have stagnated world-wide and many fish stocks are overexploited, aquaculture is expected to continue to grow (FAO 2016).

With the introduction of intensive fish farming, we became responsible for the whole life cycle of the fish, and the farmer must care well for his animals if they are to survive, grow, and stay healthy (Chap. 14). We have, therefore, moved from a hunting culture to a culture of care, which also raises new ethical concerns and responsibilities about how we handle the fish and how well the fish fare through life. In this chapter, we present a brief history of how the relatively young fields of fish welfare science and legislation have emerged, and explore their concomitant challenges and controversies.

1.2 The Origin of Animal Welfare Science

The concern for the welfare of intensively produced mammals and birds prepared the ground for concern for fish welfare (Fraser 2008; Phillips 2009; Broom 2011). After the Second World War, efficient intensive livestock production with large numbers of animals living at high stocking densities became increasingly common in the United States and Europe. This also led to more public concern about animal welfare. It has been claimed that Ruth Harrison's book, "Animal Machines" published in 1964, initiated the animal welfare movement and the science of animal welfare. Harrison raised questions such as:

> How far have we the right to take our domination of the animal world? Have we the right to rob them of all pleasures in life to make more money more quickly out of their carcasses? Have we the right to treat living creatures solely as food-converting machines? (Harrison 1964, p. 37)

She argued for better standards for animal welfare and described vividly how animals were reared in "factory farms". The book resulted in a public outcry, and only 6 weeks after its publication the British government appointed a commission, chaired by Professor F.W. Rogers Brambell, "*to examine the conditions in which livestock are kept under intensive husbandry and advise whether standards ought to be set in the interest of their welfare, and if so what they should be.* (Brambell 1965, page 1)" After several farm visits and expert interviews they published their report in 1965: "... to *enquire into the Welfare of Animals Kept under Intensive Livestock Husbandry Systems*". It was the first systematic evaluation of animal welfare in intensive farming systems, and it included many recommendations for improvements (Brambell 1965). In an appendix to the report, we find an essay by Professor W.H. Thorpe of Cambridge University, on "The assessment of pain and stress in animals". He formulated important questions that formed the agenda for the study of animal welfare in the decades that followed (Thorpe 1965).

As a direct result of the Brambell Report, the UK government appointed a Farm Animal Welfare Advisory Committee (FAWAC; called the Farm Animal Welfare

Council (FAWC) from 1979). However, in the course of the following years few of the recommendations in the report or FAWAC were followed up (Broom 2011). In a 1979 press release, FAWC gave a list of five essential conditions that all farm animals should be provided with. This list was later refined and named the "Five freedoms" (Webster 2005) (Box 1.1). These have been widely acknowledged and adopted by professional groups and animal protection NGOs, as a framework for animal welfare assessment and legislation (Welfare Quality 2009).

In his essay on "The history of animal welfare science" Professor Donald Broom writes: "In the 1960s, the emphasis of discussions was on what people should do, i.e. on animal protection rather than on animal welfare. In the 1970s and early 1980s, the term animal welfare was used but not defined and not considered scientific by most scientists" (Broom 2011). During the 1970s, the animal welfare area consisted of two branches: an ethical and philosophical branch, which may be labelled "animal ethics", by and large initiated by the moral philosophers Peter Singer and Tom Regan (Singer 1975, 1981; Regan 1983). These philosophers discussed *inter alia* whether or not animals had moral rights, or intrinsic value, over and above their subjective experiences constituting welfare. Their writings inspired a range of animal welfare/animal protection initiatives, and NGOs and (often more extreme) animal rights groups. The other branch, which may be called welfare science, was more science and ethology based. Animal welfare scientists mostly came from university zoology departments (Broom 2011), but also from animal science groups at agricultural universities and some from veterinary faculties. Whilst animal ethicists have been concerned about moral/ethical (i.e. prescriptive) issues (concerning what is acceptable), such as questioning our moral "rights" to kill or use/exploit animals, animal welfare scientists have historically been concerned with the descriptive questions (about what is de facto the animal welfare situation). A problem for welfare science has been how to define and measure welfare. Since subjective experiences cannot be measured directly, welfare scientists have often adopted a pragmatic approach, for example by studying measurable biological parameters that were assumed to correlate with welfare as actually experienced by animals (Broom 2011; Lund et al. 2007). Some scientists have taken this one step further by more or less equating animal welfare with some measure of physiological functioning. A problem for such a definition of animal welfare is that it misses the relationship with the ethical questions that are associated with animal welfare and animal suffering (Dawkins 2006; Diggles et al. 2011; Torgersen et al. 2011; Mellor 2016). New knowledge about motivational systems, animal decision-making, and basic and behavioural needs has been contributing to a gradual change in the scientific views of animals from instinct-driven "automata" to goal-directed agents that have needs that can be satisfied or frustrated (Dawkins 1980, 1990; Broom 2011). New knowledge about stress physiology, behaviour, and health is, of course, relevant for animal welfare (Moberg and Mench 2000), even though at that time (the 1970s and 1980s) veterinarians were reluctant to talk about animal feelings (Broom 2011).

Box 1.1 The Five Freedoms

(1) freedom from thirst, hunger or malnutrition;
(2) appropriate comfort and shelter;
(3) prevention, or rapid diagnosis and treatment, of injury and disease;
(4) freedom to display most normal patterns of behaviour;
(5) freedom from fear.

In a press statement FAWC (1979) published five demands for farm animal welfare conditions (see list above). They have since been refined and labelled "the five freedoms" (Webster 2005). On its web page, FAWC states: "The welfare of an animal includes its physical and mental state and we consider that good animal welfare implies both fitness and a sense of well-being. Any animal kept by man, must at least be protected from unnecessary suffering. We believe that an animal's welfare, whether on farm, in transit, at market or at a place of slaughter should be considered in terms of 'five freedoms'".

The "five freedoms" are often said to originate from the Brambell report (e.g. Broom 2011), but those referred to were the following five very modest minimum requirements for the ability of calves to move their body in industrial housing systems: "the animals should at least have the ability to stand up, lie down, turn around, groom themselves and stretch their limbs". This has later been referred to as "Brambell's Five Freedoms", even though they were not labelled "five freedoms" in the Brambell report.

http://webarchive.nationalarchives.gov.uk/20121010012427/http://www.fawc.org.uk/freedoms.htm

1.2.1 *The Welfare Definition Problem*

"Animal welfare" is a concept used by people involved in the protection of animals, and most people probably agree that welfare is related to the quality of life as experienced by the individual animal (Bracke et al. 1999). In order to implement welfare standards and welfare monitoring schemes, we need to know how it can be measured. However, despite the growing importance of animal welfare and a common understanding of what is meant by the term, the concept of animal welfare has been surprisingly difficult to define scientifically, and there is still no consensus (Duncan 1996; Webster 2005; Fraser 2008; Lerner 2008; Phillips 2009; Broom 2011).

The Brambell report did not define welfare, but described it as follows:

> Welfare is a wide term that embraces both the physical and mental well-being of the animal. Any attempt to evaluate welfare therefore must take into account the scientific evidence available concerning the feelings of animals that can be derived from their structure and functions and also from their behaviour. (Brambell 1965 (p. 9), also cited in Duncan 2006)

This view is probably close to what most animal welfare scientists still believe today.

One of the first and widely used definitions of animal welfare was presented by Professor Donald Broom (1991): "*The welfare of an individual is its state as regards its attempts to cope with its environment.*", where coping means "*having control of mental and bodily stability and that welfare is poor when coping ability is low*". Broom's definition has been said to identify animal welfare with some sort of physiological functioning, but Broom later stressed that aversive feelings, such as pain and fear, and hedonic feelings, such as pleasure and comfort, are parts of an evolved coping strategy, and that feelings are essential parts of welfare (Broom 2011, see also Chap. 13).

There are three common ways of defining animal welfare: in terms of *physiological functioning*, of *natural living* and of *feelings* (Fraser 2008, 2009). An important controversy has centred around the questions of whether animal welfare should primarily refer to biological functioning or to feelings (Duncan 1996; Turnbull and Kadri 2007; Torgersen et al. 2011; Mellor 2016) and whether animal welfare should be restricted to conscious animals only. Broom (2011) held that all animals could fit in his definition, whilst being conscious was more relevant to deciding which/when animals should be protected. Many biologists have argued that feelings and subjective experiences are not accessible to scientific enquiry, and they have therefore proposed restricting welfare to physiological functioning, e.g. related to health and other measurable parameters (Arlinghaus et al. 2007a; Diggles et al. 2011). Animal rights activists have emphasized that a natural environment and natural species-specific behaviour should be important aspects or components of animal welfare, and have questioned the ethics of killing and restricting animals' freedom of expression of normal behaviour (Regan 1983). Besides, Singer's preference utilitarianism (greatest good for the greatest number) (Singer 1975, 1981), and Regan's rights views (animals are subjects of a life with inalienable rights), the field of animal ethics also encompasses the telos concept proposed by Bernard Rollin, referring to an animal's "nature" and its right to live according to its nature (Rollin 1989).

Some scientists have claimed that fishes lack the essential brain structures that enable them to consciously experience pain or other feelings (Rose 2002; Rose et al. 2012; Arlinghaus et al. 2007a; Key 2016). The consciousness issue has therefore persisted for much longer in relation to fish compared to other higher vertebrates (in which consciousness is nowadays generally accepted, Chap. 8). This "fish pain controversy" will be discussed in more detail below.

The main criticism of defining animal welfare in terms of feelings is that animals' subjective experiences are not accessible to us. However, this problem can be solved if we can use proxies like behaviour, appearance, and health indicators as correlates of welfare. To be able to experience pleasure and pain, an animal needs to have some kind of conscious qualitative experience and we are not attributing welfare to organisms believed to lack consciousness, e.g. fungi or plants. This is because if we agree that unconscious organisms, like bacteria, experience welfare, the concept loses its meaning, as well as its moral, political, and social implications (Torgersen et al. 2011).

1.3 Emerging Concerns for Fish Welfare

Fish have been a part of animal protection legislation for a long time, even before the concept "fish welfare" was used. The first *Cruelty to Animals Act*, which prohibits painful experiments on *all animals other than invertebrates*, was passed in the United Kingdom in 1876, and was followed by the *Protection of Animals Act* in 1911, which defined "*domestic animal*" and "captive animal" as any animal "*of whatsoever kind or species, and whether a quadrupled or not*", including birds, fish, and reptiles.

The first report that specifically concerned fish welfare was probably the "*Report of the panel of enquiry into shooting and angling*", commissioned by the RSPCA (Medway 1980). The Medway report concluded that: "*In the light of evidence . . .. , it is recommended that, where considerations of welfare are involved, all vertebrate animals (i.e. mammals, birds, reptiles, amphibians and fish) should be regarded as equally capable of suffering to some degree or another, without distinction between 'warm-blooded' and 'cold-blooded' members*" (p. 52). It suggested "*that every angler should review his appreciation of the sport in the light of evidence presented on the perception of pain. Panel members believe that many anglers are concerned to promote the welfare of fish and will welcome advice on methods of lessening the likelihood of suffering among fish (p.53)*". The report further discusses the welfare consequences of angling and gives several suggestions as to how to mitigate these, e.g. through the use of barbless hooks.

During the 1980s, the intensive farming of Atlantic salmon started to grow into a significant industry in Scotland and Norway, and by 1990 the industry was experiencing major problems of bacterial diseases and sea lice. In a 1992 report on "*The Welfare of Farmed Fish*", Peter Lymbery, on behalf of the NGO Compassion in World Farming, was the first that raised concern about the poor welfare conditions in the salmon farming industry, especially the slaughter process, and argued that urgent action was needed to stop the widespread suffering of farmed fish (Lymbery 1992). Two years later, the RSPCA also showed concern for fish welfare in a report written by Steve Kestin on "*Pain and stress in fish*" (Kestin 1994). This attention to fish welfare and pain raised ethical questions about human activities that harm fish. In response, the Angling Governing Bodies Liaison Group and the British Field Sports Society ordered a second opinion from Dr. T.G Pottinger. His "*Fish welfare literature review*" (Pottinger 1995) was intended to "*assess the current state of knowledge regarding two areas of key importance in fish welfare: physiological stress and pain perception, with particular reference to the relationship between angling practices and fish welfare*" (p. 7). By 1995 several studies had already been performed on the responses to stress, performance, and health of fish produced for aquaculture and stock enhancement. These studies showed that the stress physiology of fish was very similar to that of mammals (Pickering 1981; Schreck 1981, 1990; Barton and Iwama 1991). However, at that time there were very few anatomical, biochemical, or behavioural studies regarding pain in fish, and Pottinger concluded: "*There is no information available in the literature at present which provides firm evidence that fish perceive pain as mammals apparently do or, conversely, that they cannot perceive pain as mammals do. On balance, it seems unlikely that fish experience pain as understood by humans. The problem of assessing exactly what a fish perceives when exposed to stimuli considered to be noxious or unpleasant in human terms may prove to be intractable.*" (p. 5).

In 1996, the British Farm Animal Welfare Council (FAWC 1996) also addressed fish welfare in a "*Report on the welfare of farmed fish*", covering Atlantic salmon (*Salmo salar*), rainbow trout (*Oncorhynchus mykiss*), and trout (*Salmo trutta*), with "*brief comments on carp (Cyprinus carpio)* and *those species of wrasse which are used for parasite control during salmon farming*". A FAWC working group carried out an extensive consultation exercise, obtained oral and written evidence from experts in salmon and trout production, and carefully examined scientific data. A number of fish farms in the United Kingdom (and a few in Norway) were visited, a seminar was held with invited experts from industry and research institutes, and opinions were collected from animal protection societies. FAWC interpreted the available scientific information somewhat differently from Pottinger, and concluded: "*we do not know what fish feel but the evidence available makes it very likely that at least some aspects of pain are felt by fish. In addition to any effect of pain, injury to a fish results in poor welfare where there is impairment of function or increased susceptibility to disease*" (p. 2). This report also made recommendations for the fulfilment of needs related to rearing, husbandry practices, and slaughter.

1.4 Fish Welfare in Politics, Laws, and Regulations

During the 1990s politicians too became increasingly concerned about animal welfare. In the European Union, animals were granted the status of "sentient beings" through the Treaty of Amsterdam (European Parliament 1997). In 1998, the EU issued *Council Directive 98/58/EC concerning the protection of animals kept for farming purposes*, which also included fish. Article 3 stated that: *Member States shall make provision to ensure that the owners or keepers take all reasonable steps to ensure the welfare of animals under their care and to ensure that those animals are not caused any unnecessary pain, suffering or injury*. The directive was later implemented in national legislation, and animal welfare, including fish welfare, became a priority research topic funded by the European Commission.

The Holmenkollen Guidelines for Sustainable Industrial Fish Farming from the Norwegian National Committee for Research Ethics (Sundli 1999) were amongst the first international protocols to suggest that ethical principles aimed at ensuring the health and welfare of fish, including humane slaughter, should govern the aquaculture industry. The World Organisation for Animal Health (OIE) later identified animal welfare, including fish welfare, as a priority area in its Strategic Plan (2001–2005) (Lund et al. 2007). In 2005, the Council of Europe adopted a recommendation on the welfare of farmed fish and in 2008, the OIE adopted guidelines for fish welfare.

The European Food Safety Authority (EFSA) also played an active role. EFSA aims to gain an in-depth understanding of the factors affecting animal welfare in general, and to provide a science-based foundation for European policies and legislation. In 2004, EFSA's independent Panel on Animal Health and Welfare (AHAW) issued scientific opinions regarding the transport and stunning/killing of farmed fish (EFSA 2004), and in 2008–2009 the panel published eight "scientific opinions" on the welfare and husbandry systems of Atlantic salmon (*Salmo salar*), carp (*Cyprinus carpio*), European sea bass (*Dicentrarchus labrax*), gilthead sea bream (*Sparus aurata*), rainbow trout (*Oncorhynchus mykiss*), European eel (*Anguilla anguilla*) (EFSA 2008a, b, c, d, e), and stunning and killing of turbot (*Scophthalmus maximus*) and Atlantic bluefin tuna (*Thunnus thynnus*) (EFSA 2009a, b). In 2009, this was followed by a *General approach to fish welfare and to the concept of sentience in fish* (EFSA 2009c). EFSA's scientific opinions focus on helping risk managers to identify methods to reduce unnecessary pain, distress, and suffering, and to increase animal welfare wherever possible.

Fish welfare is also considered in national legislation and recommendations, e.g. the New Zealand Animal Welfare Act (1999), the Australian Capital Territory Animal Welfare Act (1992), the Queensland Government Animal Care and Protection Act (2001), the Norwegian Animal Protection Act (1974), and the Norwegian Animal Welfare Act (2009). The Norwegian legislation gives fish a level of protection that is similar to that of other vertebrates. These laws were also followed by a range of regulations that specify both functional and specific demands regarding the individual stages of fish production.

International organizations have also issued recommendations and guidelines regarding fish welfare, and codes of practice adopted by the industry include measures to safeguard fish welfare. The Code of Conduct issued by the Federation of European Aquaculture Producers (FEAP) built on these documents and had a strong focus on fish welfare. Other standards and certification schemes for sustainable aquaculture that include fish welfare have been issued by the Aquaculture Stewardship Council (ASC), GLOBALGAP aquaculture standard, Best Aquaculture Practices (a division of the Global Aquaculture Alliance), and the RSPCA Assured (Freedom Food) (see also Chap. 13).

1.5 The Rise of Fish Welfare Science

> Drawing on all branches of biology, including behavioural ecology and neuroscience, the science of animal welfare asks three big questions: Are animals conscious? How can we assess good and bad welfare in animals? How can we use science to improve animal welfare in practice?
>
> Marian Stamp Dawkins (2006, p. 77)

Before 2000, the concept of "fish welfare" was not used much by fish scientists. A search in Google Scholar for scientific publications containing the phrase "fish welfare" produces only 16 papers and reports published before 1990, and 51 in the period 1990–1999, most of which were not really about fish welfare. However, after the turn of the millennium, "fish welfare" gradually came to be used by more and more scientists working in aquaculture and fish farming. This is illustrated by a rise to 1120 "fish welfare" papers and reports between 2000 and 2009, and 4940 hits between 2010 and 2019.

However, even if not labelled "fish welfare", the rapidly growing fish farming sector, together with a long tradition and growing practice of rearing juvenile fish for stock enhancement and sea ranching, had already led to the publication of a substantial body of studies relevant to fish welfare, including several on topics like fish production and husbandry, stress physiology, nutrition, health, diseases, and vaccines (e.g. Pickering 1981; Schreck 1981, 1990; Pottinger 1995; Wendelaar Bonga 1997; Iwama et al. 1997). Ethologists have also been studying fish behaviour for decades, addressing welfare relevant topics such as sensory biology, learning and cognition, aggression, territorial competition, and reproductive behaviour (e.g. Huntingford and Toricelli 1993; Pitcher 1992; Godin 1997; Huntingford et al. 2006).

The sharper focus in the 1990s on animal welfare in general, created a need for more scientific knowledge about fish welfare, that led to more funding opportunities from national research funds, and fish biologists willingly turned to animal welfare science. For example, in 2001 a Norwegian Strategic Institute Programme was funded to develop animal welfare competence at the two main Norwegian fisheries research institutes and the Norwegian University of Life Sciences (Damsgård et al. 2006). During the following decade, several Norwegian scientists led national and EU-funded projects on fish welfare that focused on basic research on many topics such as stress tolerance, farming environment, welfare indicators and assessment, sterile fish, coping mechanisms, individual variation, genomics, and health (Cordis 2019).

In 2002, an influential "Briefing Paper" on fish welfare was published by the Fisheries Society of the British Isles (FSBI 2002). This became an agenda and framework for scientific studies of fish welfare. The paper examined suffering and pain in fish, human–fish interactions, responses to stressors, the assessment of welfare, and much more. The authors offered no clear definition of welfare, but considered physiological function, feelings, and a natural life as different aspects of the welfare concept. However, it did not offer an opinion regarding what is acceptable fish welfare. The authors of the FSBI paper also pointed to gaps in our understanding of the concept of fish welfare, what it means and how it might be measured. The single most important gap was a lack of understanding of the mental capabilities of fish and of whether and how measurable states (such as physical injury and physiological and behavioural responses to challenge) generate subjective states of well-being and suffering. An updated version of this briefing paper was published by the same authors, plus the ethicist Peter Sandøe. The new version also included a section on "science, ethics and welfare" (Huntingford et al. 2006). In the 2000s, several review papers were published that focused on various aspects of fish welfare such as pain and consciousness (Chandroo 2004; Chandroo et al. 2004; Broom 2007), the welfare of farmed fish (Conte 2004; Ashley 2007), psychological stress and welfare, (Galhardo and Oliveira 2009), animal ethics (Lund et al. 2007), welfare assessment (Håstein et al. 2005; Volpato 2009), and recreational fishing (Cooke and Sneddon 2007). In 2008, the first textbook on fish welfare was published by British fish welfare scientists (Branson 2008).

An important role in the establishment of a fish welfare science in Europe was the EU COST action network Cost 867 WELFISH (2006–2011). The network included more than 100 participating stakeholders from 26 countries, including Canada, New Zealand, and Australia (Van de Vis et al. 2012). Its main objectives were to improve our knowledge of fish welfare, to formulate a set of guidelines embodying a common and scientifically sound understanding of farmed fish welfare, and to construct a range of targeted operational welfare indicator protocols to be used by the industry. The COST Action focused on the five main farmed species in Europe: Atlantic salmon, rainbow trout, sea bass, sea bream, and carp.

1.5.1 The Fish Pain Controversy

In the course of the past few decades, the question "do fish experience pain?" has been receiving a growing amount of attention. This is an ongoing controversy in fish welfare science, probably since it is an essential topic for welfare concern and legislation. The accepted definition of human pain is "*An unpleasant sensory and emotional experience associated with actual or potential tissue damage, or described in terms of such damage*" (IASP 1979, p. 247). According to this definition, pain comprises both sensory and negative affective aspects, implying it is a conscious experience. A review paper by Rose (2002) claimed that fish lack the essential brain regions and the neural basis of consciousness and pain perception, "*making it untenable that they can experience pain*". Rose also claimed that "*Because the experience of fear, similar to pain, depends on cerebral cortical structures that are absent from fish brains, it is concluded that awareness of fear and conscious experiences is impossible for fishes*" (p. 1). If the conclusions drawn by this paper were generally accepted there would scarcely be any point in promoting fish welfare. However, Rose still had some concern for fish and accepted that fish display "*nonconscious neuroendocrine and physiological stress responses to noxious stimuli. Thus, avoidance of potentially injurious stress responses is an important issue in considerations about the welfare of fishes*". This would appear to leave room for a definition of fish welfare in terms of "physiological functioning".

The same year, however, Lynne Sneddon from the Roslin Institute in the United Kingdom published an influential study on nociception in the trigeminal nerve of the rainbow trout, *Oncorhynchus mykiss*. She documented the presence of A-delta and C-fibres that convey nociceptive information to the fish brain (Sneddon 2002). Subsequently, she showed that rainbow trout possess nociceptors that respond to mechanical pressure, high temperatures, and acetic acid, reporting greatly increased opercular beat rate and a delay in resuming feed intake after exposure to acetic acid or bee venom injected into the lip, compared to a saline injection (Sneddon et al.

2003a, b). Since then, empirical evidence and supportive arguments have been accumulating for the ability of fish to experience pain (Sneddon et al. 2003a, b, Chandroo 2004; Chandroo et al. 2004; Dunlop and Laming 2005; Dunlop et al. 2006; Sneddon 2006; Broom 2007; Brown 2015; Ashley et al. 2007; Braitwaite 2010; Sneddon et al. 2014, 2018; Woodruff 2017; see also this volume, Chap. 10). However, Rose and others have not been convinced and they have tried to refute the studies by Sneddon and others (Pen and Rose 2007) referring to their brain morphology argument (Rose 2007; Arlinghaus et al. 2007a; Diggles et al. 2011; Rose et al. 2012; Key 2015, 2016). However, neurobiologists have recently started to seriously question their logic and arguments (Merker 2016; see also many other responses to Key (2016) in Animal Sentience 3(1) and Chap. 10).

1.5.2 What Next?

In the course of the past decades, the "fish welfare" concept has become fully adopted in Europe by fish farmers, technology companies, animal advocates (NGOs), scientists, politicians, authorities, consumers, and even amongst some fishermen. The rest of the world is also following, and we can find "fish welfare" sessions in most aquaculture conferences all over the world. Many scientists are keen to include welfare measures in their field of science, perhaps also because fish welfare may increase opportunities for funding.

The growing importance of aquaculture and stronger legal protection of animals, in general, makes it reasonable to expect that concern for fish welfare will be growing in the near future. Modern aquaculture is a highly science based but still relatively young industry, and the need for more knowledge is evident. Fish welfare is a multifaceted science and consists of fields such as physiology and health, water chemistry and technology, cognitive sciences, and neurobiology, not to mention also philosophy and ethics. Fish, especially zebrafish, have become the most used research animals (Chap. 16), and fishes are also some of the most popular pets all over the world (Chap. 15), all of which points to growing awareness of and concern for fish welfare. The following chapters present a considerable amount of new knowledge about fish welfare-related topics, but they also make it clear that our overall level of knowledge about the basic welfare needs and physiological and behavioural traits and functioning of the multitude of fish species is still very limited.

As concern for fish welfare seems to be related to dwindling wild populations (Pauly et al. 1998), combined with an increase in intensive farming practices, we can now perhaps anticipate a similar pattern concerning cephalopods, crustaceans, and insects (Carere and Mather 2019; Mellor 2019). Growing concern about falling populations of some insect species and implications, for example for bird populations on farmlands, particularly pastures, together with the rise of intensive insect farming is likely to follow a similar pattern as has already taken place for common livestock species and now, apparently, for fish. These developments will pose challenging questions for science too. How are we to measure brain activity that may be related to consciousness, for example when we stun animals before slaughter? This is a challenge in fish, and will be even more so in "lower" animals like crustaceans and cephalopods, and in due course, insects will presumably raise the same question: Are they mere automata, with or without feelings (Mellor 2019)?

References

Arlinghaus R, Cooke SJ, Schwab A, Cowx IG (2007a) Fish welfare: a challenge to the feelings-based approach, with implications for recreational fishing. Fish Fish 8:57–71. https://doi.org/10.1111/j.1467-2979.2007.00233.x

Arlinghaus R, Cooke SJ, Lyman J, Policansky D, Schwab A, Suski C, Sutton SG, Thorstad EB (2007b) Understanding the complexity of catch-and-release in recreational fishing: an integrative synthesis of global knowledge from historical, ethical, social, and biological perspectives. Rev Fish Sci 15:75–167. https://doi.org/10.1080/10641260601149432

Ashley PJ (2007) Fish welfare: current issues in aquaculture. Appl Anim Behav Sci 104 (3–4):199–235

Ashley PJ, Sneddon LU, McCrohan CR (2007) Nociception in fish: stimulus-response properties of receptors on the head of trout *Oncorhynchus mykiss*. Brain Res 1166:47–54. https://doi.org/10.1016/j.brainres.2007.07.011

Barton BA, Iwama GK (1991) Physiological changes in fish from stress in aquaculture with emphasis on the response and effects of corticosteroids. Annu Rev Fish Dis 1:3–26

Beveridge MCM, Little DC (2002) History of aquaculture in traditional societies. In: Costa-Pierce BA (ed) Ecological aquaculture. Blackwell Science, Oxford, pp 3–29

Bracke MBM, Spruijt BM, Metz JHM (1999) Overall welfare reviewed. Part 3: welfare assessment based on needs and supported by expert opinion. Neth JAgric Sci 47:307–322

Brambell FWR (1965) Report of the technical committee to enquire into the welfare of animals kept under intensive livestock husbandry systems. Her Majesty's Stationery Office, London

Branson EJ (ed) (2008) Fish welfare. Blackwell Publishing, 300 p

Broom DM (1991) Animal welfare: concepts and measurement. J Anim Sci 69:4167–4175

Broom DM (2007) Cognitive ability and sentience: which aquatic animals should be protected? Dis Aquat Org 75:99–108

Broom DM (2011) A history of animal welfare science. Acta Biotheor 59:121–137. https://doi.org/10.1007/s10441-011-9123-3

Brown C (2015) Fish intelligence, sentience and ethics. Anim Cogn 18(1):1–17

Carere M, Mather J (2019) The welfare of invertebrate animals. Animal welfare book series, vol 18. Springer

Chandroo K (2004) Can fish suffer?: perspectives on sentience, pain, fear and stress. Appl Anim Behav Sci 86:225–250. https://doi.org/10.1016/j.applanim.2004.02.004

Chandroo KP, Yue S, Moccia RD (2004) An evaluation of current perspectives on consciousness and pain in fishes. Fish Fish 5:281–295. https://doi.org/10.1111/j.1467-2679.2004.00163.x

Conte F (2004) Stress and the welfare of cultured fish. Appl Anim Behav Sci 86:205–223. https://doi.org/10.1016/j.applanim.2004.02.003

Cooke SJ, Sneddon LU (2007) Animal welfare perspectives on recreational angling. Appl Anim Behav Sci 104:176–198. https://doi.org/10.1016/j.applanim.2006.09.002

Cordis (2019). https://cordis.europa.eu/search/result_en?q=fish+welfare

Damsgård B, Juell J, Braastad, BO (2006) Welfare in farmed fish Fiskeriforskning Report 5/2006

Dawkins MS (1980) Animal suffering: the science of animal welfare. Chapman and Hall, London

Dawkins MS (1990) From an animal's point of view: motivation, fitness and animal welfare. Behav Brain Sci 13:1–31

Dawkins MS (2006) A user's guide to animal welfare science. Trends Ecol Evol 21:77–82

Diggles BK, Cooke SJ, Rose JD, Sawynok W (2011) Ecology and welfare of aquatic animals in wild capture fisheries. Rev Fish Biol Fish 21:739–765. https://doi.org/10.1007/s11160-011-9206-x

Duncan IJH (1996) Animal welfare defined in terms of feelings. Acta Agric Scand Suppl 27:29–35

Duncan I (2006) The changing concept of animal sentience. Appl Anim Behav Sci 100:11–19

Dunlop R, Laming P (2005) Mechanoreceptive and nociceptive responses in the central nervous system of goldfish (*Carassius auratus*) and trout (*Oncorhynchus mykiss*). J Pain 6:561–568. https://doi.org/10.1016/j.jpain.2005.02.010

Dunlop R, Millsopp S, Laming P (2006) Avoidance learning in goldfish (Carassius auratus) and trout (Oncorhynchus mykiss) and implications for pain perception. Appl Anim Behav Sci 97:255–271. https://doi.org/10.1016/j.applanim.2005.06.018

EFSA (2004) Welfare aspects of the main systems of stunning and killing the main commercial species of animals. EFSA J 45:1–29

EFSA (2008a) Scientific Opinion of the Panel on Animal Health and Welfare on a request from the European Commission on animal welfare aspects of husbandry systems for farmed Atlantic salmon. EFSA J 736:1–31

EFSA (2008b) Scientific Opinion of the Panel on Animal Health and Welfare on a request from the European Commission on animal welfare aspects of husbandry systems for farmed common carp. EFSA J 843:1–28

EFSA (2008c) Scientific Opinion of the Panel on Animal Health and Welfare on a request from the European Commission on animal welfare aspects of husbandry systems for farmed European seabass and gilthead seabream. EFSA J 844:1–21

EFSA (2008d) Scientific Opinion of the Panel on Animal Health and Welfare on a request from the European Commission on animal welfare aspects of husbandry systems for farmed trout. EFSA J 796:1–22

EFSA (2008e) Scientific Opinion of the Panel on Animal Health and Welfare on a request from the European Commission on animal welfare aspects of husbandry systems for farmed European eel. EFSA J 809:1–17

EFSA (2009a) Scientific Opinion of the Panel on Animal Health and Welfare – Species-specific welfare aspects of the main systems of stunning and killing of farmed tuna. EFSA J 1072:1–53

EFSA (2009b) Scientific Opinion of the Panel on Animal Health and Welfare – Species-specific welfare aspects of the main systems of stunning and killing of farmed tuna. EFSA J 1073:1–34

EFSA (2009c) General approach to fish welfare and to the concept of sentience in fish. Scientific opinion of the panel on animal health and welfare. EFSA J 954:1–27

European Parliament (1997) Treaty of Amsterdam. www.europarl.europa.eu/topics/treaty/pdf/amst-en.pdf

FAO (2016) The state of world fisheries and aquaculture 2016. Contributing to food security and nutrition for all. Rome, 200 p. http://www.fao.org/aquaculture/en/

FAWC (1979) Press statement. http://www.fawc.org.uk/pdf/fivefreedoms1979.pdf

FAWC (1996) Report on the welfare of farmed fish. The Farm Animal Welfare Council, Surbiton, Surrey

Fraser D (2008) Understanding animal welfare. Acta Vet Scand 50(Suppl 1):S1

Fraser D (2009) Assessing animal welfare: different philosophies, different scientific approaches. Zoo Biol 28:507–518

FSBI (2002) Fish Welfare. Briefing Paper 2, Fisheries Society of the British Isles, Granta Information Systems, Sawston, Cambridge

Galhardo L, Oliveira RF (2009) Psychological stress and welfare in fish. ARBS Annu Rev Biomed Sci 11:1–20

Godin J-GJ (1997) Behavioural ecology of teleost fishes. Oxford University Press, 384 p

Harrison R (1964) Animal machines – the new factory farming industry. Vincent Stuart, London, 186 p

Hart PJB, Reynolds JD (2002) Handbook of fish biology and fisheries, vol 2. Wiley, 428 p

Håstein T, Scarfe AD, Lund VL (2005) Science-based assessment of welfare: aquatic animals. Rev Sci Tech 24:529–547

Huntingford FA, Toricelli P (1993) Behavioural ecology of fishes. Ettore Majorama Life Sciences Series, vol 11. Harwood Academic, Chur, 326 p

Huntingford FA, Adams C, Braithwaite VA, Kadri S, Pottinger TG, Sandøe P, Turnbull JF (2006) Current issues in fish welfare. J Fish Biol 68:332–372

IASP (International Association for the Study of Pain) (1979) Pain terms: a list with definitions and notes on usage. Pain 6:247–252

Iwama GK, Pickering AD, Sumpter JP, Schreck CB (1997) Fish stress and health in aquaculture. Cambridge University Press, Cambridge

Kestin SC (1994) Pain and stress in fish. Royal Society for the Prevention of Cruelty to Animals. Amended. RSPCA, Horsham, West Sussex, 36 p

Key B (2015) Fish do not feel pain and its implications for understanding phenomenal consciousness. Biol Philos 30:149–165. https://doi.org/10.1007/s10539-014-9469-4

Key B (2016) Why fish do not feel pain. Anim Sent 3(1). https://animalstudiesrepository.org/animsent/vol1/iss3/1/

Lerner H (2008) The concepts of health, well-being and welfare as applied to animals. A philosophical analysis of the concepts with regard to the differences between animals. Linköping studies in arts and science No. 438. Dissertations on Health and Society No. 13. Linköpings Universitet, Department of Medical and Health Sciences. Linköping 2008

Lund V, Mejdell CM, Röcklinsberg H, Anthony R, Håstein T (2007) Expanding the moral circle: farmed fish as objects of moral concern. Dis Aquat Org 75:109–118

Lymbery P (1992) The welfare of farmed fish. Compassion in World Farming. Petersfield, Hampshire, 23 p

Lymbery P (2002) In too deep—the welfare of intensively farmed fish. Compassion in World Farming, Petersfield

Medway L (1980) Report of the panel of inquiry into shooting and angling (1976–1979). Panel of Enquiry into Shooting and Angling, Horsham, 58 p

Mellor DJ (2016) Updating animal welfare thinking: moving beyond the "five freedoms" towards "A lifeworth living". Animals 6:21. https://doi.org/10.3390/ani6030021

Mellor DJ (2019) Opinion: welfare-aligned sentience: enhanced capacities to experience, interact, anticipate, choose and survive. Animals 9(7):440. https://doi.org/10.3390/ani9070440

Merker B (2016) Drawing the line on pain. Anim Sent 30:23. https://pdfs.semanticscholar.org/ef9d/a66d1fc0aae06ef22d43fc2784c28a3703f7.pdf

Moberg GP, Mench JA (eds) (2000) The biology of animal stress: basic principles and implications for animal welfare. CAB International, Wallingford. 384 p

O'Connor S, Ono R, Clarkson C (2011) Pelagic fishing at 42,000 years before the present and the maritime skills of modern humans. Science 334(6059):1117–1121

Pauly D, Christensen VV, Dalsgaard J, Froese R, Torres F Jr (1998) Fishing down marine food webs. Science 279(5352):860–863

Pen O, Rose JD (2007) Anthropomorphism and "mental welfare" of fishes. Dis Aquat Org 75:139–154

Phillips C (2009) The welfare of animals. The silent majority. Springer, 220 p

Pickering AD (ed) (1981) Stress and fish. Academic, London

Pitcher TJ (1992) Behaviour of teleost fishes, 2nd edn. Chapman and Hall, 717 p

Pottinger TG (1995) Fish welfare literature review. Institute of Fresh Water Ecology, IFE Report No. WI/T11063f7/1, 82 p. http://nora.nerc.ac.uk/id/eprint/7223/1/Fish_Welfare_Literature_Review_-_TG_Pottinger_-_1995.pdf

Regan T (1983) The case for animal rights. Routledge & Kegan Paul, London. 425 p

Rollin BE (1989) Studies in bioethics. The unheeded cry: animal consciousness, animal pain and science. Oxford University Press, New York, NY, 330 p

Rose JD (2002) The neurobehavioral nature of fishes and the question of awareness and pain. Rev Fish Sci 10:1–38

Rose JD (2007) Anthropomorphism and mental welfare of fishes. Dis Aquat Org 75:139–154
Rose JD, Arlinghaus R, Cooke SJ, Diggles BK, Sawynok W, Stevens ED, Wynne CDL (2012) Can fish really feel pain? Fish Fish:1–35. https://doi.org/10.1111/faf.12010
Schreck CB (1981) Stress and compensation in teleostean fishes: response to social and physical factors. In: Pickering AD (ed) Stress and fish. Academic, London, pp 295–321
Schreck CB (1990) Physiological, behavioural, and performance indicators of stress. Am Fish Soc Symp 8:29–37
Singer P (1975) Animal liberation. A new ethics for our treatments of animals. Harper Collins, New York, NY, p 311
Singer P (1981) The expanding circle: ethics and sociobiology. Farrar, Straus & Giroux, New York, 208 p. http://www.stafforini.com/docs/Singer%20-%20The%20expanding%20circle.pdf
Sneddon LU (2002) Anatomical and electrophysiological analysis of the trigeminal nerve in a teleost fish, *Oncorhynchus mykiss*. Neurosci Lett 319:167–171
Sneddon LU (2006) Ethics and welfare: pain perception in fish. Bull Eur Assoc Fish Pathol 26:7–10
Sneddon LU, Braithwaite VA, Gentle MJ (2003a) Do fishes have nociceptors? Evidence for the evolution of a vertebrate sensory system. Proc Biol Sci 270:1115–1121. https://doi.org/10.1098/rspb.2003.2349
Sneddon LU, Braithwaite VA, Gentle MJ (2003b) Novel object test: examining nociception and fear in the rainbow trout. J Pain 4(8):431–440
Sneddon LU, Elwood RW, Adamoc SA, Leach MC (2014) Review: defining and assessing animal pain. Anim Behav 97:201–212
Sneddon LU, Lopez-Luna K, Wolfenden DCC, Leach MC, Valentim AM, Steenbergen PJ, Bardine N, Currie AD, Broom D, Brown C (2018) Fish sentience denial: muddying the waters. Anim Sentience 3(21):1
Sundli A (1999) Holmenkollen guidelines for sustainable aquaculture (adopted 1998). In: Svennevig N, Reinertsen H, New M (eds) Sustainable aquaculture: food for the future? A.A. Balkema, Rotterdam, pp 343–347
Thorpe WH (1965) The assessment of pain and distress in animals. Appendix III in Report of the technical committee to enquire into the welfare of animals kept under intensive husbandry conditions, F.W.R. Brambell (chairman). H.M.S.O., London
Torgersen T, Bracke M, Kristiansen TS (2011) Reply to Diggles et al. (2011): Ecology and welfare of aquatic animals in wild capture fisheries. Rev Fish Biol Fish 21:767–769
Torrissen O, Olsen RE, Toresen R, Hemre GI, Tacon AGJ, Asche F, Hardy RW, Lall S (2011) Atlantic Salmon (*Salmo salar*): the "super-chicken" of the sea? Rev Fish Sci 19:257–278
Turnbull JF, Kadri S (2007) Safeguarding the many guises of farmed fish welfare. Dis Aquat Org 75:173–182
Van de Vis H, Kiessling A, Flik G, Mackenzie S (eds) (2012) Welfare of farmed fish in present and future production systems. Springer, Dordrecht. 302 p
Volpato GL (2009) Challenges in assessing fish welfare. ILAR J 50(4):329–337
Webster J (2005) Animal welfare limping towards eden. Blackwell Publishing, UFAW Animal Welfare Series. 283 p
Welfare Quality (2009) Assessment protocols for cattle, pigs and poultry; Welfare Quality Consortium, Lelystad, The Netherlands
Wendelaar Bonga SE (1997) The stress response in fish. Physiol Rev 77:591–625
Woodruff ML (2017) Consciousness in teleosts: there is something it feels like to be a fish. Anim Sentience 2(13):1

Chapter 2
Ethics and the Welfare of Fish

Bernice Bovenkerk and Franck Meijboom

Abstract To what extent fish can experience suffering and enjoyment is not just an empirical question, but one that also calls for ethical reflection. This is firstly, because animal welfare research is value laden and secondly, because the empirical evidence requires a normative framework in order to become action guiding in practices involving fish, such as aquaculture. In this chapter, we describe the role of ethics and different ethical theories that have been applied in animal ethics and that are relevant for discussions on fish welfare. We particularly focus on utilitarian, rights based, relational, and virtue ethical animal ethics theories. We furthermore argue that fish welfare is a term that combines moral norms and biological concepts. After all, when we implement fish welfare measures we have already made certain normative choices. We illustrate the integration between ethics and science in seven steps, from implementing fish welfare at the farm level, to weighing welfare against other values, defining and measuring welfare, to the questions of why welfare is morally relevant and what this means for the moral status of fish. We then consider the question of whether fish should be attributed to moral status and hence whether their welfare should be taken into account in our moral deliberations. However, not all moral concerns regarding our treatment of fish can be addressed by focussing on welfare. We discuss a number of concerns beyond welfare that need to be taken into consideration in a moral discussion on how to relate to fish: does the killing of fish constitute a moral harm? and how should we morally evaluate the process of domesticating fish in aquaculture? The chapter concludes by pointing out a number of moral issues in four practices involving fish: aquaculture, wild fisheries, experimentation, and recreation.

Keywords Animal ethics · Welfare · Moral status · Harm of death · Domestication

B. Bovenkerk (✉)
Philosophy Group, Wageningen University and Research, Wageningen, The Netherlands
e-mail: bernice.bovenkerk@wur.nl

F. Meijboom
Faculty of Veterinary Medicine, Department of Animals in Science and Society, Utrecht University, Utrecht, The Netherlands
e-mail: F.L.B.Meijboom@uu.nl

T. S. Kristiansen et al. (eds.), *The Welfare of Fish*, Animal Welfare 20,
https://doi.org/10.1007/978-3-030-41675-1_2

2.1 Introduction

From an ethical point of view, fish form an interesting case. They form a borderline case, between on the one hand mammals—about which broad consensus exists, both on the basis of common sense and scientific research, that they are sentient—and on the other hand other natural entities, such as rocks, about which we are certain that they are not sentient. While almost everyone now assumes that mammals experience pain, not everyone is convinced that fish do. This leads them to treat fish differently than other animals. Whether or not fish indeed suffer less than mammals is in the first place an empirical question and we need scientific research in the fields of neurophysiology, physiology, and ethology to answer it. At the same time, as this chapter will make clear, it is a question that cannot be answered without ethical reflection. This is, firstly, because scientific research is value laden (Longino 1990). Secondly, as we will show, this is because the empirical evidence requires a normative framework in order to become action guiding in practices involving fish, such as aquaculture.

We will start this chapter by briefly describing the role of ethics and different ethical frameworks, or theories, that have been applied in animal ethics and that are relevant for discussions on fish welfare. Next, we will address the question of what we mean by fish welfare. As we will argue, fish welfare is a term that combines moral norms and biological concepts. When we implement fish welfare measures we have already made certain normative choices. However, not all moral concerns regarding our treatment of fish can be addressed by focussing on welfare. We will discuss a number of concerns beyond welfare that need to be taken into consideration in a moral discussion on how to relate to fish. Finally, we will illustrate our arguments by briefly pointing out the moral aspects of a number of practices involving fish: aquaculture, wild fisheries, experimentation on fish, and recreational uses of fish.

2.1.1 Ethics Is Dynamic

Ethics is the systematic reflection on morality, i.e. the set of norms and values that a person or group considers to be important and action guiding. In daily life, we often answer moral questions and make moral decisions, for example while feeding animals and implicitly considering that one *should* care for animals. In other cases, there is much more unclarity or debate on what one should do. In these situations ethical reflection is important. Therefore, we need to understand the process of moral judgement formation.[1] In this process, the purpose of ethical theorizing is twofold:

[1]We will use the terms ‘moral’ and ‘ethical’ interchangeably.

Firstly, theories justify the basis of morality and thereby seek to answer questions such as 'why should we be moral?' and 'what is the goal of ethics?'. Is it, for example to enable peaceful coexistence in society or to protect the vulnerable? Secondly, theories aim to give guidance in practical moral problems or dilemmas by offering principles and values that help us make decisions about the right action to take or the good character to cultivate.

A distinction can be made between normative theories, which we will discuss below, and meta-ethical theories. The latter seeks to answer questions such as 'are moral judgments objective or subjective and are they universal or relative to culture?'. On a meta-ethical level, we embrace coherentist moral theories, which do not believe there is one ultimate foundation on which a theory must be built, but rather that a theory is valid when it achieves coherence between principles. In particular, we hold that forming a moral judgement goes by way of reaching a 'reflective equilibrium'—a balance between three 'pillars', namely our considered intuitions or moral emotions, morally relevant facts of the case at hand, and moral principles (Daniels 1979). When we deliberate about the right thing to do in a specific case, we need to move back and forth between these three pillars until we reach a balance. We often start with a moral intuition, for example that something is morally problematic about the case at hand—say, keeping a fish in a small round fishbowl. We then need to test our intuitions by squaring them with the facts of the case—are small round fishbowls bad for fish?—and relating them to moral principles—for example, a principle of respect for animal welfare. However, the principles themselves can be tested by our intuitions; if principles were to lead to very counter-intuitive implications, we have reason to consider whether it is necessary to refine or change our principles. By moving back and forth between these pillars we reach a considered moral judgement. Such a judgement has normative power and is action guiding, although it remains a temporary judgement.

From this perspective, ethics is dynamic and this means that our judgements can change, for example when new information comes to light, when we are confronted with new situations, or when ethical theories are refined after discussion between ethicists. Proponents of different normative theories may reach different conclusions on a specific case, because they have different decision criteria; in fact, they may even have different views on who belongs to the moral community in the first place, or in other words, on whose interests we need to take into account in our moral deliberations. However, there can be points of convergence between different moral theories as well; this will lead to well-established moral judgements that are widely shared.

2.1.2 Different Animal Ethics Theories

Normative theories provide a frame that answers questions such as 'what is just and right?' and 'how should one act in the light of the available options?'. The two most influential normative theoretical frames are utilitarianism and duty-based theories (such as Kantianism). Two other theories that have recently been applied to animal ethics are relational or care ethics and virtue ethics. We will briefly explain these four theories here. Utilitarianism is a forward-looking theory, as it only looks at the possible consequences of our actions. Utilitarians argue that we should achieve the best possible balance of happiness, well-being, or some other intrinsic value, over unhappiness or suffering, for all those affected by our moral decision. This entails that when we have to make a moral decision, we need to weigh the prospective consequences of different courses of action and make a calculation about which of the courses of action will lead to the best outcome. There are many different versions of utilitarianism. For instance, the specific version of utilitarianism that famous animal ethicist Peter Singer supports in *Practical Ethics* (2011) and *Animal Liberation* (1975) is preference utilitarianism, meaning that we have an obligation to weigh the preferences of different entities against one another.[2] This view is quite different from the version of utilitarianism that has been leading in discussions about the implementation of animal welfare measures at the farm. This so-called *animal welfarism* argues that the only thing that morally matters regarding the treatment of animals is what the effects of certain measures are on the welfare of all involved animals (and human beings) (Schmit 2011). This approach has been criticized for favouring only marginal reforms in animal husbandry and animal testing, rather than questioning the validity of these practices as such (Harfeld et al. 2016; Haynes 2008).

Criticism to animal welfarism often derives from rights-based theories, which tend to take a more abolitionist position on animal use. According to Tom Regan (1983), for example all beings that are subjects-of-a-life[3] have inherent value and we should treat them with respect for this value; this means, amongst other things, that we should not use them as mere instruments or means, but also always treat them as ends in themselves; a view that is clearly based on Kant. This principle of respect for inherent value is absolute, as inherent value does not admit of degrees. According to this theory we are not allowed to sacrifice an individual's integrity, right to be free from bodily harm, or autonomy in order to achieve good consequences for others. This means that there is a presumption against animal farming or animal testing in rights-based theories. Even though rights-based theories can take into account the prospective consequences of our actions, they are not only forward looking. They also place value on duties that are derived from our past actions; for example if we have promised we would do something we should stick to this promise. Also, the purpose and intention behind an action are relevant for the assessment of the action.

[2]Although recently, Singer seems to have shifted back to the hedonistic version of utilitarianism of his earlier writing.

[3]A being is a subject of a life, which can be understood as being able to experience one's life subjectively, when they are sentient but also possess a certain form of self-awareness, memory, beliefs, perception of the future, and preference autonomy. Of course, an important question in this context is whether fish could be considered subjects-of-a-life. While Regan at the time of writing his seminal work *The Case for Animal Rights* did not consider them as such, scientific research into the physiology of fish has advanced since then.

If our actions accidentally or unintentionally lead to good consequences, Kantians do not necessarily deem the action morally right.

Relational animal ethics—also sometimes termed contextual ethics or care ethics—rejects the abstract and rationalist nature of utilitarian and Kantian principles in favour of a more context-sensitive understanding of morality and a focus on social relationships (Donovan 2006). Relational ethics particularly focuses on relationships of care for vulnerable others, grounded in the understanding that each one of us could end up in a situation in which we need care. According to relational ethics, rational argumentation, like that of Regan and Singer, overlooks the centrality of feelings of sympathy or empathy that we can have towards animals. It is through these feelings that people come to change their behaviour, and not solely through rational argument (Gruen 2010). Our obligations to animals are determined by the specific relationships we have with them. For example, we have more responsibilities towards animals in our care than to animals in the wild, because through our act of domesticating them we have made commitments to them (Palmer 2010). Relational animal ethicists hold that we cannot make moral decisions regarding our treatment of animals without taking into account social and political contexts. Moreover, as Donovan (2006) has argued, relational animal ethicists do not only seek to theorize about animals, but to enter a kind of dialogue with them in the sense that they try to not drown out the 'voice' of the animal but try instead to include the animal's point of view in their ethical deliberations.

Similarly, virtue ethics rejects the abstract and universalistic modes of reasoning of theories such as utilitarianism and Kantianism. For virtue ethics the central question is not 'what is the right action to undertake?', but rather 'what makes me a good person?'. In other words, virtue ethics is not action oriented but character oriented. Virtue animal ethicists regard animals as individuals with whom we share a common life (Gruen 2010). If we treat animals badly, we are displaying the wrong character traits. Virtuous character traits are, for example sensitivity and compassion and we do not cultivate these traits when we routinely harm animals.

For all the above-mentioned theories the intentions and purposes behind animal use are relevant. For example, it is generally deemed more justified to kill fish for consumption than for recreational use and it may be deemed more justified for people with no other means of sustenance to kill and eat fish than it is for people who have other alternatives. From a utilitarian point of view, this is because the interests involved in recreation are not as important as the interests involved in consumption. From a rights-based perspective this is because not only the consequences, but also the intentions of an actor should be morally assessed. For a relational ethicist, this is because we need to take the social context into account: if someone who is barely surviving kills a fish for consumption this is done out of necessity and not out of cruelty or dominance. For a virtue ethicist, intentions behind actions are relevant, because they tell us about someone's character. If someone kills a fish simply for pleasure, this betrays a cruel disposition. Against the background of these normative theories, we will now look at fish welfare.

2.2 Welfare of Fish and Its Moral Dimensions

When we speak about fish welfare, it is important to realize that we are not merely talking about a biological category that we can measure. As we will argue below, welfare is a concept that combines biological aspects and moral dimensions.[4]

2.2.1 *Defining Animal Welfare*

As Haynes (2011, 112) argues, 'animal welfare is an evaluative concept, like product quality and building safety'. This means that the discussion about the welfare of animals cannot be seen independently from normative assumptions. For instance, innovations in order to slaughter fish 'humanely' or the question of what housing system makes fish suffer the least (e.g. Van de Vis et al. 2003) require more than empirical evidence to come to decisions. Moral considerations are at play here too. With regard to animal welfare and housing systems, we need to ask how to balance values related to animal welfare to other legitimate values that play a role. For example, how do we weigh the value of public health against that of animal welfare? For public health reasons it is best to transport live fish to specialized slaughter facilities, but this transport leads to stress (Manuel et al. 2014) in the fish and may be detrimental to their welfare. This interplay between biological views and moral norms raises the question of what we mean when we speak about fish welfare in the first place. From this general claim that moral dimensions play a role in the welfare debate, it is important to stress that the moral questions are not restricted to problems of implementation. It is also linked to the level of defining and assessing animal welfare.

The definition of animal welfare has changed through time from only denoting balanced biological functioning to also including an animal's subjective experiences. While at first good welfare meant the absence of negative experiences, more recently the presence of positive emotions and the capacity to carry out natural or species-specific behaviour are also included in the definition of welfare (Ohl and van der Staay 2012; Duncan 2006). However, an emphasis on the negative aspects is still found in authoritative definitions of animal welfare such as the five freedoms of the Farm Animal Welfare Council.[5] According to this definition, we can establish that an animal's welfare is met if it is free from hunger and thirst, discomfort, pain, injury or disease, and fear and distress, and if it is free to express normal behaviour. Only the last freedom potentially entails positive experiences. Within this welfare concept,

[4]We argue for this position in more detail in Bovenkerk, B. and Meijboom, F.L.B. (2013). 'Fish Welfare in Aquaculture: Explicating the chain of interactions between science and ethics' in *Journal of Agricultural and Environmental Ethics*, vol 26 (1): 41–61, special issue on fish welfare.

[5]See http://webarchive.nationalarchives.gov.uk/20121010012427/http://www.fawc.org.uk/freedoms.htm (accessed on July 3, 2018).

a moral evaluation of animal welfare needs to be made when some of these freedoms conflict with each other. For example, dehorning dairy cattle may come with some sort of pain, but is often argued for with reference to the prevention of future injuries. Consequently, one has to weigh the relative weight of these freedoms in this case. The moral dimension is, however, not the result of this specific five freedom definition. Also, other views on the concept of animal welfare lead to similar ethical issues. For instance, a more recent dynamic view on animal welfare states that an animal is in a state of welfare if it has the ability and capacity to adapt to its environment and experiences it as positive (Ohl and Van der Staay 2012; Nordquist et al. 2017). Here again, the definition includes moral assumptions and needs a normative view on how to deal with situations in which different parts of animal welfare conflict. An example is the situation in which exploring a new environment may be stressful for an animal, but may also entail positive emotions in the long run.

To structure the variety of animal welfare concepts and the related moral dimensions, we follow Fraser (2003). He defines three groups of views on welfare: function-, feeling-, and nature-based views. Applied to fish, function-based views are about the ability of fish to cope with farming conditions. Feeling-based views assume that fish have subjective feelings and that these are constitutive of their welfare (Duncan 1996). Nature-based views regard fish welfare as the ability of fish to display natural or species-specific behaviour. These views are not necessarily mutually exclusive, but they can certainly conflict in specific contexts. For example, robust species of fish can cope with the stress caused by handling (EFSA 2008), but this does not exclude negative emotions as a result of the handling. One's ethical theoretical framework often determines which of the three views one emphasizes. For example, a utilitarian who focuses on sentience and strives for maximizing overall welfare, would be more likely to support the feeling-based view. Someone who argues from an ecocentric theory, which gives a central place to natural ecosystems, would be more inclined to reason from the nature-based view on welfare. However, in most practical discussions on fish welfare, we see an emphasis either on function-based parameters or on the absence of pain. This is understandable, as questions surrounding fish welfare arise in the context of aquaculture and in this context ability to cope with farm conditions is important. Moreover, while there is an increasing consensus amongst fish biologists and physiologists that fish can feel pain (Braithwaite 2010), this dispute is still being disputed (see for instance Rose et al. 2014). Only recently, research is being carried out into what constitutes positive experiences for fish. For example, research is being done on environmental enrichment in fish tanks and on preferred substrates (Manuel et al. 2015; Galhardo et al. 2009).

2.2.2 *Measuring Animal Welfare*

The question of how to implement fish welfare in aquaculture assumes not only that we know what welfare is, but also that we know how to measure it. At first sight, this seems to be a purely empirical question, but in fact, this also involves an interaction between empirical science and ethics. In any scientific study value assumptions and judgements are made at several points, from the formulation of the research question, to the determination of the test set-up and the interpretation of results (Longino 1990). For instance, we can ask what we are in fact measuring when we perform a preference test. Do we test short-term or long-term preferences? Or do we only find the least bad option out of two evils? A more fundamental question is to what extent preferences are indicative of welfare in the first place. Furthermore, there is a difference between the experience and assessment of welfare on an individual- or on the group-level. For example, on fish farms fish welfare can be assessed by measuring the amount of cortisol in the water. This gives the farmer information about the welfare on the level of the group, but of course, there can be large differences between the welfare of individual fish. This is a relevant distinction for ethical assessment and invites debate about the question whether or not the system works if the farmer cannot offer individualized care. Or perhaps group welfare is what the farmer ought to be striving for in the first place? Furthermore, we can measure the state of an animal's welfare at a specific point in time or over the course of the animal's whole lifespan. Also, the question of whether an animal experiences acute or chronic discomfort has implications for how we assess its welfare. During the assessment of welfare, we, therefore, make implicit value choices about what we deem important about animals' welfare.

Measuring fish welfare is even more complex than measuring the welfare of mammals, since we cannot use our own experiences as a frame of reference as fish's physiology is so different from ours. Little is known yet about preferences and experiences of fish. Moreover, there are many different species of fish and if we found out what preferences or experiences one species has this would not automatically translate to other fish species. We need to bear in mind that most research has been carried out on fish species that are of particular interest to humans, such as trout and salmon. There are over 30,000 species of fish and the differences between them can be as big as the difference between—say—an elephant and a mouse. This raises the question of whether welfare indicators can be translated between different species of fish. It also demonstrates the enormity of the task at hand if we want to find out more about fish welfare.

We have now given some examples of normative aspects of defining and measuring fish welfare, but before we claim that we need to implement, weigh, define, and measure welfare, we have already made two important steps. First, the assumption that welfare matters morally. Only from this starting point, it matters whether animals can experience pain or pleasure. There are theories, however, that give a less central place to pain or pleasure, such as virtue ethics. Other theories do not focus on the interests of individual animals at all, but rather on collectives, such as ecosystems or species. From an ecocentric viewpoint, avoiding suffering is not what counts, but rather the survival and flourishing of an ecosystem or species. In this view, suffering is simply part of life and has an important function, namely survival. Second, the focus on the welfare of fish indicates we already assume that fish have moral status. This point asks for some additional elaboration.

2.3 Do Fish Have Moral Status?

Implementing fish welfare implicitly assumes that fish matter from a moral point of view. Discussions about housing conditions, sustainable aquaculture, or humane slaughter all raise the question of how we should treat fish and this implies that the interests of fish matter from a moral perspective. Another way of saying this is that fish have moral status. But what exactly do we mean when we speak about moral status of animals? And what does the attribution of moral status imply for the way we treat them? In case of conflicts between the interests of different animals, or animals and humans, how should we adjudicate between these interests? As we will show, a theory of moral status does not yet tell us how we should weigh different duties in practice.[6] This requires a normative theory. This means that when we encounter practical questions about how to treat fish, for example in aquaculture, we need to be aware that we cannot find answers without adopting a specific moral framework.

In animal ethical discussions, moral status functions as an umbrella concept that encompasses both moral considerability and moral significance (Gruen 2010; Goodpaster 1978). Lori Gruen explains it as follows:

> To say that a being deserves moral consideration is to say that there is a moral claim that this being has on those who can recognize such claims. A morally considerable being is a being who can be wronged in a morally relevant sense. (Gruen 2010, np)

We could say, then, that moral considerability gives a being an entry ticket into the moral community. Moral significance, on the other hand, says something about the relative weight of a being's interests. Gruen explains the difference below:

> That non-human animals can make moral claims on us does not in itself indicate how such claims are to be assessed and conflicting claims adjudicated. Being morally considerable is like showing up on a moral radar screen—how strong the signal is or where it is located on the screen are separate questions. (Gruen 2010, np)

Determination of an animal's moral significance sheds light on the question of how we should treat an animal in a particular situation, but it does not fully determine this treatment. This is because other considerations may enter into our decision-making process. What considerations these are in turn depends on what specific normative theory one holds. For instance, a relational animal ethicist holds that we have a stronger duty of care to our pet goldfish than we have to a fish in the wild. Both fish may have the same moral considerability and significance, but our moral judgement about how to treat them is different because towards the pet fish we have made a commitment that we have not made to the wild fish. In order to know how we should decide when conflicts of interest arise, we, therefore, need more input than just a position on moral considerability and significance. Furthermore, even if two animal

---

[6]We argue for this position in more detail in B. Bovenkerk & F. Meijboom (2012). 'The Moral Status of Fish. The importance and limitations of a fundamental discussion for practical ethical questions in fish farming' in *Journal of Agricultural and Environmental Ethics* vol. 25, iss. 6, pp. 843–860.

ethicists were to grant moral considerability to an animal on the same basis, for example its capability to suffer, they could still reach different conclusions about how to treat the animal, because their arguments are based on different normative theories. For instance, two animal ethicists can agree that salmon has moral status because it has the capacity to suffer and enjoy, but they can still disagree about the moral acceptability of genetically modifying the salmon. For a welfarist it might be allowed as long as the welfare of the salmon is not harmed, while for a Kantian it may be morally problematic, because it would not respect the salmon's inherent value.

There is no theory neutral answer to the question of whether fish, or animals in general, have moral status. Ethicists with different theoretical backgrounds justify the moral considerability and significance of animals on different grounds. Most animal ethicists do use a similar strategy, however, which is basing moral status on the possession of a certain property or group of properties. Candidates for such properties are, most commonly, sentience or capacity to suffer, conscious experience, the possession of desires, self-reflective agency, or autonomous activity. For Singer (1975) animals belong to the moral community in as far as they are sentient, or have the ability to experience pain and pleasure. In his view, what matters in morality is the possession of interests and his theory starts from the basic principle that equal interests should be treated equally. Only sentient beings have interests, because only to them it matters how we treat them: A rock does not have an interest in not being kicked, while a mouse does. In the Kantian view of Tom Regan, as we saw above, animals are attributed moral status if they are subjects-of-a-life, which is to say that they can experience their life subjectively due to certain characteristics, including sentience, but also a certain amount of self-awareness. Relational or virtue ethicists, on the other hand, place less strict demands on the cognitive capacities that beings need to possess before they count morally; animals are simply recognized as belonging to our moral community either because they are embodied like us and they can be vulnerable, or because they are the kind of creatures we can have empathy with.

How do these views on moral status relate to fish? As we saw, for utilitarianism and Kantianism cognitive capacities of animals are important, in particular sentience or a certain amount of self-awareness. Many fish have a nervous system and nociceptors, but this does not tell us yet whether they subjectively or consciously experience sensations such as pain (Braithwaite 2010). Conscious pain perception would require a signal to be sent from the nociceptors to the brain and some researchers doubt whether this happens, because fish brains are so different to mammalian brains (Rose 2002; Arlinghaus et al. 2002. Rose et al. 2014). Yet, an increasing consensus amongst fish researchers now seems to be that fish can consciously experience pain (Braithwaite 2010; Roques et al. 2010, Chap. 10). The possession of more complex cognitive capacities is harder to establish, but research performed with for instance cod indicates that in this species a declarative memory is present (Nilsson et al. 2008, Chap. 8) and for other species studies show that they are able to generate complex representations of their environment rather like a mental map (Braithwaite and De Perera 2006; Ebbesson and Braithwaite

2012). Evidence is also found that different species of fish, in particular groupers and moray eels, cooperatively hunt (Bshary et al. 2006).

More research needs to be done into fish cognition and we need to bear in mind that such research is complicated by the fact that fish are anatomically quite different from us. It takes a lot of imagination to devise tests to establish whether fish can do things like act intentionally or whether they have a sense of the future. Moreover, such test results need to be interpreted and such interpretations are often not value neutral. We need empirical research to find out whether the animals experience pain or stress or have other cognitive capacities, but for the interpretation of this research we also need moral reflection. Especially because of the large knowledge gap we still have about fish it is necessary to reflect on our normative presuppositions. We encounter factual uncertainties and the relevance of these uncertainties depends on one's moral principles and values. For example, if mere sentience is sufficient for the attribution of moral status, information about intentionality in fish is less relevant, than if we also think fish need to possess more complex cognitive abilities in order to be part of the moral community. Moreover, we need reflection from the field of the philosophy of mind to help us determine what we can know about animal consciousness and what we should understand by concepts such as animal awareness, consciousness, and mind.

2.4 Limits of the Animal Welfare Concept

The discussion on moral status shows that we—based on scientific evidence and in line with most of the theories of moral status—have good reasons to consider fish morally considerable for their own sake. This stresses the importance of the attention to welfare of fish. However, it also indicates that the ethical dimensions related to our interactions with fish cannot be reduced to a discussion about welfare only. Such a reduction would result in two problems. First, an exclusive emphasis on animal welfare tends to hide the plurality of views on the moral position of animals. Quite often animal welfare seems to serve as an overarching concept that can be embraced by different people who hold a variety of moral positions. On the one hand, we grant that the broadly shared importance of welfare serves the important function of enabling discussion by a common frame of reference to groups with otherwise opposing views. On the other hand, this means that all manner of considerations and values are translated into welfare terms, even if these considerations are in fact not about welfare at all. This is a consequence of a strategic use of animal welfare arguments, because they are broadly considered as legitimate, while less consensus exists about other moral concerns. This leads advocates of, for example relational ethical or rights-based views to restate their arguments in terms of welfare, while in fact their concerns address considerations about relationships or rights (Leuven, J., 2015, The role of philosophical theory in political activism: animal advocacy and the political turn, unpublished manuscript). For instance, in the debate on the early separation of cow and calf implicit relational- or rights-based arguments are put forward as animal welfare concerns (Ventura et al. 2013). This means that the debate on animal welfare has to deal with a large variety of questions and this muddies both the conceptual scientific and moral discussions. Animal welfare scientists are then called upon to answer questions that actually arise from public views on sustainable animal farming and the relationship between farmer and animal, rather than on the long-term effects on the calves' social behaviour.

Second, when animal welfare has become an all-encompassing concept in public deliberations on the just treatment of animals, it results in a lack of attention for anything other than welfare. Surely, many issues can ultimately be framed in terms of animal welfare. Yet, public discussions on, for example tail docking in dogs for aesthetic reasons or keeping wild animals in circuses show that we do not see the full picture if we only approach these from an animal welfare perspective. Some people are against keeping animals in a circus because this violates their intrinsic value or because they oppose using animals for amusement and not only because it may be bad for their welfare (Brando 2016). Others argue that when we dock a dog's tail we violate the dog's integrity, even if the dog does not suffer from this (Bovenkerk et al. 2001). These are animal rights and virtue ethical considerations that cannot be reduced to a discussion on animal welfare. Therefore, we need to be aware of the limits of animal welfare and keep a broader perspective on the ethical debate on the human interactions with fish. Otherwise, we miss many important considerations. To further elaborate this point, we address two issues that we consider to be 'beyond welfare', namely killing of fish and domestication.

2.5 Is It Morally Harmful to Kill Fish?[7]

In many practices involving fish, such as aquaculture, wild fishing, and recreational fishing killing fish plays a central role. If we attribute moral status to fish, this does not only mean that we have to take into account their welfare in these practices, but it may also mean that killing them constitutes harm, even if this killing would be done painlessly. In other words, the ethics related to killing fish is not restricted to the question of 'how' fish should be killed, but also includes the question of whether killing as such is a moral problem and harms[8] the fish. This latter question, in other words, focuses on whether it would still be harmful if we were to kill fish painlessly.

[7]The arguments in this section are laid out in more detail in Bovenkerk, B. and Braithwaite, V. (2016). 'Beneath the Surface: killing of fish as a moral problem'. In: F. Meijboom and E. Stassen (eds) *The end of animal life: a start for ethical debate. Ethical and societal considerations on killing animals.* Wageningen: Wageningen Academic Publishers, pp. 227–250.

[8]We should mention here that when we speak about 'the harm of death', we are talking about harm in a moral sense. Of course death harms a fish in the sense that its body is damaged—in the same sense as a plant can be harmed when it is cut—but is this a harm that matters morally?

2.5.1 A Preference to Stay Alive

This leads to the question of what arguments have been put forward for the view that killing animals is harmful. Some argue it is wrong to kill animals if they have a preference for staying alive (Singer 1980).[9] The next question then is whether (some species of) fish have such a preference. According to Singer, an animal can form a preference to stay alive only when it has the capacity to be aware of itself as a distinct entity existing over time (Singer 1980). This question can also be approached from the other side: Some argue death is only a harm to those animals that have a preference not to die (e.g. Bracke 1990; Cigman 1981). This would imply that the animals need to have a concept of death. It has been argued that this requires language or second-order beliefs or intentions (Davidson 1982; Bracke 1990). From a rights theory perspective, it has been argued that a being can only have a right to life if it has a desire to live and that only beings who have an awareness of their desire actually have a desire to live (Tooley 1972). Tooley thinks this requires self-consciousness. Similarly, Cigman (1981) takes self-consciousness as necessary, because she thinks death is only a harm for beings with the capacity for categorical desires. Life as a categorical desire answers the question of whether or not 'one wants to remain alive' (Cigman 1981, 58). Desires like wanting to raise children or writing a book are categorical desires, because they give us reasons to go on living.

This discussion on the harm of death in terms of preferences or desires suggests that fish—until evidence proves otherwise—do not fulfil the right criteria to be able to speak about a preference to stay alive or avoid death. This, however, does not imply that therefore killing fish does not include a moral harm. One can raise doubts about the framing of the harm of death in terms of a desire for continued life. We can wonder whether we value continued life because it is desirable or whether we desire continued life because it is valuable. If we value life and therefore desire it, then perhaps the desire itself is not the decisive factor, but rather the value that we place on life.

2.5.2 Foregone Opportunities

This connects to an alternative argument, the so-called 'foregone opportunities account' according to which death is morally harmful for animals because it deprives them of future happiness or goods (DeGrazia 2002). Animals derive pleasure from certain goods in their life and they have an interest in the continuation of these goods.

[9]When confronted with avoidance behaviour of animals that are in danger, such as the struggling for survival of a fish on a hook, at first sight we might interpret this as a fear of death. Singer (1980), however, warns us against taking this to mean a preference for continued existence. Rather, we should interpret this as a desire to stop the pain or the threatening situation and of course this desire can also come about by killing the animal.

According to DeGrazia (2002, 61), 'death forecloses the valuable opportunities that continued life would afford'. In other words, life is *instrumentally* valuable for animals to the extent that they can have valuable experiences that make their lives worth living. According to Kaldewaij (2006, 61) a benefit of this view is 'that it can explain the magnitude of the harm of death: death takes away the possibility of ever experiencing, doing or accomplishing anything you value again'. One could object that animals are not aware of these foregone opportunities. However, this view on the harm of death does not require that individuals are aware of their lost opportunities. A being, it is argued, can have an interest in continued life, without actively being interested in it (Višak 2013). As long as the animal has the ability to have experiences that matter to it and that it would be deprived of when dead, it can be harmed by death in this account. As animal welfare scientists have shown, animals—including fish—do not just have simple desires such as eating when they are hungry and sleeping when they are tired, but they actually derive pleasure from acts such as eating and mating and it could be argued that this makes their life worth living (Duncan 2006).[10]

2.5.3 *The Harm of Death: Reason for Ethical Assessment*

The view that death is more than a welfare issue and that killing is harmful to a fish, does not straight forwardly lead to all manner of prohibitions. The implications of this view depend on how the harm of death should be weighed compared to other harms or benefits that are linked to fish consumption, sports fishing, or other activities where fish are routinely killed, such as in the aquarium industry or animal experimentation. At this stage, we need input from ethical theory again. Utilitarians make a calculus, weighing the total amount of happiness, pleasure, or preferences that an act yields against the total amount of unhappiness, displeasure, or unfulfilled preferences. In such a calculus, if people need to eat fish to survive this outweighs the death of a number of fish. This would particularly be the case for people in poor countries or for Inuit, who may have no realistic alternatives to eating fish, whereas people from wealthy countries can resort to alternative sources of protein. While some argue that it would be more sustainable if people ate more fish, thereby contributing less to climate change than eating meat (Kiessling 2009, but see Röcklinsberg 2012, p. 10 for a critical discussion of this viewpoint), this would not justify the vast numbers of fish being caught for consumption (including by-catch) today (estimated to be between 9.7×10^{11} and 2.7×10^{13} individuals)

[10]While the foregone opportunities account seems rather plausible, it does raise a troubling question, namely whether we can really be deprived of something if we do not exist anymore. After all, when we are dead, we do not know what we are missing. This problem has spurned a philosophical debate too complex to discuss within the scope of this chapter. This debate centres on the question of whether you can be harmed by something even if you do not experience this harm and no consensus has as of yet been reached in this debate (Nagel 1991; Silverstein 1980).

(Mood and Brook 2012). Arguing from a rights-based perspective one could claim that even if fish have a right to life, this right can be trumped. Rights are not absolute, so when another being's life is on the line, killing of fish might be justified. This gives rise to the question of how much the right to life for fish counts vis-a-vis the rights of other animals (including humans).

According to a broadly shared intuition, it is worse to kill a human being or another mammal than to kill a fish. What could this intuition be based on? DeGrazia (2002) argues that life is instrumentally valuable for the goods that it brings a being. However, different species can have different interests in life if they differ—qualitatively or quantitatively—in the goods that are valuable for them. Assuming that this reasoning is convincing, what does it tell us about the moral acceptability of killing fish? This question is by no means settled, but depends on an assessment of basic, serious, and peripheral interests of humans in killing fish for consumption, recreation, or experimentation and weighing these against the basic interest of fish in survival, or in other words what they stand to lose when killed.

2.6 The Domestication of Fish

A moral issue in the context of aquaculture that moves our discussion beyond welfare concerns regards the domestication of fish. Whether or not this is done intentionally, keeping fish in captivity and selecting them for favourable traits, leads to a change in their behaviour and genetic make-up. A formerly wild species then becomes domesticated.[11] For example, at the advent of aquaculture, many fish were nervous and became stressed by contact with humans, but after a couple of decades of selecting for fish that were easier to handle, their genetic make-up has changed and they can deal much better with human proximity. While on the one hand domestication might be beneficial both to the farmer and the fish, it also raises moral issues. We want to illustrate this with an example involving the farming of naturally aggressive fish species. Placing aggressive fish in high-density conditions could lead to attacks and hence to welfare problems in the fish that are attacked. Even if everyone agrees on the importance of the value of welfare and agrees that this kind of housing leads to welfare problems, then it is still not directly evident how one

[11]We build on the definition of domestication given by Swart and Keulartz (2011) who make a distinction between wild and domesticated animals on the basis of two characteristics: the degree to which an animal has adapted to its human environment and the degree to which it is dependent on humans. The more an animal has adapted and the more dependent it is on humans, the more domesticated it is. We use this definition because it remains neutral on the human intentions by which animals were domesticated (i.e. domestication can be the product of unintended and unforeseen selection pressure that we have put onto animals) and emphasizes the fact that wildness and domesticity are matters of degree. Regarding the part of the definition about adaptation, we assume that this adaptation is generally passed on to the next generation and that the genetic make-up of domesticated animals has changed.

should deal with this problem. First, scientists could try to select and cross less aggressive specimens, in fact changing the species to become less aggressive (thereby domesticating them). Second, they could examine what stocking density of these fish would lead to less aggression and change the density accordingly. A third option could be the claim that given these welfare problems these fish should not be kept under farming conditions in the first place. The underlying moral question, that determines how we assess the different options, is whether we should adjust the animal to its farming surroundings or whether we should adjust the farm to the animal.

Another example is the case of piscivorous fish. Because of the high costs, both economically and environmentally, of feeding kept fish with wild-caught fish, it is deemed preferable to switch the fish to more plant-based diets. In aquaculture, we see that many carnivores are in fact slowly being turned into herbivores. At first sight changing these fish' constitutions by domestication seems very efficient and practical, but this move does cause resistance. In response to livestock farming, there has been moral and societal discussion about the consequences of animal domestication and it is likely that aquaculture will face a similar reaction. Part of this discussion focuses on the harmful side effects of adapting animals' genomes. For example, salmon in aquaculture are three times more likely to become deaf than their wild counterparts, due to an ear malformation caused by abnormally fast growth of the fish (Reimer et al. 2017). However, there are also moral objections to changing animal genomes when animals' welfare is not obviously harmed (see Bovenkerk and Nijland 2017 for an overview). Some argue that such changes violate an animal's integrity (Rutgers and Heeger 1999; Bovenkerk et al. 2002) or that it treats animals as if they are mere things (Brom 1997), or that it objectifies or commodifies them (Bos et al. 2018). These objections all revolve around a view of what animals (in this case fish) should be like; they assume a certain 'natural' species norm that is disregarded. If a predator is turned into a herbivore, the species' integrity has been violated. A herbivore catfish is then somehow less of a catfish. The fish is used instrumentally to achieve our goals without respecting its own goals in life. Most of these moral objections have a Kantian or a care-ethical background, and whether they are convincing to someone will depend at least in part on the ethical framework she espouses.

2.7 Practices Involving Fish: Ethical Aspects

In the foregoing sections, we have discussed animal ethics theories pertaining to fish welfare. Also, we have shown that not all ethical discussions about our treatment of fish can be captured under discussions about welfare. In this section, we will point out a number of moral issues in four practices involving fish: aquaculture, wild fisheries, experimentation, and recreation.

2.7.1 Aquaculture

In 2014, the consumption of fish raised in aquaculture facilities has surpassed consumption of fish from wild fisheries[12] and it is projected that in 2030 aquaculture will generate nearly two-thirds of the global fish supply for consumption.[13] Of course, there are different types of aquaculture facilities—large or small-scale, at sea, in ponds or on land in recirculation systems, commercial or subsistence farms—and each comes with its own moral questions. Certification systems such as the Aquaculture Stewardship Council (ASC) label for responsibly farmed fish tend to focus on social and environmental sustainability rather than animal welfare concerns, although recent efforts have been made to include the latter.[14] Welfare issues at fish farms revolve around stocking density, water quality, transport stress, feeding strategies, slaughter, and negative side effects of breeding for desirable traits, such as growth rate. If we take the Animal Welfare Council Five Freedoms as a measure for farmed fish welfare, it becomes apparent that certain freedoms can be in tension with each other. For example, if we think it is important for fish' welfare that the fish have the freedom to carry out natural or species-specific behaviour, we encounter a dilemma in the case of predatorial fish, such as the African Catfish. Do we let them carry out their natural tendencies or do we want to protect the potential victims' welfare? Moreover, at fish farms it is expedient to sort fish of different sizes, but this might run counter to natural living conditions of the fish. Which aspect of fish welfare is deemed more important is dependent on one's background ethical theory; an ecocentrist might find it more important to closely mimic natural conditions, whereas a utilitarian might in the first place want to reduce pain and suffering, for example.

Another issue to consider is that public perception of fish welfare can conflict with the perceptions of farmers or fish biologists. For example, for the public, welfare during slaughter appears to be very important, while fish biologists focus more on water quality. This difference may be understood if we consider that animal welfare is not a purely objective biological term, but is a combination of moral and biological norms. Perhaps the general public focuses on the severity of discomfort at one point in time (i.e. the moment of slaughter) whereas biologists tend to perceive of welfare as a cumulative notion over time, e.g. the whole life of the fish. The realization that different notions of animal welfare may inform the public or farmers' or biologists' views on how to humanely farm fish, and that none of these notions is a priori better, might help to avoid unnecessary polarization between these groups. Moreover, if we do focus on humane slaughter it is important to note that a lot is still unclear, as only for a small number of the 362 fish species farmed worldwide in 2016 specifications to achieve effective stunning are available (Chap. 14). Stunning

[12]http://www.fao.org/3/a-i5692e.pdf (accessed 2/7/2018).

[13]http://www.worldbank.org/en/news/press-release/2014/02/05/fish-farms-global-food-fish-supply-2030 (accessed 2/7/2018).

[14]https://www.asc-aqua.org/the-principles-behind-the-asc-standards/ (accessed on 2/7/2018).

devices have been created that render fish unconscious before slaughter, and these can use, for example percussion or electric field exposure (Chap. 14). In general, fish is exposed to air prior to the application of percussive stunning. Exposure to air also occurs prior to electrical stunning after dewatering. Another approach in which the exposure to air can be minimized or avoided, is electrical stunning in water. Various studies show that percussion and electrical stunning in water and outside the water can induce an immediate stun in fish (Chap. 14). Reported studies show that neither of the two approaches is necessarily better (Chap. 14). Obviously, their assessment depends on which physiological or behavioural measurements are used (This underlines our point made above that on the level of measuring welfare value choices have to be made.

As pointed out above, besides welfare issues, other moral issues come up in discussions about aquaculture, raising questions such as 'are we entitled to domesticate fish and thereby change their genetic make-up?', and 'are we allowed to kill fish for consumption in the first place?'. Also, concerns are voiced about increasing intensification of aquaculture, including the fear that we might run into the same kind of objections to the objectification and instrumentalization of animals as we have witnessed in response to conditions in the livestock production sector.

2.7.2 *Wild Fisheries*

An often-heard reason why some people who give up eating meat choose to still eat fish (so-called pescetarians) is that at least fish have had a good life in the wild. While there is some merit to this position, it disregards suffering that also takes place in the wild, and the suffering fish inevitably experience when they are caught and slaughtered. The main animal welfare issues in wild fisheries revolve around the last moments of the fish' lives (Chap. 17). A recent discussion about methods to catch fish focussed on the welfare implications of pulse fishing. In this technique, a low-frequency electric pulse is applied to the water, which startles bottom dwelling fish such as shrimp and flatfish (Rijnsdorp et al. 2016). From a sustainability perspective, pulse fishing appears to have benefits as fishermen have to use less fuel, it leads to less by-catch and disturbs the sea bottom less than other intensive fishing styles that use trawling. However, discussion exists about the animal welfare aspects; some argue that the fish barely sense the electrical pulse, while others argue that the shock is sometimes so severe that it can break the fish' spine (particularly in the case of larger specimens of cod) (Rijnsdorp et al. 2016). This shows our point in the discussion above, that trade-offs may have to be made between animal welfare and environmental sustainability and therefore value choices have to be made when we want to implement animal welfare measures.

In contrast to fish killed in aquaculture, particularly when they are stunned before slaughter, wild-caught fish will necessarily experience welfare problems from the way they are caught. They are driven together in a net, sometimes they are dragged and compressed for hours, and when they are hauled up from deep water at a high speed the pressure difference can force their internal organs out of their orifices (Braithwaite 2010). On board they will die of suffocation, freezing on ice, or being eviscerated. In all of these methods, it takes considerable time for the fish to lose consciousness and sensibility, sometimes up to 5 h. Work is being done to develop stunning devices for wild-caught fish, but this is a time-consuming and costly process. This raises the moral question of who should be responsible for investing in such measures; fisheries, the government, or consumers?

2.7.3 *Experimental Use of Fish*

The number of fish used in animal experimentation is increasing. Even though fish are vertebrates and using them for experimentation purposes is therefore subject to ethical review (Chap. 16), there appears to be a common conception that using fish in research is less problematic than using mice or other mammals. Sometimes fish are even regarded as a replacement alternative to mice or rats. This conception could be based on the fact that less is known about pain and suffering in fish than in mammals. However, one would be conducting the fallacy of ignorance when assuming that just because we do not know what a fish experiences, it, therefore, experiences less than other animals. It has been argued that fish are not sentient animals, due to the difference in brain structure to mammalian brains (Rose et al. 2014). The next step is to argue that it is more morally permissible to use less cognitively complex animals (such as zebrafish) than more complex animals (such as dogs) in experimentation. Even though it is reasonable to assume that consciousness comes in degrees, and more conscious animals may often have richer experiences, it is not self-evident that cognitive complexity will always make suffering worse. Even though there are forms of mental suffering that fish will not experience—for example suffering from an existential crisis—it may also be possible that fish may experience some kinds of more acute suffering as worse than for example humans. Yeates (2011) casts doubts on the idea that more complexity necessarily leads to more pain. In fact, more cognitively complex animals can in some cases cope better with pain, if the pain is short and the animals realize the pain will be over quickly. On the other hand, when they realize the pain is chronic, they might not cope as well, as they know the pain will continue.[15]

Besides causing discomfort, a morally problematic aspect of animal experimentation is that the animals are routinely killed after the experiment (cf. Franco and Olsson 2016). In animal experimentation committees (AECs) it is generally assumed that painless killing is morally unproblematic. At least, the fact that the animals are killed is not meant to be part of the ethical assessment. However, if our arguments

[15]These arguments are explained in more detail in Bovenkerk, B. & Kaldewaij, F. (2014). 'The Use of Animal Models in Behavioural Neuroscience Research', in: G. Lee, J. Illes, and F. Ohl (eds), Current Topics in Behavioural Neuroscience. Berlin: Springer, pp. 17–46.

above about the harm of death for fish are valid, painless killing is not morally neutral and should become a separate concern for AECs.[16]

2.7.4 *Recreational Fishing*

While hunting animals such as deer or boar raise public moral concern in many countries, recreational fishing seems to be an accepted activity. Recreational angling is an extremely popular pastime, with approximately 47.1 billion fish that are caught by recreational fishermen annually (Cooke and Cowx 2004). This sheer number raises moral concerns on its own, but in principle many people do not seem to find fishing problematic, especially when they release the fish back into the water. About two-thirds of fish caught this way are released back into the water. However, this 'catch and release' system of fishing that is practiced in many countries (Bartholomew and Bohnsack 2005, Chap. 19), raises several moral concerns. Fish that are severely wounded by the hook often die a slow and painful death after they have been put back into the water. This raises the question of whether it would be better practice in these cases to kill the fish quickly while it is still captive or to give it another chance to survive. In the latter case, it can be recaptured and if this happens multiple times over several days, there is a strong chance the fish will become chronically stressed, potentially altering the stress physiology of the fish such that the fish becomes immunocompromised (Barton 2002). This increases the chance that the wound where the hook pierced the fish's skin becomes infected, or the overall capacity for the fish to cope with future capturing and handling, or other environmental challenges such as the threat of predation, becomes impaired. Again, it depends on one's normative framework on how one deals with this dilemma and whether one puts more emphasis on fish welfare or on survival.

2.8 Conclusion

In this chapter, we argued that questions about fish welfare cannot be answered without ethical reflection and that one's ethical framework will influence how welfare is assessed. Empirical evidence requires a normative framework in order to become action guiding in practices of aquaculture. We furthermore argued that in a moral discussion on how to relate to fish we also need to take into consideration concerns beyond welfare. We discussed two of these: the question of whether killing fish—even if painlessly—constitutes a moral harm, and the question of how we

[16]Of course, one could argue that in assessing for replacement alternatives to an experiment, the fact that animals are killed is indirectly assessed. However, replacement in practice does not seem to have the highest priority for those carrying out experiments. See Franco et al. (2018).

should deal with the inevitable consequence of farming fish that they become domesticated. Some of the moral issues that we raised are highlighted in our discussion of practices involving fish: how should we deal with conflicting notions of animal welfare in aquaculture? What trade-off should we make between fish welfare and other values in wild fisheries, such as sustainability? Who is responsible for improving fish welfare during slaughter? Is it justified to assume that it is worse to use mammals for experimental purposes than fish? Is a catch-and-release system in recreational angling justified? While we have not provided clear-cut answers to these difficult questions, we hope to have given the reader enough ethical background to continued reflection on them.

Acknowledgement We would like to thank Victoria Braithwaite and Frederike Kaldeway for their collaboration on book chapters (Bovenkerk and Braithwaite 2016; Bovenkerk and Kaldewaij 2014; Bovenkerk and Meijboom 2012; Bovenkerk and Meijboom 2013) that have served as input for this chapter.

References

Arlinghaus R, Cooke SJ, Schwab A, Cowx IG (2002) Fish welfare: a challenge to the feelings-based approach, with implications for recreational fishing. Fish Fish 8(1):57–71

Bartholomew A, Bohnsack JA (2005) A review of catch-and-release angling mortality with implications for no-take reserves. Rev Fish Biol Fish 15:129–154

Barton BA (2002) Stress in fishes: a diversity of responses with particular reference to changes in circulating corticosteroids. Integr Comp Biol 42:517–525

Bos J, Bovenkerk B, Feindt P, Van Dam Y (2018) The quantified animal: precision livestock farming and the ethical implications of objectification. Food Ethics 2:77–92

Bovenkerk B, Braithwaite V (2016) Beneath the surface: killing of fish as a moral problem. In: Meijboom F, Stassen E (eds) The end of animal life: a start for ethical debate. Ethical and societal considerations on killing animals. Wageningen Academic Publishers, Wageningen, pp 227–250

Bovenkerk B, Kaldewaij F (2014) The use of animal models in behavioural neuroscience research. In: Lee G, Illes J, Ohl F (eds) Current topics in behavioural neuroscience. Springer, Berlin, pp 17–46

Bovenkerk B, Meijboom F (2012) The moral status of fish. The importance and limitations of a fundamental discussion for practical ethical questions in fish farming. J Agric Environ Ethics 25 (6):843–860

Bovenkerk B, Meijboom FLB (2013) Fish welfare in aquaculture: explicating the chain of interactions between science and ethics. J Agric Environ Ethics 26(1):41–61, special issue on fish welfare

Bovenkerk B, Nijland H (2017) The pedigree dog breeding debate in ethics and practice: beyond welfare arguments. J Agric Environ Ethics 30(3):387–412

Bovenkerk B, Brom FWA, Van den Bergh BJ (2001) Brave new birds. The use of 'animal integrity' in animal ethics. Hastings Centre Rep 32(1): 16–22, reprinted in Armstrong SJ, Botzler RG (eds) (2003) The animal ethics reader. Routledge, London, pp 351–358

Bovenkerk B, Brom FWA, Van den Bergh BJ (2002) Brave new birds. The use of 'animal integrity' in animal ethics. Hastings Cent Rep 32(1):16–22

Bracke MBM (1990) Killing animals, or, why no wrong is done to an animal when killed painlessly. MA thesis

Braithwaite V (2010) Do fish feel pain? Oxford University Press, Oxford
Braithwaite V, de Perera TB (2006) Short-range orientation in fish: how fish map space. Mar Fresh Water Behav Physiol 39(1):37–47
Brando S (2016) Wild animals in entertainment. In: Bovenkerk B, Keulartz J (eds) Animal ethics in the age of humans. Blurring boundaries in human-animal relationships. Springer, Dordrecht, pp 295–318
Brom FWA (1997) Onherstelbaar verbeterd: biotechnologie bij dieren als een moreel probleem. Van Gorcum, Assen
Bshary R, Hohner A, Ait-el-Djoudi K, Fricke H (2006) Interspecific communicative and coordinated hunting between groupers and giant moray eels in the Red Sea. PLoS Biol 4(12):e431
Cigman R (1981) Death, misfortune and species inequality. Philos Public Aff 10(1):47–64
Cooke SJ, Cowx IG (2004) The role of recreational fisheries in global fish crises. Bioscience 54:857–859
Daniels N (1979) Wide reflective equilibrium and theory acceptance in ethics. J Philos 76 (5):256–282
Davidson D (1982) Rational animals. Dialectica 36:318–327
DeGrazia D (2002) Animal rights: a very short introduction. Oxford University Press, New York, NY
Donovan J (2006) Feminism and the treatment of animals: from care to dialogue. Signs J Women Cult Soc 31(2):305–329
Duncan IJH (1996) Animal welfare defined in terms of feelings. Acta Agric Scand. Sect A Anim Sci 27(Suppl):29–35
Duncan IJH (2006) The changing concept of animal sentience. Appl Anim Behav Sci 100(1–2.) (October):11–19
Ebbesson LOE, Braithwaite VA (2012) Environmental impacts on fish neural plasticity and cognition. J Fish Biol 81:2151–2174
EFSA (2008) Scientific opinion of the panel on animal health and welfare on a request from the European Commission on animal welfare aspects of husbandry systems for farmed fish: carp. EFSA J 843:1–28
Franco NH, Olsson A (2016) Killing animals as a necessary evil? The case of animal research. In: Meijboom F, Stassen E (eds) The end of animal life: a start for ethical debate. Ethical and societal considerations on killing animals. Wageningen Academic Publishers, Wageningen, pp 187–201
Franco NH, Olsson A, Sandøe P (2018) How researchers view and value the 3Rs – an upturned hierarchy? PLoS One 13(8):e0200895. https://doi.org/10.1371/journal.pone.0200895
Fraser D (2003) Assessing animal welfare at the farm and group level: the interplay of science and values. Anim Welf 12:433–443
Galhardo L, Almeida O, Oliveira R (2009) Preference for the presence of substrate in male cichlid fish: effects of social dominance and context. Appl Anim Behav Sci 120(3–4):224–230
Goodpaster KE (1978) On being morally considerable. J Philos 75(6):308–325
Gruen L (2010) The moral status of animals. In Zalta EN (ed) The Stanford encyclopedia of philosophy (Fall ed.) http://plato.stanford.edu/archives/fall2010/entries/moral-animal/
Harfeld JL, Cornou C, Kornum A, Gjerris M (2016) Seeing the animal: on the ethical implications of de-animalization in intensive animal production systems. J Agric Environ Ethics 29 (3):407–423
Haynes RP (2008) Animal welfare: competing conceptions and ethical implications. Springer, Dordrecht
Haynes RP (2011) Competing conceptions of animals welfare and their ethical implications for the treatment of non-human animals. Acta Biotheor 59:105–120
Kaldewaij F (2006) Animals and the harm of death. In: Kaiser M, Lien M (eds) Ethics and the politics of food. Wageningen: Wageningen Academic, pp. 528–532. Reprinted in Armstrong SJ, Botzler RG (eds) The animal ethics reader. Routledge, New York

Kiessling A (2009) Feed – the key to sustainable fish farming. In: Fisheries, sustainability and development. Fifty-two authors on co-existence and development of fisheries and aquaculture in developing and developed countries. Royal Swedish Academy of Agriculture and forestry (KSLA), Halmstad, pp 303–323

Longino H (1990) Science as social knowledge: values and objectivity in scientific inquiry. Princeton University Press, Princeton

Manuel R, Boerrigter J, Roques J, van der Heul J, van den Bos R, Flik G, van de Vis H (2014) Stress in African catfish (*Clarias gariepinus*) following overland transportation. Fish Physiol Biochem 40(1):33–44

Manuel R, Gorissen M, Stokkermans M, Zethof J, Ebbesson LOE, van de Vis H, Flik G, van den Bos R (2015) The effects of environmental enrichment and age-related differences on inhibitory avoidance in Zebrafish (*Danio rerio* Hamilton). Zebrafish 12(2):152–165

Mood A, Brook P (2012) Estimating the number of farmed fish killed in global aquaculture each year. Fishcount, London. http://tinyurl.com/qxao6o7

Nagel T (1991) Mortal questions. Cambridge University Press

Nilsson J, Kristiansen TS, Fosseidengen JE, Fernö A, van den Bos R (2008) Learning in cod (Gadus morhua): long trace interval retention. Anim Cogn 11(2):215–222

Nordquist RE, van der Staay FJ, van Eerdenburg FJCM, Velkers FC, Fijn L, Arndt SS (2017) Mutilating procedures, management practices, and housing conditions that may affect the welfare of farm animals – implications for welfare research. Animals 7(2):12. https://doi.org/10.3390/ani7020012

Ohl F, van der Staay FJ (2012) Animal welfare: at the interface between science and society. Vet J 129(1):13–19

Palmer C (2010) Animal ethics in context. Columbia University Press, New York

Regan T (1983) The case for animal rights. University of California Press, Berkeley

Reimer T, Dempster T, Wargelius A, Fjelldal PG, Hansen T, Glover KA, Solberg MF, Swearer SE (2017) Rapid growth causes abnormal vaterite formation in farmed fish otoliths. J Exp Biol 220:2965–2969

Rijnsdorp A, De Haan D, Smith S, Strietman WJ (2016) Pulse fishing and its effects on the marine ecosystem and fisheries. An update of the scientific knowledge. Wageningen University and Research Report. http://edepot.wur.nl/405708

Röcklinsberg H (2012) Fish for food in a challenged climate: ethical reflections. In: Potthast T, Meisch S (eds) Climate change and sustainable development. Ethical perspectives on land use and food production. Wageningen Academic, Wageningen, pp 326–334

Roques JAC, Abbink W, Geurds F, van de Vis H, Flik G (2010) Tailfin clipping, a painful procedure: studies on Nile tilapia and common carp. Physiol Behav 101(4):533–540

Rose JD (2002) The neurobehavioral nature of fishes and the question of awareness and pain. Rev Fish Sci 10:1–38

Rose JD, Arlinghaus R, Cooke SJ, Diggles BK, Sawynok W, Stevens ED, Wynne CDL (2014) Can fish really feel pain? Fish Fish 15(1):97–133

Rutgers LJE, Heeger FR (1999) Inherent worth and respect for animal integrity. In: Dol M (ed) Recognizing the intrinsic value of animals: beyond animal welfare. Van Gorcum, Assen, pp 41–52

Schmit K (2011) Concepts of animal welfare in relation to positions in animal ethics. Acta Biotheor 59(2):153–171

Silverstein HS (1980) The evil of death. J Philos 77:414–415

Singer P (1975) Animal liberation. A new ethics for our treatment of animals. New York Review, New York

Singer P (1980) Animals and the value of life. In: Regan T (ed) Matters of life and death. Random House, New York

Swart JAA, Keulartz J (2011) Wild animals in our backyard. A contextual approach to the intrinsic value of animals. Acta Biotheor 59(2):185–200

Tooley M (1972) Abortion and infanticide. Philos Public Aff:37–65

Van de Vis H, Kestin S, Robb D, Oehlenschlager J, Lambooij B, Munkner W, Kuhlmann H, Kloosterboer K, Tejada M, Huidobro A, Ottera H, Roth B, Sorensen NK, Akse L, Byrne H, Nesvadba P (2003) Is humane slaughter of fish possible for industry? Aquac Res 34:211–220
Ventura BA, von Keyserlingk MAG, Schuppli CA, Weary DM (2013) Views on contentious practices in dairy farming: the case of early cow-calf separation. J Dairy Sci 96 (9):6105–6116. https://doi.org/10.3168/jds.2012-6040
Višak T (2013) Killing happy animals: explorations in utilitarian ethics. Palgrave McMillan, London
Yeates JW (2011) Brain-pain: do animals with higher cognitive capacities feel more pain? Insights for species selection in scientific experiments. Large animals as biomedical models: ethical, societal, legal and biological aspects. In: Hagen K, Schnieke A, Thiele F (eds) Large animals as biomedical models: ethical, social, legal and biological aspects. Europaische Akademie, Bad-Neuenahr-Ahrweiler

Websites

http://webarchive.nationalarchives.gov.uk/20121010012427/http://www.fawc.org.uk/freedoms.htm. Accessed on 3/7/2018
http://www.fao.org/3/a-i5692e.pdf. Accessed on 2/7/2018
http://www.worldbank.org/en/news/press-release/2014/02/05/fish-farms-global-food-fish-supply-2030. Accessed 2/7/2018
https://www.asc-aqua.org/the-principles-behind-the-asc-standards/. Accessed on 2/7/2018

Chapter 3
The Diverse World of Fishes

Anders Fernö, Otte Bjelland, and Tore S. Kristiansen

Abstract When we try to improve the welfare of fish in aquaculture, public and private aquaria, and experimental research we need to take into account how fish live in their natural environments. There is enormous diversity in the world of fishes. Each species has adapted to specific habitats and co-existing species, with their anatomical, physiological, and behavioural traits in concert enabling them to survive, grow and reproduce. Fish species have different life histories with regard to longevity, rate of growth, age of reproduction, and number of reproductions and offspring. The great diversity of ways in which fish eat and avoid being eaten results in a wide range of patterns of activity and movement, while reproductive behaviour shows remarkable variation, ranging from mass spawning to long-term pair bonds. Fish may live as solitary individuals, in shoals, or huge schools. They use a set of sensors to obtain an integrated view of their environment, and communicate using various sensory channels. To behave in an adaptive way, they require the mechanisms to do what they need to do to survive and prosper. Physiological adaption to environmental variations is crucial, but rapid changes in the environment caused by human activities can impair welfare and even cause selective mortality, leading to genetic changes throughout entire populations. Fish may be classified into proactive and reactive species with different basic "personalities", and only species with the appropriate personality are suitable for farming.

Keywords Habitat · Life history · Feeding · Predation · Shoaling · Communication · Mechanisms · Human activities

A. Fernö (✉)
Department of Biology, University of Bergen, Bergen, Norway
e-mail: Anders.Ferno@uib.no

O. Bjelland · T. S. Kristiansen
Institute of Marine Research, Bergen, Norway

T. S. Kristiansen et al. (eds.), *The Welfare of Fish*, Animal Welfare 20,
https://doi.org/10.1007/978-3-030-41675-1_3

3.1 Introduction

Humans interact with fishes in many ways that may impact their welfare. A wealth of fish species live in habitats that in many respects differ from the situations to which they are exposed to artificial environments. So in order to improve the welfare of farmed fish, we need to know how fish live in their natural environments and about their basic needs. Welfare decisions should be based on the ecological lifestyle specific to the species, and we, therefore, need to know how fish have adapted in a multitude of ways to different habitats and how they cope with environmental variations within their chosen habitat. Fishes need to know what is the right thing to do and they must have the mechanisms that make it possible to do so.

A majority of all vertebrates are fish (Helfman et al. 2009), and more than 34,000 fish species have been described (https://www.fishbase.de), with a much longer phylogenetic distance between some fish species than between frog and man (Johns and Avise 1998). Hundreds of millions of years of evolutionary history have produced an enormous variety of species living in habitats that range from mountain lakes to the deep sea and from small seasonal ponds to global species that span the oceans. Fish make up the most diverse group of vertebrates (Ohno 1999), displaying an amazing array of anatomical, physiological, and behavioural adaptations that in concert have shaped organisms that are able to survive, grow, and reproduce in the widest range of habitats.

Fish can be divided into ancient (*Agnatha*), cartilaginous (*Chondrichthyes*), and bony fish (*Osteichthyes*, teleosts). The evolutionary history of ancient and modern fish separated about 500 million years ago (Nelson 2006), but they still share common features that suit them for life in water. Ray-finned fish (*Actinopterygii*) contain more copies of many genes than other vertebrates (van de Peer et al. 2002). A large number of genes were produced during a genome duplication event early in the evolution of *Actinopterygii* before the teleost radiation, and many lineages within the *Actinopterygii* have also experienced more recent genome duplications that may have provided a powerful toolkit for diversification (van de Peer et al. 2002).

So a fish is not just a fish! It is true that fish live in water (mostly) and that they swim (not always) and that they can be eaten if cooked well (though some fish taste better raw and some are poisonous). Fish come in all sizes, with body shapes that range from round, to elongated, to flat. The maximum lifetime of a species can range from months to hundreds of years. Fish can live as solitary individuals or in groups. Some species are always swimming, while others stay almost motionless. Fish feed on everything from zooplankton to scales and mucus on other fish. The young of some species help to rise other young (Taborsky 1984), while in others, the young cannibalise their siblings (Gilmore et al. 1983).

And a fish species is not merely a fish species! (1) Populations within a species may differ in many ways, such as the response to predators (Brown et al. 2005). (2) Fish may look and behave quite differently at different life stages (Helfman et al. 2009). (3) Males and females differ in many ways that ultimately have to do with the difference in reproductive investment between the sexes (Fleming and Huntingford 2012). (4) Fish of different sizes may pursue different strategies, with small fish making “the best out of a bad job” (Maynard Smith 1982). (5) Individuals can have different personalities and coping styles with some fish being more active and exploratory than others (Øverli et al. 2007). (6) Individual fish with the same coping style may also be genetically different, and thus display different morphology and behaviour (Helfman et al. 2009). (7) The range of variability increases further as fish with a given genetic make-up interact with the environment during the ontogeny (phenotypic plasticity, Ghalambor et al. 2007).

However, even an individual of a given species, population, sex, size, and coping style that has been shaped by its environment does not always behave in the same way. (viii) A fish that encounters an environment that varies in space and time must possess the behavioural flexibility to cope with situations it has earlier experienced as well as with new events. African catfish (*Clarias gariepinus*), for instance, are usually aggressive towards conspecifics, but during the dry season the fish are confined to small pools at extremely high densities, and are not aggressive (Hecht and Uys 1997). As the situation in the wild can change from one moment to the next more rapid behavioural changes are also needed. If a predator suddenly emerges, a fish need to cease all its other activities and escape. Fish thus continuously tune their behaviour to the risks and opportunities of the external world.

Any given appearance, organ or behaviour of a fish has ultimately evolved because it increases Darwinian fitness. Individual selection in biology is key to explaining how a species has adapted to the environment and corresponds to the law of gravity in physics. However, a fish cannot optimise a single trait but must make trade-offs. It cannot hide when searching for food, concentrate on predators, and prey at the same time or spawn when escaping. Meanwhile, other species, including humans, also adapt their behaviour and strategies to exploit the evolved “habits” of a fish species, and change the rules of the game (Hoffmeyer 2008).

This chapter attempts to unravel the huge diversity in the world of fish. Species with different life histories live in a wide range of habitats either alone or in shoals in which the fish communicate with one another. The various ways in which fish eat, avoid being eaten and reproduce are linked to patterns of activity and movement. Fish must have the physiological, neurobiological, and behavioural toolboxes to do what they need to do. The ability to physiologically adapt to changes in the environment is crucial, and rapid changes in the environment caused by human activities present serious challenges. The personality of a fish species may influence whether it is suitable for farming. This chapter looks at many fish species. Only a few species are used in aquaculture and in experimental research, but more species interact with fishing gears and even more with humans in public and private aquaria.

3.2 A Diversity of Habitats

Modern bony fish are found in virtually all aquatic niches and even in terrestrial environments (e.g. lungfishes, Helfman et al. 2009), and they face different challenges in different habitats. Some environmental factors are consumable and thus density dependent (e.g. food and dissolved oxygen), while others are non-consumable and density independent (e.g. temperature). Some habitats are extremely cold while others are hot. In the first place, fishes need to adapt to the average situation in a habitat, but the *variability* of the environment is also crucial, and fish exposed to environmental variations must be able to sense and cope with the changes this brings. The *diversity* of the habitat is also important, and a diverse environment opens up for specialisations, with each species occupying its own niche. Coevolution affects the fitness of different strategies and the composition and functions of ecosystems.

One crucial environmental factor is salinity. Some 41% of all fish species live in freshwater and 58% in seawater, and only 1% of all species move between fresh water and the sea in the course of their life cycle (Helfman et al. 2009). Some species are capable of tolerating a wide range of salinities, and some can even complete their life cycle in both salt and freshwater (e.g. three-spined stickleback, Gasterosteus aculatus; Östlund-Nilsson et al. 2007). Such euryhaline species are found throughout the piscine phylogenetic tree, and range from elasmobranchs such as the bullshark (*Carcharhinus leucas*; Heupel and Simpfendorfer 2008) to perch-like fishes such as the striped bass (*Morone saxatilis*; Rulifson and Dadswell 1995) and barramundi (*Lates calcarifer*; Jerry 2014). The desert pupfish (*Cyprinodon macularius*) takes this to an extreme, coping with salinities ranging from 0 to 70 ppt (Walker 1961).

Other species also live in environments that in the first place seem to be unnecessarily challenging. Atlantic salmon (*Salmo salar*) parr stay in rivers with little food and low temperature, but seem to be willing to pay the price in order to live in an environment with low risk of predation (Gross 1987). The Magadi tilapia (*Alcolapia grahami*) could be the "hottest" fish on earth (Wood et al. 2016), living in highly alkaline, hyper saline waters at temperatures above 40 °C, and with the highest metabolic rates recorded in any fish. One extreme strategy if the environment undergoes dramatic changes is to simply die after passing one's genes on to the next generation. Some species of killifish in the African genus *Nothobranchius*, for instance, live in temporary bodies of waters in pools that form during the rainy season. Before the fish die when the habitat dries out they deposit their eggs in the muddy bottom, and the young hatch at the onset of the next rainy season (Terzibasi et al. 2008).

About 41% of all fish species live in freshwaters, which means that nearly half of all fish species occupy less than 1% of the world's water supply (Helfman et al. 2009). Tropical lakes and rivers are ecosystems with an enormous number of species and specialisations, and in the Amazon basin we currently know of 5600 species of fish, and more are discovered every year (Albert et al. 2011). But what is actually the point of this huge number of species? Would just a few not be sufficient to utilise the various food resources in a habitat? But this reasoning is built on the notion that there exists a master plan, with every species playing a necessary role in the ecosystem. Evolution is, however, a blind process and the emergence of new species is not driven by any specific needs. Moreover, the species of a community differ along many dimensions of the environment. Lake Malawi in East Africa is home to more than 1000 species of mainly endemic cichlid fishes (Genner et al. 2007). A combination of adaptions to different habitats within the lake, a diversification of the feeding apparatus, and sexual selection based on diversification of colour patterns could explain such explosive speciation (Kocher 2004). Fish species living in warmer water were found to have a 1.6 times faster rate of microevolution than close relatives inhabiting cooler water (Wright et al. 2011). The high diversity in the tropics has been explained by the temperature dependence of ecological and evolutionary rates with the relatively high temperatures in the tropics generate a high diversity because "the Red Queen runs faster when she is hot" (Brown 2014).

Certain marine habitats also host a huge number of species. Tropical reefs have a diverse fish fauna that includes 30–40% of all known fish species (Moyle and Cech 2004). In contrast, only 2% of all fish species occupy the epipelagic zone in the open sea ("the twilight zone", the upper 200 m). The habitat diversity here is low and the availability of nutrients limited. Yet, some species in upwelling regions such as anchovies (*Engraulidae*) are very abundant and important for fisheries. Few modern fish species have managed to establish themself in the deep sea below the epipelagic zone, and here ancient groups of teleosts dominate (Moyle and Cech 2004). Fishes in the mesopelagic zone (200–1000 m depth) are the most abundant groups of fish, and mesopelagic fish are actually the most abundant vertebrates in the biosphere (Nelson 2006, Box 3.1). The bathypelagic zone (1000–4000 m depth) constitutes a very large habitat, with 75% of all the water in the oceans found below 1000 m, and stable conditions throughout millions of years have resulted in a diverse fish fauna (Moyle and Cech 2004). In the dark and uniform environment with little available energy the bathypelagic fishes live a simple, passive life, and the brain size and visual and taste lobes of the brain are greatly reduced (Kotrschal et al. 1998).

Box 3.1 Fish in the Mesopelagic Zone in the Deep Sea

Mesopelagic fish live at 200–1000 m depth and migrate almost 1000 meters up and down on a daily basis. During these migrations they experience huge variations in light level, temperature and pressure, but manage apparently this quite well as the biomass of mesopelagic fish has been suggested to be around 10 billion tonnes (Irigoien et al. 2014)! The scattered small individuals avoid trawls and are difficult to catch (Kaartvedt et al. 2012), so this enormous protein resource is not yet available for commercial exploitation. Scientists are now trying to develop more effective fishing techniques. Image: *Polyipnus triphanos*. Maximum length 4.5 cm. Attribution: Thomas Gloerfelt-Tarp, Wikipedia Commons.

3.3 Life History

Fish can live very different lives with great variations in length of life, age, and size at maturity as well as number of times a fish reproduce during its life and number of offspring. The tiny coral reef fish Sign Eviota (*Eviota sigillata*), for example completes its entire life cycle within eight weeks, and has the shortest lifespan of any vertebrate (Depczynski and Bellwood 2005), whereas a Greenland shark (*Somniosus microcephalus*) could live for at least 272 years (Nielsen et al. 2016). The degree of flexibility for life-history traits seems to be relatively low, but the juvenile growth history can affect key life-history trade-offs (Taborsky 2006). Human impacts can also influence life-history trajectories (see below).

Reproduction has a cost that results in trade-offs between reproduction, growth, and survival (Roff 1984). A key factor is the predictability of the environment. In an environment with an unpredictable, nonselective, mortality a fish will allocate a larger portion of its resources to reproductive activities (an "r strategist"; Adams 1980). Conversely, the optimal allocation of resources in an environment with predictable, selective mortality will aim at increasing individual fitness, frequently through competitive ability (a "K strategist"). In a predictable environment a fish could collect energy during its whole life and transfer this energy into sexual products and just reproduce once ("Big bang") and still have a good chance of success. However, the aquatic environment with moving water masses is often unpredictable and the outcome of reproduction uncertain. For example, the year class of haddock (*Melanogrammus aeglefinus*) in the North Sea varies by a factor of more than 100 (Cook and Armstrong 1986). One way to handle unpredictability could be to reproduce only when the conditions are suitable, but this is not easy for fish to know. Even for scientists it is extremely difficult to predict the strength of a year class, which creates problems for managing fish stocks. Another solution is to play the lottery several times. To do so requires a long lifespan and to reproduce many times ("Bet-hedging") in order to succeed at least once. Herring (*Clupea harengus*), which live in an unpredictable environment may spawn as often as about 15 times (Fernö et al. 1998), and cod (*Gadus morhua*) spawn as many as 17 times in the course of a single reproductive season (Kjesbu 1988). The third solution is to produce a large number of eggs and hope for the best. Most fish are "gamblers", and the sunfish (*Mola mola*) can produce 300 million eggs—the highest number of all vertebrates (Pope et al. 2010). The number of eggs that a fish spawns is one outcome of maximization of individual fitness, with an optimal relationship between the number and size of the eggs. No birth control here! In a predictable environment competition is often strong, and it pays to invest a lot into each offspring in order to enable it to successfully compete with the offspring from other fish. Atlantic salmon (*Salmo salar*) that spawns in a predictable environment in rivers with a low risk of predation produces relatively few large eggs and may spawn only once.

3.4 Ontogeny and Metamorphosis

Some fish species change gradually in the course of their life, whereas others undergo tremendous changes between different stages (Balon 1975). Many fishes emerge from an egg as a small larva with little resemblance to the juvenile or adult stage (Helfman et al. 2009). Due to these great changes in size, fish usually occupy different niches and play different roles in different stages of life. Depth and habitat preferences, as well as prey choice and social behaviour, often change markedly during the ontogeny. Atlantic salmon pass through distinct phases, with parr in freshwater and smolts migrating to the sea where the adult salmon remain until they return to spawn in freshwater (Huntingford 1993). Halibut (*Hippoglossus hippoglossus*) change from bilateral symmetric, pelagic larvae to adult demersal fish with two eyes on the right side (Sæle et al. 2006). The larvae and adult fish of the sea lamprey (*Petromyzon marinus*) are so different that they were long believed to be separate species. Newborn sharks, on the other hand, are small replicas of the adults (Helfman et al. 2009). Although several aspects of development are genetically hard wired, development is a critical period in early life, during which the environment may irreversibly influence the phenotype by epigenetic processes (Pittman et al. 2013). When the different phenotypes correctly predict the future environment, this could confer a fitness advantage. Farmed fish are gradually shaped by their environment, whereas in capture-based aquaculture the fish are suddenly placed in a new environment they must cope with.

3.5 Survive, Feed, and Grow

The lives of most fish are terminated by being eaten by a predator. An aggregation of more than 50 million juvenile cod, for example was wiped out in only five days by predatory whiting (*Merlangius merlangus*, Temming et al. 2007). Fish have evolved a wide range of antipredatory behaviours (Chap. 4). If a predator evolves a more effective way to catch prey, the prey could respond by evolving an even more effective way to avoid it. In this evolutionary arms race the prey may be able to stay just ahead, because the risk of death exerts a stronger selection pressure than a meal does for the predator (the life-dinner principle; Dawkins and Krebs 1979). Fish can avoid contact with predators by using spatial refuges in various spatial and temporal scales (Ahrens et al. 2012), as capelin (*Mallotus villosus*), for example move into thermal refuges that are too cold for cod (Rose and Leggett 1990). Fish that co-occur with their predators can avoid detection (via primary defence mechanisms such as camouflage and hiding) or decrease the risk of capture after being detected (via secondary defence mechanisms like escape and protection by other organisms). The hard fin rays carried by percomorph fish are a smart solution that makes small fish that lack heavy armour difficult to swallow and painful to attack. Fish need not always escape as soon as a predator is detected as this would mean high energy and lost opportunity costs but should rather stay where they are when the cost of fleeing exceeds that of remaining (the economic cost-benefit model, Ydenberg and Dill 1986). Hide-and-seek is an ongoing process in the deep. Fish in the mesopelagic and bathypelagic zones use bioluminescence to hide from predators from below using photophores that mimic the light from the surface to hide or confuse predators by emitting light (Haddock et al. 2010).

Parasites are a special type of extremely small predators. The importance of parasites has long been underestimated, but many small and rapidly evolving enemies can be an even stronger threat than a few large ones. Parasites have powerful effects on the vitality and life history of fish (Barber and Rushbrook 2008) and can even manipulate the behaviour of their hosts to make them susceptible to the next host in the life cycle (Gopko et al. 2017). Some fish species are in fact parasites themselves (Leung 2014). Farmed fish are seldom predated on except by parasites and cannibalism, but their reactions to various disturbances reflect the responses to natural predators (Chap. 4). Farmed fish habituate to repeated events without any consequences, and in public aquaria, fish do not react to even the most spectacular behaviour of people on the other side of the glass.

It is crucial to avoid being eaten, but fish themselves also need to eat. A fish makes three feeding decisions (Hart 1993): (1) *When to feed* both in a life-history perspective and on a shorter time scale. One important decision is if a fish should continue to feed after having satisfied its immediate needs. Energy allocation to storage is triggered by environmental variations (Fischer et al. 2011), and food insecurity is a driver of obesity also in humans (Nettle et al. 2017). (2) *Where and how to search for food* taking into account habitat profitability, trade-offs, and competition from other fish. Some fish just wait for the prey to come (Box 3.2). An interesting example of deception in observed in the large predatory cichlid *Nimbochromis livingstonii* from Lake Malawi that lures its prey by feigning death. The predator lies on its side on the bottom looking like a dead fish, and when scavengers investigate its body the predator engulfs them (McKaye 1981). (3) *What to feed on*. Fish exploit a wide variety of food types (Chap. 4), but most species are carnivores that hunt live prey. Fish are gape-limited and cannot consume prey that is too large, but it can be divided up by taking bites out of it (sharks), by spinning around their long body axis while holding on to their food in order to tear chunks out of it (rotational feeding, eels) or by passing a knot forward along the body and pressing the knot against the prey to lever off a piece of flesh (hagfish, Helfman et al. 2009). Most species are generalists/opportunists but some are specialists. Competition generally makes different species specialise on different prey, but multiple predators can feed on the same prey type provided that none of the predators deplete the prey population (Ahrens et al. 2012). Even individuals within a species may specialize in a particular type of prey (Smith and Skúlason 1996). Box 3.3 provides an interesting example of how frequency-dependent natural selection influences how individuals feed.

Box 3.2 The Frogfish—A Voracious Predator in Disguise

The frogfish *Histrio histrio* does not swim well but is a master of camouflage and ambush. It conceals its position by adapting its colour to the background and attracts prey by a lure. When the prey is close enough it gapes and sucks using rapid muscles, and water and everything else is sucked into the mouth (Pietsch and Grobecker 1987). The image reproduced with permission from David Cook, FishBase.

Box 3.3 Frequency-Dependent Natural Selection in the Way Individuals Feed

The cichlid *Perissodus microlepis* eats scales from other fish. The jaws are curved to the left or right with "left handed" forms attacking the right side of the prey and "right handed" the left side. Left- and right-handed individuals are equally common and their numbers are kept in equilibrium. If the population of left-handed fish was to exceed that of the "right-handers", the prey would

(continued)

Box 3.3 (continued)
become more wary of attacks from the right, and right-handed would then be at an advantage (Hori 1993). The fish have a naturally stronger side for attacking prey fish, and they learn to use the dominant side through experience (Takeuchi and Oda 2017). The picture reproduced from Takeuchi and Oda 2017, Fig. 1A, Scientific Reports (http://creativecommons.org/licenses/by/4.0/).

Feeding is a prerequisite for growth. Most fishes show indeterminate growth with a continual increase in length, but growth rates may slow considerably as fish age (Helfman et al. 2009) and a fish does not grow indefinitely. The sizes of adult fish ranges from less than 10 millimetres (the cyprinid fish *Paedocypris progenetica*, Kottelat et al. 2006) to 12 metres (the gigantic whale shark, *Rhincodon typus*; Helfman et al. 2009); Fig. 3.1). Most bony fish are born very small and their weight may increase by as much as 60 million times during their lifespan from larvae to fully grown (e.g. sunfish; Pope et al. 2010). Large fish are less likely to be preyed on, so why are many fish so small? But staying small can be beautiful! Rapid growth demands a lot of energy and thus a high rate of activity that increases the risk of detection by predators. Nor is it always easier to catch a small fish. An object with less mass requires less energy to change direction than an object with more mass, and a smaller fish has thus a shorter turning radius and can avoid predator attacks more effectively than a larger fish (Abrahams 2006).

3.6 Body Shape and Movements

A typical fish is streamlined so it can cut through the water with the least effort, but the body shape is linked to the activity level and way of swimming (Webb 1984). Cruising specialists like yellowtail (*Seriola quinqueradiata*) have a torpedo-like

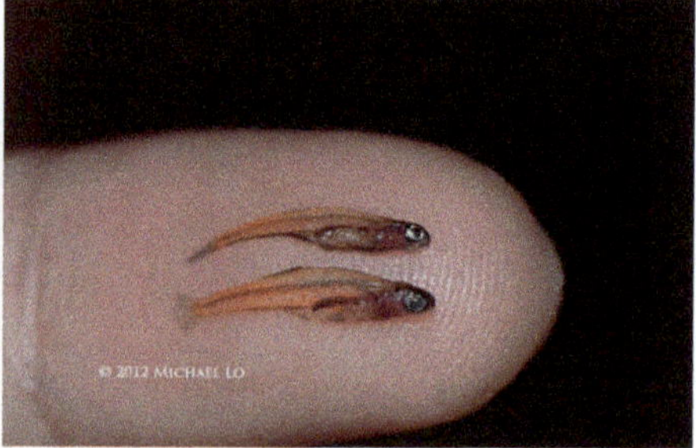

Fig. 3.1 The biggest (whale shark, *Rhincodon typus*, >12 m) and one of the smallest (*Paedocypris micromegethes*, around 10 mm) fish species. Image whale shark attribution Pixabay. Image *Paedocypris micromegethes* attribution Michael Lo

Fig. 3.2 A sunfish drifting through the sea does not seem to have a very functional body shape, but the shape is the outcome of a long evolutionary history so it certainly knows what it should look like! Attribution: By U.S. National Oceanic and Atmospheric Administration [Public domain], via Wikimedia Commons

body form and a thin tail root. Species relying on fast acceleration like the pike (*Esox lucius*) have a broad posterior body to generate thrust. Species living in habitats demanding rapid manoeuvring have a short body and fins placed in different positions generating forces in different directions. However, like the common carp (*Cyprinus carpio*) most species are generalists that perform all the functions of swimming reasonably well. Some species have a really strange body shape (Fig. 3.2). The shape is a product of natural selection but may also be modified by the environment. An extreme example of phenotypic plasticity is found in the crucian carp (*Carassius carassius*), which responds to chemical cues emitted by piscivorous fishes by increasing its body depth (Brönmark and Pettersson 1994), thereby reducing the risk of predation by gape-limited piscivores (Nilsson et al. 1995).

Fish species have different activity rhythms. To keep still without moving decreases the risk of detection, but may make it difficult to come in contact with food and mates, although prey can be attracted by lures in camouflaged lie-in-wait predators such as anglerfish (*Lophiiformes*; Nelson 2006). A fish that moves has the choices of when and where to move. Some fish move only during certain life stages or time periods. Coral reef fish are relatively stationary as adults and colonise new habitats through their planktonic eggs and larvae (Cowen and Castro 1994). Diel rhythms often reflect a trade-off between prey availability and predation risk (Reebs 2002). About one-half to two-thirds of all fish species are diurnal, one-quarter to one-third are nocturnal, and about 10% are crepuscular (Helfman et al. 2009). Cartilaginous fish are generally active at night, wrasses (*Labridae*), mullets (*Mugilidae*), and carps (*Cyprinidae*) are day active and mesopelagic fish migrate vertically during the transition period between day and night.

Approximately 2.5% of all fish species make long migrations at least once during their life cycle (Binder et al. 2011). The migrations are undertaken in order to exploit spatially and temporally changing food resources (Hoar 1953; Arnold and Cook 1984; Russell 2008), for reproductive purposes (Harden Jones 1968; Russell 2008) or due to changes in environmental factors such as temperature (Binder et al. 2011). Technical advances in aquatic telemetry have revolutionised our understanding of fish migrations by relating movements to their physiology and the environment (Hussey et al. 2015), but we still do not fully understand how fish navigate during long-distance migrations over hundreds of kilometres. Fishes live in a three-dimensional world and they also migrate vertically, and diel vertical migrations in the aquatic environment probably represent the largest movement of biomass on earth (Hays 2003). As the light level and temperature change rapidly in the vertical plane, a fish will encounter a quite different environment even following short ascents or descents. It is easier to find the optimal place in the vertical than horizontal dimension because of the short distances and low energetic migration costs and a more predictable environment in time and place. Fishes typically ascend before dusk and descend at dawn, which can be explained in terms of multiple trade-offs between predator avoidance (less risk of being detected by visually feeding predators in low light levels) and feeding rates (Clark and Levy 1988) and bioenergetic advantages (saving energy in low temperature, "*Hunt warm - Rest cool*", Sims et al. 2006).

3.7 Reproduction

Apart from 20 known asexual (parthenogenic) species, a fish must come into contact at least once with another of its own species in order to reproduce. Even simultaneous hermaphrodites exchange sexual products during egg trading (Connor 1992). In most fish the sex is fixed for life, but there are exciting exceptions. In some species, the fish are transvestites (simultaneous hermaphroditism) while in others, individuals change sex (sequential hermaphroditism). When the competition for females is strong, as in wrasses, the fish switch from female to male, with the sex change being under social control (Warner and Swearer 1991). In a few species where the fecundity of females is strongly related to size as in clownfishes (*Amphiprioninae*), the fish change from male to female (Sadovy de Mitcheson and Liu 2008). The two sexes often look rather similar, but an extreme difference is the rudimentary males of deep-water anglerfish that live attached to the much larger females and depend on her for supply of energy (Pietsch 2005). Some species even lack males! In the asexually reproducing Amazon molly (*Poecilia formosa*) offspring are true clones of their mothers (Warren et al. 2018). Sperm from males of sympatric sexual species triggers embryonic development, but the sperm DNA is excluded from the developing egg. Although *P. formosa* has existed for approximately 500,000 generations, there are no widespread signs of genomic decay. A total triumph for the female sex!

Fish reproductive behaviour shows remarkable variations that range from mass spawning to long-term pair bonds, and courtship takes many forms (Chap. 4). Alternative reproductive tactics within a species are common. Conditional strategies with unequal fitness based on the individual's condition are most frequent, but in some species the strategies have equal fitness resulting in a mixed evolutionarily stable strategy (Mixed ESS; Knapp and Nett 2008). Territorial bluegill sunfish (*Lepomis macrochirus*) males that provide parental care and sneaking males that only fertilise the eggs, for instance, enjoy equal fitness. Females tend to invest more energy in reproductive activities than males, resulting in competition for females (Fleming and Huntingford 2012). Mate selection can be based on the resources of the male such as the size of the territory or good genes that are reflected in large size, high dominance status, or intense colour patterns. Female swordtails (*Xiphophorus hellerii*) prefer males with a long sword-like tail (Basolo 1990), although this increases swimming energy expenditure and the risk of predation (Basolo and Alcaraz 2003; Basolo and Wagner Jr 2004). But in fact, such epigamic characters are selected precisely because they are costly and thus honest (see below). To make fish reproduce in farms and aquaria often represents a challenge, and in several species viable young can only be produced by artificial fertilisation (Chap. 4).

Parental care is costly and most species provide no care for their offspring (Sargent and Gross 1993). An association between parents and offspring after fertilisation may improve offspring survival, but a parent that continues to invest in its offspring does so at the expense of its own potential future reproductive success. Such care is usually consisting of defence of the eggs or fanning oxygen-rich water (Lindström and St.Mary 2008). It is not usual to care for the young after hatching, perhaps because of the difficulty of keeping them together in the moving medium. Most teleosts are egg layers, but several hundred species, such as the guppy (*Poecelia reticulata*), are live-bearers with internal fertilisation and females giving birth to live offspring. As the cost of parental care for future reproduction is higher for females, the males usually take care of the offspring (Sargent and Gross 1993). Male parental care has not only evolved by natural selection but also by sexual selection with females selecting males based on how good fathers they could be (Forsgren 1997). Previous breeding experience could influence the parental care, and in the biparental convict cichlid *Amatitlania siquia* the breeding experience of the female affects the parental care strategies of both parents (Santangelo 2015). Filial cannibalism, i.e. consumption of their own young is widespread in fishes but varies among individuals and is influenced by the personality of the fish (Vallon et al. 2016).

3.8 Alone or Together

In contrast to the situation in the high densities found in aquaculture, wild fish have the choice between swimming alone or in groups. About 25% of all fish species live in shoals or schools (polarised and synchronised shoals, Pitcher and Parrish 1993) for their entire life after the larval stage. A fish has to find the optimal balance between getting food and minimising the risk of predation (Pitcher and Parrish 1993). Living alone means that you get to consume all the food you detect yourself, but first you must find it yourself and you are on your own when a predator attacks. Small fish are generally at higher risk of predation than large fish, and shoaling species are often small, and many species change from shoaling to swimming alone at a certain size and age.

Shoaling reduces the risk of predation in several ways, such as more rapid detection of predators, a dilution effect with lower probability of being the target and confusion generated by multiple targets. Fish can live in enormous anonymous schools of millions of individuals like the herring (Nøttestad et al. 2004) or in small shoals where the individuals could recognize each other and compete and cooperate. Bigger schools offer better protection from predators, and it has been shown experimentally that banded killifish (*Fundulus diuphunus*) prefer to shoal with the larger of two conspecific shoals after a simulated avian predator attack (Krause and Godin 1994). However, this is true only up to a certain number of fish in a shoal (Rieucau et al. 2015). Very large groups that are found at high population densities may, in fact, increase predatory and disease mortality, and it has even been suggested that this "suicidal behaviour" may create a density-dependent regulation of fish populations and explain the stability of marine ecosystems (Maury 2017). A dense ball of prey fish can be an easy target. Killer whales (*Orcinus orca*) outsmart the predator defence and herd herring from deep waters to the surface (Similä and Ugarte 1993; Nøttestad et al. 2002) and produce sounds that induce schooling herring to cluster, thereby increasing the efficiency of their tail slaps (Simon et al. 2006). The herring thus seem to be the losers in this arms race. So why do they not make a desperate attempt to escape by a flash expansion (Pitcher and Parrish 1993) spreading in all directions? Perhaps they have become so stressed and exhausted that they just give up against these powerful cooperating predators. Not to form shoals at all might actually be an effective strategy for dealing with the most powerful predator of all—Man—and scattered mesopelagic fish are difficult to exploit using traditional fishing gears (Kaartvedt et al. 2012; Box 3.1). The positions of the individual fish within a school may also influence how vulnerable they are. Individuals closer to the margins of a group have a much higher risk of predation than those nearer the centre (Duffield and Ioannou 2017). Schools of sardines (*Sardinella aurita*) swim away from approaching sailfish (*Istiophorus platypterus*), and all attacks are thus directed at the rear of the school, resulting in higher predation risk for individuals in these peripheral positions (Krause et al. 2017). Collective manoeuvres (Pitcher and Parrish 1993; Box 3.4) can reduce the risk of predation. A prerequisite for coordinated avoidance is an ability to efficiently transfer information, and rapid waves of agitation crossing large anchovy schools enable them to react more rapidly to attacks by sealions (Gerlotto et al. 2006).

Although competition for food increases with the group size (Rieucau et al. 2015), shoaling may make feeding more effective by improving sampling efficiency and providing more time for feeding (Pitcher and Parrish 1993) as well as by enabling cooperative hunting (Handegard et al. 2012). Schooling fish also gain a hydrodynamic advantage (Hemelrijk et al. 2014) and can learn from others (Laland et al. 2011). Farmed species that normally form schools may be stressed at low densities (Pickering 1981) and could benefit from an improved group structure through reduced aggression (Solstorm et al. 2016).

Box 3.4 How Can the Fish Perform Coordinated Group Manoeuvres?
When we observe a school we wonder how the fish manage to keep together, whether some individuals are leaders and how the school manages to perform co-ordinated collective manoeuvres. In mammalian societies, there is usually a social hierarchy so we tend to explain structure in terms of hierarchy. But unlike mammals, shoaling fish typically live in anonymous groups without any leader. The collective behaviour is instead the outcome of individual decision rules and self-organisation (Parrish et al. 2002). Self-organisation is nature's decentralised way of organising a system from the inside based on local interactions that produce global patterns. Simple local rules determine swimming direction and the distance between fish, which in turn creates collective patterns. The individual decision rules are tuned by natural selection to produce adaptive collective patterns.

In the wild, fish do not only encounter conspecifics but also come in contact with other fish species and animal groups. In many cases, these species do not interact with their non-conspecifics, some of which will be either predators or prey. But mutually beneficial relationships do exist. A fish shoal may comprise several species. More individuals in the shoal mean earlier detection of an approaching predator and less risk to the individual fish (Pitcher and Parrish 1993). Many species use larger fish of other species as a protection from predators (Helfman et al. 2009). Cleaner fish remove parasites, and the host and the cleaner communicate by various signals to reduce aggression and facilitate cleaning (Bshary 2011).

3.9 Fish Can "Talk" with Each Other

It is important to understand how fish communicate in order to understand how farmed fish are influenced by human activities. To be a true signal a stimulus should provide some advantage to the sender and not only be a by-product like sounds generated during swimming and gas released from the swim bladder. However, communication is result of a coevolutionary arms race between senders playing the role of manipulators and receivers playing the role of mind readers (Krebs and Dawkins 1984). The displays could give static information about the species and sex but also dynamic information about the state of the individual (Helfman et al. 2009). A signal usually releases an immediate response but it may also inform or manipulate the receiver or influence motivation, for instance in order to synchronise spawning.

A fish uses a set of sensors to obtain an integrated overview of the world surrounding it, and draws up a balance of all pieces of information to enable it to anticipate future conditions. Staying ahead of what is about to happen can mean the difference between life and death. Fish must interpret signals from conspecifics, as well as those that indicate whether a predator is in hunting mode. Fish living in groups need to communicate about where to go, lurking predators and their willingness to spawn. Important senses are vision, smell, taste, hearing, sense of vibrations, touch and various types of nociceptors (touch, heat, acid, see Chap. 9). Different species live in different sensory worlds with their senses adapted to their habitat. Deep-sea fish, for instance, have large eyes, while fish in completely dark caves are blind but have well-developed chemical senses (Helfman et al. 2009). Each sensory modality has its own advantages and limitations. When reacting to prey and predators a fish needs to utilise all information it can obtain, but when signalling to conspecifics a fish can communicate through the most effective sensory channel. Critical characteristics of the stimuli are range, velocity, information about direction and distance, and information content. Visual stimuli, for example provide rapid, detailed information about the distance and direction to an object as well as about its movements (Schuster et al. 2006) but have restricted range and depend on being in direct line of sight.

Visual signals are important for fish and consist of a morphological structure or colour pattern combined with a particular posture or movement. When threatened, visual signals may be hidden within seconds by folding fins or turning off colours. Besides their important physiological functions, carotenoids play a key role in visual signalling, in particular as sexual signals (Svensson and Wong 2011). Carotenoid pigments provide honest signals because they are expensive to produce and maintain (Kodric-Brown 1998). The adipose fin indicates the social status of male Arctic char (*Salvelinus alpinus*) (Haugland et al. 2011), and female Atlantic salmon select males based on the relative size of the adipose fin (Järvi 1990). The sea is not a silent world and the soundscapes underwater have long been under-appreciated. Many species produce acoustic signals by stridulation by pharyngeal teeth or fin rays, or by muscles associated with the swim bladder. The calls of the black drum (*Pogonias cromis*) are so loud that they can be heard inside houses beside seawater canals (Locascio and Mann 2011).

Aquatic animals often use chemical information for decisions related to reproduction. Three-spined stickle-backs prefer mates with an odour type that differs from their own, which maintains major histocompatibility complex (MHC) diversity and provides offspring with better immune defence (Milinski et al. 2005). Damaged-released cues known as chemical alarm cues or Schreckstoff are found in many fish species and cause other fish in the shoal to escape or freeze (Brown et al. 2006). An interesting possibility that could be tested is whether wounds caused by sea lice increase the general stress level of farmed salmon. Olfactory signals may be consciously controlled. Swordtail females respond to male chemical cues, and the males urinate more often when females are present and situate themselves just upstream of females when courting (Rosenthal et al. 2011). Hydrodynamic imaging using the lateral line may enable fish to form mental maps in order to navigate (Burt de Perera 2004), and pressure waves recorded by the lateral line organ also enable information to be exchanged between schooling fish (Partridge and Pitcher 1980). Some species such as knifefishes (*Notopteridae*) and elephant fishes (*Mormyridae*) have a fascinating secret electric language based on signals that we are unable to sense

(Hagedorn and Heiligenberg 1985). The knifefish *Eigenmannia virescens* emits an electric organ discharge of constant frequency and waveform that varies with sex and age (Kramer 1999). When two individuals with similar wave frequencies meet, one fish shifts its frequency upward and the other downward, in order to distinguish between their own signal and those of others (the jamming avoidance response). The frequency rises in subdominant fish and is thought to represent a submissive signal (Hagedorn and Heiligenberg 1985), which indicates that the fish is aware of its own relative strength. Some fish even seem to communicate with humans (Box 3.5). But if we wish to understand fish we need to know if they are telling the truth! (Box 3.6).

Box 3.5 Mantas Talk to People

The devil rays (family *Mogulidae*) have a brain weight compared to similar-sized mammals, and the brain mass of the giant manta ray (*Manta birostris*) is the highest of all fish species studied (Ari 2011). The regions of the brain that account for this enlargement are the telencephalon and the cerebellum, which are responsible for many higher functions in mammals. The brain is kept warm by a network of blood vessels (Alexander 1996) making mantas less dependent on the drop in temperature they experience when they dive to deep waters (Braun et al. 2014). Mantas seem to have an ability to build up a cognitive map of their environment based on long-term memory (Ari and Correia 2008). Csilla Ari (2015) tells us that mantas are curious and voluntarily approach humans, and even "ask for help" when they become entangled in fishing lines. This indicates high social and cognitive abilities, and the behaviour of giant manta rays when interacting with a mirror may, in fact, indicate self-awareness (Ari and D'Agostino 2016). Attribution: By Moesmand (Own work) [CC BY-SA 3.0 (https://creativecommons.org/licenses/by-sa/3.0)], via Wikimedia Commons.

Box 3.6 Is a Fish Always Telling the Truth?

What keeps a fish from lying about its strength and state? Scientists used to interpret signals as manipulation of the receiver, but in fact, most signals are honest. An effective display often has high costs that discourage cheating (Zahavi 1975), and the interplay between social and physiological costs may maintain a close relationship between an animal's abilities and its ornamentation (Tibbetts 2014). Bluff can also be risky, as short-term escalations during interactions could reveal the cards being played, and a fish must give away information when trying to obtain information from its counterpart.

However, a fish may bluff when the cost of escalation could be high. False eye spots on the gill covers of the firemouth cichlid (*Cichlasoma meeki*) make a frontally displaying fish look like a large and dangerous individual that the receiver cannot risk ignoring (Radesäter and Fernö 1979). The picture reproduced from Radesäter and Fernö 1979, Fig. 1, Behavioural Processes with permission from Elsevier.

3.10 The Underlying Mechanisms

If we want to understand the diversity of fishes we must not forget the mechanisms involved. Fish generally behave adaptively, but to do so they need to possess the relevant proximate mechanisms and tools. It is crucial to distinguish between proximate and ultimate explanations (Box 3.7; Fig. 3.3; Chap. 4).

Box 3.7 Proximate and Ultimate Mechanisms
There are two kinds of questions and answers that lie on different levels:

Why 1?

The proximate factors. How do the fish do it? Which mechanisms are involved? Which stimuli release the response?

Why 2?

The ultimate factors. Which function does an organ or a behaviour have?

(continued)

Box 3.7 (continued)

It is of crucial importance to keep these questions separate. A question like the following makes no sense: Do fish make diel vertical migrations as a response to variations in the light level or do they descend during the day to avoid predation? Both answers are of course correct but lie on different levels.

It is not always easy to understand the ultimate factors that underlie why farmed fish do what they do. The confined fish are not evolutionarily adapted to the extreme densities in tanks and net pens, and mechanisms could go wrong resulting in maladaptive behaviour (Fernö et al. 2011). Aquaculture scientists often concentrate on proximate mechanisms, and welfare is on the proximate level.

Behaviour and physiology are inextricably linked (Sloman et al. 2006). Physiological mechanisms enable fish to maintain a stable internal environment, which is particularly challenging when the external environment undergoes rapid changes. For instance, when Atlantic salmon smolts migrate from freshwater to the sea, the osmotic challenge is suddenly reversed. From having to cope with an external environment of lower osmolarity than the internal environment, they now find themselves in water with higher osmolarity. The smolts cannot immediately

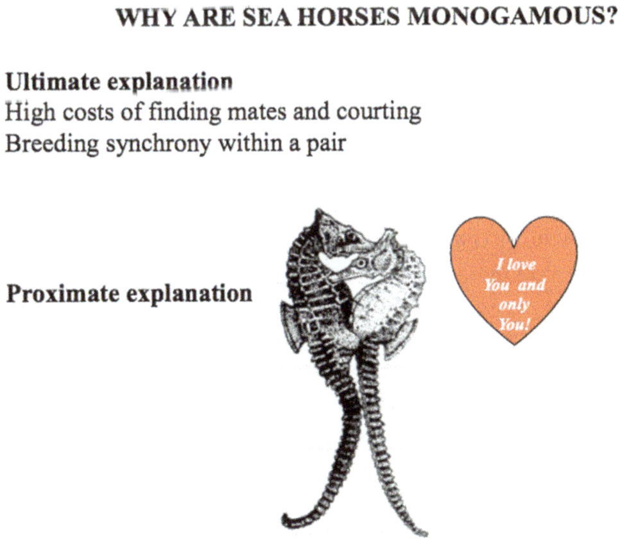

Fig. 3.3 The male and female sea horse *Hippocampus whitei* are linked by a strong pair bond that persists even if reproduction fails (Vincent and Sadler 1995). We do not know the exact nature of this bond, but in order to emphasise the difference between ultimate and proximate mechanisms and assuming that fish experience emotions (Chaps. 7 and 9), we have taken the liberty of interpreting the driving force behind this pair bond as love. Sea horse illustration reproduced from Vincent and Sadler 1995, Fig. 1 with permission from Elsevier

compensate for this, and physiological constraints make them react to a predator at a reduced distance and to school less effectively (Handeland et al. 1996). This is not the theoretically optimal way to behave, but given the constraints the smolts make the best out of a bad job. Species that encounter large variations in environmental factors in the wild (environmental generalists) should be expected to adapt more easily to artificial environments than species that are environmental specialists. Cobia (*Rachycentron canadum*) encounter wide variations in temperatures and salinities in the natural environment (Nguyen et al. 2013), and farmed cobia grow rapidly even in high temperatures (Sun and Chen 2014).

The metabolism of a fish is related to its lifestyle. Its metabolic rate limits its scope of activity and thereby the ability to perform certain types of behaviour, but a high metabolic rate demands a great deal of energy. Some species have a slower pace of life than others with low activity and metabolism, which is linked to life-history strategies such as slow growth rate and a long lifespan (Réale et al. 2010). The resting metabolic rate is an intrinsic trait of an individual and even among individuals of the same species, age and sex the metabolic rate may differ by a factor of as much as two or three (Metcalfe et al. 2016). Bold bluegill sunfishes have greater metabolic scope for activity than shy individuals (Binder et al. 2016). Group living means competition for food, and individuals of qingbo carp (*Spinibarbus sinensis*) with a high metabolic rate and high demand for food are least sociable (Killen et al. 2016a). Golden grey mullets (*Liza aurata*) with high aerobic capacity maximise food intake at the leading edge of schools, where they need to cope with greater drag (Killen et al. 2012).

3.11 How Are Fish Influenced by Human Activities?

Fish thus face many challenges to cope with within their natural environment, and modern man does not make their life easier. Human impacts often have strong adverse effects on fish populations, and a number of large fishes are now ecologically extinct in many ecosystems (McCauley et al. 2015). Anthropogenic environmental changes may also alter evolutionary trajectories by changing which individuals that have the greatest fitness (Killen et al. 2016b) and may make fish approach their tolerance level and thereby impair welfare. Behavioural changes may be detectable before physiological changes are manifested, and changes in behaviour may thus be used to identify critical thresholds (Beitinger 1990). Anthropogenic contaminants may have a negative effect on cognition, and have been found to impair conditioned learning (Purdy 1989) and spatial discrimination learning (Levin et al. 2003).

Climate change has raised the water temperature, and fish are generally "cold-blooded" with the same body temperature as the water. Ectothermy is energetically more economical than maintaining a high and constant temperature (van de Pol et al. 2017), but temperature is the environmental factor that induces most phenotypic plasticity in fish and affects a number of physiological and developmental processes

(Pittman et al. 2013). Temperature affects metabolic rates and gastric evacuation (He and Wurtsbaugh 1993; Andersen 2001; Angilletta et al. 2002) as well as swimming speed and feeding ecology (Stoner 2004; Claireaux et al. 2006). However, ectotherms are not at the mercy of ambient temperature but can migrate towards warmer or colder environments, and global warming has already led to changes in the abundance and distribution of many fish species (Rijnsdorp et al. 2009). Mackerel (*Scomber scombrus*), for instance, have now moved further north and become established in the Norwegian Sea, and the rapid increase in the size of the mackerel stock has increased intra- and interspecific competition for food (Ólafsdóttir et al. 2016). Climate change has also triggered an enormous natural light experiment in the Arctic by reducing the extent and thickness of ice (Varpe et al. 2015). Less ice means that more light enters the water column and dramatically increasing the amount of light the fish receive. This enables them to detect prey more easily and forage more efficiently, which in turn influences the whole ecosystem.

The chemical and physical characteristics of the water are crucial because of the intimate relationship between the ambient water and the body fluids in the gills needed for respiratory and excretory functions. Severe oxygen depletion events in coastal waters have become increasingly common as a result of anthropogenic inputs of nutrients and organic matters (Wu 2002). The oxygen concentration affects metabolism, activity, schooling, and antipredatory behaviour (Pollock et al. 2007; Domenici et al. 2007), and severe levels of hypoxia influence growth and survival (Smith and Able 2003). The concentration of CO_2 in the water is rising in line with the increases in atmospheric CO_2 (Doney 2010). Elevated CO_2 concentrations have many negative effects on fish eggs and larvae (Heuer and Grosell 2014), and they also influence the distribution of fish (Fost et al. 2016). Coral reefs are severely threatened both by ocean acidification and rising temperatures, and most warm-water coral reefs will probably have disappeared by 2040–2050 (Hoegh-Guldberg et al. 2017). Pollution can also have negative effects on fish (Chap. 20). Small plastic debris has become a rapidly increasing problem due to the material itself and to chemical pollutants that attract to it, and fish that consume such material can suffer liver toxicity and pathology (Rochman et al. 2013). In intensive rearing, it is crucial to prevent poor water quality that restricts growth and impairs welfare.

Industrial fishing could result in a serious fall in the abundance of exploited species (McCauley et al. 2015). K-selected species are most sensitive to overfishing (Adams 1980). While developed countries are improving their management of fish stocks, the situation is becoming worse in developing countries (Ye and Gutierrez 2017). Local fisheries targeting manta rays for human consumption have continued for centuries, but in the past decade the growing markets for these gill rakers that are believed to cure various diseases have significantly increased the fishing effort (O'Malley et al. 2016). Manta rays have life-history characteristics that make them particularly vulnerable to overfishing (Couturier et al. 2012). Moreover, harvesting can exert a strong selection pressure and has driven many marine species to become smaller, grow more slowly, become less fecund and reproduce at smaller sizes (Jørgensen et al. 2007), and may also change the personality and activity level of the fish (Biro and Post 2008; Alós et al. 2012; Diaz and Sih 2017).

3.12 Do Fish Species Have Different Personalities?

We often imagine the sea as a dangerous place where fish all the time must prioritise not being eaten. The basic state ought therefore be to be shy and careful. But being on the defensive means fewer opportunities to encounter food and mates. So perhaps many fish species are actually less defensive than we usually think? In fact, modelling work has suggested that if an organism does not accurately know its risk of mortality, it should act as if the risk is less than the mean risk, and this can be viewed as being optimistic (McNamara et al. 2012). Individual members of any given species have different coping styles with some being more exploratory (proactive) and others more passive (reactive, Øverli et al. 2007; Chap. 12), so might it not be possible to classify fish species into proactive and reactive species with different basic personalities? A combination of conscious and accidental selection has resulted in personality differences between dog breeds (Starling et al. 2013), and it would be strange if the different selection pressures that fish species are exposed to should not have resulted in different basic personalities. Comparative approaches to animal personality face some problems, but Carter and Feeney (2012) identified four "clusters" of behaviourally similar coral reef species with regard to the bold–shy axis. The three teleost species bluegill sunfish, European crucian carp (*Carassius langsdorfii*) and goldfish (*Carassius auratus*) were also found to differ in their reactions to a novel environment, thus demonstrating that emotional reactivity and boldness are species specific (Yoshida et al. 2005). It has been suggested that differences in boldness and activity are the result of a trade-off between growth and mortality (Mittelbach et al. 2014), and the personality may be linked to habitat use and exploratory behaviour (Réale et al. 2010). A fish species that experiences low predation pressure and feeds on diverse and dispersed prey should be expected to have a positive exploratory attitude to unknown objects, as it would otherwise miss the opportunity to find food and gather information about its environment (Welton et al. 2003), while species that are at high risk of predation, feeding on prey that is easy to localise and identify should be expected to be less outgoing. We, therefore, suggest that fish species could be categorised along a proactive and reactive axis. The personality of a species could determine how well it adapts to farms, aquaria, or experimental tanks (see Box 3.8 and Chaps. 7 and 12).

Box 3.8 The Personality of a Species May Determine Whether It Is Suitable for Farming

Proactive individuals are risk prone and have low sensitivity to environmental stressors (Castanheira et al. 2017), so proactive fish species may be more exploratory and "optimistic" and therefore more suitable for farming than reactive species. On the other hand, they may be less flexible and more aggressive (Sih et al. 2004; Chap. 12). The pace of life is also a part of a species personality (Réale et al. 2010), and a high activity level can set limits

(continued)

Box 3.8 (continued)
on how small artificial environments a fish can tolerate. Social species may be able to endure higher densities than solitary species, but on the other hand they may be more negatively affected by various disturbances.

References

Abrahams M (2006) The physiology of antipredator behaviour: what you do with what you've got. In: Behaviour and physiology of fish. Elsevier, Amsterdam, pp 79–108
Adams PB (1980) Life history patterns in marine fishes and their consequences for fisheries management. Fish Bull 78:1–12
Ahrens RNM, Walters CJ, Christensen V (2012) Foraging arena theory. Fish Fish 13:41–59
Albert JS, Petry P, Reis RE (2011) Major biogeographic and phylogenetic patterns. In: Historical biogeography of neotropical freshwater fishes. University of California Press, Berkeley, pp 21–58
Alexander RL (1996) Evidence of brain-warming in the mobulid rays, *Mobula tarapacana* and *Manta birostris* (*Chondrichthyes*: *Elasmobranchii*: *Batoidea*: *Myliobatiformes*). Zool J Linnean Soc 118:151–164
Alós J, Palmer M, Arlinghaus R (2012) Consistent selection towards low activity phenotypes when catchability depends on encounters among human predators and fish. PLoS One 7:e48030
Andersen NG (2001) A gastric evacuation model for three predatory gadoids and implications of using pooled field data of stomach contents to estimate food rations. J Fish Biol 59:1198–1217
Angilletta MJ, Niewiarowski PH, Navas CA (2002) The evolution of thermal physiology in ectotherms. J Therm Biol 27:249–268
Ari C (2011) Encephalization and brain organization of mobulid rays (*Myliobatiformes*, *Elasmobranchii*) with ecological perspectives. Open Anat J 3:1–13
Ari C (2015) Shark tales. X-Ray Mag 69:80–82
Ari C, Correia JP (2008) Role of sensory cues on food searching behavior of a captive *Manta birostris* (*Chondrichtyes*, *Mobulidae*). Zoo Biol 27:294–304
Ari C, D'Agostino DP (2016) Contingency checking and self-directed behaviors in giant manta rays: do elasmobranchs have self-awareness? J Ethol 34:167–174
Arnold GP, Cook PH (1984) Fish migration by selective tidal stream transport: first results with a computer simulation model for the European continental shelf. In: Mechanisms of migration in fishes. Plenum Press, New York, pp 227–261
Balon EK (1975) Terminology of intervals in fish development. Can J Fish Aquat Sci 32:1663–1670
Barber I, Rushbrook BJ (2008) Parasites and fish behaviour. In: Fish behaviour. Science Publishers, Enfield, pp 525–561
Basolo AL (1990) Female preference for male sword length in the green swordtail, *Xiphophorus helleri* (*Pisces*: *Poeciliidae*). Anim Behav 40:332–338
Basolo AL, Alcaraz G (2003) The turn of the sword: length increases male swimming costs in swordtails. Proc R Soc Lond Ser B 270:1631–1636
Basolo AL, Wagner WE Jr (2004) Covariation between predation risk, body size and fin elaboration in the green swordtail. Biol J Linn Soc 83:87–100
Beitinger TL (1990) Behavioural reactions for the assessment of stress in fishes. J Great Lakes Res 16:495–528
Binder TR, Cooke SJ, Hinch SG (2011) Fish migrations – the biology of fish migration. In: Encyclopedia of fish physiology – from genome to environment. Academic Press, San Diego, pp 1921–1927

Binder TR, Wilson ADM, Wilson SM, Suski CD, Godin J-GJ, Cooke SJ (2016) Is there a pace-of-life syndrome linking boldness and metabolic capacity for locomotion in bluegill sunfish? Anim Behav 121:175–183

Biro PA, Post JR (2008) Rapid depletion of genotypes with fast growth and bold personality traits from harvested fish populations. Proc Natl Acad Sci U S A 105:2919–2922

Braun CD, Skomal GB, Thorrold SR, Berumen ML (2014) Diving behavior of the reef manta ray links coral reefs with adjacent deep pelagic habitats. PLoS One 9:e88170

Brönmark C, Pettersson LB (1994) Chemical cues from piscivores induce a change in morphology in crucian carp. Oikos 70:396–402

Brown JH (2014) Why are there so many species in the tropics? J Biogeogr 41:8–22

Brown C, Gardner C, Braithwaite VA (2005) Differential stress responses in fish from areas of high- and low-predation pressure. J Comp Physiol 175:305–312

Brown GE, Ferrari MCO, Chivers DP (2006) Learning about danger: chemical alarm cues and threat-sensitive assessment of predation risk by fishes. In: Fish cognition and behavior. Blackwell, Oxford, pp 59–74

Bshary R (2011) Machiavellian intelligence in fishes. In: Fish cognition and behavior, 2nd edn. Wiley, Oxford, pp 277–297

Burt de Perera T (2004) Fish can encode order in their spatial map. Proc R Soc Lond Ser B 271:2131–2134

Carter AJ, Feeney WE (2012) Taking a comparative approach: analysing personality as a multivariate behavioural response across species. PLoS One 7:e42440

Castanheira MF, Conceicão LEC, Millot S, Rey S, Bégout M-L, Damsgård B, Kristiansen T, Höglund E, Øverli Ø, Martins CIM (2017) Coping styles in farmed fish: consequences for aquaculture. Rev Aquac 9:2–41

Claireaux G, Couturier C, Groison A-L (2006) Effect of temperature on maximum swimming speed and cost of transport in juvenile European sea bass (*Dicentrarchus labrax*). J Exp Biol 209:3420–3428

Clark CW, Levy DA (1988) Diel vertical migrations by juvenile sockeye salmon and the antipredation window. Am Nat 131:271–290

Connor RC (1992) Egg-trading in simultaneous hermaphrodites: an alternative to tit-for-tat. J Evol Biol 5:523–528

Cook RM, Armstrong DW (1986) Stock-related effects in the recruitment of North Sea haddock and whiting. ICES J Mar Sci 42:272–280

Couturier LIE, Marshall AD, Jaine FRA, Kashiwagi T, Pierce SJ, Townsend KA, Weeks SJ, Bennett MB, Richardson AJ (2012) Biology, ecology and conservation of the *Mobulidae*. J Fish Biol 80:1075–1119

Cowen RK, Castro LR (1994) Relation of coral-reef fish larval distributions to island scale circulation around Barbados, West-Indies. Bull Mar Sci 54:228–244

Dawkins R, Krebs JR (1979) Arms races between and within species. Proc R Soc Lond Biol Soc 205:489–511

Depczynski M, Bellwood D (2005) Shortest recorded vertebrate lifespan found in a coral reef fish. Curr Biol 15:288–289

Diaz PB, Sih A (2017) Behavioural responses to human-induced change: why fishing should not be ignored. Evol Appl 10:231–240

Domenici P, Lefrancois C, Shingles A (2007) Hypoxia and the antipredator behaviours of fishes. Proc R Soc Lond Ser B 362:2105–2121

Doney SC (2010) The growing human footprint on coastal and open-ocean biogeochemistry. Science 328:1512–1516

Duffield C, Ioannou CC (2017) Marginal predation: do encounter or confusion effects explain the targeting of prey group edges? Behav Ecol 28:1283–1292

Fernö A, Pitcher TJ, Melle V, Nøttestad L, Mackinson S, Hollingworth C, Misund OA (1998) The challenge of the herring in the Norwegian Sea: making optimal collective spatial decisions. Sarsia 83:149–167

Fernö A, Huse G, Jakobsen PJ, Kristiansen TS, Nilsson J (2011) Fish behaviour, learning, aquaculture and fisheries. In: Fish cognition and behavior, 2nd edn. Wiley, Oxford, pp 359–404
Fischer B, Dieckmann U, Taborsky B (2011) When to store energy in a stochastic environment. Evolution 65:1221–1232
Fleming IA, Huntingford F (2012) Reproductive behaviour. In: Aquaculture and behavior. Wiley, Oxford, pp 286–321
Forsgren E (1997) Female sand gobies prefer good fathers over dominant males. Proc R Soc Lond Ser B 264:1283–1286
Fost BA, Ferreri CP, Braithwaite VA (2016) Behavioral response of brook trout and brown trout to acidification and species interactions. Environ Biol Fish 99:983–998
Genner MJ, Seehausen O, Lunt DH, Joyce DA, Shaw PW, Carvalho GR, Turner GF (2007) Age of cichlids: new dates for ancient lake fish radiations. Mol Biol Evol 24:1269–1282
Gerlotto F, Bertrand S, Bez N, Gutierrez M (2006) Waves of agitation inside anchovy schools observed with multibeam sonar: a way to transmit information in response to predation. ICES J Mar Sci 63:1405–1417
Ghalambor CK, McKay JK, Carroll SP, Reznick DN (2007) Adaptive versus non-adaptive phenotypic plasticity and the potential for contemporary adaptation in new environments. Funct Ecol 21:394–407
Gilmore RG, Dodrill JW, Linley PA (1983) Embryonic development of the sand tiger shark, *Odontaspis taurus* Rafinesque. Fish Bull 81:201–225
Gopko M, Mikheev VN, Taskinen J (2017) Deterioration of basic components of the anti-predator behavior in fish harboring eye fluke larvae. Behav Ecol Sociobiol 71:68
Gross MR (1987) Evolution of diadromy in fishes. Am Fish Soc Symp Ser 1:14–25
Haddock SHD, Moline MA, Case JF (2010) Bioluminescence in the sea. Annu Rev Mar Sci 2:443–493
Hagedorn M, Heiligenberg W (1985) Court and spark: electric signals in the courtship and mating of gymnotoid fish. Anim Behav 33:254–265
Handegard NO, Boswell KM, Ioannou CC, Leblanc SP, Tjøstheim DB, Couzin ID (2012) The dynamics of coordinated group hunting and collective information transfer among schooling prey. Curr Biol 22:1213–1217
Handeland SO, Järvi T, Fernö A, Stefansson SO (1996) Osmotic stress, antipredatory behaviour and mortality of Atlantic salmon (*Salmo salar* L.) smolts. Can J Fish Aquat Sci 53:2673–2680
Harden Jones FR (1968) Fish migration. Edward Arnold, London
Hart P (1993) Teleost foraging: facts and theories. In: Behaviour of teleost fishes, 2nd edn. Chapman and Hall, London, pp 253–284
Haugland T, Rudolfsen G, Figenschou L, Folstad I (2011) Is the adipose fin and the lower jaw (kype) related to social dominance in male Arctic charr *Salvelinus alpinus*? J Fish Biol 79:1076–1083
Hays G (2003) A review of the adaptive significance and ecosystem consequences of zooplankton diel vertical migrations. Hydrobiologia 503:163–170
He E, Wurtsbaugh WA (1993) An empirical model of gastric evacuation rates for fish and an analysis of digestion in piscivorous brown trout. Trans Am Fish Soc 122:717–730
Hecht T, Uys W (1997) Effect of density on the feeding and aggressive behaviour in juvenile African catfish, *Clarias gariepinus*. S Afr J Sci 93:537–541
Helfman G, Collette BB, Facey DE, Bowen BW (2009) The diversity of fishes: biology, evolution and ecology, 2nd edn. Wiley, Oxford
Hemelrijk CK, Reid DAP, Hildenbrandt H, Padding JT (2014) The increased efficiency of fish swimming in a school. Fish Fish 16:511–521
Heuer M, Grosell M (2014) Physiological impacts of elevated carbon dioxide and ocean acidification on fish. *American Journal of Physiology. Regulatory*. Integr Comp Physiol 307:1061–1084
Heupel MR, Simpfendorfer CA (2008) Movement and distribution of young bull sharks *Carcharhinus leucas* in a variable estuarine environment. Aquat Biol 1:277–289
Hoar WS (1953) Control and timing of fish migration. Biol Rev 28:437–452

Hoegh-Guldberg O, Poloczanska ES, Skirving W, Dove S (2017) Coral reef ecosystems under climate change and ocean acidification. Front Mar Sci 80:1737–1742
Hoffmeyer J (2008) Biosemiotics: an examination into the signs of life and the life of signs. University of Scranton Press, Chicago, IL
Hori M (1993) Frequency-dependent natural selection in the handedness of scale-eating cichlid fish. Science 260:216–219
Huntingford FA (1993) Development of behaviour in fish. In: Behaviour of teleost fishes, 2nd edn. Chapman and Hall, London, pp 57–83
Hussey NE, Kessel ST, Aarestrup K, Cooke SJ, Cowley PD, Fisk AT, Harcourt RG, Holland KN, Iverson SJ, Kocik JF, Mils Flemming JE, Whoriskey FG (2015) Aquatic animal telemetry: a panoramic window into the underwater world. Science 348:1221–1231
Irigoien X, Klevjer TA, Røstad A, Martinez U, Boyra G, Acuña JL, Bode A, Echevarria F, Gonzalez-Gordillo JI, Hernandez-Leon S, Agusti S, Aksnes DL, Duarte CM, Kaartvedt S (2014) Large mesopelagic fishes biomass and trophic efficiency in the open ocean. Nat Commun 5:3271
Järvi T (1990) The effects of male dominance, secondary sexual characteristics and female mate choice on the mating success of male Atlantic salmon (*Salmo salar*). Ethology 84:123–132
Jerry DR (2014) Biology and culture of Asian Seabass Lates calcarifer. CRC, Boca Raton
Johns GC, Avise JC (1998) A comparative summary of genetic distances in the vertebrates from the mitochondrial cytochrome b gene. Mol Biol Evol 15:1481–1490
Jørgensen C, Enberg K, Dunlop ES, Arlinghaus R, Boukal DS, Brander K, Ernande B, Gardmark A, Johnston F, Matsumura S, Pardoe H, Raab K, Silva A, Vainikka A, Dieckmann U, Heino M, Rijnsdorp AD (2007) Managing evolving fish stocks. Science 318:1247–1248
Kaartvedt S, Staby A, Aksnes DL (2012) Efficient trawl avoidance by mesopelagic fishes causes large underestimation of their biomass. Mar Ecol Prog Ser 456:1–6
Killen SS, Marras S, Steffensen JF, McKenzie DJ (2012) Aerobic capacity influences the spatial position of individuals within fish schools. Proc R Soc Lond Ser B 279:357–364
Killen SS, Fu C, Wu Q, Wang Y-X, Fu S-J (2016a) The relationship between metabolic rate and sociability is altered by food deprivation. Funct Ecol 30:1358–1365
Killen SS, Adriaenssens B, Marras S, Claireaux G, Cooke SJ (2016b) Context dependency of trait repeatability and its relevance for management and conservation of fish populations. Conserv Physiol 4:1–19
Kjesbu OS (1988) The spawning activity of cod (*Gadus morhua* L.). J Fish Biol 34:195–206
Knapp R, Nett BD (2008) Parasites and fish behaviour. In: Fish behaviour. Science Publishers, Enfield, pp 411–433
Kocher TD (2004) Adaptive evolution and explosive speciation: the cichlid model. Nat Rev Genet 5:288–298
Kodric-Brown A (1998) Sexual dichromatism and temporary color changes in the reproduction of fishes. Am Zool 38:70–81
Kotrschal K, van Staaden MJ, Huber R (1998) Fish brains: evolution and environmental relationships. Rev Fish Biol Fish 8:373–408
Kottelat M, Britz R, Hui TH, Kai-Erik Witte K-E (2006) *Paedocypris*, a new genus of Southeast Asian cyprinid fish with a remarkable sexual dimorphism, comprises the world's smallest vertebrate. Proc R Soc Lond Ser B 273:895–899
Kramer B (1999) Waveform discrimination, phase sensitivity and jamming avoidance in a wave-type electric fish. J Exp Biol 202:1387–1398
Krause J, Godin J-GJ (1994) Shoal choice in the banded killifish (*Fundulus diuphunus*, Teleostei, Cyprinodontidae): effects of predation risk, fish size, species composition and size of shoals. Ethology 98:128–136
Krause JE, Herbert-Read F, Seebacher P, Domenici ADM, Wilson S, Marras MBS, Svendsen D, Strömbom JF, Steffensen S, Krause PE, Viblanc P, Couillaud P, Bach PS, Sabarros PS,

Zaslansky P, Kurvers RHJM (2017) Injury-mediated decrease in locomotor performance increases predation risk in schooling fish. Proc R Soc Lond Ser B 372:20160232
Krebs JR, Dawkins R (1984) Animal signals: mind-reading and manipulation. In: Behavioural ecology: an evolutionary approach, 2nd edn. Blackwell, Oxford, pp 380–402
Laland KN, Atton N, Webster MM (2011) From fish to fashion: experimental and theoretical insights into the evolution of culture. Proc R Soc Lond Ser B 366:958–968
Leung TLF (2014) Fish as parasites: an insight into evolutionary convergence in adaptations for parasitism. J Zool 294:1–12
Levin ED, Chrysanthis E, Yacisin K, Linney E (2003) Chlorpyrifos exposure of developing zebrafish: effects on survival and long-term effects on response latency and spatial discrimination. Neurotoxicol Teratol 25:51–57
Lindström K, St.Mary CM (2008) Parental care and sexual selection. In: Fish behaviour. Science Publishers, Enfield, pp 377–409
Locascio JV, Mann DA (2011) Localization and source level estimates of black drum (*Pogonias cromis*) calls. J Acoust Soc Am 130:1868–1879
Maury O (2017) Can schooling regulate marine populations and ecosystems. Prog Oceanogr 156:91–103
Maynard Smith J (1982) Evolution and the theory of games. Cambridge University Press, Cambridge
McCauley DJ, Pinsky ML, Palumbi SR, Estes JA, Joyce FH, Warner RR (2015) Marine defaunation: animal loss in the global ocean. Science 347:247
McKaye KR (1981) Death feigning: a unique hunting behavior by the predatory cichlid, *Haplochromis livingstoni* of Lake Malawi. Environ Biol Fish 6:361–365
McNamara JM, Trimmer PC, Houston AI (2012) It is optimal to be optimistic about survival. Biol Lett 8:516–519
Metcalfe NB, van Leeuwen TE, Killen SS (2016) Does individual variation in metabolic phenotype predict fish behaviour and performance? J Fish Biol 88:298–321
Milinski M, Griffiths S, Wegner K, Reusch T, Haas-Assenbaum A, Boehm T (2005) Mate choice decisions of stickleback females predictably modified by MHC peptide ligands. Proc Natl Acad Sci U S A 102:4414–4418
Mittelbach GG, Ballew NG, Kjelvik MK (2014) Fish behavioral types and their ecological consequences. Can J Fish Aquat Sci 71:927–944
Moyle PB, Cech JJ (2004) Fishes: an introduction to ichthyology, 5th edn. Prentice-Hall, Upper Saddle River, NJ
Nelson JS (2006) Fishes of the world. Wiley, Hoboken, NJ
Nettle D, Andrews C, Bateson M (2017) Food insecurity as a driver of obesity in humans: the insurance hypothesis. Behav Brain Sci 40:e105
Nguyen THD, Wenresti GG, Nitin KT, Truong H (2013) Cobia cage culture distribution mapping and carrying capacity assessment in Phu Quoc, Kien Giang province. J Viet Environ 4:12–19
Nielsen J, Hedeholm RB, Heinemeirer J, Bushnell PG, Christiansen JS, Olsen J, Ramsey CB, Brill RW, Simon M, Steffensen KF, Steffensen JF (2016) Eye lens radiocarbon reveals centuries of longevity in the Greenland shark (*Somniosus microcephalus*). Science 353:702–704
Nilsson PA, Brönmark C, Pettersson LB (1995) Benefits of a predator-induced morphology in crucian carp. Oecologia 104:291–296
Nøttestad L, Fernö A, Axelsen BE (2002) Digging in the deep: killer whales´ advanced hunting tactic. Polar Biol 25:939–941
Nøttestad L, Fernö A, Misund OA, Vabø R (2004) Understanding herring behaviour: linking individual decisions, school patterns and population distribution. In: The Norwegian Sea ecosystem. Tapir, Trondheim, pp 227–262
O'Malley MP, Townsend KA, Hilton P, Heinrichs S, Stewart JD (2016) Characterization of the trade in manta and devil ray gill plates in China and South-east Asia through trader surveys. Aquat Conserv Mar Freshwat Ecosyst 27:394–413

Ohno S (1999) Gene duplication and the uniqueness of vertebrate genomes circa 1970–1999. Cell Dev Biol 10:517–522
Ólafsdóttir AH, Slotte A, Jacobsen JA, Oskarsson GJ, Utne KR, Nøttestad L (2016) Changes in weight-at-length and size-at-age of mature Northeast Atlantic mackerel (*Scomber scombrus*) from 1984 to 2013: effects of mackerel stock size and herring (*Clupea harengus*) stock size. ICES J Mar Sci 73:1255–1265
Östlund-Nilsson S, Mayer I, Huntingford FA (2007) Biology of the three-Spined stickleback. CRC, Boca Raton
Øverli Ø, Sørensen C, Pulman KGT, Pottinger TG, Korzan W, Summers CH, Nilsson E (2007) Evolutionary background for stress-coping styles: relationships between physiological, behavioral, and cognitive traits in non-mammalian vertebrates. Biobehav Rev 31:396–412
Parrish JK, Viscido SV, Grünbaum D (2002) Self-organized fish schools: an examination of emergent properties. Biol Bull 202:296–305
Partridge BL, Pitcher TJ (1980) The sensory basis of fish schools: relative role of lateral line and vision. J Comp Physiol 135:315–325
Pickering AD (1981) Stress and fish. Academic, London
Pietsch TW (2005) Dimorphism, parasitism, and sex revisited: modes of reproduction among deep-sea ceratioid anglerfishes (*Teleostei*: *Lophiiformes*). Ichthyol Res 52:207–236
Pietsch TW, Grobecker DB (1987) Frogfishes of the world. Systematics, zoogeography, and behavioral ecology. Stanford University Press, Stanford
Pitcher TJ, Parrish JK (1993) The functions of shoaling behaviour. In: The behaviour of teleost fishes, 2nd edn. Chapman and Hall, London, pp 363–439
Pittman K, Yúfera M, Pavlidis M, Geffen AJ, Koven W, Ribeiro L, Zambonino-Infante JL, Tandler A (2013) Fantastically plastic: fish larvae equipped for a new world. Rev Aquac 5(Suppl. 1):224–267
Pollock MS, Clarke LMJ, Dube MG (2007) The effects of hypoxia on fishes: from ecological relevance to physiological effects. Environ Rev 15:1–14
Pope E, Hays G, Thys T, Doyle T, Sims D, Queiroz N, Hobson V, Kubicek L, Houghton JR (2010) The biology and ecology of the ocean sunfish *Mola mola*: a review of current knowledge and future research perspectives. Rev Fish Biol Fish 20:471–487
Purdy JE (1989) The effects of brief exposure to aromatic hydrocarbons on feeding and avoidance behaviour in coho salmon, *Oncorhynchus kisutch*. J Fish Biol 34:621–629
Radesäter T, Fernö A (1979) On the function of the “eye-spots” in agonistic behaviour in the firemouth cichlid (*Cichlasoma meeki*). Behav Process 4:5–13
Réale D, Dingemanse NJ, Kazem AJN, Wright J (2010) Evolutionary and ecological approaches to the study of personality. Phil Trans R Soc B Biol Sci 365:3937–3946
Reebs SG (2002) Plasticity of diel and circadian activity rhythms in fishes. Rev Fish Biol Fish 12:349–371
Rieucau G, Fernö A, Ioannou CC, Handegard NO (2015) Towards of a firmer explanation of large shoal formation, maintenance and collective reactions in marine fish. Rev Fish Biol Fish 25:21–37
Rijnsdorp AD, Peck MA, Engelhard GH, Möllmann C, Pinnegar JK (2009) Resolving the effect of climate change on fish populations. ICES J Mar Sci 66:1570–1583
Rochman CM, Hoh E, Kurobe T, Teh SJ (2013) Ingested plastic transfers hazardous chemicals to fish and induces hepatic stress. Sci Rep 3:3263
Roff DA (1984) The evolution of life history parameters in teleosts. Can J Fish Aquat Sci 41:989–1000
Rose GA, Leggett WC (1990) The importance of scale to predator-prey spatial correlations: an example of Atlantic fishes. Ecology 71:33–43
Rosenthal GG, Fitzsimmons JN, Woods KU, Gerlach G, Fisher HS (2011) Tactical release of a sexually-selected pheromone in a swordtail fish. PLoS One 6:e16994
Rulifson RA, Dadswell MJ (1995) Life history and population characteristics of striped bass in Atlantic Canada. Trans Am Fish Soc 124:477–507

Russell ES (2008) Fish migrations. Biol Rev 12:320–337
Sadovy de Mitcheson Y, Liu M (2008) Functional hermaphroditism in teleosts. Fish Fish 9:1–43
Sæle Ø, Smáradóttir H, Pittman K (2006) Twisted story of eye migration in flatfish. J Morphol 267:730–738
Santangelo N (2015) Female breding experience affects parental care strategies of both parents in a monogamous cichlid fish. Anim Behav 104:31–37
Sargent RC, Gross MR (1993) William's principle: an explanation of parental care in teleost fishes. In: Behaviour of teleost fishes. Chapman and Hall, London, pp 333–361
Schuster S, Wöhl S, Griebsch M, Klostermeier I (2006) Animal cognition: how archer fish learn to down rapidly moving targets. Curr Biol 16:378–383
Sih A, Bell A, Johnson JC (2004) Behavioural syndromes: an ecological and evolutionary overview. Trends Ecol Evol 19:372–378
Similä T, Ugarte F (1993) Surface and underwater observation of cooperatively feeding killer whales in Northern Norway. Can J Zool 71:1494–1499
Simon M, Ugarte F, Wahlberg M, Miller LA (2006) Icelandic killer whales *Orcinus orca* use a pulsed call suitable for manipulating the schooling behaviour of herring *Clupea harengus*. Bioacoustics 16:57–74
Sims DW, Wearmouth VJ, Southall EJ, Hill JM, Moore P, Rawlinson K, Hutchinson N, Budd GC, Righton D, Metcalfe JD, Nash JP, Morritt D (2006) Hunt warm, rest cool: bioenergetic strategy underlying diel vertical migration of benthic shark. J Anim Ecol 75:176–190
Sloman KA, Wilson RW, Balshine S (2006) Behaviour and physiology of fish. Elsevier, Amsterdam
Smith KJ, Able KW (2003) Dissolved oxygen dynamics in salt marsh pools and its potential impacts on fish assemblages. Mar Ecol Prog Ser 258:223–232
Smith TB, Skúlason S (1996) Evolutionary significance of resource polymorphisms in fishes, amphibians, and birds. Annu Rev Ecol Syst 27:111–133
Solstorm F, Solstorm D, Oppedal F, Olsen RE, Stien LH, Fernö A (2016) Not too slow and not too fast: water currents affect group structure, aggression and welfare in post-smolt Atlantic salmon *Salmo salar*. Aquac Environ Interact 8:339–347
Starling MJ, Branson NJ, Thomson PC, McGreevy PD (2013) "Boldness" in the domestic dog differs among breeds and breed groups. Behav Process 97:53–62
Stoner AW (2004) Effects of environmental variables on fish feeding ecology: implications for the performance of baited fishing gear and stock assessment. J Fish Biol 65:1445–1471
Sun L, Chen H (2014) Effects of water temperature and fish size on growth and bioenergetics of cobia (*Rachycentron canadum*). Aquaculture 426–427:172–180
Svensson PA, Wong BBM (2011) Carotenoid-based signals in behavioural ecology: a review. Behaviour 148:131–189
Taborsky M (1984) Broodcare helpers in the cichlid fish *Lamprologus brichardi*: their costs and benefits. Anim Behav 32:1236–1252
Taborsky B (2006) The influence of juvenile and adult environments on life-history trajectories. Proc R Soc Lond Ser B 273:741–750
Takeuchi Y, Oda Y (2017) Lateralized scale-eating behaviour of cichlid is acquired by learning to use the naturally stronger side. Sci Rep 7:8984
Temming A, Floeter J, Ehrich S (2007) Predation hot spots: large scale impact of local aggregations. Ecosystems 10:865–876
Terzibasi E, Valenzano DR, Benedetti M, Roncaglia P, Cattaneo A, Domenici L, Cellerino A (2008) Large differences in aging phenotype between strains of the short-lived annual fish *Nothobranchius furzeri*. PLoS One 3:e3866
Tibbetts EA (2014) The evolution of honest communication: integrating social and physiological costs of ornamentation. Integr Comp Biol 54:578–590
Vallon M, Grom C, Kalb N, Sprenger D, Anthes N, Lindström K, Heubel KU (2016) You eat what you are: personality-dependent filial cannibalism in a fish with paternal care. Ecol Evol 6:1340–1352

van de Peer Y, Taylor JS, Joseph J, Meyer A (2002) Wanda: a database of duplicated fish genes. Nucleic Acids Res 30:109–112

van de Pol I, Flik G, Gorissen M (2017) Comparative physiology of energy metabolism: fishing for endocrine signals in the early vertebrate pool. Front Endocrinol 8:1–18

Varpe Ø, Daase M, Kristiansen T (2015) A fish-eye view on the new Arctic lightscape. ICES J Mar Sci 72:2532–2538

Vincent ACJ, Sadler LM (1995) Faithful pair bonds in wild seahorses, *Hippocampus whitei*. Anim Behav 50:1557–1569

Walker BW (1961) The ecology of the Salton Sea, California, in relation to the sportfishery. Fish Bull 113:1–204

Warner RR, Swearer SE (1991) Social control of sex change in the bluehead wrasse, *Thalassoma bifasciatum* (*Pisces*: *Labridae*). Biol Bull 181:199–204

Warren WC, García-Pérez R, Xu S, Lampert KP, Chalopin D, Stöck M, Loewe L, Lu Y, Kuderna L, Minx P, Montague MJ, Tomlinson C, Hillier LW, Murphy DN, Wang J, Wang Z, Garcia CM, Thomas GCW, Volff J-N, Farias F, Aken B, Walter RB, Pruitt KD, Marques-Bonet T, Hahn MW, Kneitz S, Lynch M, Schartl M (2018) Clonal polymorphism and high heterozygosity in the celibate genome of the Amazon molly. Nat Ecol Evol 2:669–679

Webb PW (1984) Body form, locomotion and foraging in aquatic vertebrates. Am Zool 24:107–120

Welton NJ, McNamara JM, Houston AI (2003) Assessing predation risk: optimal behaviour and rules of thumb. Theor Popul Biol 64:417–430

Wood CM, Brix KV, De Boeck G, Bergman HL, Bianchini A, Bianchini LF, Maina JN, Johannsson OE, Kavembe GD, Papah MB, Letura KM, Ojoo RO (2016) Mammalian metabolic rates in the hottest fish on earth. Sci Rep 6:26990

Wright SD, Ross HA, Keeling DJ, McBride P, Gillman LN (2011) Thermal energy and the rate of genetic evolution in marine fishes. Evol Ecol 25:525–530

Wu RSS (2002) Hypoxia: from molecular responses to ecosystem response. Mar Pollut Bull 45:35–45

Ydenberg RC, Dill LM (1986) The economics of fleeing from predators. Adv Study Behav 16:229–249

Ye Y, Gutierrez NL (2017) Ending fishery overexploitation by expanding from local successes to globalized solutions. Nat Ecol Evol 1:0179

Yoshida M, Nagamine M, Uematsu K (2005) Comparison of behavioral responses to a novel environment between three teleosts, bluegill *Lepomis macrochirus*, crucian carp *Carassius langsdorfii*, and goldfish *Carassius auratus*. Fish Sci 71:314–319

Zahavi A (1975) Mate selection - a selection for a handicap. J Theor Biol 53:205–214

Chapter 4
Fish Behaviour: Determinants and Implications for Welfare

Felicity A. Huntingford

Abstract This chapter starts by spelling out different ways of answering the question "why does an animal behave in this way?", often referred to as Tinbergen's four question (Tinbergen, N., The study of instinct. Clarendon Press, Oxford, 1951). These are: what causes the behaviour? How does it develop? What are its consequences for fitness? and Through what stages did it evolve? Examples of answers to these different questions are given in relation to the behaviour of wild fish, looking at: use of space; foraging and feeding; avoiding predators; aggression and fighting and courtship. Farmed fish are relatively little domestication compared to farmed terrestrial species and they bring with them to the fish farm many of the behavioural characteristics of farmed fish; this is discussed in the context of each of the broad behavioural systems mentioned above. Problems for the welfare of farmed fish arising from expression of the natural behaviour of the fish concerned are discussed and behaviourally-based solutions suggested.

Keywords Aggression · Aquaculture · Cannibalism · Causes of behaviour · Courtship · Development of behaviour · Feeding · Functions of behaviour · Orientation · Predation · Welfare

4.1 Introduction: Questions About Fish Behaviour

Niko Tinbergen, the Nobel prize winning co-founder of modern behavioural biology, recognised that the question "why does an animal behave in this way?" could mean four quite different things (Tinbergen 1951). These are summarised in Box 4.1. The causes of behaviour are equivalent to the proximate explanations of behaviour discussed in Chap. 3 and its functions to the ultimate explanations. The behaviour of wild fish has been very extensively studied and this chapter describes

F. A. Huntingford (✉)
Institute of Biodiversity, Animal Health and Comparative Medicine, University of Glasgow, Glasgow, Scotland, UK
e-mail: Felicity.Huntingford@glasgow.ac.uk

T. S. Kristiansen et al. (eds.), *The Welfare of Fish*, Animal Welfare 20,
https://doi.org/10.1007/978-3-030-41675-1_4

briefly some important aspects of the behaviour of fish in their natural environments. It then gives examples of some answers to Tinbergen's questions and explores their implications for welfare of farmed fish.

Box 4.1 Tinbergen's Four Questions
When confronted with a large benthic prey item such as a gammarid, three-spined stickleback (*Gasterosteus aculeatus*) may or may not ingest it. A biologist attempting to understand why this is the case could be asking:

Causes: What mechanisms make the fish eat or not eat a particular prey item at a particular time, while rejecting other prey?

Development: When and how does eating large prey appear in the fishes' behavioural repertoire as it grows from a fertilised egg to an independent animal?

Functions: What are the consequences for fitness of eating or refusing a large gammarid?

Phylogenetic history: Through what series of changes has the capacity for eating large benthic prey evolved?

4.2 Broad Behavioural Systems in Fish

For convenience, behaviour is often classified in terms of its broad biological function, including use of space; feeding; avoiding predation; aggression and courtship. Some of these are discussed in Chap. 3, but are described again briefly here.

4.2.1 Use of Space

The movements of wild fish through their 3-dimensional world can be loosely classified into movements within a localised area or home range, periodic dispersal from such home ranges and migration (Huntingford et al. 2012a and refs. therein). Cleaner wrasse (for example, *Labroides* spp.) occupy small home ranges centred around particular coral heads that serve as cleaning stations for client species with ectoparasites that need removing (Oates et al. 2010). Saithe (*Pollachius virens*) sometimes focus their activity around floating fish farms, which provide both shelter and food (Skilbrei and Ottera 2016). Many fish species are able to return to their home area following displacement; for example displaced black rockfish (*Sebastes cheni*) can find their way home from as much as 4 km away (Mitamura et al. 2012). Spontaneous dispersal often occurs when local conditions deteriorate and ceases once favourable conditions are encountered; juvenile Atlantic salmon (*Salmo salar*) abandon shallow feeding stations when water levels fall, moving into nearby pools (Huntingford et al. 1999). Migration refers to directed movements from one well-defined area to another, often with a return migration to the original location. Some examples of fish migration on different scales are presented in Box 4.2.

Box 4.2 Examples of Fish Migrations on Different Temporal and Geographic Scales

Daily movements through relatively short distances: Brown surgeonfish (*Acanthurus nigrofuscus*) move each day between night time shelters and offshore feeding sites some 1.5 km away (Mazeroll and Montgomery 1995).

Daily vertical migration: Many fish show regular daily vertical migrations for the purpose of feeding, avoiding predators and/or tracking favourable temperatures. For example, Atlantic salmon tagged before migrating to sea swim near the surface most of the time, but move into slightly deeper water during daytime (Fig. 4.1; Gudjonsson et al. 2015).

Annual migration over very long distances: Many fish species migrate on a seasonal basis between feeding and spawning areas, often hundreds of kilometres apart; examples include Norwegian cod, eels and many salmonids (Metcalfe et al. 2008) and lumpfish (Kennedy et al. 2016).

4.2.2 *Foraging and Feeding: What Fish Eat, How and When?*

Fish exploit many kinds of food (Jobling et al. 2012a and refs therein; Box 4.3), which they gather in a variety of ways. Algae may be cropped or grazed and benthic invertebrates may be pick up individually or sifted en masse from water or substratum. Several groups of fish (lumpfish, *Cyclopterus lumpus*, for example Imsland et al. 2015a) act as cleaners, nipping at and removing ectoparasites from the skin, mouth and gill cavities of larger fish. Some piscivorous fish eat parts of their prey without killing them, taking flesh (piranhas), blood (*Trichomycterid* catfish) or scales (several cichlid and blenny species), but most swallow their prey whole, often using suction or ram feeding.

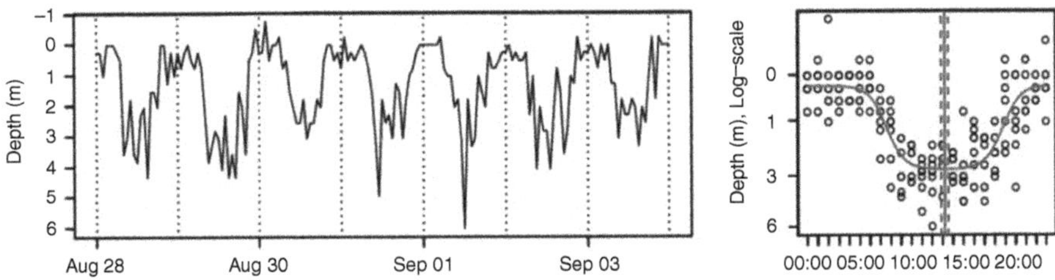

Fig. 4.1 Daily vertical movements in tagged Atlantic salmon migrating off the coast of Iceland over a 4-day period in autumn. *LHS* Original measurements (dashed lines represent midnight), *RHS* Summarised data across the 4 days, with line of best fit and estimated solar noon (vertical line. Gudjonsson et al. 2015)

Box 4.3 Some Examples of What Fish Eat

Herbivory: Fish feed on a wide range of different types of plants. Some African cichlids feed on phytoplankton (*Tilapia esculenta*), some on algae (*Pseudotropheus zebra*, which scrapes algae off rocks and *P. fuscus*, which feeds on filamentous algae) and some on macrophytes (*Oreochromis rendalli*).

Carnivory: Fish feed on a wide variety of animals, including zooplankton (ciscos, or lake herring, in large North American lakes), macro-invertebrates (pufferfish) and other fish (pike *Esox lucius* and many species of shark). Piscivory often involves cannibalism, which usually occurs early in development, as in young cod, but may also persist throughout life, as in pike.

Detritivory: Many fish, including the red-eyed mullet, *Liza haematocheila*, which is farmed in China, and the mrigal carp (*Cirrhinus mrigala*), feed on the remains of dead plants or animals; detritus is included to a lesser extent in the diet of many other cyprinids and cichlids.

Omnivorous: Many fish eat a mixture of plant and animal food. This is the case for the Mozambique cichlid (*Oreochromis mossambicus*), channel catfish (*Ictalurus punctatus*) and many carp species.

Within each broad dietary category, fish have specific, age-dependent dietary preferences; young fish of several *Tilapia* species switch from feeding selectively on zooplanktonic prey, which are rich in the protein needed for early growth, to eating less nutritious phytoplankton or detritus (Beveridge and Baird 2000). Fish display so-called "nutritional wisdom", selecting from a variety of food types the combination that best meets their physiological needs. Given access to three feeders each containing a pure macronutrient (protein, lipid and carbohydrate), carnivorous fish such as rainbow trout (*Oncorhynchus mykiss*), sea bass (*Dicentrarchus labrax*) and seabream (*Diplodus puntazzo*) select diets composed predominantly of protein, whereas omnivorous species such as goldfish (*Carassius auratus*) select lipid-rich diets (Fortes da Silva et al. 2016).

Fish are not always equally likely to eat, even when offered identical food, because of changes in appetite. Such changes may be irregular. For example, the tendency to eat increases during a fast as the intestinal tract empties and nutrient reserves are depleted; appetite falls as fish become sated during a meal and as nutrient reserves build-up. Appetite shifts may also be rhythmic, mapping onto environmental cycles with tidal, daily, lunar and/or annual periods (Jobling et al. 2012b and refs therein). For example, in juvenile Atlantic salmon held in cages and fed by an appetite-based feeding system (Sect. 4.4.2), daily feeding rhythms are apparent, changing from a dawn peak in summer to a predominantly midday peak in winter (Fig. 4.2a). Daily food intake decreases during autumn and winter, before increasing in spring, relating mainly to seasonal changes in temperature. Food intake also falls for a few days after stressful events such as vaccination and grading (Fig. 4.2b; Noble et al. 2007a).

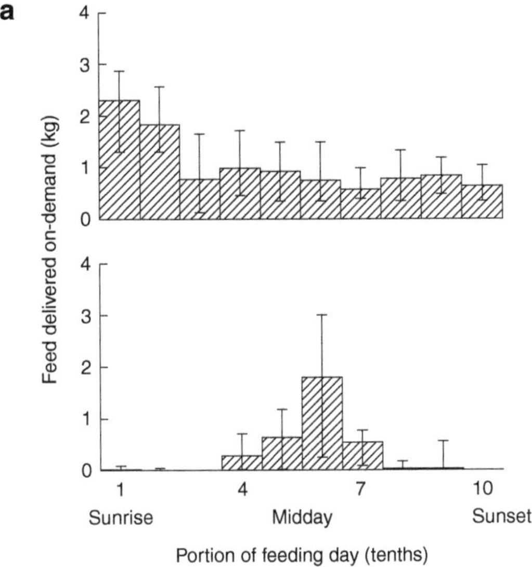

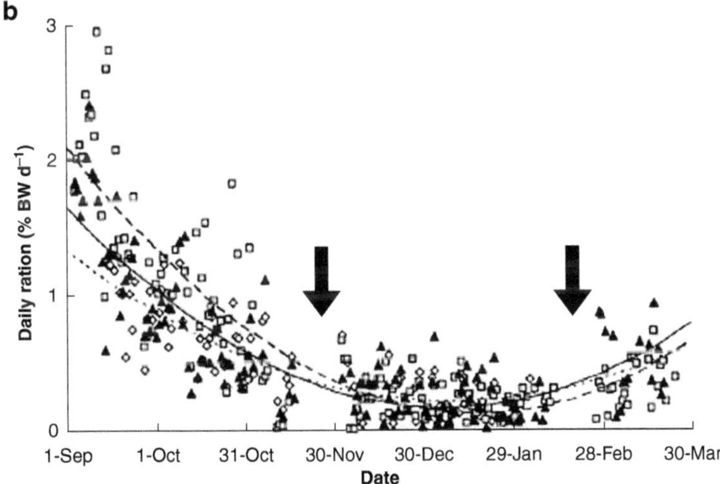

Fig. 4.2 Food intake patterns in Atlantic salmon fed to appetite. (**a**) The amount of food delivered (median and interquartile range) in each of 10 sections of the feeding day, with significant appetite peaks at dawn and midday. *Top* Dawn peak in autumn. *Bottom* Midday peak in winter. (**b**) Daily ration over one winter. Arrows represent husbandry events (vaccination and grading), after which the fish stopped feeding for a few days (modified and reproduced with permission from Noble et al. 2007a)

4.2.3 Avoiding Predation

Wild fish are eaten by many different kinds of predators and, as a consequence, show a variety of protective responses (Huntingford et al. 2012b and refs therein). Fish may avoid predation by resting at times when predators are active or by inhabiting places where predators are uncommon. Newly hatched salmonids disperse at night, when the risk of attack by visually hunting predators is low (Fraser et al. 1994). Juvenile perch (*Perca fluviatilis*) congregate in hypoxic water, which is avoided by their predators (Vejrik et al. 2016). In many species, particularly cyprinids such as carp, damaged skin releases an alarm substance that causes other fish to move away from and often subsequently avoid places where a conspecific has been injured.

Many species of fish live in loose shoals (for example, sticklebacks) or tight, polarised schools (for example, Atlantic herring *Clupea harengus*), both of which give protection against predators. If an approaching predator is detected early, prey fish often move gradually away towards shelter, but if it is close, a rapid escape response is triggered, taking the fish out of the line of attack. Individual fish of a given species are not necessarily equivalent in their response to the threat of predation, striking and consistent individual differences having been described for many species (Huntingford et al. 2013 and refs therein). Such individual variation has important implications for production and welfare of cultured fish and is discussed further in Chap. 12.

One way in which prey fish protect themselves against predation, while at the same time carrying on with other important activities, involves predator inspection, a remarkable behaviour during which they swim, cautiously but persistently, towards a potential predator (Pitcher 1992). In the process, they gather information about the actual risk that the potential predator poses, based on its species, size, hunger state and feeding preferences (Kelley 2008 and refs therein). If the risk is high, other activities are suppressed in favour of fleeing or hiding; if the risk is low, normal behaviour carries on, perhaps with an added degree of caution. As an example of the broadcast fear created by the presence of potential predators, juvenile damselfish (*Pomacentrus amboinensis*) show a dramatic increase in metabolic rate when they see known predator, a response that is not given to neutral, non-predatory fish (Fig. 4.3; Hall and Clark 2016). Such stressful experiences can suppress important activities; for example an osprey moving overhead induces strong fright responses in

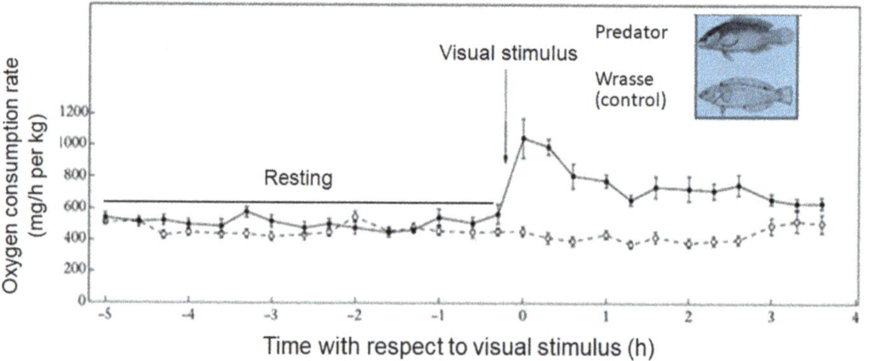

Fig. 4.3 Stress-related effects of exposure to a predator. Oxygen consumption (mean and standard error) in damselfish before and after a brief exposure to a harmless species and a potential predator (modified and reproduced with permission from Hall and Clark 2016)

pumpkinseed sunfish (*Lepomis gibbosus*) and suppresses parental behaviour (Gallagher et al. 2016).

4.2.4 Aggression and Fighting

Many fish compete for key resources (food, shelter or mates) by means of aggressive behaviour (Damsgård and Huntingford 2012 and refs therein). Aggressive encounters usually start with an exchange of increasingly vigorous displays. They are often resolved without overt fighting, but fighting does happen and can result in injury (Maan et al. 2001). Winners of fights may gain direct access to contested resources, but may also acquire a territory or a dominant position within a social group, which will give them preferential access to resources in the future. Defence of feeding territories is relatively rare among fish (seen in less than 10% of species), but in 78% of fish species, males defend territories on which they breed (Grant 1997). Within groups, the same individuals may fight on successive occasions and over time, if one loses consistently, it starts to avoid fights and becomes subordinate, while the consistent winner emerges as dominant. Following such a series of defeats, subordinate Arctic charr (*Salvelinus alpinus*) takes on a darker colouration, avoids moving and makes no attempt to feed (Øverli et al. 1998). In larger groups, a series of pairwise dominance–subordinance relationships may generate more-or-less stable dominance hierarchies that determine access to resources. Such hierarchies have been described in, for example Atlantic salmon, Arctic charr, rainbow trout, as well as in gilthead seabream (*Sparus aurata*) and Nile tilapia (*Oreochromis niloticus*; Damsgård and Huntingford 2012).

4.2.5 Courtship

While all aspects of reproductive behaviour are relevant to captive rearing, this chapter concentrates on courtship, a more or less complex exchange of behavioural signals between partners that, when successful, culminates in spawning (Fleming and Huntingford 2012 and refs therein). In the majority of fish species, courtship and mating take place in the context of a temporary association between ripe males and females. For example, male lumpfish (*Cyclopterus lumpus*) establish nest sites in crevices to which they attract ripe females during a courtship that involves males brushing the female with their fins and quivering their body against the female (Fig. 4.4). Females lay their eggs in the nest and then leave; having fertilised several egg clutches, the males care for these until they hatch (Goulet et al. 1986). Courtship without pair bonding is also seen in Atlantic salmon, Atlantic cod (*Gadus morhua*) and the Mozambique tilapia (*Tilapia mosamicus*; Fleming and Huntingford 2012). In rare cases, courtship marks the beginning of a more permanent association between partners; in seahorses a permanent, exclusive pair bond is formed between

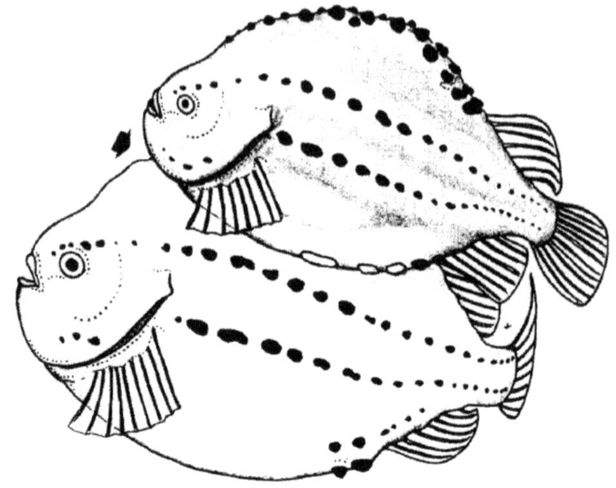

Fig. 4.4 Courtship in the lumpfish. The male (above) places the base of his gills against the larger female's back and in this position quivers against her (modified and reproduced with permission from Goulet et al. 1986)

a single female that provides the eggs and a single male into whose brood pouch these are placed, fertilised and nurtured (Kvarnemo and Simmons 2004).

4.3 Tinbergen's Questions in Relation to Fish Behaviour

This section looks at how Tinbergen's questions have been applied to the behavioural systems outlined above. Relatively little is known about the phylogenetic history of behaviour (although increasingly accurate molecular-based phylogenetic trees and advances in the comparative analysis are changing this), so discussion is confined to questions relating to causation, development and function. In the text, answers are illustrated mainly with reference to aggressive behaviour, supplementary information for other systems being presented in Table 4.1.

4.3.1 The Causes of Fish Behaviour

In general, what causes a fish to behave in a particular way at a particular time is the combined result of the external stimuli it detects and its internal state. In the case of prey choice in sticklebacks (Box 4.1), whether a fish eats a particular prey item depends on sensory cues emanating from potential prey (very small, very large and immobile prey are unattractive) and its current internal state (an empty stomach and low reserves predispose the fish to eat).

In the context of fighting, relevant stimuli come from the sight, sound or smell of a rival conspecific. Some stimuli make the fish more likely to fight; breeding male goldfish release reproductive hormones in their urine and these olfactory cues

Table 4.1 Some examples of answers to Tinbergen's questions for other behavioural systems

Causation: external stimuli
Movement through space: Young reef-dwelling fish settling from the plankton use sounds to locate appropriate reef habitats (Simpson et al. 2005). Different reefs have a different bouquet of smells and settling juveniles also use these to locate suitable reefs (Døving et al. 2006; Gerlach et al. 2007).
Feeding: Olfactory stimuli are important in the control of feeding in fish. The amino acid cysteine stimulates feeding in many carnivorous fish, whereas arginine often inhibits feeding; whether food containing both chemicals is eaten depends on their proportions (Raubenheimer et al. 2012).
Avoiding predation: Rapid escape responses to the approach of a predator are triggered by mechanosensory stimuli acting on the lateral line system; inactivating this system impairs the escape response in Atlantic menhaden (*Brevoortia tyrannus*; Higgs and Fuiman 1996).
Courtship: Male Atlantic cod use their swim bladder to produce drumming during courtship, which attracts females. Just before spawning, the sounds change from "grunts" to "hums", suggesting that they may stimulate gamete release (Brawn 1961a, b).
Causation: internal processes
Movement through space: Migration from freshwater to seawater in salmonids (Munakata et al. 2012) and autumn migration from coastal to offshore areas in Atlantic cod (Comeau et al. 2001) coincide with rising levels of thyroxine, which increases metabolic rate and promotes strong swimming.
Feeding: Food intake in fish is controlled by a series of cues signalling current energy reserves and the progress of digestion and absorption of individual meals. These cues activate complex, well-documented brain systems that adjust food intake to need (Jobling et al. 2012b).
Avoiding predation: Food-deprived juvenile Atlantic salmon spend less time sheltering from predators than do well-fed fish (Vehanen 2003). Growth hormone reduces predator avoidance in rainbow trout, via increased metabolic demand and appetite (Johnsson et al. 1996).
Courtship: Seasonal changes in the production of sex hormones track the appearance and disappearance of sexual behaviour in both males and females. Breeding male Mozambique tilapia have high levels of androgen (11-keto-testosterone or 11KT) at the start of the nesting period when they are establishing territories and courting females (Goncalves and Oliveira 2011).
Development: the role of inheritance
Movement through space: Lake whitefish (*Coregonus clupeaformis*) come in two forms, one preferring shallower and the other deeper water; hybrids show intermediate-depth preference (Rogers et al. 2002). There are distinct migratory and non-migratory stocks of Atlantic cod and among migratory stocks different substocks travel in different directions and for different distances (Metcalfe et al. 2008 and refs therein).
Feeding: Goldfish find citric acid and proline aversive, whereas common carp shows a positive preference for these substances; hybrids between a female goldfish and a male carp are also attracted to the substances, suggesting a taste preference inherited through the paternal line (Kasumyan and Døving 2003).
Avoiding predation: Naïve paradise fish inspect the first potential predator they encounter (Gerlai and Hogan 1992). Naïve guppies (*Poecilia reticulata*) from populations with a high predation risk respond to a predator more strongly than do those from low-risk sites (Odell et al. 2003).
Courtship: In the pygmy swordtail (*Xiphophorus nigrensis*) a Y-linked locus controls courtship behaviour through an effect on body size; large males use only frontal display during courtship, while small males switch from display to sneaking if larger males are present (Zimmerer and Kallman 1989). In male guppies, attractiveness to females is partly dependent on Y chromosome-linked colour patterns (Hughes et al. 2005).

(continued)

Table 4.1 (continued)

Development: the role of the environment
Movement through space: The ability of salmonids to return to their natal streams to spawn after travelling thousands of kilometres on the high seas depends on imprinting to olfactory (and possibly magnetic) cues at the start of the downstream smolt migration (Odling-Smee et al. 2006 and refs therein).
Feeding: Naïve archerfish (*Toxotes chatareus*), which dislodge terrestrial insects from overhanging leaves by spitting a jet of water, are initially very poor at judging direction and distance of prey, but improve rapidly with practice or even by simply watching other, proficient archerfish hunting (Schuster et al. 2006).
Avoiding predation: Encounters with predators have marked effects on subsequent anti-predator behaviour in fish through various forms of learning, based on direct experience (predator-naïve Nile tilapia develop effective protective responses after a series of simulated chases by a predator, and these effects persist for many weeks; Mesquita and Young 2007) or of observational learning (zebrafish learn effective routes of escape from an approaching trawl by social learning; Lindeyer and Reader 2010).
Courtship: In two closely related sympatric cichlids from Lake Victoria (*Pundamilia pundamilia* and *P. nyerere*) early sexual imprinting modifies subsequent mate preferences; females that have been cross-fostered onto mothers of the alternative species response more strongly to the courtship of heterospecific males than do those fostered onto females of their own species (Verzijden and ten Cate 2007).
Function of behaviour: cost–benefit trade-offs
Movement through space: A common pattern for pelagic fish is to stay in deep water during the day, ascending into shallower water at dusk and spending the night close to the surface. Such diurnal vertical migrations represent a continuous trade-off between the need to feed (which is easier in shallow, well-lit water), the need to avoid predators (which is easier in deeper, darker water) and the need to conserve energy (which depends on water temperature, which in turn varies with depth and time of day; Huntingford et al. (2012a, b) and refs therein).
Feeding versus avoiding predators: Feeding decisions in fish are adjusted more or less continually to current predation risk. Well-fed Brazilian catfish (*Pseudoplatystoma corruscans*) show rapid escape and freezing when exposed to alarm substance, but food-deprived fish simply move away without freezing (Giaquinto and Volpato 2001). Fish at the front of a school are more vulnerable to predators than are those in the centre, but they gain more food; hungry fish position themselves near the front of a group, whereas satiated fish move towards the rear (Krause et al. 1992).
Courtship: When courting in the absence of a predator, male pipefish (*Syngnathus typhle*) prefer more active females (potentially healthier mates), but in the presence of a predator this is reversed, with less active and therefore less conspicuous partners being favoured (Billing et al. 2007).

stimulate other males to attack (Sorensen et al. 2005). Other stimuli make fish less likely to attack; urinary androgens from dominant male Mozambique tilapia suppresses aggression in rivals (Saralva et al. 2017) and the darkened eye ring of subordinate Atlantic salmon inhibits attack by dominant companions (O'Connor et al. 1999). The displays exchanged during a fight between two breeding male cichlids (*Cichlasoma centrarchus*), involve visual cues (brightly coloured fins and flared gill covers), mechanical cues (water currents generated by tail beating) and auditory cues (sounds produced by grinding pharyngeal teeth). Short sounds with long gaps stimulate attack, whereas long, rapidly repeated pulses inhibit aggression.

The balance between such attack-eliciting and attack-inhibiting cues determines whether a given male gets into a fight (Schwarz 1974a, b).

The internal state of both participants is also important. In contests over food, up to a point fish with lower energy reserves are more likely to fight than are their well-fed counterparts and fish often fight more fiercely after a period of food deprivation (Coho salmon, *Oncorhynchus kisutch*; Damsgård and Dill 1998). When fish fight over breeding opportunities, endocrine status is important; in general the high levels of circulating androgens, as experienced by breeding males, increase their aggressiveness (Damsgård and Huntingford 2012 and refs therein). Reproductive hormones have direct effects on the brain mechanisms that control aggression, but also act indirectly by altering structures and signals used in fights; the hooked snout or kype of breeding male salmon and the bright red chest of the breeding male three-spined stickleback provide examples. One important hormonal influence on a fishes' tendency to fight comes from the physiological stress response. In general, an acute rise in cortisol levels increases aggression, although this depends on circumstance (Serra et al. 2015; Manuel et al. 2016); chronic stress, as experienced by subordinate fish, for example usually suppresses aggression (Damsgård and Huntingford 2012).

4.3.2 The Development of Fish Behaviour

Provided young fish grow up in broadly favourable conditions, changing patterns of gene expression during development build the machinery required for performing behaviour (sense organs, nervous system, endocrine glands and muscles). As a consequence, fish are often able to show normal species-specific behaviour the first time this is required, without any specific environmental inputs and experiences; in such cases, the behaviour concerned is sometimes described as 'innate'. This word is useful in drawing attention to the fact that quite complex behaviour can be hardwired in this way, but it does not mean that the behaviour concerned is inaccessible to environmental modification. This is clearly not the case; as illustrated in Chap. 5, behavioural development is subject to strong environmental influences from the point of fertilisation, and even before. Such influences range from general effects (of water quality, for example) to highly specific learning opportunities and they interact with inherited differences to determine just how fish behave at a given age. In the case of prey choice in sticklebacks (Box 4.1), a preference for moving prey develops regardless of experience, so might be called innate. However, many other aspects of prey choice depend on past encounters with specific prey types.

In many fish, aggression appears early in development, changing in form and frequency as the fish get older and their circumstances alter. For example, soon after the yolk sac has been absorbed, young rainbow trout first show aggressive behaviour, in the form of simple actions such as chasing, nipping and fleeing. Threatening a rival by raising fins and assuming a head down postures appear slightly later in development and become more common, largely replacing direct attack; during this

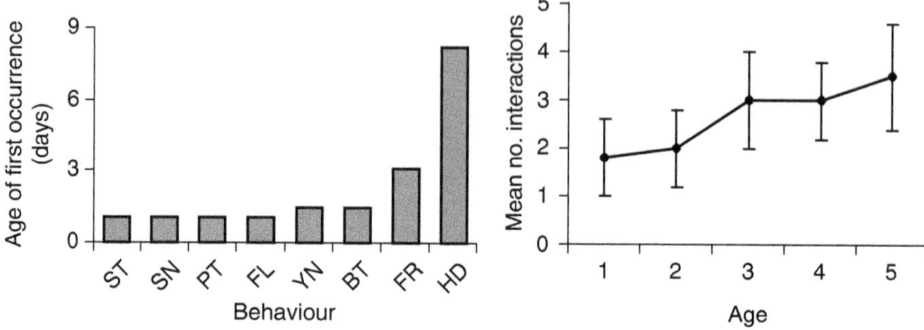

Fig. 4.5 Changing form and frequency of aggressive behaviour with age. *LHS* Age of first appearance of different components of the aggressive repertoire of young rainbow trout. *ST* stationary, *SN* snap, *PT* pursuit, *FL* flight, *YN* yawn, *BT* bite, *FR* dorsal fin raise, *HD* head down display, *RHS* Mean and standard deviation number of aggressive interactions in relation to a number of days since 50% of larvae were free swimming. 1: days 1–6. 2: days 7–12. 3: days 13–18. 4: days 19–24. 5: days 25–30 (plotted from Tables I and IV in Cole and Noakes 1980)

time the frequency of aggressive interactions increases markedly (Fig. 4.5; Cole and Noakes 1980). A number of lines of evidence bear witness to the role of inherited effects in the development of differences in aggressive behaviour in fish (Box 4.4).

Box 4.4 Some Evidence for Inherited Differences in Aggression Among Fishes

- Normal aggressive responses are often shown by fish reared without the opportunity for social interaction; like normally reared fish, male cichlids (*Astatotilapia burtoni*) reared in social isolation attack the black eyebar of other males (Fernald 1980).
- Population differences in aggressiveness persist when fish from different sites are reared in identical conditions in many species, including brown trout (Lahti et al. 2001) and coho salmon (Rosenau and McPhail 1987).
- Strains of Siamese fighting fish (*Betta splendens*) selectively bred for victory in cockfight-like contests are more aggressive than wild-type fish reared in comparable conditions (Verbeek et al. 2007).
- Hybrids between the relatively unaggressive lake charr (*Salvelinus namaycush*) and the aggressive brook charr (*Salvelinus fontinalis*) show intermediate levels of aggression (Ferguson and Noakes 1982).
- Reciprocal crosses implicate elements on the Y chromosome in the inheritance of strain differences in aggressiveness in guppies (Farr 1983).

Such inherited influences interact with a variety of environmental effects. To illustrate an effect of general conditions, adult male zebrafish (*Danio rerio*) exposed briefly as fry to a period of anoxia become more aggressive towards potential rivals and more likely to win fights than are fish reared in normoxic conditions; they also have higher testosterone levels (Ivy et al. 2017). When male three-spined sticklebacks exposed to olfactory cues from a predator later encounter a rival in the presence of a predator they attack less than do naive fish (Herczeg et al. 2016). In terms of social experience, Nile tilapia that have been reared with companions that lack a dorsal fin (so do not show normal displays) are slower to engage in fights (Barki and Volpato 1998). On a shorter timescale, experience of losing makes the mangrove rivulus (*Rivulus marmoratus*) less likely to initiate a fight in subsequent encounters; experience of winning makes them more likely to launch in with an attack rather than a display (Hsu and Wolf 2001). More complex aspects of aggressive encounters, such as adjusting fight intensity to the value of the contested resource (Sect. 4.3.3) sometimes have to be learned. Round gobies (*Neogobius melanostomus*) fight more fiercely over good quality shelters, but only if they have had prior experience of shelters of different quality (McCallum et al. 2017).

4.3.3 The Functions of Fish Behaviour

Biologists are interested in the consequences of particular actions for Darwinian fitness. Some are beneficial (sticklebacks get valuable nutrients if they eat large benthic prey); others are detrimental (large prey take time and energy to ingest and feeding on benthic prey reduces vigilance). Fish adjust their behaviour flexibly to the balance of such positive and negative consequences. Undisturbed sticklebacks prefer large benthic prey, but when predators are about they switch to zooplanktonic prey, which are small but can be eaten, albeit at a reduced rate, while simultaneously watching out for predators (Ibrahim and Huntingford 1989).

In the case of aggression, the benefit of winning a fight lies in the acquisition of valuable resources. Immature fish compete for shelter (as in the case of overwintering juvenile rainbow trout; Gregory and Griffith 1996) and food (dominant Nile tilapia gain the lion's share of available food; Vera Cruz and Brown 2007). In adult fish, the disputed resource is often access to breeding opportunities and mates; female pipefish (*Syngnathus typhle*) actively compete for males with empty pouches (Berglund and Rosenqvist 2003), while in Atlantic salmon males that fight readily fertilise more eggs (Weir et al. 2004).

In terms of costs, fighting takes time away from other important activities, such as feeding (territorial coho salmon spend more time fighting and less time feeding than do non-territorial fish; Puckett and Dill 1985) or looking out for predators (male cichlids *Nannacara anomala* fail to detect predators at a distance when they are fighting; Jakobssen et al. 1995). Taking part in a fight also uses up energy; in fighting male cichlid fish (*Aequidens rivulatus*), respiration rate increases by 33% during fights. In intense fights, both winner and loser suffer injury to mouth, fins, tail and flank (Maan et al. 2001).

Any factor that shifts the balance between such positive and negative consequences alters what happens when rivals meet. Increasing the benefits of winning makes a fight more likely to occur and to escalate; levels of aggression in medaka (*Oryzias latipes*) are lowest when food is dispersed in space, but clumped in time, so that they cannot be monopolised in a cost-effective way (Robb and Grant 1998). Conversely, increasing the costs of fighting makes animals less aggressive. Breeding male three-spined sticklebacks reduce levels of aggression in the presence of a predator, though males with a clutch of eggs in their nest do so less than those with an empty nest (Ukegbu and Huntingford 1988). The fitness equation will also be influenced by probability of winning and fish have elaborate behavioural mechanism for assessing their own fighting ability compared to that of a potential rival. This may be based on immediate condition of both parties (relative body size is an important predictor of fight outcome; Enquist et al. 1990), on past experience of taking part in fights (previous losers are less likely to fight; Hsu and Wolf 2001) or on observing the outcome of fights between other fish (Grosenick et al. 2007).

4.4 How the Natural Behaviour of Fish Is Expressed in Culture Systems

Compared to most terrestrial farmed animals, there has been little domestication of aquaculture species, so farmed fish share the natural behavioural traits of their wild counterparts. This section considers whether and how such traits are expressed on fish farms; Sect. 5.5 considers their implications for fish welfare.

4.4.1 How Farmed Fish Use Space

Patterns of space use in culture systems have been most fully investigated for Atlantic salmon in sea cages. During the day, the fish tend to swim actively around the perimeter of the cage in a natural-looking school, with speed and direction dependant on stocking density and time of day; during the night, activity falls and the fish disperse. They rarely use the whole of the cage, so the density they experience can be much higher than theoretical densities based on the volume of water available to them. The depth at which farmed salmon swim reflects their natural responses to light and temperature. Like wild salmon, in general, farmed fish descend into deeper water at dawn and ascend at dusk, especially during the summer when surface light is more intense. Superimposed on their response to light levels, the salmon concentrate their activity at the highest temperature available in daytime; again, this happens particularly during the summer when the range of available temperatures is greater (Oppedal et al. 2007 and refs therein). At low densities, juvenile cod in net pens do not show coordinated movement, but instead just “mill about”, mainly in the lower half of the pen (possibly related to their naturally benthic habits) and they are markedly more active during the daytime (Fig. 4.6). As stocking

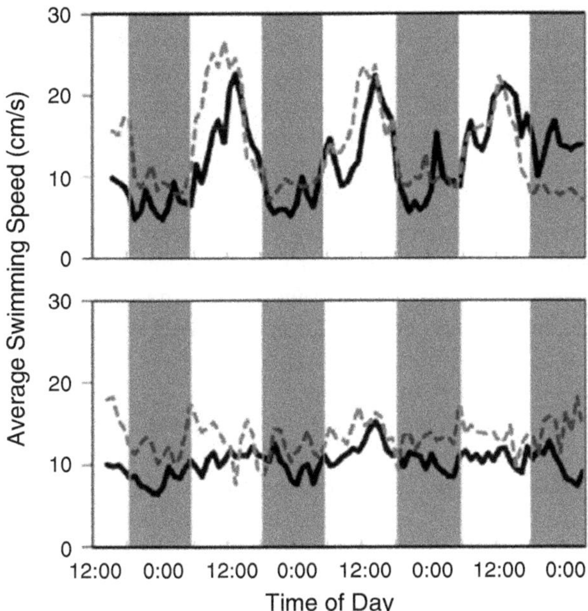

Fig. 4.6 Use of space by fish in culture. Swimming speed in juvenile cod in net pens averaged over 4 consecutive weeks in relation to the time of day (dark columns represent nighttime) and at two different stocking densities (*top ca* 20 kg/m^3 *bottom ca* 32 kg/m^3; modified and reproduced with permission from Rillahan et al. 2011)

density increases, the cod start to swim in synchronised, polarised schools, and are equally active throughout the day (Fig. 4.6; Rillahan et al. 2011). In the wild, lumpfish are frequently found attached to floating seaweed using a suction disc below their mouth. In culture, during the day, juvenile lumpfish forage actively throughout the depth of their holding pen, attaching themselves to various substrates at dusk and detaching themselves again at dawn (Imsland et al. 2015b).

4.4.2 *How Farmed Fish Feed*

4.4.2.1 How Food Is Delivered

The way in which food is presented to cultured fish varies, but it is usually delivered from above and falls gradually through the water column to the bottom of the tank or pen. Movement is imparted to the falling food by water currents. Food is often delivered manually, with various devices to distribute it over the water surface, but may also be delivered by automatic feeders, most of which are computer controlled (Le François et al. 2010).

4.4.2.2 The Type of Food Delivered

Larval feeds often include live prey such as rotifers and *Artemia* larvae, but after the larval phase, fish are usually given formulated feed in the form of crumbs, flakes or pellets. These differ in texture, appearance and often in taste and smell from the natural prey of the species concerned (Jobling 2010). Many important cultured species such as tilapias and cyprinids (including the common carp, *Cyprinus carpio*) are herbivorous or omnivorous, making provision of appropriate food relatively easy. Fishmeal (derived from reduction fisheries for small marine fish) is an ideal source of nutrients for high value piscivorous farmed fish. However, there is a real potential for overfishing stocks of such forage fish, with knock-on detrimental consequences for the marine ecosystems (Lekang et al. 2016). This has led to a successful search for other sources of nutrients for farmed fish, which are now fed diets containing a significant proportion of non-fish material. For established aquaculture species, nutritional needs are well known and formulated diets provide for all of these, so while cultured fish do not have the opportunity to be selective about their food, arguably they do need to do so.

In spite of the availability of formulated feeds, fish in intensive culture systems may experience a choice of food types, due to the presence of natural prey in the ponds, pens or cages in which they are held. Species that are cultured extensively, for example pond-reared carps and tilapias, usually have a wide range of natural prey from which to choose, as well as any additional feedstuffs they may be given as supplements (Jobling 2010). In nature, juvenile lumpfish (*Cyclopterus lumpus*) are opportunistic feeders, taking a variety of animal prey approximately in proportion to availability. When held in salmon cages for sea lice control, lumpfish have access to a range of potential food in addition to the salmon ectoparasites they are supposed to eat; here too they also feed opportunistically, concentrating on the abundant fragments of formulated feed (Fig. 4.7; Imsland et al. 2015a).

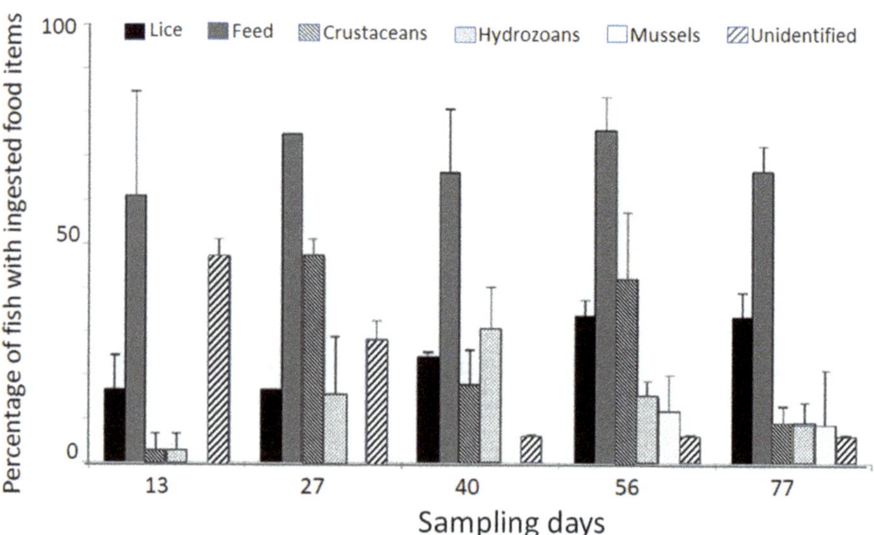

Fig. 4.7 Diet choice in lumpfish cultured in net pens with Atlantic salmon. Percentage (mean and standard deviation) of lumpfish that had eaten different food types during an 77-day feeding trial (modified and reproduced with permission from Imsland et al. 2015a)

Cannibalism is a natural part of the behavioural repertoire of many fish species, at some life history stages at least. It can be considered as a form of aggression, since one animal injures another member of the same species, or as predation; However conceptualised, cannibalism occurs during the culture of many species (Raubenheimer et al. 2012; Naumowicz et al. 2017 and refs therein). For example, in jundiá (*Rhamdia quelen*), a cultured omnivorousa catfish-like species, cannibalism is common in young fish, especially at night. Cannibalistic jundiá swallow prey that are on average 12% of their body weight (the observed maximum being ca 19%) and between 80% and 100% of fish that are small enough to be swallowed do actually fall prey to larger conspecifics, regardless of the availability of alternative food (Costenaro-Ferreira et al. 2016).

4.4.2.3 How Much Are Fish Fed and When

When fish are fed by hand, timing is constrained by logistics, but the feed delivery itself is often continued until surface signs of feeding cease, to some extent matching delivery to appetite. Automatic feeders can deliver predetermined amounts at times that are based on what is known of natural nutritional needs and appetite fluctuations of the species concerned. Such fluctuations are often difficult to predict, for example if appetite is suppressed as a result of disturbance. To allow for this, farmed fish are sometimes fed by demand feeders (Bégout et al. 2012 and refs therein). These include self-feeding devices that require the fish themselves to activate a trigger to release feed (used for tilapia and rainbow trout, for example) and various interactive feedback systems that adjust the amount delivered to appetite of the fish concerned (used for Atlantic salmon and rainbow trout, for example).

4.4.3 Avoiding Predation

Even though cultured fish are to a large extent protected from predators, in spite of the best efforts of farmers farmed fish sometimes receive direct predatory attacks that may cause stress, injury and death. Farmed salmonids in sea cages may be attacked by seals, otters, mink, grey herons and cormorants (Huntingford et al. 2012b and refs. therein). Bluefish (*Pomatomus saltatrix*) congregate around seabass and sea bream farms, break into cages and attack fish in them (Sanchez-Jerez et al. 2008). Some farmed fish, for example larval jundiá, are attacked by cannibals and show appropriate predator avoidance, moving away and remaining inconspicuously at the water surface (Costenaro-Ferreira et al. 2016).

Even if farmed fish do not receive any direct predatory attacks, they may well be exposed to cues that signal potential predation risk. For African catfish, alarm substance released from fish injured during fights causes transient stress responses in nearby fish (Van de Nieuwegiessen et al. 2009). Routine husbandry practices such as capture and handling in a sense mimic a predatory attack and these too induce

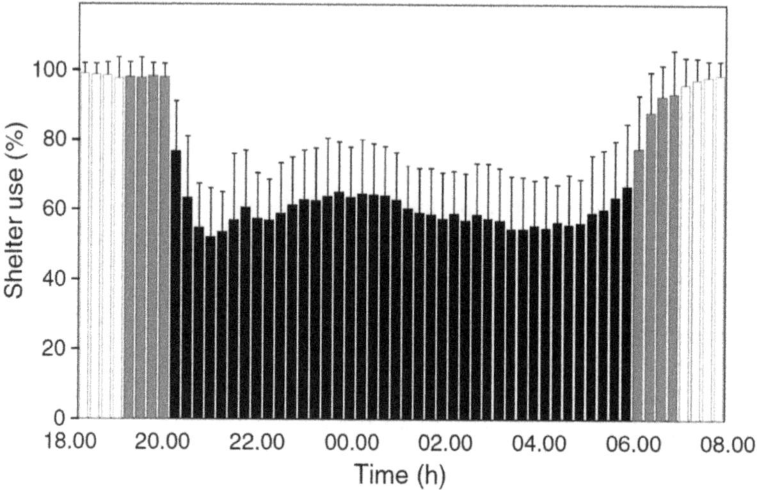

Fig. 4.8 Use of shelters by cultured burbot. Mean and standard deviation percentage of fish making use of shelter (halved plastic pipes, 8 per tank) in relation to the time of day. Black bars indicate night time, white bars day time and grey bars dusk and dawn (modified and reproduced with permission from Wocher et al. 2011)

stress responses. After being chased with a net, jundiá show increased cortisol levels, accompanied by reduced activity and fewer social interactions (Abreu et al. 2016). In nature, burbot (*Lota lota*) often hide during the day to avoid predation. When provided with shelters, cultured burbot also hide during the day, which reduces stress levels and promotes growth (Fig. 4.8; Wocher et al. 2011).

4.4.4 Aggression

Many farmed fish species attack each other and the frequency of aggressive interactions in production systems can be surprisingly high (Damsgård and Huntingford 2012 and refs. therein). Particularly just before feeding, larval cod held at high densities nip at the tail of smaller companions at a rate of *ca* 3 attacks per fish per hour, which multiplies up to a fair number of attacks across the day (Forbes 2007). In juvenile cod held in production conditions, more than 80% of fish on low rations have aggression-induced damage to their anterior fins; even when feed is delivered to excess, fin damage still occurs at a non-trivial level (Fig. 4.9; Hatlen et al. 2006). Some other examples are shown in Box 4.5.

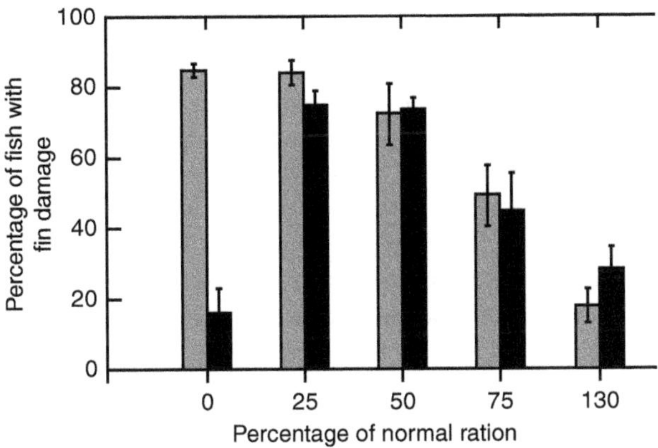

Fig. 4.9 Indicators of aggression in farmed cod. The percentage (mean and standard error) of 50 g fish with damage to the anterior dorsal fin when held at different rations. Grey bars: Fish sampled at the start of the observation period. Black bars: Fish sampled 55 days later (modified and reproduced with permission from Hatlen et al. 2006)

Box 4.5 The Frequency of Aggressive Interactions in Fish in Culture Conditions

- In cultured juvenile koi carp attacks on the abdomen and fins of rivals occur at a rate of over 1/fish/min (Jha et al. 2005).
- In adult rainbow trout, nips and chases occur at the rate of 0.75 attacks/fish/min in fish fed 1 meal per day and 0.30 attacks/fish/min in fish given 3 meals per day (Noble et al. 2007c).
- In juvenile sea bream held in groups of 15 fish, dominance interactions occur at a rate of 62 per hour (Papadakis et al. 2016). Attack rate is 2 per min in dominant fish and 0.6 per min in subordinates (Goldan et al. 2003).

In farmed fish as in wild fish, aggressive interactions often develop into stable dominance–subordinance relationships. Approximately 10% of (dominant) fish are responsible for 60% of attacks among juvenile halibut (*Hippoglossus hippoglossus*; Greaves and Tuene 2001) and in larval yellow perch (*Bidyanus bidyanus*) held at densities of 25 and 50 fish/m^3, one or two (subordinate) fish are chased continually (Rowland et al. 2006). Within groups of rainbow trout, a subset of dominant fish, identified from their relatively undamaged dorsal fins, monopolise localised food (Moutou et al. 1998). Taking part in a fight causes complex changes in the fish involved, including physiological stress response, particularly in losers. In the wild, this will normally be short lived, but in production systems fish that lose fights may be unable to escape from their victors, in which case they will experience chronic stress, which is of concern from a welfare perspective.

4.4.5 Courtship

Ensuring the reliable supply of fry needed for sustainable aquaculture invariably interferes with natural reproductive behaviour (Fleming and Huntingford 2012 and refs therein). At the most extreme, eggs and sperm are collected by stripping, followed by artificial fertilisation. This method, which is used for intensively farmed species such as Atlantic salmon and rainbow trout, precludes all the normal behavioural components of reproduction, including competition, courtship, and mate choice. For some species (tilapia and channel catfish in some systems), single males and females are paired artificially in enclosures or tanks where spawning takes place spontaneously, allowing natural courtship, but with no competition and little scope for mate choice. One of the least intrusive ways of generating fry for aquaculture is to hold large, mixed sex groups of fish in tanks or ponds where mating occurs spontaneously and fertilised eggs are collected for use. This is the case for marine pelagic spawners such as cod, seabass and sea bream and, in some cases, for tilapia. Such group spawning allows competition, natural courtship and mate choice. In Nile tilapia held in breeding ponds, 33% of males (usually among the largest in the group) fertilise more than 70% of eggs (Fessehaye et al. 2006); for cod spawning in groups, 10% of males fertilise 90% of eggs (Herlin et al. 2008).

4.5 Welfare in Cultured Fish

4.5.1 The Tricky Concept of Fish Welfare

There is controversy surrounding the question of fish welfare, partly arising from the fact than welfare can be defined in different ways. When defined in terms of how well an animal functions, good welfare requires it to be adapted to its current environment, with all its biological systems working appropriately. Feelings-based definitions equate good welfare with an animal's being free from negative experiences such as pain, fear and hunger and having access to positive experiences such as social companionship. This is predicated on an assumption that the animal concerned is sufficiently sentient to experience positive and negative emotions. Definitions based on what is natural equate welfare with an animal's being able to show the same kinds of behaviour in captivity as it would in the wild. These definitions emphasise different important aspects of a complex topic and none is inherently correct or incorrect. What welfare means and how it can be measured are discussed in more detail in Chaps. 1, 2, 9, and 13. The material presented in this chapter mostly fits comfortably within the least controversial definition of welfare in terms of effective functioning. However, some examples concern the ability of farmed fish to show their natural behavioural repertoire and in particular to the concept of behavioural need.

4.5.2 *Natural Behaviour and the Welfare of Farmed Fish*

Effects on fish welfare of many necessary husbandry practices such as transport, handling, delivery of medical treatments and slaughter have been extensively reviewed (Ashley 2007). Of interest in this chapter are welfare problems arising specifically from the expression of natural behavioural traits in farmed fish. Some problems arise because fish in culture show natural responses in circumstances where this is inappropriate or harmful. Others arise because cultured fish fail to show natural responses when this would be appropriate or beneficial. In either case, problems can arise from the mechanisms that control the behaviour concerned, how it develops or the way that it has been shaped by natural selection. Examples are given for each of these cases, which are not mutually exclusive, and often suggest similar strategies for mitigation.

4.5.2.1 Natural Responses to External Stimuli

Right Response, Wrong Context In some cases, problems arise in aquaculture because cultured fish are exposed to relevant stimuli and show natural responses to them, but because of the context in which the behaviour is shown this has adverse effects on welfare. For example, when Atlantic salmon are held in sea cages and food is provided at the water surface, their natural response to spatial cues (location of food and gradients of temperature and light) can result in aggregations of fish that are sufficiently high to cause collisions and seriously deplete oxygen concentrations (Johansson et al. 2006). This can be mitigated by strategically positioned underwater light sources, the natural response to which causes fish to spread out more evenly throughout the cage (Juell et al. 2003). By keeping salmon away from the water surface, where the infective stages of sea lice congregate, submerged lights also protect them against louse infestation. The number of lice per fish falls from *ca* 7 with surface lights to *ca* 1 with submerged lights (Frenzl et al. 2014).

In the context of anti-predator responses, many husbandry practices send out signals that may be interpreted by the fish as indicating the presence of a predator. These might include scents (humans release serine from their skin, which in nature signals the presence of mammalian predators; Idler et al. 1956), visual cues (people and objects loom over the water surface) and mechanosensory cues (farmed fish are exposed to noise from boats, pumps and other farm equipment). Responding to such cues in nature protects fish against danger, but in aquaculture escape responses can cause damage when fish collide with each other or with the walls of their enclosure. For example, swimming into tank walls by larval striped trumpeter *Latris lineata* results in a high incidence of jaw malformations (Cobcroft and Battaglene 2009).

Physiological stress is a natural and adaptive response to challenge, but in culture effective coping is rarely possible and prolonged stress in response to real or perceived threats can have adverse effects on welfare. For example, Atlantic salmon show increased plasma cortisol concentrations and reduced appetite after handling

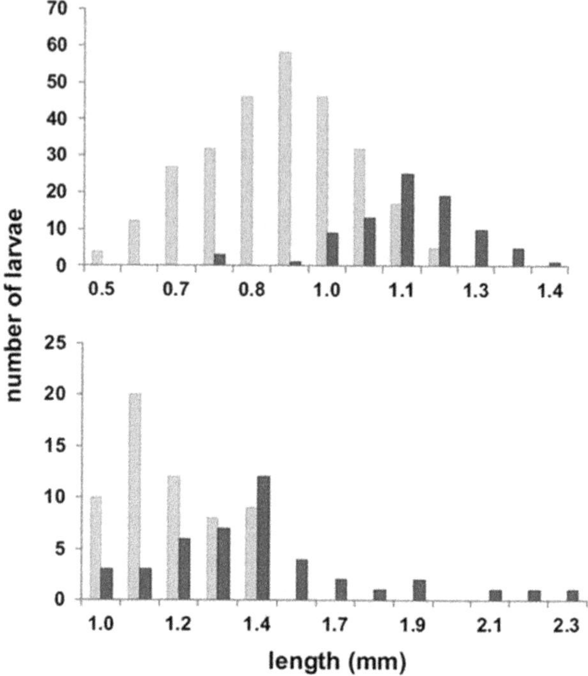

Fig. 4.10 Size distribution of larval pikeperch on the light (light bars) and dark (dark bars) side of a grading mesh at 22 (*top*) and 34 (*bottom*) days post hatch (reproduced with permission from Tielmann et al. 2016)

and confinement (Pankhurst et al. 2008). These are a component of many husbandry practices, including size grading, which normally involves capturing and confining fish before passing them through a grid, often in air. As an example of mitigation, self-grading involves allowing fish to swim voluntarily through an underwater grid, sometimes in response to simple directional stimuli. For example, up to the age of 28 days post hatching (dph) pikeperch larvae show a very strong tendency to move towards light. This positive phototactic response promotes successful self-grading in pikeperch up to 22 dph, but not from 28 dph onwards (Fig. 4.10; Tielmann et al. 2016).

Lack of Relevant Stimuli in Culture Systems In other cases, welfare problems arise because natural stimuli that are important in guiding fish behaviour are absent in culture systems. For many fish, visual cues are important for locating and identifying food; poor visibility compromises how well-cultured fish can feed and hence their welfare. Larval Atlantic cod survive poorly when held at low light intensities even with abundant food, because they are unable to capture prey efficiently (Puvanendran and Brown 2002). In older fish, at least initially, feed, though nutritionally appropriate, may fail to fit natural preferences (chemical attractants may be missing, for example), in which case the pellets may be rejected. Adding natural feeding stimulants to formulated feed can help; combining mussel extracts (containing a known feeding stimulant) with feed formulated with rapeseed (containing unattractive compounds) makes this more palatable to turbot (*Psetta*

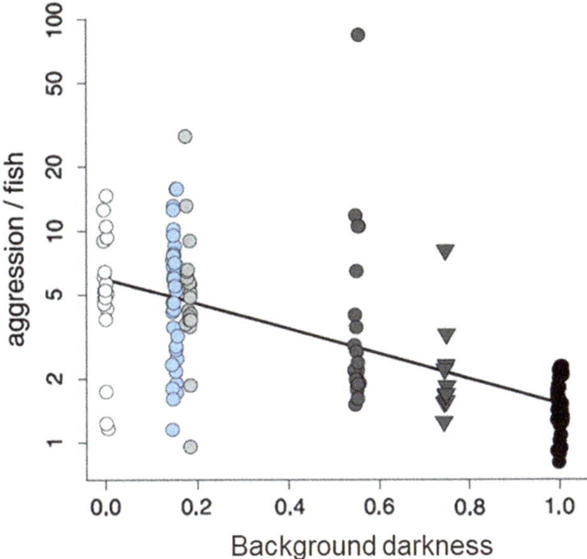

Fig. 4.11 Using natural responses to improve welfare in farmed fish: number of aggressive acts per fish per 10 mins in Coho salmon held in tanks in groups of 10, in relation to selected background colour. Degree of shading of symbols represents darkness of background colour (modified and reproduced with permission from Gaffney et al. 2016)

maxima) and promotes better growth (Nagel et al. 2014). Absence of natural stimuli in culture conditions does not always have negative consequences for welfare. For example, bright skin colour stimulates aggression in many species of fish. Since the skin tends to darken in fish resting on a dark background, this suppresses attack by dominant companions, suggesting a strategy for reducing aggression in production systems. Given a series of pairwise choices, Coho salmon (*Oncorhynchus kisutch*) show a strong preference for black backgrounds, and the darker the selected background, the less aggression they show (Fig. 4.11; Gaffney et al. 2016).

4.5.2.2 Natural Expression of Internal Processes

Motivational Systems in Farmed Fish Some welfare problems in aquaculture arise as a result of the activation of the systems that control behaviour in a context that would be appropriate in nature, but does not promote welfare in culture systems. In the case of feeding, failure to take account of natural changes in appetite may cause over- or under-feeding, with adverse consequences for production, fish welfare and environmental protection (Jobling et al. 2012b and refs therein). Such problems can be avoided by delivering feed (by hand or from computerised automatic feeders) in quantities and on timescales that match natural appetite patterns. Alternatively, farmers may use one of the varieties of commercially available demand feeders that match feed delivery to current appetite (Sect. 4.4.2). Use of such feeders promotes fast, uniform growth and better welfare in several farmed species (Attia et al. 2012 and refs therein).

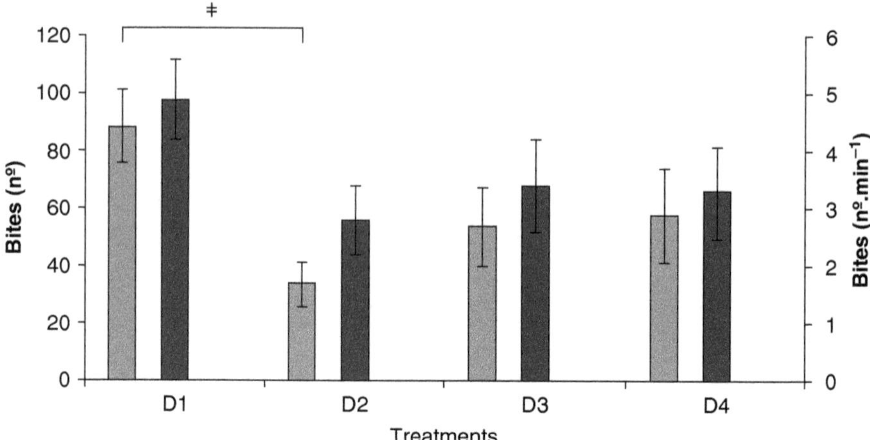

Fig. 4.12 Reducing aggression by manipulating motivational state. Number of bites during the whole observation period (mean and standard error. Grey bars) and frequency of bites (bites/min) in the period after the first observed bite (mean and standard error. Black bars) in *Brycon amizonicus* fed for 7 days on diets supplemented with different amounts of tryptophan (TRP). *D1* control: 0.47% TRP. D2: 0.94% of TRP. D3: 1.88% TRP and D4: 3.76% of TRP. Asterisk represents a significant difference at $p < 0.05$ (Wolkers et al. 2012)

The motivational basis of fighting over food includes low energy reserves (Sect. 4.3.1) and levels of aggression in production systems can be reduced by ensuring that fish have sufficient food. Since demand feeders potentially allow all fish in a unit, even subordinates, to feed to satiation, their use can lower levels of aggression. In Atlantic salmon at both parr and post-smolt stage, fish fed using an interactive feedback system (matching food release to appetite) show less energetically expensive scramble competition than do fish fed the same rations at predetermined times by a conventional feeder. In post-smolts, overt aggression is rare in demand-fed fish at all times, but in conventionally fed fish attacks increase dramatically during meals (Noble et al. 2007a, b). Another promising strategy for reducing aggression among farmed fish is to tap into the internal processes that underpin the resolution of fights. The experience of participating in, and particularly losing, a fight increases the production of the neurotransmitter serotonin, which, among other behavioural effects, inhibits aggression. Diets supplemented with tryptophan, a natural precursor of serotonin, reduce aggression in several farmed species, including rainbow trout (Lepage et al. 2005), Atlantic cod (Hoglund et al. 2005) and martinx (*Brycon amazonicus*; Wolkers et al. 2012; Fig. 4.12).

In many species of fish the behavioural precursors of spawing, including courtship, are long and complex and serve a number of important functions, including fine-tuning the fish's endocrine state and synchronising gamete production and release. Cultured fish inherit these mechanisms and if the conditions in which brood stock are held do not allow them to be expressed, spawning may be prevented or compromised. In pairs of Nile tilapia held in visual isolation, males have smaller gonads relative to their body size and show less courtship and females spawn less than do those allowed visual contact with other fish (Castro et al. 2009). It has proved notoriously difficult to induce full maturation in cultured eels (*Anguilla anguilla*), whose complex life cycle includes migration from freshwater growing areas to breeding grounds in the Sargasso Sea. Exposing groups of cultured eels in the pre-migrant stage to 9 weeks simulated migration in freshwater and then seawater stimulates gonadal development to levels equivalent to wild fish; gonad weight in relation to body weight is 44% higher in males allowed to "migrate" than in control fish (Mes et al. 2016). One might say in this context that eels have a physiological need to swim.

The Question of Behavioural Need The concept of animals having behavioural needs that must be met for good welfare is based on a number of assumptions about the interacting external stimuli and internal systems that control behaviour. At one extreme, behaviour patterns such as escaping and hiding from a predator are triggered by particular aversive stimuli from which, in nature, the animal removes itself through the appropriate behaviour. When such responses are not shown because no predator is present, there is no reason to suppose that the animal concerned is experiencing an unfulfilled behavioural need. Where behaviour is activated by specific causal factors within the animal (a nutritional deficit in the case of feeding), the simple consequences of that behaviour (acquisition of the necessary nutrients) may remove the relevant causal factor, so bringing the behaviour to an end. In such a case, if nutrients were gathered by simply ingesting and swallowing a nutritionally adequate pellet of food rather than locating, hunting and capturing prey, once again there is no reason to suppose that the animal is experiencing an unfulfilled behavioural need.

However, it is possible that natural selection has produced animals that are motivated not just to achieve particular outcomes, but also to perform species-specific actions. In other words, reinforcing aspects of performance per se may be built into the mechanisms that control important but possibly costly behaviour. In the case of feeding, in addition to requiring nutrients, fish could in some sense need to perform specific foraging responses. This might be the case when a predatory fish hunts prey that is difficult and dangerous to catch, meaning that there are inhibitions to be overcome. The author has seen hungry pike showing strong signs of fear while preparing to attack a stickleback, an attractive morsel except for its sharp spines. In such a case, the adaptive internal mechanism that controls foraging might well include some sort of positive reinforcing effect of performance, independent of its beneficial nutritional consequences. The author has also observed pike still stalking prey even when their stomachs are literally full to overflowing, suggesting that something other than simply meeting a nutritional need is going on. According to this scenario, the welfare of some farmed fish might conceivably be compromised in spite of good nutritional status where there is no opportunity to perform the full foraging sequence required of wild fish. Possibly relevant examples among farmed fish include sea bream, which naturally feed on hard-bodied prey that requires chewing. When fed standard pellets they "play" with these, often eventually breaking them up so that much feed is wasted; this can be prevented by feeding the fish with especially hard pellets (Andrew et al. 2004). Farmed juvenile cod, which also feed naturally on hard prey, are notorious for picking and chewing at irregularities in the net walls of their cages, in the process making holes through which escapes occur; 50% of cod escapes are caused by such behavioural factors (Jackson et al. 2015). Net biting is related to the presence of food (increasing when food is released outside the net) and to food deprivation (Zimmermann et al. 2012; Hansen et al. 2012). This behaviour can be reduced but not prevented altogether by keeping the fish well fed, and also by environmental enrichment in the form of stimulating objects placed within the cage (Zimmermann et al. 2012).

In the context of migration, it could be that juvenile salmonids undertake long journeys simply because they are tracking widely dispersed food, in which case so long as plenty of food is provided, keeping them in cages would not create a behavioural need. On the other hand, it could be that, regardless of feeding, they are motivated by an internally generated drive to swim. Certainly, numerous studies have documented beneficial effects of sustained swimming at moderate speeds for health and welfare in many species of fish (Solstorm et al. 2016). In such a case, provided fish are held in cages that are sufficiently large to allow continuous swimming, this behavioural need can be met even though fish are confined. A third possibility is that migrating fish, like some migrating birds, are somehow driven to swim in a particular direction for a particular distance to a particular location. In this case, by definition, fish held in cages would have an unfulfilled behavioural need.

Although there is nothing rewarding about losing a fight, other aspects of aggressive encounters, such as an exchange of aggressive displays, might be rewarding to the animals, over and above the benefits gained from winning a fight. Certainly, fights are costly and dangerous and animals might well need some kind of motivational nudge to participate, even when the fitness sums (Sect. 4.3.3) add up in favour of a fight. One striking feature of aggressive interactions is the complex steps that fish take to gain information about potential rivals, from comparing relative size, strength and stamina before and during a given encounter, to remembering the results of past encounters and performing complex computations based on observations of fights between other individuals. It seems that fish are very strongly motivated to reduce uncertainty by gathering information about how much their status as territory owner or dominant group member is threatened in any given context. Similar issues are raised by encounters with potential predators, in which prey fish take considerable risks to gain information about the threat posed by a potential predator, which allows them to function effectively in a dangerous world. Here too, natural selection has produced animals that are highly motivated to gain information in the face of risk. It may be that fish that have been alerted to the presence of a rival or a predator, but are unable to gather any information about the actual threat posed (as might well happen in culture) experience an unfulfilled behavioural need.

It is important to note that the points raised in the previous paragraphs are largely speculative and we know very little about whether captive fish do indeed have behavioural needs and nothing about how they experience (or what they feel about) the motivational states involved.

4.5.2.3 Behavioural Development in Aquaculture Systems

Changing space use and feeding habits during development, particularly dramatic when planktonic larvae of marine species leave the water column and settle onto the substratum, place heavy demands on culture systems and, unless accommodated, often compromise welfare, particularly for new aquaculture species. In the summer flounder (*Paralichthys dentatus*), large, early metamorphosing fish attack and eat smaller, late metamorphosing companions, a problem for both production and welfare. This can be reduced by exposing larvae to low-salinity water (a key cue that triggers metamorphosis) during late development; this synchronises settlement, reduces variability in size at settlement and cuts down on cannibalism after metamorphosis (Gavlik and Specker 2004).

The fact that some behavioural differences (in aggression, for example, Box 4.4) are inherited, means that the culture conditions that optimise health and welfare may not be the same for all individuals of a cultured fish species. It also raises the possibility of selective breeding for behavioural traits that promote welfare. For example, targeted selective breeding for low-stress responsiveness has been successful for a variety of species, including rainbow trout (Quillet et al. 2014) and sea bass (Vandeputte et al. 2016).

The fact that all aspects of fish behaviour are influenced by environmental events from a very early stage is also critical. To give just one example, the highly simplified conditions experienced by cultured fish result in poor brain development and impaired ability to form and use mental maps (Salvanes et al. 2013; Chap. 5). This contributes to the poor survival of cultured fish released for restocking, which raises welfare concerns. For fish that spend their whole life in culture systems, it could be that small brains and poor memories make them better suited to the simple physical environment of production systems. However, it is also possible that poor ability to use landmarks may compromise effective space use even in the simple conditions provided by most production systems. In any event, poor brain growth and limited spatial memory can be mitigated by quite simple environmental enrichment (Chap. 5).

4.5.2.4 How Behaviour Has Been Shaped by Natural Selection; Its Functions

The flexible adjustments that wild fish make to the costs and benefits of the behavioural options available to them potentially contribute to the ease with which they can adapt to culture conditions. Sometimes, however, if the selective advantage of showing a given behaviour pattern is very strong, this can cause problems. For example, the adjustments that fish make to perceived predation risk can result in suppression of feeding, with negative effects for both production and welfare. This can be mitigated by culturing fish that are relatively resistant to stress and/or by developing low-stress husbandry practices such as passive grading.

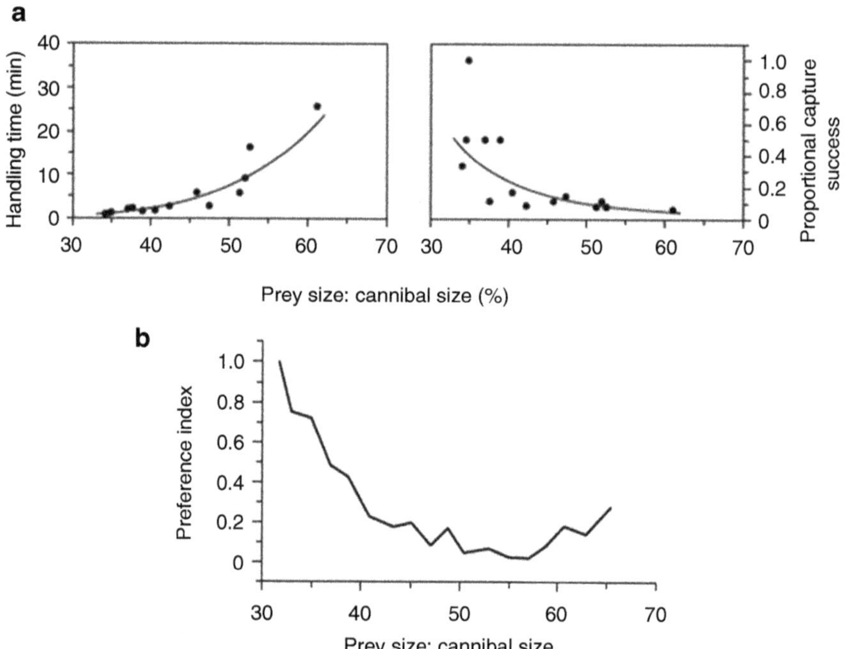

Fig. 4.13 A cost: benefit approach to cannibalism. (**a**) Handling time (*LHS*) and capture success (*RHS*) in juvenile barramundi in relation to the relative size of prey and cannibal. (**b**) An index of preference for prey of different relative sizes in juvenile barramundi (approx. 40 days post hatching at the start of the study; modified and reproduced with permission from Ribeira and Qin 2015)

In other cases, we can make use of such adaptive flexibility to solve welfare problems in aquaculture; since the mechanisms that control behaviour are themselves the result of natural selection, many of the cases discussed in Sect. 4.3.3 are also relevant here. For example, reducing the benefits or increasing the costs of fighting over food by using demand feeders cuts levels of aggression and allows subordinate fish to feed and grow well (Noble et al. 2007a, b; Attia et al. 2012). Increasing the costs of fighting by exposing fish to a current so that performing aggressive acts becomes more energetically costly also reduces levels of aggression (Huntingford and Kadri 2012 and refs therein).

As discussed above, cannibalism is part of the natural feeding repertoire of many species of fish, especially early in development; in culture this causes problems for welfare and production (Raubenheimer et al. 2012; Naumowicz et al. 2017). From a functional perspective, fish of the same species represent a highly nutritious food source and the larger the prey fish, the more nutrients it can provide. Set against this, the time taken to catch and consume another fish increases with relative prey size, so that profitability falls. In barramundi (*Lates calcarifer*), which can be ferocious cannibals, handling time increases and the probability of capture decreases with relative prey size (Fig. 4.13a), so profitability falls dramatically. Theory predicts that cannibals should feed preferentially on smaller conspecifics and this is exactly what they do (Fig. 4.13b; Ribeira and Qin 2015). From this cost–benefit perspective, cannibalism can be reduced by decreasing its benefits, for example by providing abundant alternative food; provision of more frequent meals does indeed cut down on cannibalism, for example in barramundi (Ribeiro et al. 2015) and spotted trout (*Cynoscion nebulosus*; Manley et al. 2016). Increasing the costs of cannibalism by minimising size differences within cohorts, so that cannibals are only exposed to relatively large, expensive prey, also reduces cannibalism in a number of species. Homogeneous size distributions can be achieved by feed management (increasing feeding frequency from one to three meals a day generates homogeneous growth and postposes the development of cannibalism; Ribeiro et al. 2015) or by regular size grading (Naumowicz et al. 2017).

4.6 Conclusions

The welfare of fish in aquaculture has received considerable research attention since the 1980s (Fig. 4.14a), with at least 460 articles showing up on a Web of Science analysis, representing many different disciplines (Fig. 4.14b). Behavioural biology has a respectable place among these contributing disciplines and this final section considers briefly what, if any, unique contribution it can make to understanding and protecting welfare in farmed fish.

Most farmed fish are effectively undomesticated and, as the examples presented in this chapter show, they naturally express many of the behavioural traits that are shown by their wild counterparts. Problems for welfare can arise from the mechanisms that control behaviour (as when low energy reserves promote aggression), how behaviour develops (as when unsynchronised larval settlement encourages cannibalism) and the way that it has been shaped by natural selection (as when the benefits of fighting over a clumped and predictable food supply outweigh the costs). By the same token, an understanding of the biology of behaviour in the framework of Tinbergen's questions can suggest solutions to such problems; providing plenty of food increases energy reserves and reduces aggression, manipulating the stimuli that trigger larval settlement synchronises metamorphosis and reduces cannibalism and feeding fish to appetite via interactive feeders removes the benefits of fighting over food.

This is not to suggest that behavioural biologists have a monopoly of wisdom in this context. Different disciplines will often converge on the same solution; both behavioural biologists and physiologists rightly suggest that promoting sustained swimming in farmed fish may improve their welfare. In addition, experienced fish farmers have extensive knowledge of what is good and bad for fish, being well aware, for example that cannibalism is more common when size differences within cohorts are large. A behavioural approach provides an additional framework through which established solutions can be understood and potentially provides tools for fine-tuning existing solutions. For example, profitability curves such as that shown

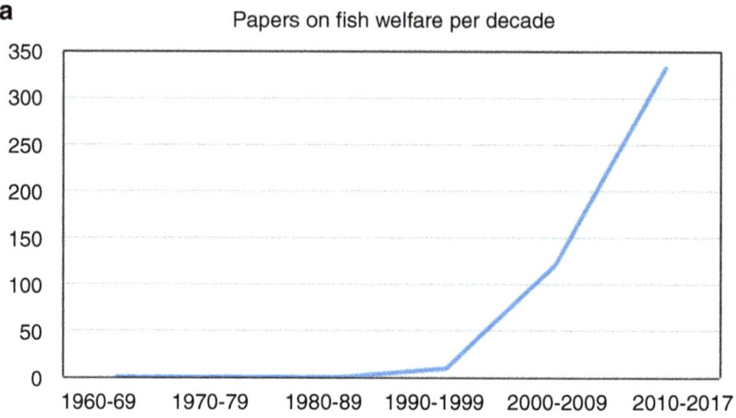

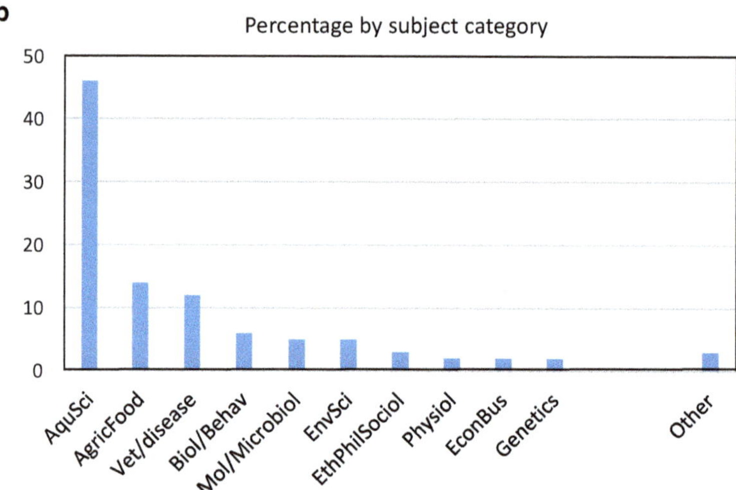

Fig. 4.14 (**a**) Number of papers on the welfare of farmed fish published per decade since 1960, from the Web of Science. (**b**) Percentage of the 435 papers published on the welfare of farmed fish, classified by Web of Science subject categories, combined where appropriate and omitting categories with <2 entries

in Fig. 4.14 can inform the grading process by identifying the exact size differential that will prevent cannibalism in a given system. It also potentially offers a strategy for finding fast-track solutions if a similar problem arises in different species.

One context in which behavioural biology may have something special to add to the debate is in broad explorations of the nature of welfare in fish. For example, biological understanding of the mechanisms that control behaviour provides a framework for discussing whether fish have behavioural needs. If they do, we need to find out whether the experience of an unfulfilled behavioural need is aversive. There is increasing interest in and understanding of cognition and sentience in fishes (Chaps. 7–9) and the difficult issue of the emotional experiences of fish is being increasingly scrutinised, using a powerful combination of behavioural biology, psychology and neuroscience. Box 4.6 provides an example. To quote the authors of that study: "...we have shown that Sea Bream exposed to stimuli that vary according to valence (appetitive, aversive) and salience (predictable, unpredictable) exhibit different behavioural, physiological and neuromolecular states that are specific to each combination of valence and salience...The neuromolecular data presented here suggest an involvement of both (the ventral and the dorsal telencephalic area) in the appraisal of emotional stimuli, which supports the occurrence of an evolutionary conserved neural substrate for the processing of emotional stimuli, given the similar role played by the mammalian homologues of these areas." (Cerqueira et al. 2017). Through such multidisciplinary studies, the next few years will surely see significant increases in our understanding of many aspects of the behaviour of fish, including their welfare.

Box 4.6 Probing the Experiences of Fish

Affective space in humans can be conceptualised in terms of valence (intrinsic attractiveness or aversiveness of events or experiences) and salience (strength or importance of events or experiences). To determine whether fish show specific emotion-like states related to these two dimensions, sea bream (*Sparus auratus*) were trained in one of four different conditions. They received either a *positive* event (food delivery) or a *negative* event (brief exposure to air) that was either always preceded by a light stimulus (*predictable*) or occurred at random with respect to a light stimulus (*unpredictable*). The subsequent responses of the fish to the light stimulus alone were distinct for the four treatments. Social behaviour was shown by fish trained with food delivery, especially when this had been predicted by the light cue. Escape attempts were shown by fish trained with air exposure, again especially when this had been predicted by the light cue. Plasma cortisol levels also differed between treatments, with lower levels in fish trained with food as opposed to exposure and lower levels in the predictable that the unpredictable condition. Immediate early genes (indicative of recent neural activity) were also differentially expressed in specific areas of the forebrain in the four treatment groups. Differences in valence (food versus exposure) and predictability were associated with unique patterns of expression of several different immediate early genes in the ventral telencephalic area (suggested homologue of the mammalian septum) and the dorsal telencephalic area (suggested homologue of the mammalian basal amygdala. Cerqueira et al. 2017).

References

Abreu MS, Giacomini ACVV, Koakosk G, Piato AL, Barcellos LJG (2016) Evaluating "anxiety" and social behavior in jundiá (*Rhamdia quelen*). Physiol Behav 160:59–65

Andrew JE, Holm J, Huntingford FA (2004) The effect of pellet texture on the feeding behaviour of gilthead sea bream (*Sparus aurata* L.). Aquaculture 23:471–479

Ashley PJ (2007) Fish welfare: current issues in aquaculture. Appl Anim Behav Sci 104:199–235

Attia J, Millot S, Di-Poi C, Begout M-L, Noble C, Sanchez-Vasquez FJ, Terova G, Saroglia M, Damsgaard B (2012) Demand feeding and welfare in farmed fish. Fish Physiol Biochem 38:107–118

Barki A, Volpato GL (1998) Early social environment and the fighting behaviour of young *Oreochromis niloticus*. Behaviour 135:913–929

Bégout M-L, Kadri S, Huntingford F, Damsgård B (2012) Tools for studying the behaviour of farmed fish. In: Huntingford F, Jobling M, Kadri S (eds) *Aquaculture and behavior*. Wiley, Chichester, pp 65–86

Berglund A, Rosenqvist G (2003) Sex role reversal in pipefish. Adv Study Behav 32:131–167

Beveridge MCM, Baird DJ (2000) Diet, feeding and digestive physiology. In: Beveridge MCM, McAndrew BJ (eds) Tilapias: biology and exploitation. Kluwer Academic, Dordrecht, pp 59–87

Billing AM, Rosenqvist G, Berglund A (2007) No terminal investment in pipefish males: only young males exhibit risk-prone courtship behaviour. Behav Ecol 18:535–540

Brawn VM (1961a) Sound production by the cod (*Gadus callarias*). Behaviour 18:239–255

Brawn VM (1961b) Reproductive behaviour of the cod (*Gadus callarias*). Behaviour 18:177–198

Castro ALS, Gonçalves-de-Freitas E, Volpato GL, Oliveira C (2009) Visual communication stimulates reproduction in Nile tilapia, *Oreochromis niloticus*. Braz J Med Biol Res 42:368–374

Cerqueira M, Millot S, Castanheira MF, Félix AS, Silva T, Oliveira GA, Oliveira CC, Martins CIM, Oliveira RF (2017) Cognitive appraisal of environmental stimuli induces emotion-like states in fish. Sci Rep 7:13181. https://doi.org/10.1038/s41598-017-13173-x

Cobcroft J, Battaglene SC (2009) Jaw malformation in striped trumpeter *Latris lineata* larvae linked to walling behaviour and tank colour. Aquaculture 289:274–282

Cole KS, Noakes DLG (1980) Development of early social behaviour of rainbow trout, *Salmo gairderi*. Behav Process 5:97–112

Comeau LA, Campana SE, Chouinard GA, Hanson JM (2001) Timing of Atlantic cod *Gadus morhua* seasonal migrations in relation to serum levels of gonadal and thyroidal hormones. Mar Ecol Prog Ser 221:245–253

Costenaro-Ferreira C, Oliveira RRB, Oliveira PLS, Hartmann GJ, Hammes FB, Pouey JLOF, Piedras SRN (2016) Cannibalism management of jundiá fry, *Rhamdia quelen*: behavior in heterogeneous batches fed on food with different particle sizes. Appl Anim Behav Sci 185:146–151

Damsgård B, Dill LM (1998) Risk-taking behavior in weight-compensating coho salmon, *Oncorhynchus kisutch*. Behav Ecol 9:26–32

Damsgård B, Huntingford F (2012) Fighting and aggression. In: Huntingford F, Jobling M, Kadri S (eds) Aquaculture and behavior. Wiley, Chichester, pp 248–285

Døving KB, Stabell OB, Östlund-Nilsson S, Fisher R (2006) Site fidelity and homing in tropical coral reef cardinalfish: are they using olfactory cues? Chem Senses 31:265–272

Enquist M, Leimar O, Ljungberg T, Mallner Y, Segerdahl N (1990) A test of the sequential assessment game - fighting in the cichlid fish *Nannacara anomala*. Anim Behav 40:1–14

Farr JA (1983) The inheritance of quantitative fitness traits in guppies, *Poecilia reticulata*. Evolution 37:1193–1209

Ferguson MM, Noakes DL (1982) Genetics of social behaviour in charrs (*Salvelinus* sp). Anim Behav 30:128–134

Fernald RD (1980) Response of male cichlid fish, *Haplochromis burtoni*, reared in isolation to models of conspecifics. Z Tierpsychol 54:85–93

Fessehaye Y, El-bialy Z, Rezk MA, Crooijmans R, Bovenhuis H, Komen H (2006) Mating systems and male reproductive success in Nile tilapia (*Oreochromis niloticus*) in breeding hapas: a microsatellite analysis. Aquaculture 256:148–158

Fleming I, Huntingford F (2012) Reproductive behaviour. In: Huntingford F, Jobling M, Kadri S (eds) Aquaculture and behavior. Wiley, Chichester, pp 286–321

Forbes H (2007) Individual variability in the behaviour and morphology of larval Atlantic cod (*Gadus morhua*). PhD Thesis. University of Glasgow, Glasgow

Fortes da Silva R, Kitagawa A, Sanchez Vazquez FJ (2016) Dietary self-selection in fish: a new approach to studying fish nutrition and feeding behaviour. Rev Fish Biol Fish 26:39–51

Fraser NHC, Huntingford FA, Thorpe JE (1994) The effect of light-intensity on the nightly movements of juvenile Atlantic salmon alevins away from the red. J Fish Biol 45A:143–150

Frenzl B, Stien LH, Cockerill D, Oppedal F, Richards RH, Shinn AP, Bron JE, Migaud H (2014) Manipulation of farmed Atlantic salmon swimming behaviour through the adjustment of lighting and feeding regimes as a tool for salmon lice control. Aquaculture 424–425:183–188

Gaffney LP, Franks B, Weary DM, von Keyserlingk MAGF (2016) Coho salmon (*Oncorhynchus kisutch*) prefer and are less aggressive in darker environments. PLoS One 11(3):e0151325. https://doi.org/10.1371/journal.pone.0151325

Gallagher AJ, Lawrence MJ, Schlaepfer SMR, Wilson ADM, Cooke SJ (2016) Avian predators transmit fear along the air–water interface influencing prey and their parental care. Can J Zool 94:863–870

Gavlik S, Specker JL (2004) Metamorphosis in summer flounder: manipulation of rearing salinity to synchronize settling behavior, growth and development. Aquaculture 240:543–559

Gerlach G, Atema J, Kingsford MJ, Black KP, Miller-Sims V (2007) Smelling home can prevent dispersal of reef fish larvae. Proc Natl Acad Sci U S A 104:858–863

Gerlai R, Hogan JA (1992) Learning to find the opponent: an ethological analysis of the behavior of paradise fish (*Macropodus opercularis*) in intra- and interspecific encounters. J Comp Psychol 106:306–315

Giaquinto PC, Volpato GL (2001) Hunger suppresses the onset and the freezing component of the anti-predator response to conspecific skin extract in pintado catfish. Behaviour 138:1205–1214

Goldan O, Popper D, Karplus I (2003) Food competition in small groups of juvenile gilthead sea bream (*Sparus aurata*). Isr J Aquacult 55:94–106

Goncalves DM, Oliveira RF (2011) Hormones and sexual behavior of teleost fishes. In: Norris DO, Lopez KH (eds) Hormones and reproduction of vertebrates 1: fishes. Elsevier, San Diego, pp 119–147

Goulet D, Green JM, Shears TH (1986) Courtship, spawning, and parental care behavior of the lumpfish, *Cyclopterus lumpus*, in newfoundland. Can J Zool 64:1320–1325

Grant JWA (1997) Territoriality. In: Godin J-GJ (ed) Behavioural ecology of teleost fishes. Oxford University Press, Oxford, pp 81–103

Greaves K, Tuene S (2001) The form and context of aggressive behaviour in farmed Atlantic halibut (*Hippoglossus hippoglossus*). Aquaculture 193:139–147

Gregory JS, Griffith JS (1996) Aggressive behaviour of under-yearling rainbow trout in simulated winter concealment habitat. J Fish Biol 49:237–245

Grosenick L, Clement TS, Fernald RD (2007) Fish can infer social rank by observation alone. Nature 445:429–432

Gudjonsson S, Einarsson SM, Jonsson IR, Gudbrandsson J (2015) Marine feeding areas and vertical movements of Atlantic salmon (*Salmo salar*) as inferred from recoveries of data storage tags. Can J Fish Aquat Sci 72:1087–1098

Hall AE, Clark TD (2016) Seeing is believing: metabolism provides insight into threat perception for a prey species of coral reef fish. Anim Behav 115:117–116

Hansen L, Dale T, Damsgaard B, Uglem I, Aas K, Bjorn PA (2012) Escape-related behaviour of Atlantic cod, *Gadus morhua*, in a simulated farm situation. Aquac Res 40:26–34

Hatlen B, Grisdale-Helland B, Helland SJ (2006) Growth variation and fin damage in Atlantic cod (*Gadus morhua* L.) fed at graded levels of feed restriction. Aquaculture 261:1212–1221

Herczeg G, Ab Ghani NI, Merilä J (2016) On plasticity of aggression: influence of past and present predation risk, social environment and sex. Behav Ecol Sociobiol 70:179–187

Herlin M, Delghandi M, Wesmajervi M, Taggart JB, McAndrew BJ, Penman DJ (2008) Analysis of the parental contribution to a group of fry from a single day of spawning from a commercial Atlantic cod (*Gadus morhua*) breeding tank. Aquaculture 274:218–224

Higgs DM, Fuiman LA (1996) Ontogeny of visual and mechanosensory structure and function in Atlantic menhaden *Brevoortia tyrannus*. J Exp Biol 199:2619–2629

Hoglund E, Bakke MJ, Øverli Ø, Winberg S, Nilsson GE (2005) Suppression of aggressive behaviour in juvenile Atlantic cod (*Gadus morhua*) by L-tryptophan supplementation. Aquaculture 249:525–531

Hsu Y, Wolf LL (2001) The winner and loser effect: what fighting behaviours are influenced? Anim Behav 61:777–786

Hughes KA, Rodd FH, Reznick DN (2005) Genetic and environmental effects on secondary sex traits in guppies (*Poecilia reticulata*). J Evol Biol 18:35–45

Huntingford FA, Kadri S (2012) Exercise, stress and welfare. In: Palsra AP, Planas JV (eds) Swimming physiology of fish. Springer, Heidelberg, pp 161–174

Huntingford FA, Aird D, Joiner P, Thorpe KR, Braithwaite VA, Armstrong JA (1999) How juvenile salmon respond to falling water levels: experiments in an artificial stream. Fish Manag Ecol 6:1–8

Huntingford F, Hunter W, Braithwaite V (2012a) Movement and orientation. In: Huntingford F, Jobling M, Kadri S (eds) Aquaculture and behavior. Wiley, Chichester, pp 87–120

Huntingford F, Coyle S, Hunter W (2012b) Avoiding predators. In: Huntingford F, Jobling M, Kadri S (eds) Aquaculture and behavior. Wiley, Chichester, pp 220–247

Huntingford F, Mesquita F, Kadri S (2013) Personality variation in cultured fish: implications for production and welfare. In: Carere C, Maestripieri D (eds) Animal personalities: behavior, physiology and evolution. University of Chicago Press, Chicago, pp 414–440

Ibrahim AI, Huntingford FA (1989) Laboratory and field studies of the effects of predation risk on foraging in sticklebacks (*Gasterosteus aculeatus*). Behaviour 109:46–57

Idler DR, Fagerland UHM, Mayoh H (1956) Olfactory perception in migrating salmon 1. L-serine, a salmon repellent in mammalian skin. J Gen Physiol 39:889–892

Imsland AK, Renolds P, Eliassen G, Hangstad TA, Nytro AV, Foss A, Vikingstad E, Elvegard TA (2015a) Feeding preferences of lumpfish (*Cyclopterus lumpus*) maintained in open net-pens with Atlantic salmon (*Salmo salar*). Aquaculture 436:47–51

Imsland AK, Reynolds P, Eliassen G, Hangstad TA, Nytro AV, Foss A, Vikingstad E, Elvegard TA (2015b) Assessment of suitable substrates for lumpfish in sea pens. Aquac Int 23:639–645

Ivy CM, Robertson CE, Bernier NJ (2017) Acute embryonic anoxia exposure favours development of a dominant and aggressive phenotype in adult zebrafish. Proc R Soc Lond B 284:20161868

Jackson D, Drumm A, McEvoy S, Jensen Ø, Mendiola D, Gabiña G, Borg JA, Papageorgiou N, Karakassis Y, Black KD (2015) A pan-European valuation of the extent, causes and cost of escape events from sea cage fish farming. Aquaculture 436:21–26

Jakobssen S, Brick O, Kullberg C (1995) Escalated fighting behaviour incurs increased predation risk. Anim Behav 49:235–239

Jha P, Jha S, Pal BC, Barat S (2005) Behavioural responses of two popular ornamental carps, *Cyprinus carpio* and *Carassius auratus* to monoculture and polyculture conditions in aquaria. Acta Ichthyolica Piscatore 35:133–137

Jobling M (2010) Feeds and feeding. In: Le François NR, Jobling M, Carter C, Blier PU (eds) Finfish aquaculture diversification. CAB International, Wallingford, pp 61–87

Jobling M, Alanara A, Kadri S, Huntingford FA (2012a) Feeding biology and foraging. In: Huntingford F, Jobling M, Kadri S (eds) Aquaculture and behavior. Wiley, Chichester, pp 121–149

Jobling M, Alanara A, Noble C, Sanchez-Vasquez J, Kadri S, Huntingford FA (2012b) Appetite and food intake. In: Huntingford F, Jobling M, Kadri S (eds) Aquaculture and behavior. Wiley, Chichester, pp 183–219

Johansson D, Ruohonen K, Kiessling A et al (2006) Effect of environmental factors on swimming depth preferences of Atlantic salmon (*Salmo salar*) and temporal and spatial variations in oxygen levels in sea cages at a fjord site. Aquaculture 254:594–605

Johnsson JI, Petersson E, Jönsson E, Björnsson BT, Järvi T (1996) Domestication and growth hormone alter anti-predator behaviour and growth patterns in juvenile brown trout, *Salmo trutta*. Can J Fish Aquat Sci 53:1546–1554

Juell JE, Oppedal F, Boxaspen K, Taranger GL (2003) Submerged light increases swimming depth and reduces fish density in Atlantic salmon *Salmo salar* in production cages. Aquac Res 34:469–477

Kasumyan AO, Døving KB (2003) Taste preferences in fishes. Fish Fish 4:289–347

Kelley JL (2008) Assessment of predation risk by prey fishes. In: Magnahagen C, Braithwaite VA, Fosgren E, Kapoor BG (eds) Fish behaviour. Science, Enfield, pp 269–302

Kennedy J, Jonsson SP, Olafsson HG, Kasper JM (2016) Observations of vertical movements and depth distribution of migrating female lumpsofh (*Cyclopterus lumpus*) in Iceland from data storage tags and trawl surveys. ICES J Mar Sci 73:1160–1169

Krause J, Bumann D, Todt D (1992) Relationship between position preference and nutritional state of individuals in schools of juvenile roach (*Rutilus rutilus*). Behav Ecol Sociobiol 30:177–180

Kvarnemo C, Simmons LW (2004) Testes investment and spawning mode in pipefishes and seahorses (Syngnathidae). Biol J Linn Soc 83:369–376

Lahti K, Laurila A, Enberg K, Piironen J (2001) Variation in aggressive behavior and growth rate between populations and migratory forms in the brown trout, *Salmo trutta*. Anim Behav 62:935–944

Le François NR, Jobling M, Carter C, Blier PU (2010) Finfish aquaculture diversification. CABI, Wallingford

Lekang OI, Salas-Bringas C, Bostock JC (2016) Challenges and emerging technical solutions in on-growing salmon farming. Aquac Int 24:757–766

Lepage O, Larson ET, Mayer I, Winberg S (2005) Serotonin, but not melatonin, plays a role in shaping dominant-subordinate relationships and aggression in rainbow trout. Horm Behav 48:233–242

Lindeyer CM, Reader SM (2010) Social learning of escape routes in zebrafish and the stability of behavioural traditions. Anim Behav 79:827–834

Maan M, Groothuis TGG, Wittenberg J (2001) Escalated fighting despite predictors of conflict outcome: solving the paradox in a South American cichlid fish. Anim Behav 62:623–634

Manley CB, Rakocinski CF, Lee PG, Blaylock RB (2016) Feeding frequency mediates aggression and cannibalism in larval hatchery-reared spotted seatrout, *Cynoscion nebulosus*. Aquaculture 437:155–160

Manuel R, Boerrigter JGJ, Cloosterman M, Gorissen M, Flik G, van den Bos R, van de Vis H (2016) Effects of acute stress on aggression and the cortisol response in the African sharptooth catfish *Clarias gariepinus*: differences between day and night. J Fish Biol 88:2175–2187

Mazeroll AI, Montgomery WL (1995) Structure and organization of local migrations in brown surgeonfish (*Acanthurus nigrofuscus*). Ethology 99:89–106

McCallum ES, Gulas ST, Balshine S (2017) Accurate resource assessment requires experience in a territorial fish. Anim Behav 123:249–257

Mes D, Dirks RP, Palstra AP et al (2016) Simulated migration under mimicked photothermal conditions enhances sexual maturation of farmed European eel (*Anguilla anguilla*). Aquaculture 452:367–372

Mesquita FO, Young RJ (2007) The behavioural responses of Nile tilapia (*Oreochromis niloticus*) to antipredator training. Appl Anim Behav Sci 106:144–154

Metcalfe JD, Righton D, Eastwood P, Hunter E (2008) Migration and habitat choice in marine fishes. In: Magnahagen C, Braithwaite VA, Fosgren E, Kapoor BG (eds) Fish behaviour. Science, Enfield, pp 187–234

Mitamura H, Uchida K, Miyamoto Y, Kakihara T, Miyagi A, Kawabata Y, Ichikawa K, Arai N (2012) Short-range homing in a site-specific fish: search and directed movements. J Exp Biol 215:2751–2759

Moutou KA, McCarthy ID, Houlihan DF (1998) The effect of ration level and social rank on the development of fin damage in juvenile rainbow trout. J Fish Biol 52:756–770

Munakata A, Masafumi A, Kazumasa I, Kitamura S, Katsumi S (2012) Involvement of sex steroids and thyroid hormones in upstream and downstream behaviors in masu salmon, *Oncorhynchus masou*. Aquaculture 362:158–166

Nagel F, von Danwitz A, Schlachter M, Kroeckel S, Wagner C, Schulz C (2014) Blue mussel meal as feed attractant in rapeseed protein-based diets for turbot (*Psetta maxima)*. Aquac Res 45:1964–1978

Naumowicz K, Pajdak J, Terech-Majewska E, Szarek J (2017) Intracohort cannibalism and methods for its mitigation in cultured freshwater fish. Rev Fish Biol Fish 27:193–208

Noble C, Kadri S, Mitchell DF, Huntingford FA (2007a) The impact of environmental variables on the feeding rhythms and daily feed intake of cage-held 1+ Atlantic salmon parr (*Salmo salar*). Aquaculture 269:290–298

Noble C, Kadri S, Mitchell DF, Huntingford FA (2007b) The effect of feed regime on the growth and behaviour of 1+ Atlantic salmon post-smolts (*Salmo salar*) in semi-commercial sea cages. Aquac Res 38:1686–1691

Noble C, Mizusawa K, Suzuki K, Tabata M (2007c) The effect of differing self-feeding regimes on the growth, behaviour and fin damage of rainbow trout held in groups. Aquaculture 264:214–222

O'Connor KI, Metcalfe NB, Taylor AC (1999) Does eye darkening signal submission in territorial contests between juvenile Atlantic salmon, *Salmo salar*? Anim Behav 58:1269–1276

Oates J, Manica A, Bshary R (2010) Roving and service quality in the cleaner wrasse *Labroides bicolor*. Ethology 116:309–315

Odell JP, Chappell MA, Dickson KA (2003) Morphological and enzymatic correlates of aerobic and burst performance in different populations of Trinidadian guppies *Poecilia reticulata*. J Exp Biol 206:3707–3718

Odling-Smee LC, Simpson SD, Braithwaite VA (2006) The role of learning in fish orientation. In: Brown C, Laland K, Krause J (eds) Fish cognition and behaviour. Blackwell, Oxford, pp 119–138

Oppedal F, Juell J-E, Johansson D (2007) Thermo- and photo-regulatory swimming behaviour of caged Atlantic salmon: implications for photoperiod management and fish welfare. Aquaculture 265:70–81

Øverli Ø, Winberg S, Damsgård B, Jobling M (1998) Food intake and spontaneous swimming activity in Arctic charr (*Salvelinus alpinus*): role of brain serotonergic activity and social interactions. Can J Zool 76:1366–1370

Pankhurst NW, Ludke SL, King HR, Peter RE (2008) The relationship between acute stress, food intake, endocrine status and life history stage in juvenile farmed Atlantic salmon, *Salmo salar*. Aquaculture 275:311–318

Papadakis VM, Glaropoulos A, Alvanopoulou M, Kentouri M (2016) A behavioural approach of dominance establishment in tank-held sea bream (*Sparus aurata*) under different feeding conditions. Aquac Res 47:4015–4023

Pitcher T (1992) Who dares, wins - the function and evolution of predator inspection behavior in shoaling fish. Netherlands J Zool 42:371–391

Puckett KJ, Dill LM (1985) The energetics of feeding territoriality in juvenile coho salmon (*Oncorhynchus kisutch*). Behaviour 92:97–111

Puvanendran V, Brown JA (2002) Foraging, growth and survival of Atlantic cod larvae reared in different light intensities and photoperiods. Aquaculture 214:131–151

Quillet E, Krieg F, Dechamp N, Hervet C, Berard A, Le Roy P, Guyomard R, Prunet P, Pottinger TG (2014) Quantitative trait loci for magnitude of the plasma cortisol response to confinement in rainbow trout. Anim Genet 45:223–234

Raubenheimer D, Simpson S, Sánchez-Vázquez J, Huntingford F, Kadri S, Jobling M (2012) Nutrition and diet choice. In: Huntingford F, Jobling M, Kadri S (eds) *Aquaculture and behavior*. Wiley, Chichester, pp 150–182
Ribeira FF, Qin JG (2015) Prey size selection and cannibalistic behaviour of juvenile barramundi *Lates calcarifer*. J Fish Biol 86:1549–1566
Ribeiro FF, Forsythe S, Qin JG (2015) Dynamics of intracohort cannibalism and size heterogeneity in juvenile barramundi (*Lates calcarifer*) at different stocking densities and feeding frequencies. Aquaculture 444:55–61
Rillahan C, Chambers MD, Howel WH, Watson WH (2011) The behavior of cod (*Gadus morhua*) in an offshore aquaculture net pen. Aquaculture 310:361–368
Robb SE, Grant JWA (1998) Interactions between the spatial and temporal clumping of food affect the intensity of aggression in Japanese medaka. Anim Behav 56:29–34
Rogers SM, Gagnon V, Bernatchez L (2002) Genetically based phenotype-environment association for swimming behavior in lake whitefish ecotypes (*Coregonus clupeaformis*). Evolution 56:2322–2329
Rosenau ML, McPhail JD (1987) Inherited differences in agonistic behavior between two populations of coho salmon. Trans Am Fish Soc 116:646–654
Rowland SJ, Mifsud C, Nixon M, Boyd P (2006) Effects of stocking density on the performance of the Australian freshwater silver perch (*Bidyanus bidyanus*) in cages. Aquaculture 253:301–308
Salvanes AGV, Moberg O, Ebbesson LOE, Nilsen TO, Jensen KH, Braithwaite VA (2013) Environmental enrichment promotes neural plasticity and cognitive ability in fish. Proc R Soc Lond B 280:20131331
Sanchez-Jerez P, Fernandez-Jover D, Bayle-Sempere J et al (2008) Interactions between bluefish *Pomatomus saltatrix* (L.) and coastal sea-cage farms in the Mediterranean Sea. Aquaculture 282:61–67
Saralva JL, Keller-Costa T, Hubbard PC, Rato A, Canario AVM (2017) Chemical diplomacy in male tilapia: urinary signal increases sex hormone and decreases aggression. Sci Rep 7:7636
Schuster S, Wohl S, Griebsch M, Klostermeier I (2006) Animal cognition: how archer fish learn to down rapidly moving targets. Curr Biol 17:378–383
Schwarz A (1974a) Sound production and associated behaviour in a cichlid fish, *Cichlasoma centrarchus*.I. male-male interactions. Z Tierpsychol 35:147–156
Schwarz A (1974b) The inhibition of aggressive behaviour by sound in the cichlid fish, *Cichlasoma centrarchus*. Z Tierpsychol 35:508–517
Serra M, Wolkers CPB, Urbinati EC (2015) Novelty of the arena impairs the cortisol-related increase in the aggression of matrinxã (*Brycon amazonicus*). Physiol Behav 141:51–57
Simpson SD, Meekan M, Montgomery J, McCauley R, Jeff A (2005) Homeward sound. Science 308:221–231
Skilbrei OT, Ottera H (2016) Vertical distribution of saithe (*Pollachius virens*) aggregating around fish farms. ICES J Mar Sci 73:1186–1195
Solstorm F, Solstorm D, Oppedal F, Olsen RE, Stien LH, Fernö A (2016) Not too slow, not too fast: water currents affect group structure, aggression and welfare in postsmolt Atlantic salmon *Salmo salar*. Aquac Environ Interact 8:339–347
Sorensen PW, Pinillos M, Scott AP (2005) Sexually mature male goldfish release large quantities of androstenedione into the water where it functions as a pheromone. Gen Comp Endocrinol 140:164–175
Tielmann M, Schulz C, Meyer S (2016) Self-grading of larval pike-perch (*Sander lucioperca*), triggered by positive phototaxis. Aquacult Eng 72–73:13–19
Tinbergen N (1951) The study of instinct. Clarendon Press, Oxford
Ukegbu AA, Huntingford FA (1988) Brood value and life expectancy as determinants of parental investment in male 3-spined sticklebacks, *Gasterosteus-aculeatus*. Ethology 78:72–82
Van de Nieuwegiessen P, Zhao H, Verreth JAJ, Schrama JW (2009) Chemical alarm cues in juvenile African catfish, *Clarius gariepinus*: a potential stressor in aquaculture? Aquaculture 286:95–99

Vandeputte M, Porte JD, Auperin B, Dupont-Nivet M, Vergnet A, Valotaire C, Claireaux G, Prunet P, Chatain B (2016) Quantitative genetic variation for post-stress cortisol and swimming performance in growth-selected and control populations of European sea bass (*Dicentrarchus labrax*). Aquaculture 455:1–7

Vehanen T (2003) Adaptive flexibility in the behaviour of juvenile Atlantic salmon: short-term responses to food availability and threat from predation. J Fish Biol 63:1034–1045

Vejrik L, Matejickova I, Juza T, Frouzova J, Seďa J, Blabolil P, Ricard D, Vasek M, Kubecka J, Riha M, Cech M (2016) Small fish use the hypoxic pelagic zone as a refuge from predators. Freshw Biol 61:899–913

Vera Cruz EM, Brown CL (2007) The influence of social status on the rate of growth, eye color pattern and insulin-like growth factor-I gene expression in Nile tilapia, *Oreochromis niloticus*. Horm Behav 51:611–619

Verbeek P, Iwamoto T, Murakami N (2007) Differences in aggression between wild-type and domesticated fighting fish are context dependent. Anim Behav 73:75–83

Verzijden MN, ten Cate C (2007) Early learning influences species assortative mating preferences in Lake Victoria cichlid fish. Biol Lett 3:134–136

Weir LK, Hutchings JA, Fleming IA, Einum S (2004) Dominance relationships and behavioural correlates of individual spawning success in farmed and wild male Atlantic salmon, *Salmo salar*. J Anim Ecol 73:1069–1079

Wocher H, Harsányi A, Schwarz FJ (2011) Husbandry conditions in burbot (*Lota lota*): impact of shelter availability and stocking density on growth and behaviour. Aquaculture 315:340–347

Wolkers CPJ, Serra M, Hoshiba MA, Urbinati EC (2012) Dietary L-tryptophan alters aggression in juvenile matrinxa *Brycon amazonicus*. Fish Physiol Biochem 38:819–827

Zimmerer EJ, Kallman KD (1989) Genetic basis for alternative reproductive tactics in the pygmy swordtail, *Xiphophorus nigrensis*. Evolution 43:1298–1307

Zimmermann EW, Purchase CF, Fleming IA (2012) Reducing the incidence of net cage biting and the expression of escape-related behaviors in Atlantic cod (*Gadus morhua*) with feeding and cage enrichment. Appl Anim Behav Sci 141:71–78

Chapter 5
The Effects of Early Life Experience on Behavioural Development in Captive Fish Species

Victoria A. Braithwaite and Ida Ahlbeck Bergendahl

Abstract Animals that are reared in constant, unchanging environments typically develop abnormal behavioural and cognitive abilities that often result in poor welfare. It is now recognized that the addition of variability through physical and social enrichment has many positive effects for captive populations of fish; these include neural stimulation, improved cognitive skills and the ability to respond to challenging situations adaptively. How fish are housed, the way that they are handled and the changes that they experience with regard to environmental stimulation can, therefore, help promote the development of behaviourally robust fish. These findings are important for a range of contexts; for fish housed and maintained in research labs (e.g. zebrafish), where the development of behaviour and cognitive skills can help create animals that are more suited for biomedical research. For fish reared in hatchery environments where the goal is to release the fish for conservation purposes (e.g. salmonids), here the development of an appropriate behavioural repertoire with the capacity to make appropriate, context-dependent decisions is known to promote post-release survival. Thus understanding how early life experiences shapes and refines adult behaviour helps us to rear fish that are best suited to live and survive in their environments whether captive, or in the wild.

Keywords Development · Enrichment · Variability · Zebrafish · Salmonid · Conservation and restocking

---

In memory of professor Victoria Anne Braithwaite, who passed away September 30, 2019, after a determined fight against cancer. She was a highly regarded scholar in animal behaviour, especially fish cognition, a greatly appreciated colleague and mentor and a beloved friend. She is deeply missed.

V. A. Braithwaite
Center for Brain, Behavior and Cognition, Penn State University, University Park, PA, USA

I. Ahlbeck Bergendahl (✉)
Department of Aquatic Resources, Swedish University of Agricultural Sciences, Drottningholm, Sweden
e-mail: ida.ahlbeck.bergendahl@slu.se

T. S. Kristiansen et al. (eds.), *The Welfare of Fish*, Animal Welfare 20,
https://doi.org/10.1007/978-3-030-41675-1_5

5.1 Early Experiences Influence the Way Adult Phenotypes Develop

As the other chapters in this volume highlight, the many ways we interact and use fish requires us to consider how our interactions affect the well-being and welfare of the fish (Huntingford et al. 2006; Branson 2008). This is particularly pertinent to the way we rear, handle and care for the fish we maintain in captivity (Braithwaite and Salvanes 2010). As an animal grows and develops, the environment it experiences changes aspects of the animal's brain, physiology and behaviour (Rabin 2003; Healy et al. 2009). This means that the rearing environments we use for juvenile fish can influence the kinds of adult phenotypes that are expressed (Ebbesson and Braithwaite 2012). In this chapter, two specific scenarios that involve rearing fishes in captivity will be considered; (1) rearing zebrafish in research facilities, and (2) rearing fish in hatcheries that will later be released for restocking purposes. These examples have been selected to highlight different species responses to captivity, and to illustrate how different kinds of captive environments influence the welfare of fishes. It is important to identify the end goal of the rearing process before judging what will promote or prevent the fishes from achieving good welfare. Thus, knowledge of what the fishes need, whether they are able to access this, and what behavioural skills are going to contribute to positive welfare need to be addressed (Huntingford et al. 2006).

When animals are reared in captive environments their experiences are very different to the conditions that they would normally experience in their natural habitat. There is now a growing recognition that constant, safe, captive environments promote certain aspects of welfare; accessibility to food is predictable, animals have a shelter that protects them from the stress of predation threat, or from unexpected, rapid fluctuations in climate. However, the development of behaviour in these unchanging environments is often compromised and can create poor welfare through under-stimulation and the development of maladaptive behaviours (Mathews et al. 2005; Richter et al. 2010). Captive environments rarely mirror the natural environment in which the animals have evolved to live, and although an animal may be safer and experience fewer challenges, captivity often leads to compromised welfare (Newberry 1995; Huntingford et al. 2006; Ashley 2007).

Some of these issues can be addressed over the long term when animals are bred in captivity and over the course of successive generations the animals adjust their behaviour and physiology as they adapt to their confinement and greater interaction with humans; a process often referred to as domestication (Richter 1952). Some of the changes that occur have a heritable basis, but others are a direct consequence of the environment that the animal lives in. In the second half of the last century, much discussion was given to the question of how an animal's behavioural phenotype arises (Breed and Sanchez 2010). The 'nature or nurture' debate focused on ideas that behaviour arose because of factors that were genetic and heritable (i.e. the 'nature') versus traits that were influenced by experience and the environment (i.e. 'the nurture'). The reality that we now understand is that most behaviours are a result of both genes and the environment, and more recently the field of epigenetics has begun to add a further twist in that some behaviours arise when animals experience a certain kind of environment, but those experiences can in turn influence gene expression, influencing some genes to become silent while promoting the expression of others (Powledge 2011).

Understanding mechanisms that affect behavioural development is important from the perspective of animal management. Rearing and maintaining captive populations of fish in a research laboratory, or on a fish farm can be designed with welfare in mind if we understand the factors that influence well-being in constant confinement situations (Turnbull and Huntingford 2012). It is also necessary to recognize the considerable differences between different species of fish, and to ensure that housing, handling and routine care are tailored to species needs (Braithwaite et al. 2013). There are also situations, however, where we breed and rear fish in captivity with the aim of translocating them into different environments—potentially even releasing fishes back into natural environments (e.g. Brockmark et al. 2007; Roberts et al. 2014). Under these conditions, the rearing environments and the experiences we provide for the fish will need to be different from those used for fishes kept exclusively in captivity. To provide positive welfare experiences for captive fishes, it is necessary to understand how captivity alters the behaviour of different fish species, and how manipulation of different factors can change the rearing environment (Salvanes and Braithwaite 2006; Johnsson et al. 2014).

What animals experience during development shapes and changes how they develop into adults. For example, in Atlantic salmon (*Salmo salar*) the experience of a social environment that includes strong competitors can trigger a specific life history trajectory that involves slow growth (Metcalfe et al. 1989; Metcalfe 1991). Salmon that select the slower-growing life history trajectory typically then experience lower levels of aggression compared to fish that become faster growers. Thus here, the social environment influences life history strategies that in turn affect multiple aspects related to the salmon's physiology and behaviour (Nicieze and Metcalfe 1999). Understanding how exposure to specific kinds of environment promotes different kinds of developmental pathways, or triggers development of certain phenotypic traits is important, because such knowledge is useful with regard to rearing fish with different rates of growth that become sexually mature at different ages.

Gaining a better understanding of the ways that natural environments alter the behaviour of fishes can also be helpful from a management perspective, because determining how fish in the wild cope with natural stressors could help to improve the resilience of fish held in captivity (Braithwaite and Salvanes 2010). Understanding how animals naturally cope with highly stressful situations might provide solutions for how captive fish cope during stressful procedures such as handling. In terms of what is known about how wild fish cope with stressors, there is a well-established literature investigating how different populations cope with environmental challenges such as exposure to high levels of predation. Poeciliid fishes living in populations that are sympatric with many predators have been shown to develop behavioural and physiological adaptations that help them cope with chronic high threat, high-stress situations (Reznick and Endler 1982; Magurran and Seghers 1994; Archard et al. 2012). This is in contrast to fish of the same species that live in locations where aquatic predators are less likely to colonize, for instance above waterfalls which act as geographic barriers to the predator species (Magurran 2005). Living in areas where there are increased threats of predation has generated many different kinds of adaptation in fishes; where there are periods of parental care, exposure to high predation environments can change the behavioural patterns of the parent fish in ways that help their offspring experience events such as chasing early on in life (Huntingford et al. 1994). During the breeding season, male three-spined sticklebacks (*Gasterosteus aculeatus*) build a nest and then provide paternal care to eggs laid by one or more females. Newly hatched fry stay close to the nest for some days after hatching, and populations of sticklebacks from areas associated with high levels of predation pressure have fathers that actively chase the fry as they move around the nest area. This behaviour is not observed in stickleback fathers from areas with lower levels of predation. Here, the experience of being chased by a father seems to prepare their offspring for life alongside many predators, and presumably priming the fry in this way helps them survive better in a high threat environment (Huntingford et al. 1994).

In addition to the environment having a direct effect on the development of behavioural phenotypes, it is also now recognized that contrasting environments influence the way that the brains of fish develop (Ebbesson and Braithwaite 2012; Chap. 6). Cichlid species found in Lake Tanganyika are well known for exhibiting a wide range of morphological and behavioural phenotypes, but studies have also found that cichlid species differ in the relative size of different brain regions; for example monogamous species develop a larger telencephalon and have greater visual acuity than polygynous species, and the telencephalon and cerebellum of cichlids living in more complex rocky habitats are relatively larger than in fish from less complex habitats (Pollen et al. 2007; Shumway 2008). The way fish use their sensory systems and brains has also been shown to be influenced by how threatening an environment is in terms of exposure to predators; Panamanian bishop fishes (*Brachyrhaphis episcopi*) exhibit a significantly stronger degree of visual laterality when they come from high, as opposed to low, predation environments. Fishes from high predation populations typically use their right eye for monitoring novelty or potentially threatening cues, and their left eye for determining the location of nearby conspecifics, whereas fish from low predation sites show no obvious laterality (Brown et al. 2004). Given the complete crossover of the optic nerves, this means the high predation fish process threatening and non-threatening information separately on the two sides of their brain (Dadda and Bisazza 2006). The capacity for parallel processing in this way is believed to improve anti-predator responses such as ability to school in an effective manner (Bisazza and Dadda 2005). In terms of how these lateralized responses develop, experiments with wild-caught bishop fishes that are bred in captivity has revealed that lab-reared offspring from high, but not low, predation fishes have a predisposition for laterality, however, the role that each eye takes appears to require specific experience for the predisposition to mirror what is seen in their wild-caught parents (Brown et al. 2007). Thus, wild-caught high predation parents produce offspring that when reared without experiencing a real predatory event still display laterality, but the specific nature of which eye is used to view threatening stimuli is not necessarily the same as seen in wild conspecifics. Whether being able to develop visually lateralized responses directly helps fish cope in stressful situations is not yet known, but if it did, creating opportunities for captive fish to become visually lateralized might improve their welfare with regard to coping with certain kinds of stressor.

The different kinds of examples described above illustrate how experience with specific kinds of environment can, over generations, and sometimes just within a generation, influence how fishes develop and ultimately affect adult physiology and behaviour. Comparing fish in the wild to fish reared in captivity, it is clear that what they experience varies considerably. The result is that different kinds of behaviour and physiology develop in captive fishes. If fish are going to be maintained in captivity throughout their life, then not promoting the development of more refined anti-predator responses is unlikely to have a negative effect on fish survival, but for fish that will ultimately be released for conservation purposes, such a behavioural deficit could be fatal. Thus, knowledge of how the captive environment affects the development of behaviour and physiology is an important component for designing rearing and housing methods, as well as determining what can be done to promote fish welfare (Branson 2008). And it is important to recognize that the kinds of behaviour that different captive fish need to develop requires that we know enough about the natural history of the species, or population, in question to understand what will be important, or relevant, for different fish (Patton and Braithwaite 2015).

5.2 Lessons from Other Taxa: Should Captive Environments Be Static or Vary?

How we house and maintain animals in captivity has been changing, and we can look at why these changes have arisen to help us learn from the mistakes made with other taxa. For many years, researchers maintained laboratory rodents in clean, clutter-free, identical cage environments. These conditions were perceived to be easy to maintain, to lower the risk of disease, and to decrease the amount of individual variation (Newberry 1995). In practice, however, they did little to allow the animal to express natural behaviours such as foraging, digging, or exploring, while housing the animals in these ways may be practical from a maintenance perspective. Such rearing methods are now criticized from a welfare perspective because they are considered to be under-stimulating (Newberry 1995). Over the last decade, there have been shifts in the way we house laboratory animals such as rats and mice; recognition that these are social animals has led to more pair or group housing of animals, and to the use of physical enrichment items to provide the animals with something to interact with. As enrichment items began to be used in rodent cages, the effects of enriched environments could be quantified and studies started to report positive effects; animals reared with enrichment express decreased anxiety-like behaviours (Roy et al. 2001; Meshi et al. 2006), they are more likely to explore (Zhu et al. 2009), and they are better at solving cognitive tasks (van Praag et al. 2000; Costa et al. 2007). The results are not always conclusive (see Cotel et al. 2012), but on the whole, most published studies report benefits from adding enrichment into rodent cages.

The positive effects of adding variability, through physical and social enrichment, have also now been demonstrated in a variety of fish species (Johnsson et al. 2014). Using similar approaches to the studies with rodents, it has been shown that providing juvenile fish with opportunities to interact with social and physical stimuli promotes neural stimulation, improves cognitive capacities and the ability to respond to challenging situations adaptively (Salvanes et al. 2013; Salvanes 2017). These findings are important for a range of contexts; for fish housed and maintained in research labs (e.g. zebrafish, *Danio rerio*), where the development of behaviour and cognitive skills can help create animals that are more suited for biomedical research. For fish reared in hatchery environments where the goal is to release the fish for conservation purposes (e.g. salmonids), the development of behavioural phenotypes that have the capacity to make appropriate, context-dependent decisions will help these fish survive after they have been released into natural streams and rivers.

Thus understanding how early life experiences shapes and refines adult fish behaviour helps us to rear fish that are better suited to live and survive in their environments whether captive, or in the wild. This means that the way we house fish, the way that they are handled, and the changes that they experience with regard to environmental stimulation can help promote the development of fish that are able to adapt and fine-tune their behavioural responses to a specific kind of environment (Olla et al. 1998; Näslund and Johnsson 2016).

5.3 Housing and Maintaining a Commonly Used Model Fish Species in Research Environments

For animals used in experiments, there has been a considerable emphasis to apply the 3 Rs; to replace, reduce and refine the way animals are used. This emphasis has resulted in changes to the kinds of animals used for scientific research. Where once the go to model species were laboratory rodents, we now find that zebrafish have become one of the most widely used biomedical model vertebrates (Lidster et al. 2017). The fish were first recognized for their potential in terms of an animal model in the 1970s and 1980s when George Streisinger, at the University of Oregon, recognized the value of their small size, fecundity, transparent eggs, and vertebrate body plan. Initially, developmental biologists began to use these fish to learn how embryos grew during their first few days of life (Westerfield 2007).

Since these early studies, zebrafish have become increasingly popular model species with numerous techniques now available to use them as models not just for development, but also for toxicology and disease research, as well as neural and behavioural development work (Lawrence 2007). A recent study has investigated a number of factors associated with the welfare of zebrafish maintained in research facilities (Lidster et al. 2017). Estimates indicate that annually, more than five million zebrafish are used globally for research (Lidster et al. 2017). Zebrafish are Cyprinids that are native to the southeastern Himalayan region. In the wild, they can be found in tropical streams, canals, ponds, ditches and flooded rice fields. In research facilities, however, they are typically maintained in commercially supplied, plain glass tanks with flow-through recirculating water systems. The recirculating systems are typically used as few research facilities are equipped to maintain their zebrafish colonies with flow-through water systems (which require much larger volumes of water and waste water treatment), even though this is a more desirable way of maintaining large populations of fish because it provides an effective way to control diseases.

Given the high numbers of fish bred for research purposes, there have been a number of studies exploring what kinds of environment allow these fish to thrive in captivity. The size and scale of the zebrafish colonies have been found to vary in terms of the numbers of fish maintained in one facility; from fewer than 500 to more than 10,000 fish maintained for breeding purposes (Lidster et al. 2017). A survey conducted by Lidster et al. (2017) gathered information from almost 100 different zebrafish research facilities to investigate how housing and handling affect the welfare of the fish. They reported that some facilities, but not all, used combinations of feed items so that the fish were given a varied diet, including some live prey such as brine shrimp which was considered a form of enrichment as it promoted natural hunting and feeding responses. In terms of physical enrichment, such as gravel as substrate and the inclusion of artificial plants, more than half of the facilities did not consider it as necessary, and only 25% of the facilities surveyed reported using any kind of physical enrichment. In addition to physical items, varying water flow by means of adding aeration to the tanks has been proposed as a form of enrichment as the changes in water flow allow the fish to change their swimming patterns. Lidster et al. (2017) reported that just over half of the facilities surveyed used aeration methods, but in most cases this was done to oxygenate the water, and was not done because it provided alternative swimming opportunities for the fish. In terms of lighting, nearly 70% of the facilities indicated that they used an abrupt off/on lighting transition, whereas only a quarter used dimming options to graduate the onset and offset of facility lighting. Sudden changes in lighting can create welfare problems because the rapid transition from dark to light can cause startle responses in the fish causing the fish to bang into the walls of their tanks.

The lack of consideration given to the rearing environments of the majority of the zebrafish facilities surveyed by Lidster et al. (2017) is not only a welfare concern, but it also raises questions about the value of the data from tests with fish reared in these ways (Ebbesson and Braithwaite 2012). It has been known for some time that rodents reared in homogeneous, or under-stimulating environments typically behave in abnormal ways, making the results from these animals difficult to replicate across studies (Würbel et al. 1998; Richter et al. 2010). Given that more biologically

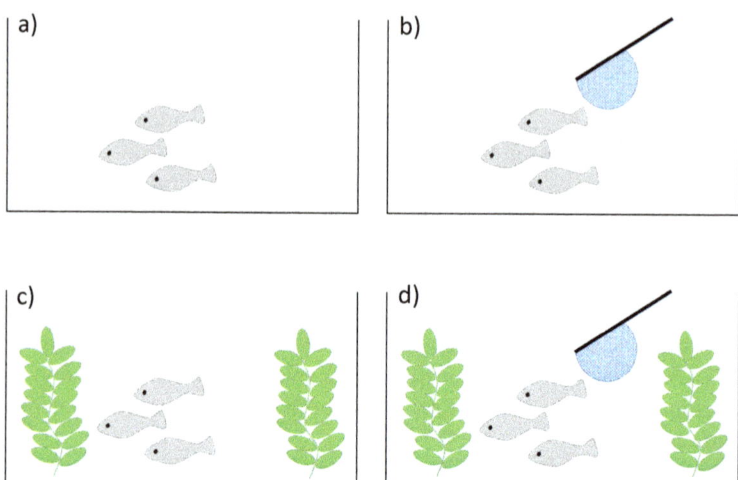

Fig. 5.1 Schematic drawings illustrating the contrasting rearing experiences given to juvenile zebrafish to study the effects of different early rearing experiences. (**a**) Typical, plain rearing conditions, (**b**) regular stressor exposure by daily chasing with a dip net, (**c**) physical complexity added using artificial plants and (**d**) a compound treatment combining chasing and plants

relevant and more reliable data can be gained from animals kept with more variable, and more stimulating environments, there is a need to consider using environmental enrichment and providing opportunities for stimulation in captive environments.

Where it has been assessed, there appear to be a number of positive effects generated by enrichment and exposure to variability in the rearing environment for zebrafish (von Krogh et al. 2010; Spence et al. 2011; DePasquale et al. 2016). For example, zebrafish reared in plain environments initially had slower learning rates in a maze task compared to fish reared with physical enrichment in the form of artificial plants (Spence et al. 2011). A more recent study has reported the effects of exposing zebrafish to different kinds of variability: (1) physical enrichment (plants in the tank), and (2) exposure to a few minutes of chasing with a dip net on a daily basis—considered a mildly stressful experience (DePasquale et al. 2016, see Fig. 5.1a–d). Here, zebrafish that had the chance to experience both kinds of variability (plants and chasing) developed lower levels of anxiety that persisted into adulthood, and the fish were also more accurate in a simple learning and memory task that required the fish to learn a correct escape route to exit a four-armed maze (DePasquale et al. 2016, Fig. 5.2a–b). Furthermore, exposure to physical enrichment has been shown to result in increased cell proliferation in the zebrafish telencephalon (von Krogh et al. 2010), and exposure to physical enrichment over several weeks also promoted an increase in overall brain size in juvenile zebrafish (DePasquale et al. 2016).

Given that the numbers of zebrafish now routinely used in biomedical research are continuing to increase, recognizing how we house and rear these fish affects their

(a)

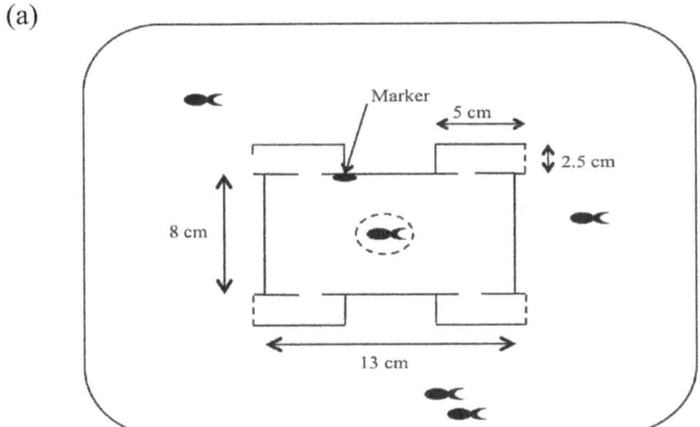

(b)

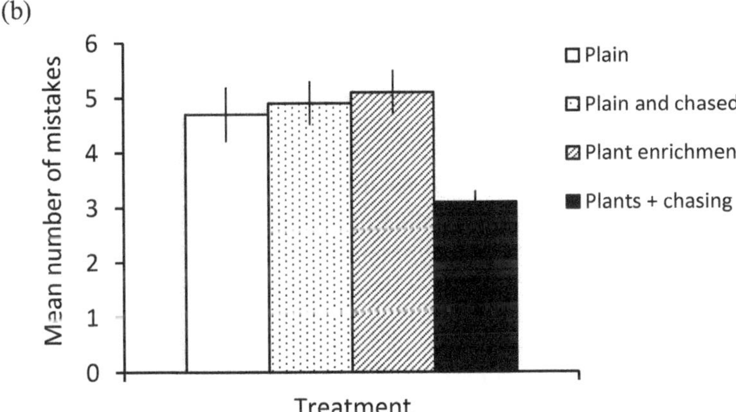

Fig. 5.2 (**a**) Schematic diagram of a four-armed maze viewed from above used to test juvenile zebrafish. Only one arm is a true exit, the other three lead to dead ends. (**b**) Accuracy in the maze comparing fish from the rearing treatments shown in Fig. 5.1. Fish reared with experience of both chasing and physical enrichment made fewer mistakes as they learn to locate the position of the open exit (see DePasquale et al. (2016) for more detail)

ability to behave across different contexts is an important observation. Paying attention to different aspects of the behavioural development of the fish is essential, because how we use and interpret the data obtained from fish reared in non-enriched, under-stimulating conditions may not be representative of a behaviourally competent animal. And as we learn more about the needs of fish and fish welfare, it is questionable whether the typical housing of this widely used species delivers good welfare to the zebrafish (Huntingford et al. 2006).

5.4 Rearing Fish in Captivity that Will Later Be Released for Conservation Purposes

Several species of fish are reared in hatcheries or in research facilities as part of conservation programs where the goal is to release fish into the wild to help increase the biomass of a threatened population (Salvanes and Braithwaite 2006; Johnsson et al. 2014). One of the biggest challenges for this approach is to overcome the mismatch between the environment in which the fish develop and the one in which they end up (Näslund and Johnsson 2016, Fig. 5.3a, b). In the hatchery, fish are reared with a high value placed on survival, however, the success of these fish once released is often poor (Olla et al. 1998). Fish reared in large numbers in hatcheries typically differ from their wild counterparts in a range of vital skills; such as foraging, predator recognition and reproductive behaviours (Sosiak et al. 1979; Ersbak and Haase 1983; Bachman 1984; Nordeide and Salvanes 1991; Fleming et al. 1997; Sundström and Johnsson 2001; Huntingford 2004; Jackson and Brown 2011). It has been proposed that fish reared in hatcheries experience environments that do not support the development of key behaviours necessary for survival in natural environments (Olla et al. 1998; Johnsson et al. 2014). The physical and social environment of the hatchery create very different conditions to those that fish in the wild experience, and several studies have investigated how changing early experiences of social and structural complexity in the hatchery alters the development of a range of behaviours considered to correlate with survival skills (Salvanes and Braithwaite 2006; Brockmark et al. 2007; Näslund and Johnsson 2016).

Researchers have tried many different ways to improve the behavioural skill set that hatchery fish develop, one approach has been to train hatchery fish prior to their release in an attempt to teach the fish about appropriate anti-predator and foraging skills (Olla et al. 1998). Successful anti-predator training, i.e. increased predator avoidance or evasion behaviour, has been documented by several authors

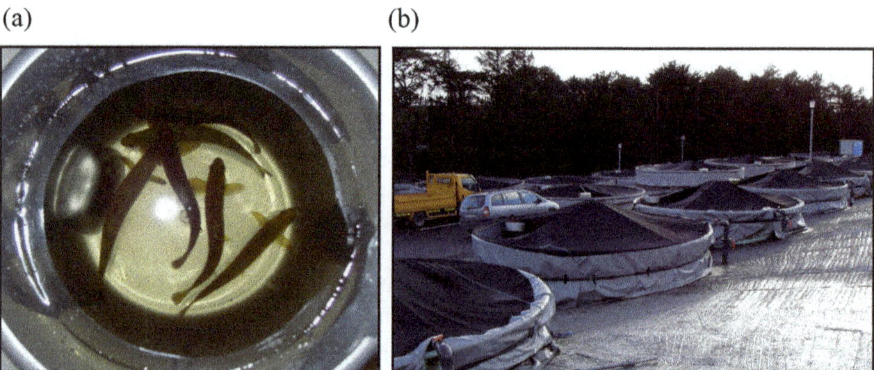

Fig. 5.3 (**a**) A sample of hatchery-reared Atlantic salmon just prior to smolting. (**b**) Outdoor hatchery facility in Ireland illustrating the large numbers of rearing tanks annually rearing thousands of salmon in highly protected, unchanging environments

(e.g. Berejikian 1995; Brown and Smith 1998; Nødtvedt et al. 1999; Mirza and Chivers 2000; Vilhunen et al. 2005; Vilhunene 2006; D'Anna et al. 2012). Exposing naive hatchery fish to live predators is a quite unforgiving way to train the fish, and certainly introduces ethical issues related to the welfare of the fish that die through predation in these captive settings (Huntingford et al. 2006). From an ethical perspective, however, utilitarian harm/benefit analyses similar to those used to justify the harm caused to animals in experiments can be used to make the case that the cost of mortality to a few fish in captivity, will benefit the many survivors, particularly when witnessing a predation event in captivity makes an individual fish more effectively recognize a real predation threat for itself after it has been released into the wild (Huntingford et al. 2006; Braithwaite and Salvanes 2010).

Training fish to learn about the threat posed by predators appears to have some success in the fish that survive this experience (Hossain et al. 2002). And, once learned, such anti-predator behaviours can spread through a population with naïve individuals learning by observing how more experienced fish react to threats (i.e. social learning; Chivers and Smith 1995; Vilhunen et al. 2005). Learning about danger through social learning has even, in some cases, been shown to be as efficient as direct predator contact (Vilhunen et al. 2005). Naïve fish can be trained in different ways, one method involves exposing hatchery fish to the odour of a predator in combination with odours from injured or killed conspecifics, so that the captive-reared fish learn to associate a predator's odour with the threat it poses (Chivers and Smith 1994; Brown and Smith 1998; Mirza and Chivers 2000; Vilhunen et al. 2005; Vilhunene 2006; Olson et al. 2012). A less common but also successful training technique has been to expose captive-reared fish to predator models that train the fish to respond aversively when they see a similar looking, but real, predator (de Oliveira Mesquita and Young 2007).

Similar attempts have been made to try and teach hatchery fish about foraging on live prey. In hatchery environments, particularly with salmonids, the fish grow best when they are fed on processed pellets made from fishmeal. The pellets, while nutritious, do little to help fish learn how to recognize, then capture and handle live prey. Attempts to train hatchery fish basic foraging skills have tried to provide direct exposure to live prey, or have allowed naïve fish to observe experienced fish feeding on live prey (Olla et al. 1998; Sundström and Johnsson 2001; Brown and Laland 2002; Brown et al. 2003a, b; Reid et al. 2010; Ellis et al. 2002; Reid et al. 2010). While a number of these training programs have been successful, several others have not (Wahl et al. 2012; Berejikian 1996; Petersson et al. 2015; Maynard et al. 1996). Where success has been poor or absent, it is not always clear whether it is a result of the training being too short, inappropriately timed training, or that the fish are not capable of learning new skills (Ahlbeck Bergendahl et al. 2017). When animals are reared in unchanging, homogeneous environments they often develop cognitive impairments, because static environments do not promote learning and memory skills (Fernö et al. 2006; Shettleworth 2010). If a fish does not learn the advantage of changing and refining its behaviour through learning and memory as it grows, then training programs that try to teach fish new skills will most likely fail (Salvanes and Braithwaite 2006). Devising environments that support continued

Fig. 5.4 (**a**) A plain tank typical for rearing juvenile Atlantic salmon, the water inlet produces an unchanging water current, and the belt feeder seen at the top left provides a constant supply of pellet food throughout daylight hours. (**b**) An enriched version of the rearing environment. By adding structure to the tank, it increases spatial complexity and allows fish to interact with objects, and additional variability can be achieved by changing the feeder position, and varying the schedule of when food is available (see Braithwaite 2005 for more detail)

learning within the captive fish populations will be important if foraging training programs are to be used.

An alternative approach to training the hatchery fish, is to rear the fish in ways that attempt to decrease the development of cognitive and behavioural deficiencies (Olla et al. 1998). If fish experience more variability in the hatchery environment, they will naturally develop ways to keep track of changes, thus learning and memory skills will be promoted as the fish grows (Fig. 5.4a, b). Several studies have shown the benefit of this kind of approach with documented benefits of increased pre- and post-survival (Heenan et al. 2009; Brockmark et al. 2010; Brockmark and Johnsson 2010; Hyvärinen and Rodelwald 2013). Changes to the environment do not need to be dramatic, rather the environment needs to be varied so that the fish learn the value of learning and adapting their behaviour. Relatively simple changes that alter where and when food is made available, or alterations can be made to the physical environment, such as the addition of rocks and artificial plants, so that the fish have to adjust their swimming behaviour (Braithwaite 2005; Rodewald et al. 2011; Bergendahl et al. 2016). Adopting this more general approach to promoting behavioural development in the hatchery has resulted in fish that develop improved learning and memory capacities and, importantly, has recently been found to improve post-release survival (Hyvärinen and Rodelwald 2013; Roberts et al. 2014; Salvanes et al. 2013).

5.5 Conclusions

Fish are held in captivity for a variety of reasons; which include research and production for restocking purposes (Näslund and Johnsson 2016). The captive environment is understandably very different from the natural habitat and environmental enrichment and variability to the husbandry routine can be used to mimic some of the features of the natural habitat. For fish maintained for research, the aim should be to promote the development of natural behaviour so that assays and tests deliver reliable data that are repeatable across studies and between research groups (Näslund and Johnsson 2016).

When fish are reared for release into the wild, it is desirable to find ways that promote the development of behavioural flexibility and competent cognitive skills as these are known to influence post-release survival. Some of these skills can be gained through specific, targeted training programs, but there also seems to be value in providing a degree of variation and use of enrichment to allow behavioural flexibility to develop in these fish from a young age. Using these different kinds of rearing methods helps fish cope better with stressful situations, particularly those that are unavoidable in captive settings (i.e. handling). Thus, rearing environments and husbandry practices that allow natural behaviours to develop can have significant, positive effects on the welfare of fish whether they will be captive throughout their lives (e.g. zebrafish), or for fish that will be released into natural environments (e.g. salmonids).

References

Ahlbeck Bergendahl I, Miller S, DePasquale C, Giralico L, Braithwaite VA (2017) Becoming a better swimmer: structural complexity enhances agility in captive-reared fish. J Fish Biol 90:1112–1117

Archard GA, Earley RL, Hanninen AF, Braithwaite VA (2012) Correlated behaviour and stress physiology in fish exposed to different levels of predation pressure. Funct Ecol 26:637–645

Ashley PJ (2007) Fish welfare: current issues in aquaculture. Appl Anim Behav Sci 104:199–235

Bachman RA (1984) Foraging behavior of free-ranging wild and hatchery brown trout in a stream. Trans Am Fish Soc 113:1–32

Berejikian BA (1995) The effects of hatchery and wild ancestry and experience on the relative ability of steelhead trout fry *Oncorhynchus mykis* o avoid a benthic predator. Can J Fish Aquat Sci 52:2476–2482

Berejikian BA (1996) Instream postrelease growth and survival of chinook salmon smolts subjected to predator training and alternate feeding strategies, 1995. In: Maynard DJ, Flagg TA, Mahnken CVW (eds) Development of a natural rearing system to improve supplemental fish quality 1991-1995. Bonneville Power Administration, Portland, OR, pp 113–127

Bergendahl IA, Salvanes AGV, Braithwaite VA (2016) Determining the effects of duration and recency of exposure to environmental enrichment. Appl Anim Behav Sci 176:163–169

Bisazza A, Dadda M (2005) Enhanced schooling performance in lateralized fishes. Proc R Soc B 272:1677–1681

Braithwaite VA (2005) Cognitive ability in fish. Behav Physiol Fish 24:1–37

Braithwaite VA, Salvanes AGV (2010) Aquaculture and restocking: implications for conservation and welfare. Anim Welf 19:139–149

Braithwaite VA, Huntingford FA, van den Bos R (2013) Variation in emotion and cognition in fishes. J Agric Environ Ethics 26:7–23

Branson E (2008) Fish welfare. Blackwell Scientific Publications, London

Breed M, Sanchez L (2010) Both environment and genetic makeup influence behavior. Nat Educ Knowl 3:68

Brockmark S, Johnsson JI (2010) Reduced hatchery rearing density increases social dominance, postrelease growth, and survival in brown trout (*Salmo trutta*). Can J Fish Aquat Sci 67:288–295

Brockmark S, Neregård L, Bohlin T, Björnsson BT, Johnsson JI (2007) Effects of rearing density and structural complexity on the pre- and post-release performance of Atlantic salmon. Trans Am Fish Soc 136:1453–1462

Brockmark S, Adriaenssens S, Johnsson JI (2010) Less is more: density influences the development of behavioural life skills in trout. Proc R Soc B 277:3035–3043

Brown C, Laland K (2002) Social enhancement and social inhibition of foraging behaviour in hatchery-reared Atlantic salmon. J Fish Biol 61:987–998

Brown GE, Smith JF (1998) Acquired predator recognition in juvenile rainbow trout (*Oncorhynchus mykiss*): conditioning hatchery-reared fish to recognize chemical cues of a predator. Can J Fish Aquat Sci 55:611–617

Brown C, Davidson T, Laland K (2003a) Environmental enrichment and prior experience of live prey improve foraging behaviour in hatchery-reared Atlantic salmon. J Fish Biol 63:187–196

Brown C, Markula A, Laland K (2003b) Social learning of prey location in hatchery-reared Atlantic salmon. J Fish Biol 63:738–745

Brown C, Gardner C, Braithwaite VA (2004) Population variation in lateralised eye use in the poeciliid *Brachyraphis episcopi*. Proc R Soc Biol Lett 271:S455–S457

Brown C, Western J, Braithwaite VA (2007) The influence of early experience and inheritance of cerebral lateralization. Anim Behav 74:231–238

Chivers DP, Smith JRF (1994) Fathead minnows, *Oimephales promelas*, acquire predator recognition when alarm substance is associated with the sight of unfamiliar fish. Anim Behav 48:597–605

Chivers DP, Smith RJF (1995) Chemical recognition of risky habitats is culturally transmitted among Fathead Minnows, Pimephales promelas (*Osteichthyes, Cyprinidae*). Ethology 99 (4):286–296

Costa DA, Cracchiolo JR, Bachstetter AD, Hughes TF, Bales KR, Paul SM, Mervis RF, Arendash GW, Potter H (2007) Enrichment improves cognition in AD mice by amyloid-related and unrelated mechanisms. Neurobiol Aging 28:831–844

Cotel M-C, Jawhar S, Christensen DZ, Bayer TA, Wirths O (2012) Environmental enrichment fails to rescue working memory deficits, neuron loss and neurogenesis in APP/PS1Kl mice. Neurobiol Aging 3:96–107

D'Anna G, Giacalone VM, Fernández TV, Vaccaro AM, Pipitone C, Mirto S, Mazzola S, Badalamenti F (2012) Effects of predator and shelter conditioning on hatchery-reared white seabream *Diplodus sargus* (L., 1758) released at sea. Aquaculture 356:91–97

Dadda M, Bisazza A (2006) Does brain asymmetry allow efficient performance of simultaneous tasks? Anim Behav 72(3):523–529

de Oliveira Mesquita F, Young RJ (2007) The behavioural responses of Nile tilapia (*Oreochromis niloticus*) to anti-predator training. Appl Anim Behav Sci 106(13):144–154

DePasquale C, Neuberger T, Hirrlinger A, Braithwaite VA (2016) The influence of complex and threatening environments in early life on brain size and behavior. Proc R Soc B 283 (1823):20152564

Ebbesson LOE, Braithwaite VA (2012) Environmental impacts on fish neural plasticity and cognition. J Fish Biol 81:2151–2174

Ellis TE, Hughes RN, Howell BR (2002) Artificial dietary regime may impair subsequent foraging behaviour of hatchery-reared turbot released into the natural environment. J Fish Biol 61:252–264
Ersbak K, Haase BL (1983) Nutritional deprivation after stocking as a possible mechanism leading to mortality in stream-stocked brook trout. N Am J Fish Manag 3:142–151
Fernö A, Huse G, Jakobsen PJ, Kristiansen TS (2006) The role of fish learning skills in fisheries and aquaculture. In: Brown C, Krause J, Laland K (eds) Fish cognition and behaviour. Blackwell, London, pp 278–310
Fleming IA, Lamberg A, Jonsson B (1997) Effects of early experience on the reproductive performance of Atlantic salmon. Behav Ecol 8:470–480
Healy SD, Bacon IE, Haggis O, Harris AP, Kelley LA (2009) Explanations for variation in cognitive ability: behavioural ecology meets comparative cognition. Behav Process 80:288–294
Heenan A, Simpson SD, Meekan MG, Healy SD, Braithwaite VA (2009) Restoring depleted coral reef fish populations through recruitment enhancement: a proof of concept. J Fish Biol 75:1857–1867
Hossain MAR, Tanaka M, Masuda R (2002) Predator-prey interaction between hatchery-reared Japanese flounder juvenile, Paralichthys olivaceus, and sandy shore crab, Matuta lunaris: daily rhythms, anti-predator conditioning and starvation. J Exp Mar Biol Ecol 267:1–14
Huntingford FA (2004) Implications of domestication and rearing conditions for the behaviour of cultivated fishes. J Fish Biol 65:122–142
Huntingford FA, Wright PJ, Tierney JF (1994) Adaptive variation in antipredator behaviour in threespine stickleback. In: Bell MA, Foster SA (eds) The evolutionary biology of the threespine stickleback. Oxford University Press, Oxford, pp 345–380
Huntingford FA, Adams CE, Braithwaite VA, Kadri S, Pottinger TG, Sandoe P, Turnbull JF (2006) Current understanding on fish welfare: a broad overview. J Fish Biol 68:332–372
Hyvärinen P, Rodelwald P (2013) Enriched rearing improves survival of hatchery reared Atlantic salmon smolts during migration in the River Tornionkoki. Can J Fish Aquat Sci 70:1386–1395
Jackson CD, Brown GE (2011) Difference in antipredator behaviour between wild and hatchery – reared juvenile Atlantic salmon (*Salmo salar*) under seminatural conditions. Can J Fish Aquat Sci 68:2157–2165
Johnsson JI, Brockmark S, Näslund J (2014) Environmental effects on behavioural development consequences for fitness of captive-reared fishes in the wild. J Fish Biol 85:1946–1971
Lawrence C (2007) The husbandry of zebrafish (*Danio rerio*): a review. Aquaculture 269:1–20
Lidster K, Readman GD, Prescott MJ, Owen SF (2017) International survey on the use and welfare of zebrafish *Danio rerio* in research. J Fish Biol 90:1891. https://doi.org/10.1111/jfb.13278
Magurran AE (2005) Evolutionary ecology: the Trinidadian guppy. Oxford University Press, Oxford
Magurran AE, Seghers BH (1994) Predator inspection behaviour covaries with schooling tendency amongst wild guppy, Poecilia reticulata, populations in Trinidad. Behaviour 128:121–134
Mathews F, Orros M, McLaren G, Gelling M, Foster R (2005) Keeping fit on the ark: assessing the suitability of captive-bred animals for release. Biol Conserv 121:569–577
Maynard DJ, Tezak EP, Berejikian BA, Flagg TA (1996) The effect of feeding spring Chinook salmon a live food supplemented diet during acclimation, 1995. In: Maynard DJ, Flagg TA, Mahnken CVW (eds) Development of a natural rearing system to improve supplemental fish quality 1991–1995, pp 98–112
Meshi D, Drew MR, Saxe M, Ansorge MS, David D, Santarelli L, Malapani C, Moore H, Hen R (2006) Hippocampal neurogenesis is not required for behavioral effects of environmental enrichment. Nat Neurosci 9:729–731
Metcalfe NB (1991) Competitive ability influences sea- ward migration age in Atlantic salmon. Can J Zool 69:815–817
Metcalfe NB, Huntingford FA, Graham WD, Thorpe JE (1989) Early social status and the development of life-history strategies in Atlantic salmon. Proc R Soc Lond B 236:7–19

Mirza RS, Chivers DP (2000) Predator-recognition training enhances survival of brook trout: evidence from laboratory and field enclosure studies. Can J Zool 78:2198–2208
Näslund J, Johnsson J (2016) Environmental enrichment for fish in captive environments: effects of physical structures and substrates. Fish Fish 17:1–30
Newberry RC (1995) Environmental enrichment: increasing the biological relevance of captive environments. Appl Anim Behav Sci 44:229–243
Nicieze AG, Metcalfe NB (1999) Costs of rapid growth: the risk of aggression is higher for fast growing salmon. Funct Ecol 13:793–800
Nødtvedt M, Fernö A, Gjosaeter J, Steingrund P (1999) Anti-oredator behaviour of hatchery-reared and wild juvenile Atlantic cod (*Gadus morhua* L.) and the effect of predator training. In: Howell BR, Moksness E, Svåsand T (eds) Stock enhancement and sea ranching. Blackwell, Oxford, pp 350–362
Nordeide JT, Salvanes AGV (1991) Observations on reared newly released and wild cod (*Gadus morhua* L.) and their potential predators. ICES Mar Sci Symp 192:139–146
Olla BL, Davis MW, Ryer CH (1998) Understanding how the hatchery environment represses or promotes the development of behavioural survival skills. Bull Mar Sci 62:531–550
Olson JA, Olson JM, Walsh RE, Wisenden BD (2012) A method to train groups of predator-naive fish to recognize and respond to predators when released into the natural environment. N Am J Fish Manag 32:77–81
Patton BW, Braithwaite VA (2015) Swimming against the current: ecological and historical perspectives on fish cognition. WIREs Cognit Sci 6:159–176
Petersson E, Valencia AC, Järvi T (2015) Failure of predator conditioning: an experimental study of predator avoidance in brown trout (*Salmo trutta*). Ecol Freshw Fish 24:329–337
Pollen AA, Dobberfuhl AP, Scace A, Igulu MM, Renn SCP, Shumway CA et al (2007) Environmental complexity and social organization sculpt the brain in Lake Tanganyikan cichlid fish. Brain Behav Evol 70:21–39
Powledge TM (2011) Behavioral genetics: how nurture shapes nature. Bioscience 61:588–592
Rabin LA (2003) Maintaining behavioural diversity in captivity for conservation: natural behavior management. Anim Welf 12:85–94
Reid AL, Seebacher F, Ward AJW (2010) Learning to hunt: the role of experience in predator success. Behaviour 147:223–233
Reznick DN, Endler JA (1982) The impact of predation on life history evolution in Trinidadian guppies (*Poecilia reticulata*). Evolution 36:160–177
Richter CP (1952) Domestication of the Norway rat and its implication for the study of genetics in man. Am J Hum Genet 4:273–285
Richter SH, Garner JP, Auer C, Kunert J, Würbel H (2010) Systematic variation improves reproducibility of animal experiments. Nat Methods 7:167–168
Roberts LJ, Taylor J, Forman DW, Garcia de Leaniz C (2014) Silver spoons in the rough: can environmental enrichment improve survival of hatchery Atlantic salmon Salmo salar in the wild? J Fish Biol 85:1972–1991
Rodewald P, Hyvärinen P, Hirvonen H (2011) Wild origin and enrichment promote foraging rate and learning to forage on natural prey of captive reared Atlantic salmon parr. Ecol Freshw Fish 20:569–579
Roy V, Belzung C, Delarue C, Chapillon P (2001) Environmental enrichment in BALB/c mice: effects in classical tests of anxiety and exposure to a predatory odor. Physiol Behav 74:313–320
Salvanes AGV (2017) Are antipredator behaviours of *Salmo salar* juveniles similar to wild juveniles? J Fish Biol 90:1785–1796
Salvanes AGV, Braithwaite VA (2006) The need to understand the behaviour of fish we rear for mariculture or for restocking. ICES J Mar Sci 63:346–354
Salvanes AGV, Moberg O, Ebbesson LOE, Nilsen TO, Jensen KH, Braithwaite VA (2013) Environmental enrichment promotes neural plasticity and cognitive behaviour in fish. Proc R Soc B 280:20131331

Shettleworth SJ (2010) Cognition, evolution, and behavior, 2nd edn. Oxford University Press, New York
Shumway CA (2008) Habitat complexity, brain, and behavior. Brain Behav Evol 72(2):123–134
Sosiak AJ, Randall RG, McKenzie JA (1979) Feeding by hatchery-reared and wild Atlantic salmon (*Salmo salar*) parr in streams. J Fish Res Board Can 36:1408–1412
Spence R, Magurran AE, Smith C (2011) Spatial cognition in zebrafish: the role of strain and rearing environment. Anim Cogn 4:607–612
Sundström LF, Johnsson JI (2001) Experience and social environment influence the ability of young brown trout to forage on live novel prey. Anim Behav 61:249–255
Turnbull JF, Huntingford FA (2012) Welfare and aquaculture: where BENEFISH fits in. Aquac Econ Manag 16:433–440
van Praag H, Kempermann G, Gage FH (2000) Neural consequences of enviromental enrichment. Nat Rev Neurosci 1(3):191–198
Vilhunen S, Hirvonen H, Laakkonen MV-M (2005) Less is more: social learning of predator recognition requires a low demonstrator to observer ratio in arctic charr (*Salvelinus alpinus*). Behav Ecol Sociobiol 57:275–282
Vilhunene S (2006) Repeated antipredator conditioning: a pathway to habituation or to better avoidance? J Fish Biol 68:25–43
von Krogh K, Sørensen C, Nilsson GE, Øverli Ø (2010) Forebrain cell proliferation, behavior and physiology of zebrafish, *Danio rerio*, kept in enriched or barren environments. Physiol Behav 101:32–39
Wahl DH, Einfalt LM, Wojcieszak DB (2012) Effect of experience with predators on the behavior and survival of muskellunge and tiger muskellunge. Trans Am Fish Soc 141:139–146
Westerfield M (2007) The Zebrafish book. A guide for the laboratory use of Zebrafish (*Danio rerio*), 5th edn. University of Oregon Press, Eugene, OR
Würbel H, Chapman R, Rutland C (1998) Effect of feed and environmental enrichment on development of stereotypic wire-gnawing in laboratory mice. Appl Anim Behav Sci 60:69–81
Zhu SH, Codita A, Bogdanovic N, Hjerling-Leffler J, Ernfors P, Winblad B, Dickins DW, Mohammed AH (2009) Influence of environmental manipulation on exploratory behaviour in male BDNF knockout mice. Behav Brain Res 197:339–346

Chapter 6
Fish Brains: Anatomy, Functionality, and Evolutionary Relationships

Alexander Kotrschal and Kurt Kotrschal

Abstract In this chapter, we provide an overview of the anatomy, functionality, and evolution of the fish nervous system. Our focus will be on the brain in the vertebrate group with the greatest variation in brain form and function, the actinopterygian bony fishes. We first describe central (CNS) and autonomic (ANS) nervous systems and then characterize the major distal components of the CNS (spinal cord, spinal nerves, cranial nerves), before we summarize the brain regions and their connections and highlight some similarities and differences between different fish taxa. The second part of this chapter is devoted to variation in fish brain anatomy, including a discussion of comparative brain anatomy evolution and brain plasticity. We finish with a summary of the evolutionary costs and benefits of brain size based on results in guppies (*Poecilia reticulata*) artificially selected for large and small brains. With respect to fish welfare, we conclude that their great brain diversity reflects the diverse cognitive needs of fishes. However, their lifelong high rates of neurogenesis should also make individuals capable to cognitively adapt to a certain range of environmental conditions.

Keywords Fish brain · Brain anatomy · Ecomorphology · Brain size · Artificial selection

A. Kotrschal (✉)
Zoological Institute, Stockholm University, Stockholm, Sweden

Behavioural Ecology, Insitute of Animal Sciences, Wageningen University, Wageningen, Netherlands
e-mail: alexander.kotrschal@wur.nl

K. Kotrschal
Department of Behavioural Biology, University of Vienna, Wien, Austria

Konrad Lorenz Forschungsstelle, University of Vienna, Grünau, Austria

Wolf Science Center, University of Veterinary Medicine, Ernstbrunn, Austria

T. S. Kristiansen et al. (eds.), *The Welfare of Fish*, Animal Welfare 20,
https://doi.org/10.1007/978-3-030-41675-1_6

6.1 Anatomy and Function of the Nervous System

The central nervous system (CNS) of vertebrates including fish consists of the brain and the spinal cord, linking with receptors and afferent organs via the motor and sensory nerves. Although most research is done on neuron properties and how they connect to each other a majority of cells in the CNS are of various other types. Glial cells, for instance, support the neurons physically, electrically insulate them, play a role in brain development and homeostasis, and may also be involved in information processing. Today we have a good understanding of how some anatomical arrangements function. Examples of especially well-understood systems are the Mauthner neurons governing a fundamental escape response, the electroreceptive system of mormyrid fishes, or the visual system in general.

6.1.1 The Central Nervous System

The CNS is arguably the most complex organ in any vertebrates' body. There are a large number of specialized cells, which are intricately connected, interacting with each other in diverse ways. In the following, we can only give a brief account of this complexity.

6.1.1.1 The Spinal Cord

The fish spinal cord is the phylogenetically oldest part of the CNS and hence, similar in structure to the spinal cords of all other vertebrates. During embryonic development, the CNS forms when neural folds roll in and fuse (Nieuwenhuys et al. 1998). In cross sections, the central area (with the cell bodies of the cord neurons) appears darker than the outer zone. Those areas are therefore called "grey matter" and "white matter", respectively. The white matter is mainly composed of ascending and descending fibres organized in distinct tracts: a dorsal somatic sensory tract, a lateral visceral sensory and visceral motor tract, and a large ventral somatomotor tract. In most fishes, the paired large axons of the Mauthner neurons descend in the ventral grey matter. Mauthner neurons are lacking in adult elasmobranchs (Bone 1977). They govern the C-start escape response (the"C" describes the typical body in the process of escaping) and function already early in ontogeny. Mauthner neuron axons decussate at the Mauthner chiasm; hence, when one cell is stimulated to fire, the C-start moves the head away from the aversive stimulus, which enables a very fast change of swimming direction. Mauthner neurons are the classic example of a hard-wired central nervous system response mechanism.

6.1.1.2 The Spinal and Cranial Nerves

Segmental dorsal and ventral nerve roots emerge from the spinal cord and, except for in lampreys, unite to form the spinal nerves, which carry motor, sensory, and autonomic signals between the spinal cord and the body. The ventral root axons of the spinal somatomotor neurons connect to the musculature, while the dorsal root contains the sensory neurons, connecting to the peripheral sensory systems. Whereas the bodies of the motor neurones form the ventral horn of the spinal cord grey matter, the sensory neurones are situated in segmental ganglia outside the cord (Fig. 6.1). Most of the cranial nerves follow the same basic pattern, but emerge from the rostral part of the spinal cord and the brain stem. They are numbered from rostral to caudal. The optic cranial nerve II deviates from this segmental arrangement pattern, as the eye develops via outpouching of the lateral neural tube (see below), and is, therefore, a brain-internal connection and should, therefore, be termed "tract" rather than "nerve". The olfactory nerve (I) connects the olfactory mucosa with the olfactory bulbs. If the bulbs are located directly at the mucosa, like in cyprinids, this is a brain-internal connection and therefore, called a "tract". The terminalis nerve conveys information from most rostral sensory systems; it is numbered 0, because it was described after the other nerves have already been numbered. The other cranial nerves (from front to back) are the oculomotor (III), trochlear (IV), trigeminal (V),

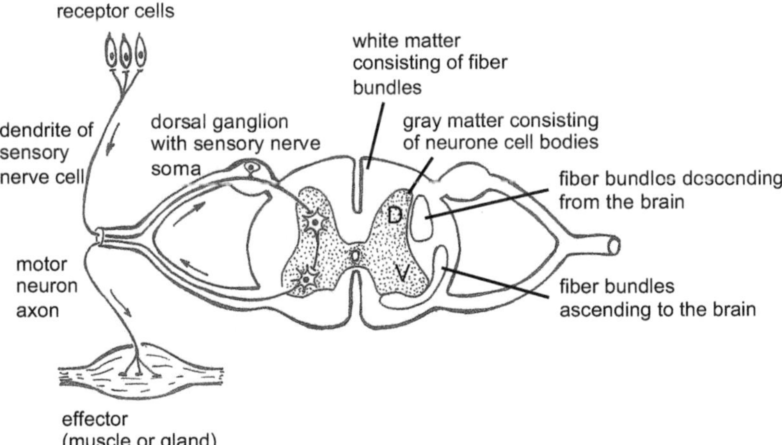

Fig. 6.1 Spinal cord and nerves (cross section) and their connections to sensory cells/organs and effectors. The central, butterfly-shaped grey matter consists mainly of nerve cells (D: sensory dorsal horn interneurones; V: ventral horn with motor neurones). Arrows give direction of potential conduction. The dendrites/axons of the sensory nerve cells, situated in the dorsal ganglia, are activated by peripheral receptors and transport the action potentials towards networks of interneurones in the grey matter, the output of which is communicated to the effectors via the axons of the ventral horn motor neurones. The local circuits of the spinal cord communicate with the brain via ascending and descending fibre bundles

abducens (VI), facial (VII), auditory (VIII), glossopharyngeal (IX), vagal (X), accessory (XI), and hypoglossal (XII) (Von Kupffer 1891).

6.1.1.3 The Brain

The brain is the enlarged anterior pole of the spinal cord, which has developed because also the major vertebrate sensory systems are located at the anterior pole of the body. For functional reasons, particularly the ventral brain areas are the phylogenetically most conservative structure; hence, its basic organization is grossly similar across all vertebrates, including fishes. However, owing to the diverse sensory orientation of fishes, the different regions can differ greatly in size and form. In fact, the fishes show the greatest variation of brain anatomy and brain function in all vertebrates (Nieuwenhuys et al. 1998). Figure 6.2 shows gross variation between the major groups of fish and Fig. 6.3 shows more detailed differences between the brains of two modern teleosts. The rostral spinal cord is continuous with the brain stem, with the ventral diencephalon as the most rostral pole, ending underneath the anterior commissure of the forebrain. Hence, the ventral brain composed of the modified rostral spinal cord; dorsally, it carries a series of prominent structures. During early ontogeny, the anterior end of the neural tube differentiates into neuromeres. Figure 6.4 shows how the three largest and most anterior neuromeres develop into the telencephalon, diencephalon, and the mesencephalon, while the seven more caudal neuromeres differentiate into the Rhombencephalon. From rostral to caudal, the three main regions are the forebrain (Prosencephalon), the midbrain (Mesencephalon) and the hindbrain (Rhombencephalon). The forebrain is divided into paired olfactory bulbs, ventrally attached to the telencephalic hemispheres, dorsally covering the Diencephalon (the "betweenbrain", consisting of Thalamus, Hypothalamus Subthalamus, Epithalamus, and Pretectum). Towards caudal, the midbrain roof is developed as paired optic lobes, followed by the Cerebellum (Metencephalon), dorsally attached to the Medulla oblongata (Myelencephalon) (Fig. 6.4; Nieuwenhuys et al. 1998; Northcutt and Davis 1983; Northcutt 1978).

The *brain stem* houses primary representation centres for all somatosensory faculties except olfaction and vision, and features a degree of variability hardly matched by other brain divisions. In unspecialized, evolutionary "mainstream" fishes, from agnathans to basic teleosts, neurone groups in the dorsal brain stem are arranged in four horizontal columns with sensory components of cranial nerves IV–XII and of two rostral and one caudal lateral line nerves terminating in the two dorsalmost columns, while motor fibres originate from ventrally located centres (Allis 1897; Webb and Northcutt 1997). The dorsal, sensory columns along the wall of the fourth ventricle process the senses of hearing, lateral line, and taste. Such somatotopic arrangement may facilitate the formation of short-loop reflexes (Kanwal and Finger 1992), and of sensomotory specializations such as the taste-dominated cyprinid palatal organ (Sibbing 1991). One additional, dorso-rostral column is found in fishes with the ability to process electrosensory information (e.g. in

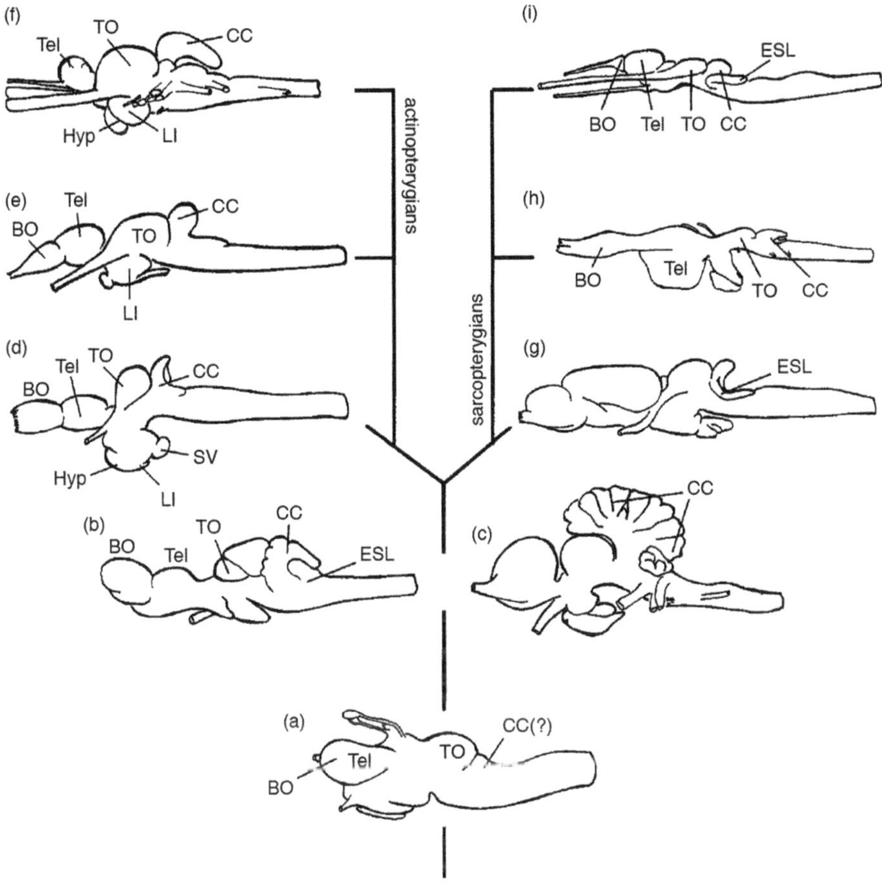

Fig. 6.2 Representative brains showing variation between major groups of fish. Forebrains are evaginated in lamprey (**a**: *Petromyzon*), in the elasmobranchs (**b**: *Acanthias*, **c**: *Cetorhinus*), lungfish (**h**: *Ceratodus*), and the coelacanth (**i**: *Latimeria*), but everted in the actinopterygian line, such as in the bichirs (**g**: *Calamoichtys*), sturgeons and neopterygians (**d**: *Acipenser*, **e**: *Amia*, **f**: *Gadus*). *BO* Bulbus olfactorius, *CC* Corpus cerebelli, *ESL* electrosensory lobe, *Hyp* Hypophysis, *LI* Lobus inferior, *SV* Saccus vasculosus, *Tel* Telencephalon, *TO* Tectum opticum. Brains are not to scale, reproduced with permission from Kotrschal et al. (1998)

Calamoichthys and *Latimeria*). The roof of the fourth ventricle is formed by a choroid plexus with varying degrees of differentiation (Weiger et al. 1988). Hypertrophy of areas within the dorsal column is associated with sensory specializations and these may form prominent bulges as in many carp- and cod-like fishes (Kotrschal et al. 1998). In addition to several ascending and descending fibre systems, the brain stem houses the reticular formation, a ventrally located system for basic maintenance and life support (Davis and Northcutt 1983).

The mesencephalic and diencephalic tegmentum continues rostrally to the brain stem with connective and integrative systems for brain structures arising from its

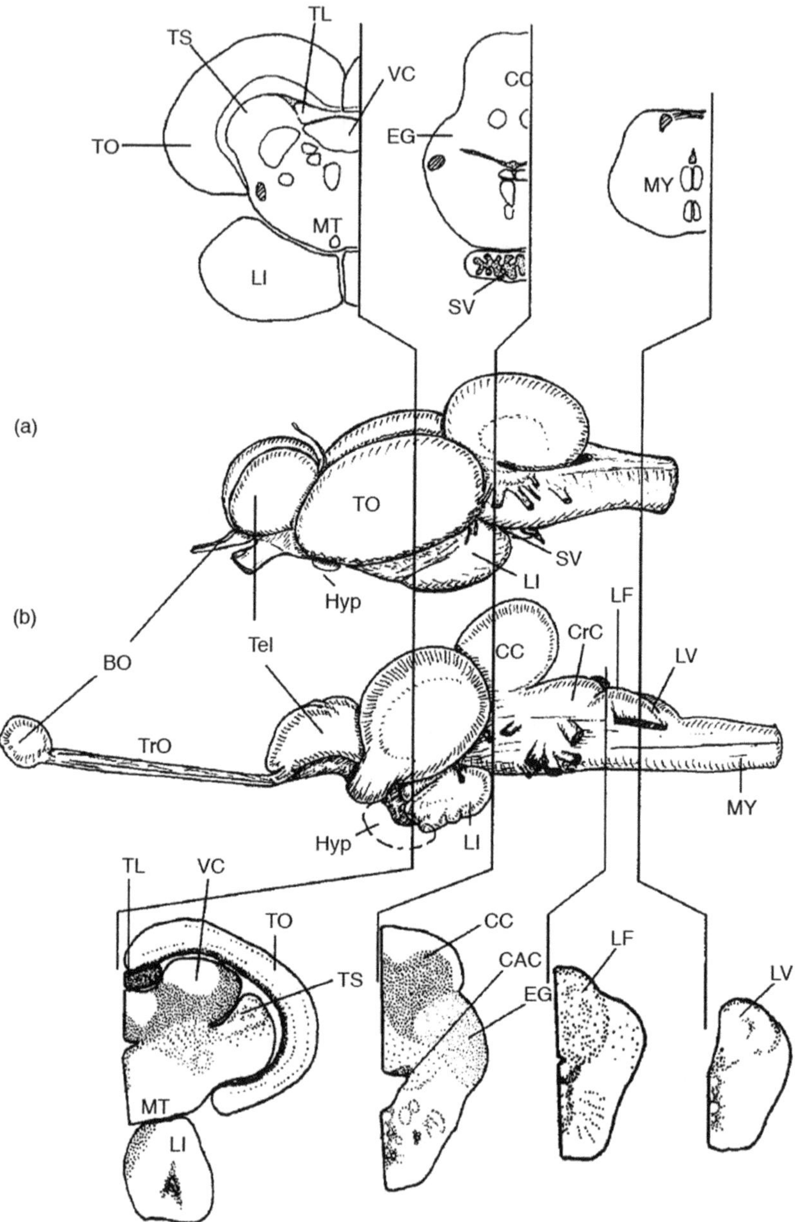

Fig. 6.3 Comparison between a perciform and a cypriniform brain: (**a**) blenny, *Blennius incognitus*; (**b**) roach, *Rutilus rutilus*. Lateral views in the middle of the page, representative cross sections at levels indicated by the vertical lines at top and bottom of the page. Note the small Bulbus olfactorius, but large Telencephalon, Tectum opticum, and Corpus cerebelli in the blenny. In the roach, the olfactory bulb is remote from the Telencephalon, and the somatosensory (taste) lobes of the brain stem, Lobus facialis, and Lobus vargus are large. *BO* Bulbus olfactorius, *CAC* Central acoustic area, *CC* Corpus cerebelli, *CrC* Crista cerebellaris, *EG* Eminentia granularis, *Hyp* Hypophysis, *LF* Lobus fascialis, *LI* Lobus inferior, *LV* Lobus vagus, *MT* Mesencephalic tegmentum, *MY* Myelencephalon, *SV* Saccus vasculosus, *Tel* Telencephalon, *TL* Torus longitudinalis, *TO* Tectum opticum, *TrO* Tractus olfactorius, *TS* Torus semicircularis, *VC* Valvula cerebelli. Redrawn after Kotrschal et al. (1998)

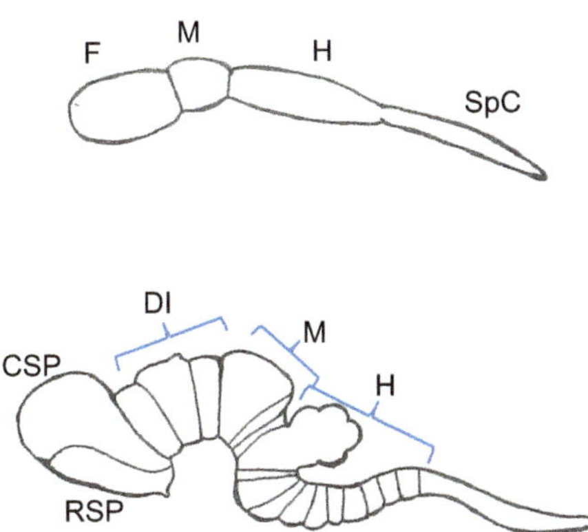

Fig. 6.4 Vertebrate brain ontogeny (mouse). Top: Early stage: The rostral (left) neural tube shows the appearance of the forebrain (F), midbrain (M), and hindbrain vesicles (H), with the developing spinal cord (SpC). Bottom: Later stage: More subdivisions appear. Forebrain: CSP: caudal secondary Prosencephalon; RSP: rostral secondary Prosencephalon; DI: prosomeres 1–3 of the Diencephalon; M: two mesomeres of the midbrain; H: isthmus region and hindbrain rhombomeres 1 to 11. Redrawn after Puelles et al. (2013)

roof: the Cerebellum, the Tectum opticum, and the forebrain (Davis and Northcutt 1983). Brain stem and tegmentum are continuous with each other and the sub-cerebellar secondary gustatory nucleus may serve as an arbitrary border. In a rostrocaudal direction, the tegmental third ventricle changes from a slit-like gap to a narrow channel before opening into the fourth ventricle. Several structures form as extensions to this ventricle. The inferior lobes of the Hypothalamus are paired, ventral diencephalic hemispheres specific to ray-finned fishes, serving as multimodal integration centres. In all vertebrates, the hypothalamic Tegmentum converts sensory inputs into hormonal and behavioural responses. In this context, the diencephalic and tegmental ventricle is lined by a number of "paraventricular organs", most of them equipped with cerebrospinal fluid-contacting neurons and distinctive ependymal cells as part of brain-internal humoral communication system based on the liquor. Ependymal cells line the ventricular cavities, they are one of the neuroglial cell types and are involved in the production of cerebrospinal fluid. Specific for the actinopterygian bony fishes is the saccus vasculosus (Fig. 6.2), an organ serving as a sensor of seasonal day length (Nakane et al. 2013). The hypothalamic Neurohypophysis serves as a central humoral command unit of physiology and behaviour. Dorsally, the choroid plexus of the third ventricle forms several extensions, such as the Saccus dorsalis with its light-sensitive and endocrine Epiphysis or other circumventricular organs.

The Cerebellum varies in extent from a small ridge in ancestral or sedentary, benthic fishes (Fig. 6.1) to a prominent structure in most modern teleosts (Fig. 6.2). Although relatively large in pelagic sharks or teleosts swiftly manoeuvring in 3D, it is not necessarily characteristic for a pelagic lifestyle per se (below). Particularly in modern electrosensory fishes, this structure may become massively enlarged, covering the entire surface of the brain in *Gnathonemus* (Maler et al. 1991). Various cerebellar subdivisions serve a variety of functions including cognitive and emotional contexts (Rodríguez et al. 2005). Corpus and Valvula cerebelli, the latter as a rostral extension beneath the optic tectum, are intimately connected and appear to play roles in spatial orientation, proprioception, motor coordination, and eye movement. The central acoustic area forms as a granular area at the ventral Cerebellum, co-varying in size with the development of the peripheral hearing apparatus (Popper and Fay 1993). Inputs from the inner ear and from lateral line fibres terminate at the Eminentia granularis, a parvocellular area on both sides of the lateral corpus (Fig. 6.2). The Crista cerebelli, caudal and in continuation with the molecular layer of the corpus, predominantly processes lateral line input (Davis and Northcutt 1983).

The Tectum opticum (TO) as the mesencephalic roof consists of paired dorsal hemispheres with a cortex-like layering of grey and white matter, separated from the Tegmentum by ventricular spaces. The TO receives projections from contralateral retinal ganglion cells; it processes the primary visual input and participates in significant bidirectional communication with the brain stem (Davis and Northcutt 1983). Tectal development varies closely with eye size, visual orientation (Vanegas and Ito 1983; Zaunreiter et al. 1985), and lateral line dependence (Schellart 1991) and is also present in ontogenetically or phylogenetically blind fishes (Voneida and Fish 1984). The Retina ontogenetically forms as part of the Diencephalon and shows considerable structural variation related to phylogeny, ontogeny, ecology, or lifestyle (Kotrschal et al. 1998).

Below the optic lobes, the Torus longitudinalis extends into the sub-tectal ventricle as a pair of longitudinal cylinders (Fig. 6.2). Its presumed functions include postural control, detection of luminance levels, and monitoring of saccadic movement. Also, it has a role as premotor centre between Telencephalon and brain stem (Wullimann 1994).

The Telencephalon arises from the rostral portion of the embryonic neural tube forming two hemispheres. In more ancestral taxa of agnathans, elasmobranchs, and sarcopterygians these develop as in the majority of vertebrates, by the bulging out (evagination) of the lateral walls and therefore, contain a central ventricle. In contrast, the actinopterygian forebrain forms by bending out (eversion) of the dorsal walls of the embryonic neural tube (Nieuwenhuys 1982). Hemispheres are therefore solid, and a T-shaped ventricle extends up to the dorsolateral surfaces of the hemispheres and separates the two halves. Centrally, the two hemispheres are closely attached to each other and may even fuse. In addition to secondary olfactory fibres, which terminate throughout the entire structure, virtually all sensory modalities project into the dorsal Telencephalon through lemniscal pathways (Finger 1980); hypothalamic and primary olfactory input ascend from the ventral forebrain. The latter also contains the Commissura anterior with a peduncle of decussating fibre tracts for a two-way flow of information between the Telencephalon and Diencephalon as well as intra-telencephalic fibres. Fish with parts of their forebrains ablated feed, grow, and behave normally in most respects, but exhibit significantly diminished rates of learning and do not engage in more complex social tasks (Salas et al. 2006; Portavella et al. 2002; Szabó 1973).

The bulbus olfactorius of all fishes evaginates from the rostral tip of the embryonic neural tube. Its ventricles are secondarily reduced or absent in advanced actinopterygians. Primary fibres from the olfactory mucosa terminate within glomerular structures of the olfactory bulb neuropil in a chemotopic way, i.e. fibres from olfactory mucosa neurons with similar receptor characteristics terminate within the same glomerulus. Large projection neurons, mitral cells, and tuft cells project into the Telencephalon and Diencephalon via medial and lateral olfactory tracts. In most species, the olfactory bulbs remain attached to the rostral telencephalon, but are attached to the olfactory mucosa in ostariophysean teleosts, connecting to the forebrain via a secondary olfactory tract. Figure 6.5 illustrates the phylogenetic change in brain regions emphasis across the fishes.

The nerve cell bodies (perikarya) of the Nervus terminals, located at the junction between olfactory bulbs and Telencephalon, send processes into the olfactory mucosa, the Diencephalon, and into most other brain areas including the Retina (Kotrschal et al. 1998; Nieuwenhuys et al. 1998). The function of this olfactoretinalis system was studied in medaka (*Oryzias latipes*). The terminal nerve gonadotropin-releasing hormone 3 (TN-GnRH3) neurons function as a gate for activating mating preferences based on familiarity. Basal levels of TN-GnRH3 neuronal activity suppress female receptivity for any male. Visual familiarization facilitates TN-GnRH3 neuron activity, which correlates with female preference for the familiarized male (Okuyama et al. 2014).

6.1.2 The Autonomic Nervous System

As part of the peripheral nervous system, the autonomic nervous system (ANS) composed of sympathetic and parasympathetic elements, governs all "involuntary" bodily functions. These fibres innervate smooth muscles, for instance around blood vessels, in the gut, spleen, urogenital tract, heart, and in teleost fishes also in pigment cells, and are vital for controlling homeostasis (Young 1931). Generally, the efferent fibres from central neurons are not directly connected with peripheral organs; they are rather linked via synapses to peripheral ganglion cells, which then innervate the target organ. Hence, the central nerve fibres are called pre-, the peripheral nerve fibres post-ganglion fibres. The autonomic nervous system in fishes is usually divided into the cranial autonomic, the spinal autonomic, and the enteric system of the gut (Bone et al. 1982). Teleosts, other fishes, and even other higher vertebrates differ from elasmobranchs in that the spinal autonomic ganglia are linked to the spinal nerves via branches carrying both pre- and post-ganglionic fibres. In teleosts, these also innervate skin melanophores (Mills 1932). Another difference lies within the vagal nerve (X) that in teleosts has both excitatory and inhibitory actions in the gut, while in elasmobranchs the vagus nerve does not control gut movements (Young 1980).

Fig. 6.5 Representative actinopterygian brains, in mid-sagittal view. Drawn from freshly perfusion fixed brains. All scale bars 5 mm. From bottom to top: *Acipenser ruthenus* (34.5 cm body length), *Amia calva* (20 cm), *Salmo trutta* (32 cm), *Tinca tinca* (16.5 cm), *Parablennius sanquinolentus* (13 cm). Note the amphibian brain-like features in the holostean (palaeozoic radiation of acinopterygians) and still, in the holostean (mesozoic radiation). Please also note the decrease in relative size, of the olfactory bulb (BO) and an increase in the size of the Telencephalon (Tel) towards the modern actinopterygian representatives. *BO* olfactory bulb, *Bst* brain stem, *Ca* anterior vommissure of the optic tract, *Cer* Cerebellum, *E/Sd* Epiphysis/Saccus dorsalis, *Hyp* Hypophysis, *Li* diencephalic inferior lobe, *Sv* Saccus vasculosus, *Tel* Telencephalon, *TO* optic tectum

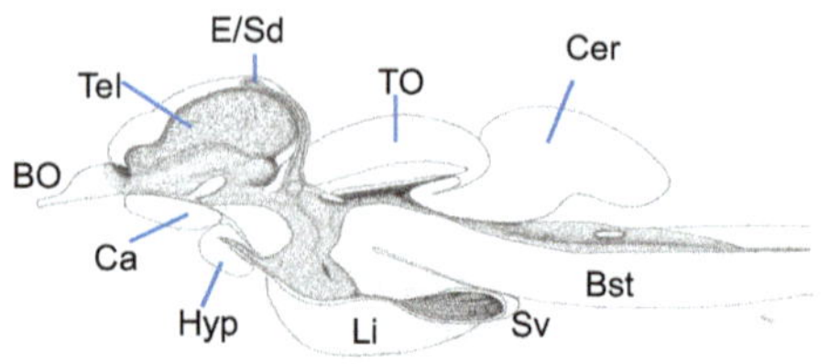

Teleostei, Perciformes, Parablennius sanquinolentus

Teleostei, Cypriniformes, Tinca tinca

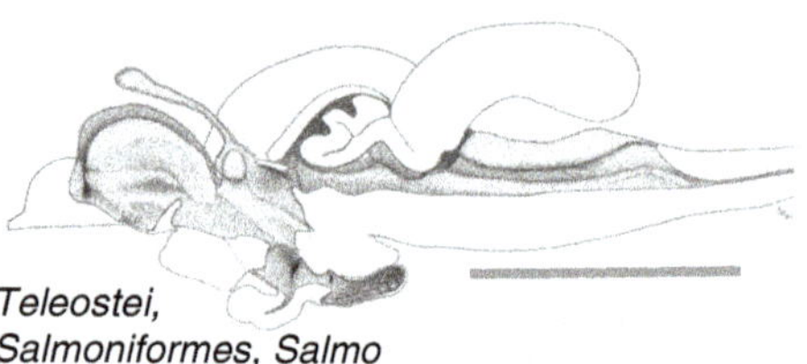

Teleostei, Salmoniformes, Salmo trutta

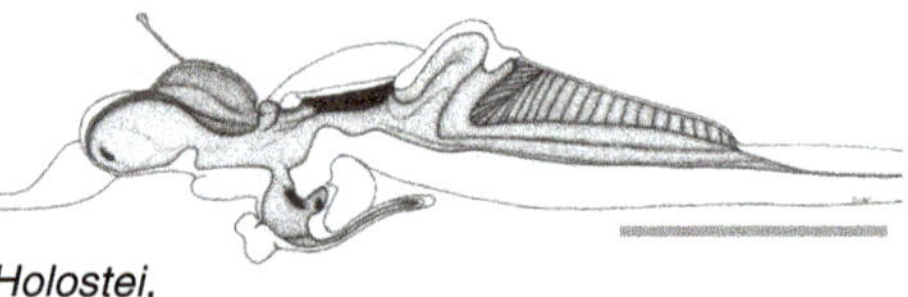

Holostei, Amiiformes, Amia calva

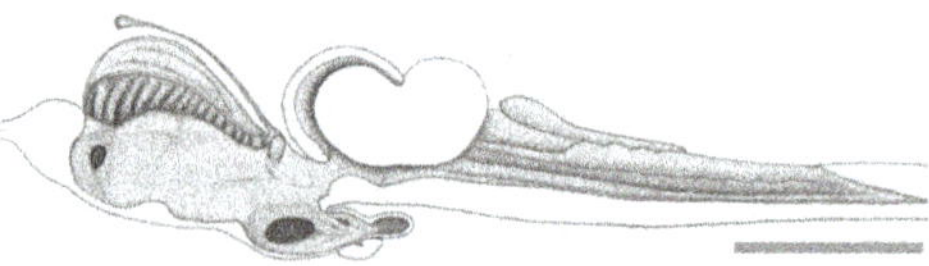

Chondrostei, Acipenseriformes Acipenser ruthenus

6.2 Variation in Brain Anatomy

In most fishes, the brain is considerably smaller than the space available and may in some cases occupy only about 6% of the brain cavity in an elasmobranch (Kruska 1988). The excess space is commonly filled with lymphatic, fatty tissue. Nevertheless, in Lake Tanganyika cichlids, skull morphometry seems to determine brain shape and constrain its evolution (Tsuboi et al. 2014a). Most neurons are relatively large in agnathans, sarcopterygians, chondrosteans, and elasmobranchs, but are small in teleosts. The apparent evolutionary decrease in cell size probably arises from size constraints during larval life, when, at only a few millimetre in length, teleost larvae are the smallest fully functional vertebrates (Kotrschal et al. 1990). Intergroup comparisons commonly based on brain size relative to body size can thus be misleading. Recent advances in using flow cytometry to quantify neuron numbers should prove useful in getting the story right (Herculano-Houzel 2009; Marhounová et al. 2019).

Brains scale negatively allometric with body size, with ontogenetically and phylogenetically small fish tending to have relatively large brains and vice versa (Brändstatter and Kotrschal 1990; Striedter 2005). There appears to be a coarse trend towards an increase in relative brain size during phylogenetic development. Agnathans, for example feature some of the relatively smallest brains, whereas those of perciforms are among the largest. The sexes generally show similar relative brain sizes with exception of the three-spined stickleback (*Gasterosteus aculeatus*), where, at similar body sizes, male brains are up to 23% larger than female brains (Kotrschal et al. 2012a). Possibly, this dramatic sexual size dimorphism is generated by the many cognitively demanding challenges that such territorial, parental males are faced with, including elaborate courtship displays, the construction of an ornate nest and a male-only parental care system (Östlund-Nilsson et al. 2007). Also, advanced courtship behaviour in cichlids (Kotrschal and Taborsky 2010) and demanding spatio-temporal orientation in females of some blennies (Costa et al. 2011) impact brain anatomy and size.

Faculties for sensory perception, central processing, and behavioural responses undoubtedly reside primarily within an organism's nervous system. The motor generators for the species-specific action patterns "innate behaviours" (Tinbergen 1951) reside in the spinal cord, whereas the brain is the command centre for selectively disinhibiting the motor generators for these action patterns in service of organized behaviour. In case of reflexes, a stimulus directly triggers a motor response via spinal cord circuits and the brain is informed only thereafter.

Adaptive radiation has produced a functional diversity of structures, shapes, and sizes rivalled by few other organs, unprecedented in non-fish vertebrates (Nieuwenhuys et al. 1998). A chief aim of evolutionary neurobiology and ecomorphology is to reveal how physical brains reflect sensory orientation, cognitive potential, and motor abilities. Viewed within a phylogenetic context, a study of this diversity can uncover how brains have adapted to the requirements of disparate habitats, ecologies, and behavioural needs. One century of ecomorphological research (Herrick 1902, 1906) has produced a large empirical database for fish brains, which we attempt to briefly summarize here.

"Fish" is a collective term for more than half of all known vertebrate taxa. Embodying more than 400 million years of vertebrate evolution, taxonomic distance within this group is immense, greatly exceeding, for example that between frog and human (Johns and Avise 1998). Fish occupy virtually every aquatic habitat, from tropical reefs to abyssal depths; some have even adopted amphibian-like lifestyles. Associated ecological and behavioural demands have fashioned basic brain designs into a vast number of species-specific variations on the theme (Nieuwenhuys et al. 1998). Recent papers based on a combination of quantitative techniques and applications of phylogenetically controlled statistical designs have illuminated the characteristics of evolutionary trends in a variety of taxa. In short, both ecology and phylogenetic distance account for brain variability. For example, when comparing the brains of sharks and teleosts, effects of evolutionary history prevail, whereas nested downwards, comparisons within the latter taxon (i.e. within the cichlids or within cyprinids) increasingly pinpoint ecology as the major covariant of morphology.

6.2.1 Comparative Studies of Brain Evolution

When brains are compared between species, this is often done to relate the detected anatomical differences to functional properties. The rationale behind this is that anatomical structure is the result of the integration of the past selection pressures. We are here adhere to the broad definition of cognition, which includes perception, learning, processing, storage, and retrieval of information (Shettleworth 2010). Although challenged by some (Chittka and Niven 2009), brain size, absolute and/or relative, is often used as a proxy for cognitive ability (Striedter 2005). Evidence for this relationship comes largely from phylogenetically controlled comparative analyses. The logic of those analyses is that more closely related species are more similar than more distantly related species (Harvey and Pagel 1991). Controlling for this phylogeny effect, comparative analyses reveal macroevolutionary patterns. In birds and mammals, relative brain size and cognitive ability are positively associated (Benson-Amram et al. 2016; MacLean et al. 2014). The majority of comparative analyses in brain anatomy is done in mammals and birds, whereas studies relating brain size to cognitive abilities across species in fish are currently lacking. In Lake Tanganyika cichlids relative brain size is positively associated with social and environmental complexity (Pollen et al. 2007; van Staaden et al. 1995), but also with the type of diet (Gonzalez-Voyer et al. 2009). Sex-specific analyses in these cichlids showed that female care-type (bi-parental or female-only) determined brain size in females, but not in males. Likewise, in pipefishes and seahorses, feeding ecology and brain size seem to be linked as longer snouts (an adaptation to more mobile prey) correlate with larger brains (Tsuboi et al. 2017). The high costs of brain tissue (Kuzawa et al. 2014) likely limit brain evolution as indicated by apparent trade-offs between brain size and other costly organs such as the gut (Lake Tanganyika cichlids, Tsuboi et al. 2014b) or fat tissue (Pacific seaweed pipefish, Tsuboi et al. 2016).

6.2.2 *Brain Plasticity*

Adaptive phenotypic plasticity (West-Eberhard 2003) can allow for faster adjustments to prevailing conditions than trans-generational adaptation (Ghalambor et al. 2007). In teleosts (Wagner 2003) and elasmobranchs (Lisney et al. 2007), life-stage specific habitat shifts trigger changes in size of the brain parts relevant for the respective ages. Brain plasticity is also commonly observed in experimental settings. Captive rearing changes brain anatomy in most fish species investigated so far, it affects olfactory bulb and Telencephalon size in Chinook salmon (*Oncorhynchus tshawytscha*; (Kihslinger and Nevitt 2006), brain, optic tectum, and Telencephalon size in Guppies (*Poecilia reticulata*; Burns et al. 2009), sometimes overall brain size in nine-spined stickleback (*Pungitius pungitius*; Gonda et al. 2011), and Telencephalon size in three-spined stickleback (*Gasterosteus aculeatus*; Park et al. 2012). Generally, brain and region sizes are smaller in captive-reared fishes, while environmental enrichment can counteract such effects and lead to size increases of brain regions. Enrichment increased cell proliferation in the Telencephalon of coho salmon (*Oncorhynchus kisutch*; Lema et al. 2005) and triggered the development of larger Cerebellum in steelhead trout (*Oncorhynchus mykiss*; Kihslinger and Nevitt 2006). Most fish grow indeterminately, and a major difference between the brain of fish and that of poikilothermic vertebrates is that fish brains show much more cell proliferation also as adults (Zupanc 2001), accounting for a pronounced phenotypic plasticity and adaptive potential also at later life stages. For instance, a change in social status correlates with increased cell proliferation rate in salmonids (Sørensen et al. 2007), a greater female availability boosts brain size in male guppies (Kotrschal et al. 2012b), and changes in rearing-group size change most brain regions in cichlids (Fischer et al. 2015).

6.2.3 *The Brain-Size-Selected Guppies*

As highlighted in Sect. 6.2.1, comparative methods are useful to unravel the macroevolutionary patterns in fish brain anatomy. However, they produce correlational results, as a causative relationship can only be established by experimental manipulation. In the following, we will summarize the findings of the first such experiments in brain evolution, done with the guppy (*Poecilia reticulata*), a small fish of the Poeciliidae that inhabits shallow streams in Trinidad and Northern Venezuela. This is a model organism in several disciplines of biology, including ecology, evolutionary biology, and behavioural biology; it was used for artificial selection on large and small relative brain size for *experimentally* testing established concepts of brain evolution. In addition, this unique model system revealed some previously unknown costs and benefits of evolving a large brain.

The guppy brain size selection lines were generated using an artificial selection design consisting of two replicated treatments (three up-selected lines and three down-selected lines). Since brain size can only be quantified after dissection, pairs were first allowed to breed at least two clutches and then sacrificed for brain quantification. The offspring from parents with large or small relative brain size were then used for starting the next generation. More specifically, to select for *relative* brain size (controlled for body size), the residuals from the regression of brain size (weight) on body size (length) of both parents were used. Three times 75 pairs (75 pairs per replicate) were started to create the first three "up" and "down" selected lines (six lines in total). Male and female residuals for each pair were summed and the offspring from the top and bottom 25% of these "parental residuals" were used to form the next-generation parental groups. Then offspring of the 30 pairs with the largest residual sums for up-selection and the 30 pairs with the smallest residual sums for down-selection were propagated for each of the following generations. Already the second generation showed a 9% difference in relative brain size between lines and the third differed by up to 14%, at unchanged body size. The larger brains were composed of more neurons (Marhounová et al. 2019), but the 11 main brain regions remained proportionally similar between the lines (Kotrschal et al. 2017). These large- and small-brained lines were compared in a range of traits to determine the costs and benefits of large brain size.

As described above a relatively larger brain seems to confer a cognitive benefit (Benson-Amram et al. 2016; MacLean et al. 2014). This was indeed the case in several tests of learning and memory in the brain size lines. For instance, using food as a reward, large-brained females were better at numerical learning (Kotrschal et al. 2013) and outperformed small-brained females in a reversal-learning test (Buechel et al. 2018). Large-brained males, in turn, were better at learning and memorizing the way through a maze when a virgin female was the reward (Kotrschal et al. 2014a). But is this simply an "academic" difference in cognitive performance or does an increase in cognitive ability matter in the life of a guppy? To test this, six large semi-natural streams were stocked with 800 guppies each (balanced over sex and brain size selection line and individually marked using green and red elastomer implants), and one pike cichlid (*Crenicichla alta*), a natural guppy predator from Trinidad, was introduced into each of those streams. In weekly censuses survival of fish was monitored and showed that large-brained females survived longer and in greater numbers (Kotrschal et al. 2015a). After 14 weeks exactly half of the large-brained females, but only 44.5% of the small-brained females were still alive. Males were eaten faster than females, but the large- and small-brained males did not differ in survival. These findings provoked two questions: Why did the large-brained females survive better and why did the large-brained males not benefit from a larger brain?

The improved survival of large- over small-brained females was likely due to their cognitive advantages that enabled them to better avoid predation. The pike cichlids sat hidden in the deepest part of the streams, striking at fish passing by. Better learning and memory should help learning about and avoiding those dangerous areas of the stream. Guppies show predator inspection (Dugatkin and Godin 1992). A change in cognitive ability may impact this behaviour in several ways and indeed, a follow-up experiment showed that larger-brained animals seem to be faster at gathering and integrating information about the predator's state because they inspected for a shorter time and also from further away (van der Bijl et al. 2015). But why did the large-brained males not show a survival advantage over small-brained males? After all, they also show predator inspection and are known to outperform small-brained males in tests of learning and memory. Potentially because large-brained males in these selection lines are more colourful than the small-brained males (Kotrschal et al. 2015c); it is still not understood why, but it is likely due to a genetic correlation between brain size and coloration. Large-brained guppy males had larger orange and iridescent spots. Pike cichlids are visual hunters with a known preference for more colourful individuals (Endler 1980). Therefore, the large-brained males' increased conspicuousness may have overridden the cognitive benefits of a larger brain.

So, a larger brain confers several benefits, related to cognition and survival. Large-brained males also seem to have a mating advantage as several traits that are known to be beneficial during mating are exaggerated in large-brained males aside the already mentioned colourfulness. They also have a longer tail fin, a known trait relevant for female choice, and also a longer gonopodium (Kotrschal et al. 2015c), which is the anal fin modified to function as intromittent organ. Guppy males often sneak up on females and attempt to mate coercively and a longer gonopodium facilitates such copulations (Houde 1987). However, several tests of mate choice did not reveal any significant mating advantage of large- over small-brained males (unpublished data). In females, brain size may also be relevant during mating, because accurate assessment of partner quality may depend on cognitive ability (Verzijden et al. 2012). Indeed, when given the choice between very colourful and less colourful males, large-brained females showed a more pronounced preference for the attractive than for the less-attractive male (Corral-Lopez et al. 2017). In contrast, small-brained and wild-type females showed no preference. In-depth analysis of optomotor response to colour cues and gene expression of opsins in the eye revealed that the observed differences were not due to differences in visual perception of colour or visual acuity (Corral-López et al. 2017), indicating that differences in the ability to process indicators of attractiveness are responsible. While brain size did not impact general sexual behaviour (Corral-López et al. 2015) in males, it did effect context-specific mate choice as large-brained males could better discriminate between differentially-sized females (Corral-López et al. 2018).

Are small-brained guppies simply "sub-optimal" guppies? This may appear so at first since they seem to be inferior to large-brained animals in so many traits. But this is probably not the case, because several classic indicators of quality showed no difference between large- and small-brained animals; those include body condition, swimming endurance, escape velocity ("C-start"), and adult body size (Kotrschal et al. 2014b). Several traits were, in fact, more prominent in small- compared to large-brained guppies indicating that costs are associated with developing a larger brain. For instance, in their first parturition, small-brained guppy females gave birth to 15% more offspring compared to large-brained females (Kotrschal et al. 2013). This indicates a trade-off between investment in brain size or fecundity. Further indications of trade-offs include a decreased innate (but not acquired) immune response in large-brained animals (Kotrschal et al. 2016), a slower juvenile speed of growth in large-brained animals (Kotrschal et al. 2015b), a smaller gut size in large-brained animals (Kotrschal et al. 2013), but slower intrinsic ageing in small-brained animals (Kotrschal et al. 2019).

In summary, three generations of artificial selection on relative brain size lead to up to 14% difference in brain size between guppies selected for large and small brains. A larger brain confers cognitive benefits as the large-brained animals outperformed the small-brained animals in several tests of learning and memory. A larger brain also seems beneficial for female mate choice, male attractiveness, and for anti-predator behaviour and survival. However, evolving a larger brain comes at some costs. Large-brained animals showed a decreased fecundity, smaller guts, slower juvenile growth rate, impaired immunity, and faster ageing. It is conceivable that guppies selected for different brain sizes may also slightly differ in their habitat preferences and requirements. However, no tests regarding habitat preferences as related to brain size have hitherto be conducted, rendering any conclusions regarding brain size-related welfare requirements speculative.

References

Allis EP (1897) The cranial muscles and cranial and first spinal nerves in *Amia calva*, vol 12. Ginn & Company, p 487

Benson-Amram S, Dantzer B, Stricker G, Swanson EM, Holekamp KE (2016) Brain size predicts problem-solving ability in mammalian carnivores. Proc Natl Acad Sci U S A 113(9):2532–2537

Bone Q (1977) Mauthner neurons in elasmobranchs. J Mar Biol Assoc U K 57:253–259

Bone Q, Marshal NB, Blaxter LHS (1982) Biology of fishes. Chapman & Hall, London

Brandstatter R, Kotrschal K (1990) Brain growth-patterns in 4 European cyprinid fish species (Cyprinidae, Teleostei) – roach (*Rutilus-rutilus*), bream (*Abramis-brama*), common carp (*Cyprinus-carpio*) and sabre carp (*Pelecus-cultratus*). Brain Behav Evol 35:195–211

Buechel SD, Boussard A, Kotrschal A, van der Bijl W, Kolm N (2018) Brain size affects performance in a reversal-learning test. Proc R Soc B 285:20172031

Burns JG, Saravanan A, Rodd FH (2009) Rearing environment affects the brain size of guppies: lab-reared guppies have smaller brains than wild-caught guppies. Ethology 115:122–133

Chittka L, Niven J (2009) Are bigger brains better? Curr Biol 19:R995–R1008

Corral-López A, Eckerström-Liedholm S, Der Bijl WV, Kotrschal A, Kolm N (2015) No association between brain size and male sexual behavior in the guppy. Curr Zool 61:265–273

Corral-Lopez A, Bloch N, Kotrschal A, van der Bijl W, Buechel S, Mank JE, Kolm N (2017) Female brain size affects the assessment of male attractiveness during mate choice. Sci Adv 3: e1601990

Corral-López A, Garate-Olaizola M, Buechel SD, Kolm N, Kotrschal A (2017) On the role of body size, brain size, and eye size in visual acuity. Behav Ecol Sociobiol 71:179

Corral-López A, Kotrschal A, Kolm N (2018) Selection for relative brain size affects context-dependent male preferences, but not discrimination, of female body size in guppies. J Exp Biol. https://doi.org/10.1242/jeb.175240

Costa SS, Andrade R, Carneiro LA, Gonçalves EJ, Kotrschal K, Oliveira RF (2011) Sex differences in the dorsolateral telencephalon correlate with home range size in blenniid fish. Brain Behav Evol 77:55–64
Davis R, Northcutt R (1983) Fish neurobiology, vol 2, Higher brain areas and functions. University of Michigan Press, Ann Arbor, MI
Dugatkin LA, Godin JGJ (1992) Predator inspection, shoaling and foraging under predation Hazard in the Trinidadian guppy, *Poecilia-reticulata*. Environ Biol Fish 34:265–276
Endler JA (1980) Natural-selection on color patterns in *Poecilia-reticulata*. Evolution 34:76–91
Finger TE (1980) Nonolfactory sensory pathway to the telencephalon in a teleost fish. Science 210:671–673
Fischer S, Bessert-Nettelbeck M, Kotrschal A, Taborsky B (2015) Rearing-group size determines social competence and brain structure in a cooperatively breeding cichlid. Am Nat 186:123
Ghalambor CK, McKay JK, Carroll SP, Reznick DN (2007) Adaptive versus non-adaptive phenotypic plasticity and the potential for contemporary adaptation in new environments. Funct Ecol 21:394–407
Gonda A, Herczeg G, Merila J (2011) Population variation in brain size of nine-spined sticklebacks (*Pungitius pungitius*) – local adaptation or environmentally induced variation? BMC Evol Biol 11:75
Gonzalez-Voyer A, Winberg S, Kolm N (2009) Social fishes and single mothers: brain evolution in African cichlids. Proc R Soc B Biol Sci 276:161–167
Harvey PH, Pagel MD (1991) The comparative method in evolutionary biology. Oxford University Press, Oxford
Herculano-Houzel S (2009) The human brain in numbers: a linearly scaled-up primate brain. Front Hum Neurosci 3:31
Herrick CJ (1902) A note on the significance of the size of nerve fibers in fishes. J Comp Neurol 12 (4):329–334
Herrick CJ (1906) On the centers for taste and touch in the medulla oblongata of fishes. J Comp Neurol Psychol 16(6):403–439
Houde AE (1987) Mate choice based upon naturally-occurring color-pattern variation in a guppy population. Evolution 41:1–10
Johns GC, Avise JC (1998) A comparative summary of genetic distances in the vertebrates from the mitochondrial cytochrome b gene. Mol Biol Evol 15:1481–1490
Kanwal JS, Finger TE (1992) Central representation and projections of gustatory systems. In: Hara TJ (ed) Fish chemoreception. Springer, pp 79–102
Kihslinger RL, Nevitt GA (2006) Early rearing environment impacts cerebellar growth in juvenile salmon. J Exp Biol 209:504–509
Kotrschal A, Taborsky B (2010) Resource defence or exploded lek?–a question of perspective. Ethology 116:1189–1198
Kotrschal K, Adam H, Brandstätter R, Junger H, Zaunreiter M, Goldschmid A (1990) Larval size constraints determine directional ontogenetic shifts in the visual system of teleosts. J Zool Syst Evol Res 28:166–182
Kotrschal K, van Staaden MJ, Huber R (1998) Fish brains: evolution and environmental relationships. Rev Fish Biol Fish 8:373–408
Kotrschal A, Heckel G, Bonfils D, Taborsky B (2012a) Life-stage specific environments in a cichlid fish: implications for inducible maternal effects. Evol Ecol 26:123–137
Kotrschal A, Rogell B, Maklakov AA, Kolm N (2012b) Sex-specific plasticity in brain morphology depends on social environment of the guppy, *Poecilia reticulata*. Behav Ecol Sociobiol 66:1485–1492
Kotrschal A, Rogell B, Bundsen A, Svensson B, Zajitschek S, Brännström I, Immler S, Maklakov AA, Kolm N (2013) Artificial selection on relative brain size in the guppy reveals costs and benefits of evolving a larger brain. Curr Biol 23:168–171
Kotrschal A, Corral-Lopez A, Amcoff M, Kolm N (2014a) A larger brain confers a benefit in a spatial mate search learning task in male guppies. Behav Ecol 26:527–532

Kotrschal A, Lievens EJ, Dahlbom J, Bundsen A, Semenova S, Sundvik M, Maklakov AA, Winberg S, Panula P, Kolm N (2014b) Artificial selection on relative brain size reveals a positive genetic correlation between brain size and proactive personality in the guppy. Evolution 68:1139–1149

Kotrschal A, Buechel S, Zala S, Corral Lopez A, Penn DJ, Kolm N (2015a) Brain size affects female but not male survival under predation threat. Ecol Lett 18:646–652

Kotrschal A, Corral-Lopez A, Szidat S, Kolm N (2015b) The effect of brain size evolution on feeding propensity, digestive efficiency, and juvenile growth. Evolution 69:3013–3020

Kotrschal A, Corral-Lopez A, Zajitschek S, Immler S, Maklakov AA, Kolm N (2015c) Positive genetic correlation between brain size and sexual traits in male guppies artificially selected for brain size. J Evol Biol 28:841–850

Kotrschal A, Kolm N, Penn DJ (2016) Selection for brain size impairs innate, but not adaptive immune responses. Proc R Soc B 283:20152857

Kotrschal A, Zeng HL, van der Bijl W, Öhman-Mägi C, Kotrschal K, Pelckmans K, Kolm N (2017) Evolution of brain region volumes during artificial selection for relative brain size. Evolution 71:2942–2951

Kotrschal A, Corral-Lopez A, Kolm N (2019) Large brains, short life: selection on brain size impacts intrinsic lifespan. Biol Lett 15:20190137

Kruska DC (1988) The brain of the basking shark (Cetorhinus maximus). Brain Behav Evol 32 (6):353–363

Kuzawa CW, Chugani HT, Grossman LI, Lipovich L, Muzik O, Hof PR, Wildman DE, Sherwood CC, Leonard WR, Lange N (2014) Metabolic costs and evolutionary implications of human brain development. Proc Natl Acad Sci U S A 111:13010–13015

Lema SC, Hodges MJ, Marchetti MP, Nevitt GA (2005) Proliferation zones in the salmon telencephalon and evidence for environmental influence on proliferation rate. Comp Biochem Physiol A Mol Integr Physiol 141:327–335

Lisney TJ, Bennett MB, Collin SP (2007) Volumetric analysis of sensory brain areas indicates ontogenetic shifts in the relative importance of sensory systems in elasmobranchs. Raffles Bull Zool 14:7–15

MacLean EL, Hare B, Nunn CL, Addessi E, Amici F, Anderson RC, Aureli F, Baker JM, Bania AE, Barnard AM, Boogert NJ, Brannon EM, Bray EE, Bray J, Brent LJN, Burkart JM, Call J, Cantlon JF, Cheke LG, Clayton NS, Delgado MM, DiVincenti LJ, Fujita K, Herrmann E, Hiramatsu C, Jacobs LF, Jordan KE, Laude JR, Leimgruber KL, Messer EJE, de A. Moura AC, Ostoji*fá* L, Picard A, Platt ML, Plotnik JM, Range F, Reader SM, Reddy RB, Sandel AA, Santos LR, Schumann K, Seed AM, Sewall KB, Shaw RC, Slocombe KE, Su Y, Takimoto A, Tan J, Tao R, van Schaik CP, Viranyi Z, Visalberghi E, Wade JC, Watanabe A, Widness J, Young JK, Zentall TR, Zhao Y (2014) The evolution of self-control. Proc Natl Acad Sci U S A 111:E2140–E2148

Maler L, Sas E, Johnston S, Ellis W (1991) An atlas of the brain of the electric fish *Apteronotus leptorhynchus*. J Chem Neuroanat 4:1–38

Marhounová L, Kotrschal A, Kverková K, Kolm N, Němec P (2019) Artificial selection on brain size leads to matching changes in overall number of neurons. Evolution 73(9):2003–2012

Mills SM (1932) The double innervation of fish melanophores. J Exp Zool A Ecol Genet Physiol 64:231–244

Nakane Y, Ikegami K, Iigo M, Ono H, Takeda K, Takahashi D, Uesaka M, Kimijima M, Hashimoto R, Arai N (2013) The saccus vasculosus of fish is a sensor of seasonal changes in day length. Nat Commun 4:2018

Nieuwenhuys R (1982) An overview of the organization of the brain of actinopterygian fishes. Am Zool 22:287–310

Nieuwenhuys R, ten Donkelaar HJ, Nicholson C (1998) The central nervous system of vertebrates. Springer, Heidelberg

Northcutt RG (1978) Brain organization in the cartilaginous fishes. In: Hodgson ES, Mathewson RF (eds) Sensory biology of sharks, skates and rays. Office of Naval Research, Washington, DC, pp 117–193

Northcutt RG, Davis R (1983) Fish neurobiology: brain stem and sense organs. University of Michigan Press, Ann Arbor, MI

Okuyama T, Yokoi S, Abe H, Isoe Y, Suehiro Y, Imada H, Tanaka M, Kawasaki T, Yuba S, Taniguchi Y (2014) A neural mechanism underlying mating preferences for familiar individuals in medaka fish. Science 343:91–94

Östlund-Nilsson S, Mayer I, Huntingford FA (2007) Biology of the three-spined stickleback. CRC Press, Boca Raton, FL

Park PJ, Chase I, Bell MA (2012) Phenotypic plasticity of the threespine stickleback *Gasterosteus aculeatus* telencephalon in response to experience in captivity. Curr Zool 58:189–210

Pollen AA, Dobberfuhl AP, Scace J, Igulu MM, Renn SCP, Shumway CA, Hofmann HA (2007) Environmental complexity and social organization sculpt the brain in Lake Tanganyikan cichlid fish. Brain Behav Evol 70:21–39

Popper AN, Fay RR (1993) Sound detection and processing by fish: critical review and major research questions (part 1 of 2). Brain Behav Evol 41:14–25

Portavella M, Vargas J, Torres B, Salas C (2002) The effects of telencephalic pallial lesions on spatial, temporal, and emotional learning in goldfish. Brain Res Bull 57:397–399

Puelles L, Harrison M, Paxinos G, Watson C (2013) A developmental ontology for the mammalian brain based on the prosomeric model. Trends Neurosci 36:570–578

Rodríguez F, Durán E, Gomez A, Ocana F, Alvarez E, Jiménez-Moya F, Broglio C, Salas C (2005) Cognitive and emotional functions of the teleost fish cerebellum. Brain Res Bull 66:365–370

Salas C, Broglio C, Durán E, Gómez A, Ocaña FM, Jiménez-Moya F, Rodríguez F (2006) Neuropsychology of learning and memory in teleost fish. Zebrafish 3:157–171

Schellart NA (1991) Interrelations between the auditory, the visual and the lateral line systems of teleosts; a mini-review of modelling sensory capabilities. Neth J Zool 42:459–477

Shettleworth SJ (2010) Cognition, evolution, and behavior, 2nd edn. Oxford University Press, Oxford

Sibbing F (1991) Food capture and oral processing. In: Nelson J, Winfield IJ (eds) Cyprinid fishes. Springer, pp 377–412

Sørensen C, Øverli Ø, Summers CH, Nilsson GE (2007) Social regulation of neurogenesis in teleosts. Brain Behav Evol 70:239–246

Striedter GF (2005) Principles of brain evolution. Sinauer Associates, Sunderland

Szabó I (1973) Path neuron system of medial forebrain bundle as a possible substrate for hypothalamic self-stimulation. Physiol Behav 10:315–328

Tinbergen N (1951) The study of instinct. Oxford University Press, New York

Tsuboi M, Gonzalez-Voyer A, Kolm N (2014a) Phenotypic integration of brain size and head morphology in Lake Tanganyika Cichlids. BMC Evol Biol 14:39

Tsuboi M, Husby A, Kotrschal A, Hayward A, Buechel S, Zidar J, Lovle H, Kolm N (2014b) Comparative support for the expensive tissue hypothesis: big brains are correlated with smaller gut and greater parental investment in Lake Tanganyika cichlids. Evolution 69:190–200

Tsuboi M, Shoji J, Sogabe A, Ahnesjö I, Kolm N (2016) Within species support for the expensive tissue hypothesis: a negative association between brain size and visceral fat storage in females of the Pacific seaweed pipefish. Ecol Evol 6:647–655

Tsuboi M, Lim ACO, Ooi BL, Yip MY, Chong VC, Ahnesjö I, Kolm N (2017) Brain size evolution in pipefishes and seahorses: the role of feeding ecology, life history and sexual selection. J Evol Biol 30:150–160

van der Bijl W, Thyselius M, Kotrschal A, Kolm N (2015) Brain size affects the behavioral response to predators in female guppies (*Poecilia reticulata*). Proc R Soc B Biol Sci 282:20151132

van Staaden MJ, Huber R, Kaufmann LS, Liem KF (1995) Brain evolution in cichlids of the African Great Lakes: brain and body size, general patterns and evolutionary trends. Zoology 98:165–178
Vanegas H, Ito H (1983) Morphological aspects of the teleostean visual system: a review. Brain Res Rev 6:117–137
Verzijden MN, Ten Cate C, Servedio MR, Kozak GM, Boughman JW, Svensson EI (2012) The impact of learning on sexual selection and speciation. Trends Ecol Evol 27:511–519
Von Kupffer C (1891) The development of the cranial nerves of vertebrates. J Comp Neurol 1:246–264
Voneida TJ, Fish SE (1984) Central nervous system changes related to the reduction of visual input in a naturally blind fish (*Astyanax hubbsi*). Am Zool 24:775–782
Wagner H-J (2003) Volumetric analysis of brain areas indicates a shift in sensory orientation during development in the deep-sea grenadier *Coryphaenoides armatus*. Mar Biol 142:791–797
Webb J, Northcutt R (1997) Morphology and distribution of pit organs and canal neuromasts in non-teleost bony fishes. Brain Behav Evol 50:139–151
Weiger T, Lametschwandtner A, Kotrschal K, Krautgartner WD (1988) Vascularization of the telencephalic choroid plexus of a ganoid fish [*Acipenser ruthenus* (L.)]. Dev Dyn 182:33–41
West-Eberhard M (2003) Developmental plasticity and evolution. Oxford University Press, Oxford
Wullimann MF (1994) The teleostean torus longitudinalis. Eur J Morphol 32:235–242
Young JZ (1931) Memoirs: on the autonomic nervous system of the teleostean fish *Uranoscopus scaber*. J Cell Sci 2:491–536
Young J (1980) Nervous control of gut movements in Lophius. J Mar Biol Assoc U K 60:19–30
Zaunreiter M, Kotrschal K, Goldschmid A, Adam H (1985) Ecomorphology of the optic system in 5 species of blennies (Teleostei). Fortschr Zool 30:731–734
Zupanc GKH (2001) Adult neurogenesis and neuronal regeneration in the central nervous system of teleost fish. Brain Behav Evol 58:250–275

Chapter 7
Inside the Fish Brain: Cognition, Learning and Consciousness

Anders Fernö, Ole Folkedal, Jonatan Nilsson, and Tore S. Kristiansen

Abstract Detecting and interpreting information about resources and dangers and behaving flexibly and effectively are essential for survival and welfare. Fish, as well as other organisms, access their surroundings through their sensory systems. They must also have a cognitive system capable of integrating and interpreting the sensory input in relation to earlier experiences, and eventually act according to this input. Learning ability and memory enable fish to detect regularities and associations and construct mental images, categories and concepts. In this way, they can adapt their behaviour to the dynamic environment and predict the near future and the consequences of their behaviour. Numerous studies have shown that many fish species have evolved good cognitive abilities, and can construct internal maps, cope with complex social relationships and retain memories for long periods. Some fish can even innovate and use tools. However, the enormous diversity within the piscine world demands that we view learning, cognition and welfare from an ecological perspective. The cognitive capacity of individual species depends on the environmental and social complexity they encounter, and also differs between populations, coping styles and sexes. It is reasonable to believe that fish are conscious and have emotions and feelings, although their subjective experiences must be very different from ours and also vary between species. Their cognitive capacity and the behavioural flexibility that enables them to cope with aquaculture environments and procedures are essential for the welfare of farmed fish.

Keywords Memory · Intelligence · Categorising · Emotions · Ecological perspective · Aquaculture · Domestication · Welfare

A. Fernö (✉)
Department of Biology, University of Bergen, Bergen, Norway
e-mail: Anders.Ferno@uib.no

O. Folkedal · J. Nilsson · T. S. Kristiansen
Institute of Marine Research, Bergen, Norway

T. S. Kristiansen et al. (eds.), *The Welfare of Fish*, Animal Welfare 20,
https://doi.org/10.1007/978-3-030-41675-1_7

7.1 Introduction

Key to all modern definitions of animal welfare is the ability to cope with challenges that the environment presents. Definitions differ in the extent to which subjects suffer when they are unable to cope, or in general whether, beyond behaviour and physiology, affects and emotions are critical for definitions of welfare. Still, the current consensus suggests that emotional systems are essential for most species to navigate through life (see Spruijt et al. 2001). The ability of fishes to have a conscious qualitative experience of emotions and experience welfare can be regarded as both a system for monitoring their current state of need and a motivational system that drives them to obtain what they need. To monitor their living conditions in relation to their state of need, fish need to sense relevant properties of the environment and their internal states. They must also have a cognitive system capable of interpreting and integrating the sensory input in relation to earlier experiences, and eventually act according to this input. Fishes also have to make trade-offs between different needs, like food and safety from predators, and on the basis of this predict and decide future actions.

Primitive organisms without an intelligent brain must rely on reflexive and automated behaviours. Presumably, most laymen believe that also fishes are not particularly intelligent. They lack facial movement and they do not produce sounds that are interpreted by humans as positive or negative as many mammalian pets do, such as purring and hissing in cats. It is, therefore, more difficult for us to relate to the mind of a fish. During the past few decades, however, it has been realised that the behaviour and cognitive abilities of fishes are rather complex and much more than just reflexive (Brown et al. 2011). We now know that fish can behave in a flexible way, learn from experience and predict results of future actions. This is critical for the ability of both wild and farmed fish to cope with a dynamic environment, as well as for fish living as pets and fish used in experiments. Fishes that are unable to adapt to a changing situation will have problems. A salmon that continues to swim in a straight course in a circular net pen will have a hard life!

As pointed out in Chap. 3, a fish is not just a fish, and both body and brain size and sensory abilities vary greatly between the more than 34,000 fish species. We should thus expect to find differences in cognitive abilities and experience of emotions both between and within species and between the often markedly different ontological states of the same individual. In this chapter, we explore the cognitive world of fishes, and discuss how cognition and emotions affect their behaviour and experience of welfare. We also look closely at the most controversial issue in fish welfare science: to what degree, if at all, are fish conscious and experience conscious feelings? And can knowledge about cognition, learning and consciousness help us to evaluate and improve the welfare of fish that are kept in artificial environments that in many respects differ from the natural environment in which their mental capacities have evolved? This chapter will generally be built on the traditional way to look at cognition and welfare in fishes, and Chap. 9 of this volume will take this a step further based on new knowledge about how the brain works.

7.2 How Do Fish Collect and Process Information?

Current theories say that cognition consists of three processes: a perception phase, a learning phase and a memory phase (Shettleworth 2010; but see Chap. 9). Fish continuously experience a multitude of stimuli from the surrounding world. To handle this huge amount of information they need to distinguish between objects and events that are relevant and useful and those that are not. Like us, fish need to know who and where prey, 'friends' and 'enemies' are, but also where and who they themselves are (e.g. stronger or weaker than a competitor), where they are moving and where they are at risk and where to hide. In an only partly predictable environment this is a considerable intellectual task.

The only access that fish and other organisms have to their surroundings is through their sensory input. To collect information, fishes have a rich toolbox of sensory organs and a relatively large brain adapted to their special habitats and the biosphere they live in (Chap. 6). Fishes interact with the outer world by physical and chemical external stimulation of the sensory organs that generate the signals they send to the nervous system (Chap. 9). To make sense of this information, they need to construct an internal representation and make inferences regarding what exists and is going on 'out there'. Fishes also interact with their inner world by sensors in the body and brain that measure various internal needs and physiological and psychological states. New evidence indicates that the brain is continuously trying to predict sensory inputs from the external world as well as proprioceptive signals of body movements and interoceptive signals from other internal processes (see Chap. 9).

7.3 Learning

What is the point of learning? If the external world were completely predictable, fixed reactions to different stimuli ought to work perfectly well. But the world is not predictable. Behaviour must continually be adapted to the dynamic environment. But even in such an environment, some patterns are repeated, and ignoring what has taken place in the past would mean neglecting relevant information. Every living organism possesses mechanisms for learning, remembering and forecasting with the objective of performing anticipatory actions in the external world (Chap. 9). Sensory information must be converted into objects, and objects must be put into categories and concepts, and interpreted as being the same, similar or quite different. Otherwise, all new objects will be experienced as different and new, and it will be impossible to learn anything and no use to remember anything. What is sensed and observed now must be put into the context of past experiences if it is to make sense, and the ability to know what is good and what is bad is dependent on learning and memory. However, learning also incurs costs. It demands neural tissue (Chap. 6) and could reduce behavioural flexibility if circumstances change, while routines might make behaviour more predictable, which could be exploited by predators.

Even so, a fish that does not modify its behaviour on the basis of learning would not be around long. It has long been known that fish can associate different cues with rewards (Bull 1928), and recent decades have seen a wealth of studies on fish learning and cognition (Brown et al. 2011). All fish species studied so far can learn and remember. The 'three-second-memory' of a goldfish (*Carassius auratus*) is a total myth. If a fish cannot be conditioned to a stimulus, this usually means that the stimulus cannot be detected. Still, it is not always obvious that fish have learnt. In an experiment where we initially attempted to teach halibut (*Hippoglossus hippoglossus*) to respond to a light signal, we used commercial fish food pellets as a reward. But the halibut displayed no response, and at first we thought that halibut are incapable of this type of associative learning. However, when we replaced pellets with shrimp, the fish reacted strongly and were easily conditioned (Nilsson et al. 2010). It turned out that the food pellets that the fish already had in excess had low affective value and did thus not provide strong enough reinforcement. So it was we who were stupid and not the halibut!

But not everything is learned equally well. Events which induce positive or negative affects that contribute significantly to survival and reproduction are more readily remembered than emotionally neutral events (Paul et al. 2005). For instance, fishes are more easily conditioned to moving than stationary objects (Wisenden and Harter 2001). Moving objects such as attacking predators or escaping prey have more important consequences for the fish than objects that do not move and should, therefore, have higher emotional value. Another example of the natural link between a signal and what it predicts is taste aversion. In rats, the taste of food or drink is easily associated with subsequent nausea after only one exposure, while an audio–visual cue is not (Garcia and Koelling 1966). The link between the taste of poisoned or spoiled food and nausea is evolutionarily important, while no association is formed when no natural link exists between the cue and nausea. Taste aversion has also been demonstrated in fish (MacKay 1974; Little 1977; Manteifel and Karelina 1996). There could, in addition, be a central inhibition to learn certain relationships. Male three-spined sticklebacks (*Gasterosteus aculeatus*) can learn to bite a rod if rewarded by the opportunity to attack a rival but cannot be trained to bite the rod to gain access to a gravid female (Sevenster 1973). Aggression and sex seem thus to be incompatible motivational systems.

Learning plays an important role in the life of a fish and it can be divided into different categories (Box 7.1). Learning influences which prey fishes select and how the prey is caught (Croy and Hughes 1991; Ibrahim and Huntingford 1992; Kristiansen and Svåsand 1992; Steingrund and Fernö 1997) and may help them to avoid predators (Kelley 2008). Fish can also learn migration routes, home ranges and territories (Helfman and Schultz 1984; Odling-Smee et al. 2011), and learning enables individual recognition in for instance hierarchies (Grosenick et al. 2007).

Box 7.1 Different Categories of Learning

Habituation—A behavioral response decrement that results from repeated stimulation not involving sensory adaptation/sensory fatigue or motor fatigue (Rankin et al. 2009). To learn not to react to something irrelevant represents the simplest form of learning but is crucial for functional behaviour. A lot of time and energy would be wasted if a fish responded to events that often occur but mean nothing, like whirling algae, raindrops hitting the surface or the appearance of organisms that mean nothing for the fish.

Classical (Pavlovian) conditioning—An association formed between an initially neutral stimulus (conditioned stimulus, CS) and a stimulus to which the animal responds naturally (unconditioned stimulus, US), which may be experienced as either positive (reward) or negative (punishment). When the animal has learnt that the probability that the US will be presented is higher after the presentation of the CS (Rescorla 1966), the CS alone releases a similar response as to the US. In this way, fish can learn to prepare for relevant events before they occur, for instance by approaching or avoiding associated cues. During *Delay conditioning* CS and US overlap in time. During *Trace conditioning,* there is a stimulus-free time interval between the CS and the US, and a 'trace' from the CS must be maintained in the working memory across the interval for an association to be formed (Lieberman 1990). Trace conditioning is, in contrast to delay conditioning, dependent on an awareness of the CS–US relationship (Clark and Squire 1998) and is sensitive to distractions (Clark and Squire 1999).

Instrumental (Operant) conditioning—An association formed between an action and its outcome. For instance, when pulling a string is followed by food delivery a cod will continue to pull, while pulling will eventually cease if it is not followed by food (Nilsson and Torgersen 2010).

Procedural learning—The development and retention of skills and habits, i.e. how to improve task performance. For instance, sticklebacks attack and deal with prey more efficiently as a result of experience (Croy and Hughes 1991).

Imprinting—The development of a strong attachment to a stimulus during a certain stage of life. Imprinting of salmonids to chemical cues in their river enables later homing behaviour (Scholtz et al. 1976).

Latent learning—Learning something with no immediate reward. For instance, a fish can learn a certain route (Odling-Smee and Braithwaite 2003) for later use. In this case the mere acquisition of new information is probably rewarding.

Learning by insight—An understanding of a problem and an idea how to solve it. The animal understands the purpose and consequences of its behaviour and can intentionally do what it takes to reach a goal, for instance by the use of tool. Some fish species can perform intentional behaviours that could be considered as tool use (Brown 2012).

7.3.1 Differences in Learning Capacity

It is not easy to draw comparisons between different animal groups with different evolutionary backgrounds by presenting them with the same task (Bitterman 1975). We assume that mammals have greater learning capacity than fish, but for instance cod (*Gadus morhua*) have a high level of behavioural sophistication and complex learning strategies (Meager et al. 2018) and are in fact in some respects as intelligent as many mammals. Although trace conditioning is more cognitively demanding than delay conditioning, as the fish must retain an awareness of and anticipate the arrival of the reward (Lieberman 1990), cod can learn even with a one- to two-minute trace interval (Nilsson et al. 2008a).

Even so, learning and cognition are species and context dependent. Fish species have been shaped by their natural environments (Chap. 3), and fishes show the widest range of variations in brain function in all vertebrates (Nieuwenhuys et al. 1998). Species live in different habitats and social structures, their investment in offspring varies, and they receive different amounts of information from their surroundings. Although few fish species have been studied, we thus ought to adopt an ecological perspective on the cognitive and learning capacities in different species. Coble et al. (1985) trained 14 species of freshwater fish to move in response to light to avoid an electrical shock and found clear differences in learning capacity, with for instance common carp (*Cyprinus carpio*) being good learners and northern pike (*Esox lucius*) poor learners, although the results may be influenced by the propensity of each species to flee in response to a stimulus. There may also be differences within a species. Fish from different populations and individuals with different personalities do not behave in the same way (Chap. 3) and should thus be expected to have different mental capacities. Huntingford and Wright (1992) found genetically based population differences in the three-spined stickleback, with predator-naive fish from a high-risk site with abundant predatory fish learning to avoid a dangerous place faster than fish from a low-risk site. Climbing perches (*Anabas testudineus*) from populations in streams learnt a route in a maze faster than perches from ponds (Sheenaja and Thomas 2011). Interestingly, pond fish learnt the route faster in a maze provided with visual landmarks than in a plain maze, while this was not the case for stream fish. A local landmark may be a more reliable cue in a relatively stable habitat like a pond, and stream fishes seem to rely more on 'egocentric' cues than visual ones. A positive relationship was observed between male boldness and a simple associative learning task in guppy *(Poecilia reticulata)* and rainbow trout (*Oncorhynchus mykiss*) (Dugatkin and Alfieri 2003; Sneddon 2003), whereas the learning rate to find a hidden food patch in a maze was lower in bold than shy brook trout (*Salvelinus fontinalis*; White et al. 2017).

Behaviour after learning has occurred could also differ between species. Conditioned fish usually show a clear response to the CS and approach a feeder or escape from the place where they had an aversive experience (Nilsson et al. 2008a; Portavella et al. 2004). However, halibut that are trace conditioned to food do not approach the CS (light) but remain on the bottom only making subtle positional changes ('Silent learning', Nilsson et al. 2010; Fernö et al. 2011). Flatfishes like halibut are lie-and-wait predators (Gibson 2005) and should thus wait for the right movement to attack the prey. A similar difference between cod and halibut is found between the rat and the lie-and-wait predator cat (van den Bos et al. 2003).

7.3.2 How Fast Should Fish Make Decisions Based on Learning?

Taking decisions comprises a complex process of assessing and weighing short- and long-term costs and benefits of different actions (van den Bos et al. 2013). That reasonable decisions are based on probability was already realised in the eighteenth century (Butler 1736). But how intelligent are fish actually? One way to find out this is to study how quickly they learn a connection or a task. But be careful now! In a classical learning experiment, the experimenter already at the start knows that there is a relationship between a stimulus and a reward, but the fish lacks this knowledge and does not even know that it is involved in a learning task. That the stimulus and a reward are present at the same time could just be a coincidence. A fish that base its future decisions on a relationship that does not exist will react when it should not react, and such mistakes could be costly. Negative fitness consequences of specific actions or events are often disproportionately more severe than positive ones (Dall 2010). It is thus reasonable to assume that a fish will require a certain number of associations between the US and the CS to 'believe' that a connection really exists. In other words, it should not learn too fast!

We could compare this with scientists who must decide whether two factors are related. A certain level of statistical significance is chosen. A significance level of 5% means that on average the conclusion in 1 out of 20 cases is that a connection exists although there is none. In statistical terminology this is known as a *Type I error*. There is, however, also the risk of making the opposite kind of error; i.e. when a connection exists, but the conclusion is that there is none—this is a *Type II error*. When a higher significance level like 1% is chosen the risk of Type I error is reduced, but the risk of Type II error increases.

Fish too may make the same two types of mistakes. The costs of Type I versus Type II errors would be expected to influence how many events where the CS and the US are linked a fish must experience in order to form an association and learn. If something potentially very dangerous takes place, the fish would be expected to learn fast as not to do so could be very costly, and antipredator defences are typically acquired quickly (Kelley 2008). When a juvenile Atlantic salmon (*Salmo salar*) is presented with alarm substance from predators at the same time as lemon odour and the next time is presented only with lemon odour, the fish displays the same decrease in swimming and feeding activity and the same increase in hiding as parr presented with the alarm cue alone (Leduc et al. 2007). After just one experience ($n = 1$) the fish thus reacts as if lemon odour means danger and behaves by giving the potential danger the benefit of the doubt. Fish can also learn to avoid electric shock after only one or a few trials (Yoshida and Hirano 2010). In contrast, to make associations involving food rewards generally demands several experiences, and a cod needs 5–7 paired events in order to learn to respond to a light-CS that announces food (Nilsson et al. 2008a). In such cases, it will not cost the fish so much not to respond, and to avoid energy and opportunity costs it may be better to be certain that a connection really exists. Hence, the costs of making the wrong decision seem to influence the number of related events needed to form an association. Interestingly, physiological modifications in the body may involve similar decision rules as rules involving the brain, but the costs of a change and thereby the critical level of experience before a change occurs can differ (Box 7.2).

Box 7.2 'Learning' at Different Levels

Other types of phenotypic plasticity than learning, such as physiological adaptations, also involve costs (Murren et al. 2015) and maybe governed by decision-making rules similar to those involved in learning. The costs of modifying neural pathways should be relatively small. Modifying the performance of an organ will usually involve higher costs, and one would thus expect that a larger number of experiences will be needed.

In accordance with this, behavioural traits exhibit greater plasticity than physiological and morphological traits. If a fish occasionally needs to swim fast this should not change the general impression that it has no need for a high-performance swimming capability, and the costs of building up such capacity could be high. But with repeated challenges, it should become clear for the fish that it must become a better swimmer and its physiology will be modified. Training improves swimming performance in fish, but not until after many days of training (Sinclair et al. 2014).

Retaining a memory may also be costly, but if the cost of forgetting is high the organism should be willing to pay the price (Dukas 1999). This could also apply to mechanisms outside of the brain. Acquired immunity is a form of phenotypic plasticity (Snell-Rood 2012). Life-history theory predicts that immune responses have evolved in the context of benefits and energetic and nutritional costs, and active downregulation of the immune system can be expected if future encounters with the pathogen are unpredictable or if the negative effects of infection are small (McKean and Lazzaro 2011). The immunological memory should thus be short when the costs of maintaining the memory are higher than the benefits. Epithelial invasion interacts mostly with the immunoglobulin IgA that declines relatively rapidly, whereas systemic invasion interacts mostly with IgG with a longer half-life (Frank 2002). So perhaps all mechanisms that show some flexibility could be located along a flexibility axis based on the costs of a change versus those of not changing.

However, other factors than pure math may modify the decision rules. (1) *Irrational behaviour*. Violations of the clauses of rationality may, in fact, be accounted for by Darwinian explanations (Huneman and Martens 2017). Irrational behaviour could be a byproduct of the selection for decision patterns that are on average adaptive in most environments or that were adaptive in past environments, whereas it is more controversial if irrationality could itself be selected for its contribution to fitness. (2) *Optimistic behaviour*. An animal that does not know the exact mortality risk could act as if the risk is less than the mean risk, and this can be viewed as being optimistic (McNamara et al. 2012), just as humans are optimistic about future scenarios that are entirely determined by chance (Langer and Roth 1975; Weinstein 1980), and such ungrounded beliefs can actually be adaptive (Smithdeal 2016). (3) *Personality*. Bolder sticklebacks make faster decisions than shyer conspecifics (Mamuneas et al. 2015). (4) *State and Control system*. When a fish is highly motivated to feed, it will be risk prone and go for anything. The output of the decision-making process is determined by an interaction between impulsive or emotionally based systems responding to immediate rewards or threats and reflective or cognitive control systems controlling long-term goals (van den Bos et al. 2013). Many decisions must be made under stressful conditions, and acute stress can hamper task-processing and promote automatic, less energetic over more explicit, energy costly behaviour (van den Bos and Flik 2015). (5) *Social environment*. Decisions may be affected by social interactions and social stress (van den Bos et al. 2013).

7.3.3 Can Fish Learn from Others?

Socially provided information enables an animal to acquire information from knowledgeable ones and avoid the costs of learning solely by individual experience, and many species of fish overcome their cognitive limitations by social learning. Still, individuals within a species often differ in their tendency to learn from others, which can make comparisons between species more challenging (Mesoudi et al. 2016). Social learning in fishes is often the result of simple local enhancement or following mechanisms, but many fish exhibit more complex forms of social learning (Laland et al. 2011). Observing and copying others could help fish to find prey, improve their antipredator response, evaluate the strength of competitors and the value of sexual mates as well as follow migration routes (Brown and Laland 2003). For instance, archer fish (*Toxotes jaculatrix*) can learn their advanced ballistic hunting technique by just watching group members performing and must, therefore, be capable of mapping and subsequently employing the shooting characteristics of other fish (Schuster et al. 2006). Helfman and Schultz (1984) performed a classic demonstration of social learning in French grunts (*Haemulon flavolineatum*) and found that daily migrations between resting and feeding grounds were maintained by guided learning. Even maladaptive behaviour can spread by social learning. Groups of guppies (*Poecilia reticulata*) were trained to take an energetically expensive route although a shorter route was available, and when all founder fish had been replaced by naive individuals, the new fish still chose the longer route (Laland and Williams 1998). The appearance of the leader may influence the decisions, and sticklebacks tend to follow a healthy looking model more than a less attractive one (Sumpter et al. 2008).

Collective memories may shape fish distributions (Macdonald et al. 2018). Information can be transmitted between generations, and young herring (*Clupea harengus*) seem to learn the migration route from older, more experienced individuals (Corten 2002). Herring return to the same overwintering area for many years but may quite suddenly choose a new area (Fernö et al. 1998). The changes coincide with new strong year classes when traditional routes followed by experienced fish are not transmitted so effectively (Huse et al. 2002). To only follow leaders above a certain threshold number reduces the probability of errors being amplified throughout a group (Ward et al. 2012). The critical ratio between experienced and inexperienced herring seems to be about 5%, and changes can thus be predicted on the basis of the age structure of the population (Huse et al. 2010). The removal of knowledgeable individuals can produce rapid changes in collective behaviour (De Luca et al. 2014), and the longer recovery times of many collapsed stocks than predicted by traditional fishery population models may be due to the breakdown of socially transmitted traditions (Petitgas et al. 2010).

The extent to which a fish relies on other fish may have to do with the costs and benefits of public versus private information. Both nine-spined (*Pungitius pungitius*) and three-spined sticklebacks use public information to locate food, but only the nine-spined stickleback uses information from others to assess food-patch quality (Coolen et al. 2003). The three-spined stickleback is better protected with plates and spines and may afford to rely to a greater extent on more reliable but riskier trial and error learning.

7.3.4 Memory: Remembering or Forgetting Experiences

Learning means that a memory is retained, and fish can remember a connection for a long time. Electric yellow cichlids (*Labidochromis caeruleus*) maintained memories for moving visual patterns over a period of at least 12 days (Ingraham et al. 2016). Cod that have learnt that a light signal means food maintained the conditioned response for at least three months (Nilsson et al. 2008a), and bamboo sharks (*Chiloscyllium griseum*) trained in visual discrimination tasks remembered the learned information for a period of up to 50 weeks (Fuss and Schluessel 2015). Rainbow fish (*Melanotaenia duboulayi*) escape more rapidly from an aversive stimulus 11 months after training (Brown 2001), and hooked carp (*Carassius carassius*) avoid the hook for at least one year (Beukema 1970). Imprinting is a way to learn for life, and Atlantic salmon smolts that migrate to the sea return several years later to the same river from which they migrated based on olfactory imprinting (Scholtz et al. 1976). However, maintaining a memory is not without costs but is an active and costly process of maintenance and repair and may require a large brain volume (Dukas 1999), so an individual's memory capacity reflects a trade-off between costs and benefits. Memories of minor importance may be forgotten rapidly, while critical memories appear to be maintained (Dukas 1999). It is also crucial to adjust predictions to a changing environment, and forgetting is adaptive if information becomes outdated (Kraemer and Golding 1997). The memory window for prey in sticklebacks in a changeable marine environment is shorter (eight days) than in sticklebacks in a more stable freshwater environment (25 days, Mackney and Hughes 1995).

7.3.5 The Predictive Brain

Although an organism must solve multidimensional problems and is never completely certain about anything (Dall 2010), the whole point of learning and memory is that fish (or we) to some extent can predict the outcome of future actions and what will take place in the future. New knowledge about how brains function shows that brains are not reactive but predictive and continuously make predictions of what we will encounter next in the world, with the coding of prediction errors by neurons in different parts of the brain representing a basic mode of brain function (Barrett and Simmons 2015). Brains are thus essentially prediction machines (Clark 2013). This could change the way we look at fishes and will be discussed in depth in Chap. 9. We know little about how events that fish experience throughout their life influence how they perceive the future. Everything that happens affects the tuning of systems and hence affects future behaviour and predictions. Animals that have been living in less preferred environments are for instance more likely to interpret a stimulus as negative than animals living in preferred environments (cognitive bias, Bateson and Mather 2007).

7.4 Cognitive Abilities and Consciousness

Animals with advanced cognitive abilities should be more capable of producing modified or new behaviours, and such abilities can demand a certain degree of consciousness, which offers animals an additional tool with which to deal with complex environments (Dawkins 1998). Perception and learning do not require animals to be conscious, which requires that information is further processed. Consciousness could be described as an awareness of internal and external stimuli (see Chap. 8). Panksepp (2005) divides consciousness into *Primary consciousness* (raw sensory/perceptual feelings and internal emotional/motivational experiences), which may have provided an evolutionary platform for the emergence of more complex layers of consciousness, *Secondary consciousness* (the capacity to have thoughts about how external events relate to internal events) and *Tertiary consciousness* (thoughts about thoughts, awareness of awareness), which is probably unique to humans. The different understandings of consciousness are reflected in the divergent conclusions reached by scientists about to what degree animals are conscious.

Fish need to use their sensory input to construct perceptual models of the world in order to explain the sensory input itself and thereby try to understand their world. Although the metabolic costs of brain tissue (Kuzawa et al. 2014; Chap. 6) could place restrictions on the evolution of consciousness, fish, in fact, do things that suggest that conscious cognition has evolved. They are able to combine detailed spatial relationships to form a mental map (Rodriguez et al. 1994), learn about relationships between group members by 'eavesdropping' using information from observations of interactions between conspecifics (Dugatkin and Godin 1992), do bookkeeping during Tit-for-Tat encounters (Milinski et al. 1990), manipulate the behaviour of others by deception (Bshary 2011) and possess numerical skills that compare favourably with those of mammals (Agrillo et al. 2017). Coral reef fish indicate hidden prey to cooperative conspecific hunting partners with the signals involved possessing all the attributes that have been proposed to infer a referential gesture possessing the hallmarks of intentionality (Vail et al. 2013), and groupers (*Plectropomus pessuliferus marisrubri*) and giant moray eels (*Gymnothorax javanicus*) perform interspecific cooperative hunting involving intentional signalling (Bshary et al. 2006). Male cichlids (*Astatotilapia burtoni*) have an impressive capability to predict the fighting abilities of competitors using known relationships to deduce unknown ones (transitive inference, Grosenick et al. 2007). Sticklebacks (*Pungitius pungitius*) compare their own foraging success with the success of other individuals to update foraging decisions (Kendal et al. 2009) using a hill climbing strategy (Laland et al. 2011)—a payoff-based copying strategy that seems to be the same as in humans (Schlag 1998). A study of learning and decision-making in cod suggests that cods have the advanced ability to trade-off between a current and a future anticipated reward (Gabagambi 2008; Box 7.3). Cleaner wrasses (*Labroides dimidiatus*) even outperform chimpanzees and orangutans in a foraging task involving a choice between two actions, both of which yield identical immediate rewards, but only one of which yields an additional delayed reward (Salwiczek et al. 2012). But this does not mean that apes are stupid and fish clever. From a biological point of view, such comparisons are only relevant when we know the ecological background of the species (see below).

Box 7.3 Cod Trade-Off Between a Current and Future Anticipated Reward

Groups of juvenile cod were trained to associate a sound signal with dry food and light flashes with shrimp. Both food types were presented 10 s after the end of the signal. Dry food and shrimp were delivered on opposite sides of the tank, and both food signals were given in the centre of the tank. Other groups

(continued)

Box 7.3 (continued)
were trained to associate the sound with shrimp and the light with dry food. Shrimp, with their more powerful odour and softer texture, provide a stronger reward than dry food.

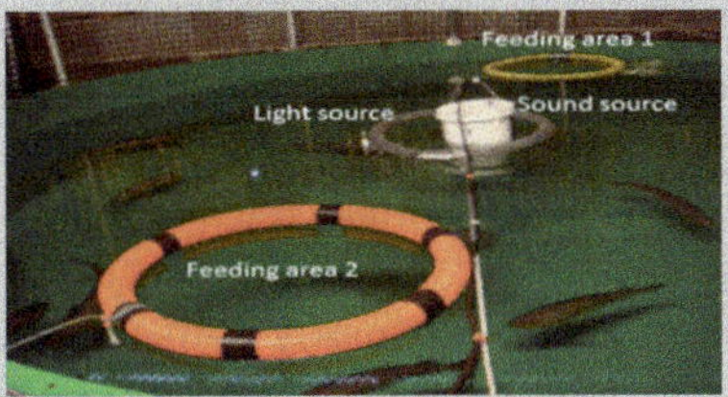

After the fish had learnt the signal–food associations, the two signal–food trials were presented together in a chain starting with the light signal followed by its associated food. Ten seconds later, while the fish were still eating, the sound signal was switched on announcing the other food type on the opposite side of the tank. Thus, the fish that were conditioned to associate the sound with shrimp were told by the sound that a higher reward would be available in the near future if they gave up the dry food and moved to the other side. And this was what they did. In contrast, for the other groups moving to the other side at the sound signal would mean giving up a high reward for a future lower one, and these fish usually ignored the sound and rather chose to stay and eat the shrimp. Thus, cod were capable of trading off between the immediate and the future reward, and gave up the current reward if the value of the future reward was higher, but not if it was lower (Gabagambi 2008). Image reproduced with permission by Nestory Peter Gabagambi.

Learning experiments may provide information on how much fish actually understand. When a CS is associated with a rewarding US such as food, the response may be directed towards the CS, a behaviour referred to as sign-tracking (Hearst and Jenkins 1974), or the response may be directed towards the location where the US reward is expected to show up, the goal, which is referred to as goal-tracking (Boakes 1977). In sign-tracking, the animal responds to the CS as if it was the US, i.e. the CS becomes a substitute for the US (stimulus substitution). Goal tracking, on the other hand, occurs when the CS is understood as an announcement signal instead of a substitute, i.e. the animal is aware of the CS–US relationship. Cod in small groups with a light CS on the opposite side of the tank to the feeder generally employ sign-tracking early in the conditioning process (Nilsson et al. 2008b). However, some individual fish initially swam towards the goal in the feeding area but eventually turned and joined the rest of the group. These individuals do not seem to behave in the best way—it would have been more rational to move directly to the goal. The reason why the individuals that had became aware of the meaning of the CS eventually joined the rest of the group to the wrong side could be that they follow leaders or that they were safer from predation in the company of others (Krause and Ruxton 2002). Later in the conditioning process, all the fish approached the goal before the arrival of food, but they did always sign-tracked to the CS first (Nilsson et al. 2008b). This may have been because they had learnt by operant conditioning and believed that it was the action of first swimming to the sign that triggered the reward. Humans do similar things. Sportsmen often perform complicated rituals before they compete. They once performed these rituals before a successful competition and to be safe they stick to what turned out to be successful. So we must not disqualify the fish because they apply sign learning!

The ultimate test of how much fish understand is if they can figure out new ways to do things. Some observations of fish that learn to do a new trick are indeed impressive. During self-feeding cod are easily trained to activate a feeder by pulling on a string with their mouth, based on initial curiosity-driven triggering rewarded by food (Nilsson and Torgersen 2010, Fig. 7.1). But some individuals learned to trigger the feeder in a quite different way by attaching the string to a tag that was attached to their dorsal side (Millot et al. 2013, Fig. 7.2). Getting the tag attached to the feeder pulley and then getting stuck first released a fright reaction, but the fish overcame this. Over time the movements became fine-tuned, and the fish attached the string to the tag by goal-directed coordinated movements, after which they swam forwards, thus activating the feeder. The original simple association between the response and the reward had by then developed into a skill representing a higher level of learning. The fish had already learnt to activate the feeder with the mouth and had thus a functional and non-aversive alternative, but an advantage with the new behaviour could be that a fish that had already triggered the feeder swam towards the pellets without first having to turn around (Fig. 7.2), during which time other fish could have eaten the pellets. The situation was totally artificial since cod do not generally use their back or fins to manipulate objects but only the mouth. The achievement may, in fact, be interpreted as *Innovative behaviour*, and only three'smart' individuals out of 56 fish learnt the trick. The behaviour could also be defined as *Tool use* with the fish using a new tool as an 'artificial limb' in a new setting. Fish lack

Fig. 7.1 Cod can be trained to activate a feeder by pulling on a string with the mouth. Photograph by Jonatan Nilsson

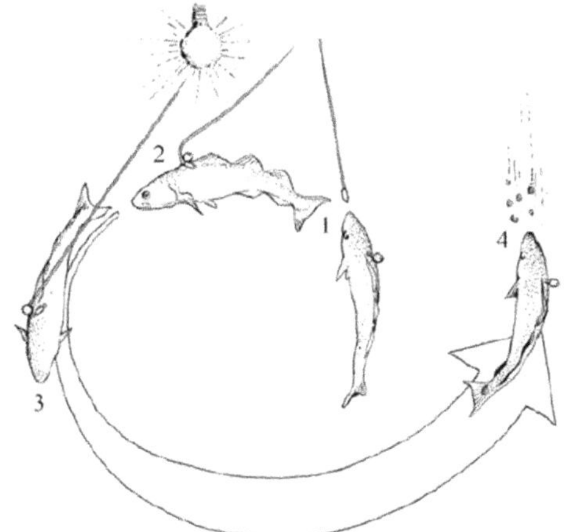

Fig. 7.2 A cod that has learnt a skill to trigger a feeder by attaching the string to a tag on the dorsal side. Reproduced from Millot et al. 2013, Fig. 1b, Animal Cognition with permission from Springer

grasping appendages, and there are clear constraints underwater with respect to the physics of tool use, and the use of tools seems generally to be confined to a limited number of fish taxa, particularly the wrasses (Brown 2012) that have been observed to use anvils to break open shellfish (Jones et al. 2011). Tool use has also been demonstrated in cartilaginous fishes, and stingrays rapidly learn to use water as a tool to extract food from a testing apparatus (Kuba ct al. 2010).

It may be argued that the complex activities fish could perform do not require any advanced cognitive abilities or understanding. Even the sophisticated interactions between cleaner fish and their clients may at least partly be explained by complex chains of key stimuli and automatic responses and simple associative learning (Bshary 2006). But although it is safe to assume that simple stimuli often trigger behaviour in fish, it is still reasonable to believe that fish have some consciousness. Conscious experience is just that something is subjectively experienced, and we must assume that also fish experience their own self-generated virtual reality (see Chap. 9). It is not a question of Either-Or. Consciousness presumably occurs on a phylogenetic sliding scale (Bekoff and Sherman 2004). Still, the cognitive abilities of fish often match or exceed other vertebrates (Brown 2015).

Many brains working together could do things that a single brain is not capable of. Individual fish can improve cognitive performance and welfare by relinquishing control to the group. Like individual brains, groups may adapt to computing 'the right thing' to do in different contexts (Couzin 2007). Swarm intelligence is defined as improved cognitive performance in groups that arises from distributed, self-organised decision-making (Ioannou 2017). Although there is no centralised control dictating how individual agents should behave, local interactions lead to the emergence of 'intelligent' global behaviour with swarms self-organising into ordered patterns. Both the speed and accuracy of decision-making increased with group size in mosquitofish (*Gambusia holbrooki*) through self-organised division of the task of vigilance combined with social information transfer (Ward et al. 2011). In golden shiners (*Notemigonus crysoleucas*) information was integrated within groups by balancing personal information and social cues, even though no individual was explicitly aware of the consensus option (Miller et al. 2013). Still, the evidence for swarm intelligence in fish is not as strong as in social insects (Ioannou 2017).

7.4.1 All Fish Do Not Have the Same Mental Capacity

Importantly, the cognitive abilities of fish depend on the environment in which they have evolved. Each species is intelligent to the extent that it successfully reproduces in an environment and adapts to changes when necessary. The question is what the added value of being aware is. There are both species and population differences in a number of cognitive traits in fish (Braithwaite et al. 2013). In cichlids (*Cichlidae*) relative brain size is positively associated with the environmental and social complexity the species encounters (van Staaden et al. 1995; Pollen et al. 2007; Bshary et al. 2014). The species trade-off neural investment in specific brain lobes depending on the environment in which they live (Kotrschal et al. 1998; White and Brown 2015), and species living in spatially complex habitats solve spatial tasks better than species in structurally simple environments. A benthic species within the three-spined stickleback complex that encounters a visually complex environment solved a maze task faster than sticklebacks from a species that live in the water column and rarely encounter fixed landmarks that can be used for spatial reference (Odling-Smee et al. 2008). Girvan and Braithwaite (1998) found differences between populations of sticklebacks in their ability to solve a spatial task, which could be related to their respective habitat, and populations of the poeciliid *Brachyraphis episcopi* from areas of low-predation pressure located a foraging patch more rapidly than fish from high-predation sites (Brown and Braithwaite 2005). In many species, males and females face different challenges that may result in sex-based cognitive differences. Guppy males that inhabit more spatially complex environments than females learned to choose between alternative routes in a complex maze after only one trial, whereas females did not learn after five trials (Lucon-Xiccato and Bisazza 2016). In a visual discrimination task, the two sexes had similar discrimination learning abilities, but decision-making speed was higher in males. There could also be relationships between personality and 'cognitive style' (how to acquire, process, store or act on information) in animals, for instance, with fast behavioural types associated with speed over accuracy as a cognitive style (Sih and Del Giudice 2012). In guppies, fast explorers made rapid, inaccurate decisions in a spatial memory task, while slow explorers took longer to make choices but were more accurate (Burns and Rodd 2008). Experience could also play a role, and experiences during early ontogeny can increase cognitive flexibility later in life (Bannier et al. 2017). Even a single environmental change can enhance cognitive abilities, and cichlid fish *Simochromis pleurospilus* subjected to a change in food ration early in life later outperformed fish kept on constant rations in a learning task later in life (Kotrschal and Taborsky 2010).

7.4.2 *Fish Compared to Humans*

The only mind you can really know is your own, which means that we can never know for certain how another organism perceives its surrounding world. However, scientists who never consider that an animal can exhibit human-like complexity can miss much of the richness of its behaviour (Bateson and Laland 2013), and demanding absolute certainty reflects a sort of double standard, as in other scientific subjects we make the best of incomplete evidence (Griffin 1998). A fish cannot tell us how it look at things and has no facial expressions that we can read-out, and its experience 'Umwelt' (von Uexküll 1921) must be very different from ours. Still, there are several similarities between the decision rules observed in fish and those in other vertebrates, including humans (Bshary et al. 2014), and fish and man may be more similar than we tend to think (Box 7.4). Some routes to action involve awareness and some do not, and most central processing is without awareness even in humans (Dawkins 2017).

Box 7.4 Are Fish and Man More Similar Than We Use to Think?
We tend to focus on differences and to ignore what is common. Imagine an organism that had evolved in a quite different world. The nervous system of such an organism could be built up in a completely different way. For example, it might do something when not stimulated and keep still when exposed to stimulation. When such an organism has initiated a certain behaviour, this could start a positive feed back loop with the animal continuing this activity forever. It could even approach danger and avoid positive stimuli. Looked upon in this way, fish and humans are rather similar.

One reason for this is that although fish and man went in different directions about 300 million years ago, we still have the same genetic roots. Functional neuroanatomy shows that the essential building blocks for spatial and emotional learning originated early in vertebrate history (Portavella et al. 2004). Another reason is that the environments in which fish and man have evolved are not so different. It is true that fish has evolved in water and man on land, and living on land posed some additional challenges, such as stronger force of gravity (maintaining posture, added risk of injury), and in the case of mammals maintaining a constant body temperature. But the basic features of these environments are in fact similar. Events take place at irregular intervals, but even so, there are certain patterns, and different events are often correlated. The environment is thus partly predictable. There are conflicts between different activities, and the presence of conspecifics opens up for the potential of cooperation as well as conflicts. Organisms adapted to similar environments will be expected to become similar in many respects. The overall structure of the behaviours and reaction patterns of fish and man are thus presumably not fundamentally different, and this may be the reason why to a certain extent we

(continued)

Box 7.4 (continued)
can imagine why and how a fish escapes or hides from a predator, when it is trading off feeding and predation and how fish interact with conspecifics to their own advantage.

7.5 Affects Giving Quality to Life

Many laymen do not believe that fish experience emotion, perhaps based on a belief that this is a more advanced ability. But to feel hungry does not appear to be particularly advanced. Hunger is a basic sensation that motivates behaviour, and when hunger becomes very strong but impossible to be mitigated, it may become endowed with negative emotions (frustration, stress). 'Emotions' refer to processes that have evolved from basic processes in the brain that enable animals to experience bad and good qualitative states (Panksepp 1994), and 'Affects' refers to behavioural and physiological responses that can vary both in terms of valence (pleasantness/unpleasantness) and intensity (Paul et al. 2005). In this chapter, we define both affects and emotions as being qualitative states of different intensity/arousal, with valence ranging from pleasure to aversion (good to bad). These terms and 'Feelings' are often used interchangeably, but we could more clearly distinguish between affects/emotions and feelings, where affects/emotions only represent an activated state, while feelings are dependent on awareness and are the conscious experiences of emotions (LeDoux 1996, 2012). For instance, the emotion fear is more primitive than the feeling of being sad. Emotional states may be accompanied by subjective feelings but this is not necessarily so (Dawkins 2008). Emotions are regarded as being functional in their own right, but they also provide the raw material for the experience of conscious feelings (Rose et al. 2014).

Affectivity goes far back in brain–mind evolution (Panksepp et al. 2017), and emotions probably serve a function in fish similar to that in other higher vertebrate species (Kittilsen 2013, see also Chap. 9). An interesting recent study by Cerqueira et al. (2017) supports that distinctive affective states occur in fish. Gilthead sea bream (*Sparus aurata*) exposed to stimuli that varied according to valence and predictability exhibited different behavioural, physiological and brain states that were characterised by the expression of early genes in brain regions homologous to regions involved in reward and aversion processing in mammals. The reward evaluating mechanisms in the brain involving neurotransmitters such as dopamine are conservative in the animal kingdom (Salas et al. 2006), and it now seems to be generally accepted that fish experience emotion. But scientists are all the same generally very careful and often use for instance the term 'Emotional value'. But is this not the same thing as emotions expressed in a seemingly neutral and economical language?

From an evolutionary point of view, it could be argued that to experience emotions is an ingenious way to make animals behave in a way that maximises survival and reproduction. The ability to perceive emotions enables an individual to assess a discrepancy between its requirements and the current environmental conditions and to avoid the bad and seeking out the good. The interplay between different emotions also offers an ideal way to solve conflicts by having a common currency permitting trade-offs during conflicts. Cabanac (1992) suggests that pleasure is the common currency that animals use to rank their priorities and needs. For instance, if hunger motivates feeding in a dangerous environment but at the same time the emotion fear motivates the need to protect oneself against predation, a fish may refrain from feeding until the hunger level is so high that the pleasure from feeding is higher than the fear of predation.

So to the million dollar question: do fish have conscious feelings of which they are aware? This is the most critical question for welfare, but it is also an extremely difficult one to answer and a controversial issue among scientists. Different scientific schools base their conclusions on different arguments. The view that fish lack conscious feelings is largely based on the fact that fish lack the neural structures that control feelings in mammals (Rose 2002; Key 2016). But this would be similar to concluding that fish cannot breathe because they have no lungs! The fish brain contains the basic structures (hippocampus, amygdala, cortical areas) which have since evolved into more elaborate structures (Broglio et al. 2005), and fish brains are in fact remarkably similar in organisation to those of other vertebrates (Bshary et al. 2014). Furthermore, Barrett (2017) claims that emotions are not housed in different parts of the brain but are constructed by systems that interact across the whole brain. Changes in the tendency to do various things, i.e. motivation, are not just the outcome of technical adjustments in a computer. Signals from the brain create waves that flow through the whole body and can dramatically change the physiology and activity of organs. Emotional experience is the brain's 'Best guess' of the causes of interoceptive signals (Seth 2013). Feelings are mental experiences of body states (Damasio and Carvalho 2013), and it would be remarkable if fish did not feel such experiences. The feeling of fear, for instance, could be the detection of a physiological signature induced by exposure to a threatening situation (Clark 2016).

Subjective interpretations of observations of fish might give us some clue. When we watch a fish we cannot read its facial expressions, but we can read its body language. Observations of fish vigorously defending their young or approaching an opponent while at the same time seemingly being frightened give us the strong impression that a robot fish should not behave in this way and that the behaviour of the fish has an emotional basis that the fish cannot be unconscious of. Of course, we could argue that such observations have no scientific value, and that a finely tuned robot would give us the same impression. But at least it explains why many scientists so strongly believe that fish experience conscious feelings.

The three approaches to the study of felt aversive emotions spontaneous responses, drug interventions and motivational tests do so far not enable us to draw definitive conclusions (Weary et al. 2017). Working from the position that animals consciously experience emotions and create test situations providing specific predictions could be a way to go (Weary et al. 2017), and reliable physiological indicators of consciousness in humans could be tested using affective paradigms

(Panksepp et al. 2017). Still, perhaps we should just admit that it is more or less impossible to definitely prove that fish experience conscious feelings. But this does not eliminate the possibility that they do! Feelings likely arise from the older regions of the brainstem (Venkatraman et al. 2017), and all vertebrates including fish are probably sensory-phenomenally and affectively conscious (Fabbro et al. 2015).

A question under debate is whether fish can feel pain and suffer. Rose (2002) and Rose et al. (2014) argue that although fish have pain receptors and react to aversive stimulation this does not mean that they experience pain because they lack the neocortex. However, the emotional brain also includes important subcortical structures (Berridge and Winkielman 2003), and Chandroo et al. (2004) and Sneddon (2011, 2015) presented strong evidence for the ability of fish to experience pain based on their cognitive capacities. Several teleost species display significant physiological and behavioural changes in response to pain over a prolonged period of time rather than instantaneous withdrawal reflex responses (see Chap. 10). Suffering maybe even more intense in simpler creatures without the mental capacities to cognitively distance themselves from their pain (Metzinger 2003). But as the importance of pain would be expected to differ based on the options available to the fish, we would expect different species to perceive pain ('the flavour of the emotion', Braithwaite et al. 2013) differently. A herring living in the pelagics can for instance not withdraw and hide if it experiences pain, so the perception of pain should be expected to be relatively weak. Eckroth et al. (2014) found that cod have a relatively weak response to different noxious stimuli including a fishing hook (but see Fernö and Huse 1983), which may be related to the eating habits including species with hard or spiky components such as crustaceans. Many sharks, skates and rays exhibit courtship biting behaviour resulting in injuries and may thus have a reduced capacity for pain, and there are in addition species-specific responses to pain in fish (see Chap. 10). There are also population differences in a number of emotional traits in fish (Braithwaite et al. 2013) and individual variations in the response to painful stimuli (Fernö and Huse 1983; Ashley et al. 2009; Weary et al. 2017).

7.6 Fish in Contact with Man-Made Stimuli

Even wild fish in the natural environment do not always do what is theoretically best. The sensory and information processing capacities of fish have limitations, their knowledge is not complete and they experience conflicts between different activities that are difficult to combine. The personality of a fish also limits behavioural plasticity and constrains its ability to act optimally in all situations (Conrad et al. 2011). However, as a starting point we assume that wild fish behave in an adaptive way. What is optimal in nature may, however, not work in a farming environment. The fish are not evolutionarily adapted to the confined environment and the extremely high densities. Some mechanisms may go wrong, the fish could learn the inappropriate types of behaviour, and they are at the mercy of collective behaviour and self-organisation (Fernö et al. 2011). The way in which animals categorise objects is preadapted and tailored to their ecological niche (Rozin and Kalat 1972), but novel objects such as fishing gear and aquaculture installations are not always easy to categorise, and learning about novel objects can produce unexpected outcomes that are not always adaptive (Fernö et al. 2011).

Wild fish that encounter fishing gear often initially categorise the novel object incorrectly, which results in maladaptive responses and the fish being caught (Fernö 1993). A cod could probably easily distinguish between a cod and a herring but will, all the same, be fooled by a metal lure. Fish may experience a net as algae that they can explore and discover too late that this is not the case. The response could, however, be modified on the basis of experience. Physical contact with baited hooks can result in learned avoidance (Fernö and Huse 1983), and fish can learn to swim through the meshes in a trawl (Özbilgin and Glass 2004). Moreover, high fishing mortality results in strong selection pressure and could thereby influence the personality of wild fish (Biro and Post 2008; Arlinghaus et al. 2017).

There are enormous variations in how different animal species react to captivity (Mason 2010), and a fish species needs to have the right personality to adapt to the farming environment. Proactive fish species may be more exploratory and 'optimistic' and less sensitive to environmental stressors than reactive species and may, therefore, be more suitable for farming (Chap. 3, Castanheira et al. 2017). How negatively an aversive stimulus is interpreted in Nile tilapia (*Oreochromis niloticus*) was found to depend on an individual's coping style with reactive individuals being more neophobic after net restraint, indicating that proactive individuals were less fearful (Martins et al. 2011). The actions of the proactive coping style are based on predictions of the environment in contrast to the reactive coping style with a more direct stimulus–response relationship (Coppens et al. 2010), and in the wild, predictable conditions are likely to favour proactive coping and unpredictable conditions reactive coping (Wingfield 2003). The predictable conditions in intensive rearing systems of Atlantic salmon may thus favour risk-taking proactive individuals (Huntingford 2004; Huntingford and Adams 2005). On the other hand, proactive individuals may have a stronger tendency to develop and follow routines (Koolhaas et al. 1999), which could be an advantage when the situation is stable as is often the case in farming, but a disadvantage if the conditions change. Proactive individuals may pay less attention to their immediate environment, and proactive rainbow trout were slower than reactive fish to alter their food-seeking behaviour in response to relocated food (Ruiz-Gomez et al. 2011), and shy but not bold brook trout continued to use cues to search for food when the environment changed (White et al. 2017). In the sole (*Solea solea*), proactive individuals tended to outcompete reactive individuals in a stable environment with feed-in excess, but the reactive ones appear to respond better in an unpredictable or variable environment (Sih et al. 2004; Mas-Muñoz et al. 2011). Coping styles are treated in more detail in Chap. 12.

Basic personality could be modified by domestication that could make fish move through the proactive–reactive axis. Genetically-based tendencies to perform different behaviours and the threshold values above which the behaviours are triggered could change within a few generations (e.g. Fernö and Järvi 1998). Selective breeding for aquaculture (e.g. Atlantic salmon) or research (e.g. zebrafish, *Danio rerio*) is usually carried out by using a set of criteria, referred to as breeding goals. Rapid growth and late sexual maturation are key goals for most farmed species, while the biomedical model species zebrafish is favoured for its short generation time. Alongside genetically driven components that directly facilitate a breeding goal, most goals are fulfilled based on various largely unknown underlying traits. For example, growth depends on the genetic composition which mitigates growth *per se*, such as levels of growth hormones (Fleming et al. 2002). But to utilise the genetic scope for growth, the fish must thrive in its environment, which is to say that adaptation is important and is promoted blindly during breeding (Huntingford 2004). After ten generations, a comparison of bred and wild-type Atlantic salmon kept under freshwater rearing conditions showed a striking 1:3 growth relationship in favour of the bred fish (Solberg et al. 2013a), whereas under natural conditions the ratio was at most 1:1.25 (Solberg et al. 2013b).

Bred Atlantic salmon are more risk prone and more socially dominant under culture conditions, while wild salmon dominate under natural conditions (Fleming and Einum 1997). Using a daily stressor regime in aquaculture tanks, wild salmon were found to be most stress sensitive, further enhancing the already high growth rate differs from that of bred fish (Solberg et al. 2013a). The effect of acute stress on feeding was similar between different families of bred salmon, but activity levels and cortisol responses differed between families and correlated positively with genetic markers for disease resistance and survival (Kittilsen et al. 2009). This highlights the importance of understanding and tailoring the genetic composition in a systems biological framework, as fish that appear to be physically similar and have the same growth rate may experience and respond to critical events quite differently. We should thus expect benefits of targeting new and more specific breeding goals to better match rearing environments and operations in aquaculture. For example, in salmon sea cages that provide a multifactorial and partly unstable environment, the fish show depth preferences along gradients of among other factors light and temperature (Oppedal et al. 2011) that seem to be genetically based (Folkedal et al., unpublished data). Such findings have a direct impact on depth-dependent sea lice infestation rate and feeding efficiency, and provide new strings to be orchestrated in selective breeding.

Learning about man-made objects and the procedures employed in aquaculture is essential for welfare—otherwise the fish will be unable to avoid aversive situations and adapt to forthcoming events. Farmed fish habituate to disturbances ('get used to it'; Grissom and Bhatnagar 2009), learn about food and feeding regimes via classical conditioning and participate in self-feeding based on operant conditioning (Fernö et al. 2011). Fish can even learn to react positively to initially aversive events. Lowering the water in the tank followed by feeding reduced the impact of subsequent handling and transportation in chinook salmon (*Oncorhynchus tshawytscha*, Schreck et al. 1995). Salmon initially reacts negatively to light flashes, but if light is associated with a food reward they learn to approach the light source (Bratland et al. 2010), and cod can learn to approach an initially aversive net if rewarded by food (Nilsson et al. 2012). Gilthead sea bream repeatedly exposed to aversive light flashes initially responded by lowering the swimming speed, in addition to escaping the light (Folkedal et al. 2018). When flashes were rewarded with food, these negative responses were eventually lost and replaced with an approach in groups, while the reduced swimming speed during flashes persisted throughout the experiment in unrewarded groups.

A farming environment deprived of critical stimuli could, however, understimulate fish and impair the capacity to learn new responses (Fernö et al. 2011). On the other hand, the high fish densities and frequent disturbances that are typical of the farming environment may be cognitively demanding, resulting in stimuli overload, and the fish could then regress to a more primitive pattern of behaviour with a shift from a high-level cognitive control of individual choices to low-level direct stimulus–response controlled schooling behaviour (Toates 2004). Unpredictable chronic stress reduces avoidance learning in zebrafish (Manuel et al. 2014). Fish can also become trapped in the collective (Fernö et al. 2011), and social behaviour in schooling salmon can overrule strong incentives such as food. Using a submerged feeding unit in a sea cage where the food pellets were available only within a one-metre vertical range, Folkedal et al. (unpublished data) found that the response of the Atlantic salmon group was all-or-none. Although the fish were hungry and could see food being available in their close vicinity, the school hesitated for up to 30 minutes before suddenly engaging in vigorous feeding activity.

It is essential for animals to prepare for forthcoming events to respond in an appropriate manner, and fish can learn to anticipate a stressor by conditioning to stimuli presented before the event (Overmier and Hollis 1990). Galhardo et al. (2011) found that the cichlid *Oreochromis mossambicus* showed a stronger stress response when subjected to an unpredictable than to a predictable confinement. However, earlier studies on the effects of predictability of aversive stimuli in animals have produced inconsistent results (Bassett and Buchanan-Smith 2007), and we could not demonstrate that the predictability of the stressor reduced the stress response of Atlantic salmon parr in tanks that were repeatedly exposed to chasing events that were either announced by a signal or not (Madaro et al. 2016). This could be due to the overall habituation of the stress response or to methodological limitations, or because predictability, in fact, has a limited effect on the stress response of salmon. The benefit of predictability of stressors may also be limited when fish have no way to avoid the stressor, as is the case in tanks. When predictability gives the opportunity to escape before danger occurs the stress response can be greatly reduced (Carpenter and Summers 2009).

Sea-ranched fish and fish escaping from aquaculture installations experience an artificial environment during the first part of their life before they encounter the natural environment, when they are faced with unfamiliar food and predators and a social environment that they have not learnt about. If the fish lack the mental tools to cope with these challenges their welfare will be impaired, and formerly stocked fish usually suffer higher natural mortality rates than wild fish (Lorenzen et al. 2012). Hatchery-reared cod show maladaptive feeding and antipredator behaviour (Kristiansen and Svåsand 1992; Steingrund and Fernö 1997; Nødtvedt et al. 1999; Meager et al. 2011). Meager et al. (2012) found multidimensional personality differences between wild and hatchery-reared cod. In hatchery fish, the proactive and reactive axes were the most important, whereas in wild fish behavioural phenotypes were more closely aligned to the boldness–shyness axis. Environmental enrichment of captive environments improves the cognitive capacity of both cod and salmon (Braithwaite and Salvanes 2005; Salvanes et al. 2013), but does not always have ecologically relevant effects (Johnsson et al. 2014). In capture-based aquaculture, the fish experience the reverse situation with a challenging transition from the wild to the farming environment (see Chap. 18).

7.7 How Can Welfare of Fish Be Evaluated?

Short-term increases and reductions in well-being ensure that fish do what is best in evolutionary terms (Spruijt et al. 2001). Organisms have not evolved to 'feel fine' regardless of the situation, and if the situation is bad a fish should feel bad. Welfare is well-being integrated over time, so short-term challenges should not impair welfare, and challenges that fish can cope with may even improve welfare. In welfare research, the traditional concept of *homeostasis*—physiological variables are kept around their 'set point'—has been replaced by *Allostasis*—maintaining stability by change (Korte et al. 2005). Both hypostimulation and allostatic overload inhibits the development of cognitive responses and functional capacity (Korte et al. 2005; McEwen and Gianaros 2011), and under-stimulated fish develop smaller brains (Marchetti and Nevitt 2003; Mayer et al. 2011). Groups of Atlantic salmon repeatedly exposed to stressful stimuli are able to cope better with subsequent aquaculture stressors compared with less exposed groups (Vindas et al. 2016).

Welfare in farmed fish is difficult to objectively define and measure. Our assessment of welfare in both fish and other animals relies on some critical assumptions. Whether we like it or not, these assumptions are based to a certain extent on our own subjective feelings. For instance, it is for instance often taken for granted that the welfare of an injured or sick fish is impaired (e.g. Dawkins 2003), but can we strictly scientifically prove that this is the case? What we do is to extrapolate from our own experience by assuming that fish feel something similar. Welfare is often used rather loosely in studies investigating production performance, but even if the fish are growing well we have no guarantee of good welfare. Wagging its tail tells us something about the welfare of a dog, but in fish this just means that the fish is moving. But behavioural and physiological indicators are important means of assessing welfare. The behaviour reflects both subjective state and the behavioural needs and is the best available tool to assess welfare (Dawkins 2004). One simple way to assess whether farmed fish have good welfare is to just watch the fish in a tank or cage. As Nobel Prize winner Niko Tinbergen wrote: Contempt for simple observation is a lethal trait in any science (Tinbergen 1963). Although there is always the risk of subjective or biased interpretations, we should not underestimate the capacity of the human brain to integrate and interpret various aspects of the observed behaviour. If the fish are swimming in a regular way without sudden reactions and frequent interactions, there is a good chance the fish are enjoying good welfare. Of course, we still need more specific quantitative behavioural indicators (Martins et al. 2012, Chap. 13).

The capacity of fish to learn opens for new ways to assess welfare. Anticipation in Pavlovian conditioning is a useful tool for assessing the balance between positive and negative affective states in the face of challenges (Spruijt et al. 2001). When Atlantic salmon parr in tanks are conditioned to food by a light signal, they respond by moving from the reference area to the CS/feeding area (Folkedal et al. 2012). However, when stressed by temperature fluctuations, hyperoxia or chasing, the parr showed no anticipation of food for longer than the time it took for the physiological responses of oxygen hyper-consumption and cortisol excretion to return to normal. Anticipation of food thus seems to be more sensitive to disturbances than other parameters, with a lack of anticipatory responses indicating impaired welfare. One way to estimate welfare could be to provoke the fish by a repeated challenge test (Box 7.5).

Box 7.5 A Challenge Test to Assess Fish Welfare

An animal that is in balance and functions well should react in an adaptive way to external events. The initial reaction of the fish to a disturbance should then not be too weak (as in sick fish) but nor should it be too strong (over-reactive, stressed fish). Furthermore, the reaction should not continue indefinitely, but wane over time and also be modified by previous experience.

A way to estimate welfare could thus be to provoke the fish by a repeated challenge test. Bratland et al. (2010) using video image analysis of the distribution of Atlantic salmon in tanks found that the fish reacted to light flashes by immediately fleeing from the feeding area. We have initiated a study to estimate day-to-day welfare in farming by locating a unit in the centre of a tank or cage, which both generates sound pulses and acoustically or visually records the behaviour, distribution and density of fish around the unit before and after stimulation.

1. The change in density close to the unit before and after the signal should give an estimate of the *Reactivity* of the fish (how fast and strongly they respond).
2. The development over time with the fish gradually returning to the area around the loudspeaker should give an estimate of the ability to *Downregulate* (calm down).

Repeated stimulations could provide information on the ability to habituate and learn, but too frequent stimulations probably reduce the response so much that the technique is used up as a welfare indicator. The challenge is also to resolve the distributional changes in sufficient detail.

7.8 Conclusions

When we approach a question scientifically we must rely on defined terms and concepts, but scientists studying the mental capacities and welfare of fish sometimes seem to try too hard to fit their observations into previously defined concepts. The terms used to analyse intelligence and feelings are not as straightforward as categorising the different parts of a plant into root, stem and leaves, but represent abstract concepts that are not easily defined. Categories such as memories, emotions or 'the self' do not correspond to brain organisation in a one-to-one fashion, and even the distinction between categories such as emotion and cognition is relative (Barrett 2009). After all, it is not easy to ask a fish if a concept is relevant for how it experiences the world. So we must be humble and not push our conceptions and reference systems onto other living beings. A way forward could be to base our research on more objective and testable concepts such as the ability to make predictions (see Chap. 9).

Fish must interpret the multitude of stimuli they sense and categorise them. Learning enables a fish to adapt its behaviour to the dynamic environment and predict what will happen. The costs of acting based on a relationship that does not exist versus the costs of missing an existing relationship seem to influence how fast fish learn. Fish seem generally to have evolved high cognitive abilities, but the scope of these abilities depends on the environmental and social complexity they encounter, and the cognitive and learning capacity differs between species as well as between populations, coping styles and sexes.

It is reasonable to believe that fish have at least a simple form of consciousness and have emotions and feelings, although their way of experiencing this may be different in kind and degree from the human experience and also to vary between fish species. Rose et al. (2014) and many others argue that we should apply a function-based approach (an animal's ability to adapt to its present environment; Huntingford et al. 2006) to fish welfare instead of a feeling-based approach. But should we really have to care about what we expose fish to if they totally lack conscious feelings (Torgersen et al. 2011, but see Dawkins 2017)? As in Court, the question is: Who is responsible for coming up with the evidence. We feel that this is up to the scientific school which claims that fish are mindless robots that react to external stimuli in a merely reflexive way. This view contrasts with many observations, and what we generally do when we lack complete knowledge is to adopt a precautionary approach.

Man-made objects are novel objects that can be difficult to categorise. How well fish cope with the situation in farming may differ between species and personality types, and proactive species and individuals are likely to generally adapt better to the situation. In Chap. 15, Thomas Torgersen makes the interesting suggestion that what to a large extent could determine the welfare of a certain species in aquaculture is whether it is able to actively respond to maladaptive conditions in the wild by swimming away. If it is, their welfare may be impaired in a farming situation, in which the fish usually have to live under suboptimal conditions, while species with limited possibilities to do something about an undesirable situation in the wild may accept the situation in the farm and adapt their physiology to their current suboptimal environment.

Acknowledgements We are grateful to Dr. Ruud van den Bos who provided very useful suggestions based on an earlier version of this chapter.

References

Agrillo C, Miletto Petrazzini ME, Bisazza A (2017) Numerical abilities in fish: a methodological review. Behav Process 141:161–171

Arlinghaus R, Laskowski KL, Alós J, Klefoth T, Monk CT, Nakayama S, Schröder A (2017) Passive gear-induced timidity syndrome in wild fish populations and its potential ecological and managerial implications. Fish Fish 18:360–373

Ashley PJ, Ringrose S, Edwards KL, Wallington E, McCrohan CR, Sneddon LU (2009) Effect of noxious stimulation upon antipredator responses and dominance status in rainbow trout. Anim Behav 77:403–410

Bannier F, Tebbich S, Taborsky B (2017) Early experience affects learning performance and neophobia in a cooperatively breeding cichlid. Ethology 123:712–723

Barrett LF (2009) The future of psychology: Connecting mind to brain. Perspect Psychol Sci 4:326–339

Barrett LF (2017) How emotions are made: The secret life of the brain. Houghton Mifflin Harcourt Publishing Company, New York

Barrett LF, Simmons WK (2015) Interoceptive predictions in the brain. Nat Rev Neurosci 16:419–429

Bassett L, Buchanan-Smith HM (2007) Effects of predictability on the welfare of captive animals. Appl Anim Behav Sci 102:223–245

Bateson P, Laland KN (2013) Tinbergen's four questions: an appreciation and an update. Trends Ecol Evol 28:712–718

Bateson M, Mather M (2007) Performance on a categorization task suggests that removal of environmental enrichment induces "pessimism" in captive European starlings. Anim Welf 16:33–36

Bekoff M, Sherman PW (2004) Reflections on animal selves. Trends Ecol Evol 19:176–180

Berridge KC, Winkielman P (2003) What is an unconscious emotion? (The case for unconscious "liking"). Cognit Emot 17:181–211

Beukema JJ (1970) Angling experiments with carp: decreased catchability through one trial learning. Netherlands J Zool 20:81–92

Biro PA, Post JR (2008) Rapid depletion of genotypes with fast growth and bold personality traits from harvested fish populations. Proc Natl Acad Sci 105:2919–2922

Bitterman ME (1975) The comparative analysis of learning. Science 188:699–709

Boakes RA (1977) Performance on learning to associate a stimulus with positive reinforcement. In: Operant-Pavlovian interactions. Erlbaum, Hillsdale, NJ, pp 67–97

Braithwaite VA, Salvanes AGV (2005) Environmental variability in early rearing environment generates behaviourally flexible cod: implications for rehabilitating wild populations. Proc R Soc Lond Ser B 272:1107–1113

Braithwaite VA, Huntingford F, van den Bos R (2013) Variation in emotion and cognition among fishes. J Agric Environ Ethics 26:7–23

Bratland S, Stien L, Braithwaite VA, Juell J-E, Folkedal O, Nilsson J, Oppedal F, Fosseidengen JE, Kristiansen TS (2010) From fright to anticipation: using aversive light stimuli to investigate

reward conditioning in large groups of Atlantic salmon (*Salmo salar*). Aquacult Int 18:991–1001
Broglio C, Gomez A, Duran E, Ocana FM, Jimenez-Moya F, Rodriguez F, Salas C (2005) Hallmarks of a common forebrain vertebrate plan: Specialized pallial areas for spatial, temporal and emotional memory in actinopterygian fish. Brain Res Bull 66:277–281
Brown C (2001) Familiarity with the test environment improves escape responses in the crimson spotted rainbowfish, *Melanotaenia duboulayi*. Anim Cogn 4:109–113
Brown C (2012) Tool use in fishes. Fish Fish 13:105–115
Brown C (2015) Fish intelligence, sentience and ethics. Anim Cogn 18:1–17
Brown C, Braithwaite VA (2005) Effects of predation pressure on the cognitive ability of the poeciliid *Brachyraphis episcope*. Behav Ecol 16:482–487
Brown C, Laland KN (2003) Social learning in fishes: a review. Fish Fish 4:280–288
Brown C, Laland K, Krause J (2011) Fish cognition and behavior, 2nd edn. Wiley-Blackwell, Oxford
Bshary R (2006) Machiavellian intelligence in fishes. In: Fish cognition and behavior. Blackwell, Oxford, pp 223–242
Bshary R (2011) Machiavellian intelligence in fishes. In: Fish cognition and behavior, 2nd edn. Wiley-Blackwell, Oxford, pp 277–297
Bshary R, Hohner A, Ait-el-Djoudi K, Fricke H (2006) Interspecific communicative and coordinated hunting between groupers and giant moray eels in the red sea. PLoS Biol 4:2393–2398
Bshary R, Gingins S, Vail AL (2014) Social cognition in fishes. Trends Cogn Sci 18:465–471
Bull HO (1928) Studies on conditioned responses in fishes. J Mar Biol Assoc U K 15:485–533
Burns JG, Rodd FH (2008) Hastiness, brain size and predation regime affect the performance of wild guppies in a spatial memory task. Anim Behav 76:911–922
Butler J (1736) The analogy of religion, natural and revealed, to the constitution and course of nature. J.J. & P. Knapton, London
Cabanac M (1992) Pleasure: the common currency. J Theor Biol 155:173–200
Carpenter RE, Summers CH (2009) Learning strategies during fear conditioning. Neurobiol Learn Mem 91:415–423
Castanheira MF, Conceicão LEC, Millot S, Rey S, Bégout M-L, Damsgård B, Kristiansen T, Höglund E, Øverli Ø, Martins CIM (2017) Coping styles in farmed fish: consequences for aquaculture. Rev Aquacult 9:23–41
Cerqueira M, Millot S, Castanheira MF, Félix AS, Silva T, Oliveira GA, Oliveira CC, Martins CIM, Oliveira RF (2017) Cognitive appraisal of environmental stimuli induces emotion-like states in fish. Sci Rep 7, article number 13181
Chandroo KP, Duncan IJH, Moccia RD (2004) Can fish suffer? Perspectives on sentience, pain, fear and stress. Appl Anim Behav Sci 86:225–250
Clark A (2013) Whatever next? Predictive brains, situated agents, and the future of cognitive science. Behav Brain Sci 36:1–73
Clark A (2016) Surfing uncertainty: prediction, action, and the embodied mind. Oxford University Press, New York
Clark RE, Squire LR (1998) Classical conditioning and brain systems: a key role for awareness. Science 280:77–81
Clark RE, Squire LR (1999) Human eyeblink classical conditioning: Effects of manipulating awareness of the stimulus contingencies. Psychol Sci 10:14–18
Coble DW, Farabee GB, Anderson RO (1985) Comparative learning ability of selected fishes. Can J Fish Aquat Sci 42:791–796
Conrad JL, Weinersmith KL, Brodin T, Saltz JB, Sih A (2011) Behavioural syndromes in fishes: a review with implications for ecology and fisheries management. J Fish Biol 78:395–435
Coolen I, van Bergen Y, Day RL, Laland KN (2003) Species difference in adaptive use of public information in sticklebacks. Proc R Soc Lond Ser B 270:2413–2419
Coppens CM, de Boer SF, Koolhaas JM (2010) Coping styles and behavioural flexibility: towards underlying mechanisms. Philos Trans R Soc B Biol Sci 365:4021–4028

Corten A (2002) The role of "conservatism" in herring migrations. Rev Fish Biol Fish 11:339–361

Couzin ID (2007) Collective minds. Nature 445:715

Croy MI, Hughes RN (1991) The role of learning and memory in the feeding behaviour of the fifteen-spined stickleback, *Spinachia spinachia* L. Anim Behav 41:149–159

Dall SRX (2010) Managing risk: the perils of uncertainty. In: Evolutionary behavioral ecology. Oxford University Press, New York, pp 194–206

Damasio A, Carvalho GB (2013) The nature of feelings: evolutionary and neurobiological origins. Nat Rev Neurosci 14:143–152

Dawkins MS (1998) Evolution and animal welfare. Q Rev Biol 73:305–328

Dawkins MS (2003) Behaviour as a tool in the assessment of animal welfare. Zoology 106:383–387

Dawkins MS (2004) Using behaviour to assess animal welfare. Anim Welf 13:3–7

Dawkins MS (2008) The science of animal suffering. Ethology 114:937–945

Dawkins MS (2017) Animal welfare with and without consciousness. J Zool 301:1–10

De Luca G, Mariani P, MacKenzie BR, Marsili M (2014) Fishing out collective memory of migratory schools. J R Soc Interface 11:20140043

Dugatkin LA, Alfieri MS (2003) Boldness, behavioral inhibition and learning. Ethol Ecol Evol 15:43–49

Dugatkin LA, Godin J-GJ (1992) Reversal of female mate choice by copying in the guppy *Poecilia reticulata*. Philos Trans R Soc B Biol Sci 249:179–184

Dukas R (1999) Costs of memory: ideas and predictions. J Theor Biol 197:41–50

Eckroth JR, Aas-Hansen Ø, Sneddon LU, Bichão H, Døving KB (2014) Physiological and behavioural responses to noxious stimuli in the Atlantic cod (*Gadus morhua*). PLoS One 9: e100150

Fabbro F, Aglioti SM, Bergamasco M, Clarici A, Panksepp J (2015) Evolutionary aspects of self- and world consciousness in vertebrates. Front Hum Neurosci 9, article number157

Fernö A (1993) Advances in understanding of basic behaviour - consequences for fish capture. ICES Mar Sci Symp 196:5–11

Fernö A, Huse I (1983) The effect of experience on the behaviour of cod (*Gadus morhua L.*) towards a baited hook. Fish Res 2:19–28

Fernö A, Järvi T (1998) Domestication genetically alters the anti-predator behaviour of anadromous brown trout (*Salmo trutta*) - a dummy predator experiment. Nord J Freshw Res 74:95–100

Fernö A, Pitcher TJ, Melle V, Nøttestad L, Mackinson S, Hollingworth C, Misund OA (1998) The challenge of the herring in the Norwegian Sea: making optimal collective spatial decisions. Sarsia 83:149–167

Fernö A, Huse G, Jakobsen PJ, Kristiansen TS, Nilsson J (2011) Fish behaviour, learning, aquaculture and fisheries. In: Fish Cogn Behav, 2nd edn. Wiley-Blackwell, Oxford, pp 359–404

Fleming IA, Einum S (1997) Experimental tests of genetic divergence of farmed from wild Atlantic salmon due to domestication. ICES J Mar Sci 54:1051–1063

Fleming IA, Agustsson T, Finstad B, Johnsson JI, Björnsson BT (2002) Effects of domestication on growth physiology and endocrinology of Atlantic salmon (*Salmo salar*). Can J Fish Aquat Sci 59:1323–1330

Folkedal O, Stien LH, Torgersen T, Oppedal F, Olsen RE, Fosseidengen JE, Braithwaite VA, Kristiansen TS (2012) Food anticipatory behaviour as an indicator of stress response and recovery in Atlantic salmon post-smolt after exposure to acute temperature fluctuation. Physiol Behav 105:350–356

Folkedal O, Fernö A, Nederlof MAJ, Fosseidengen JE, Cerqueira M, Olsen RE, Nilsson J (2018) Habituation and conditioning in gilthead sea bream (*Sparus aurata*): Effects of aversive stimuli, reward and social hierarchies. Aquac Res 49:335–340

Frank SA (2002) Immunology and evolution of infectious disease. Princetown University Press, Princetown and Oxford

Fuss T, Schluessel V (2015) Something worth remembering: Visual discrimination in sharks. Anim Cogn 18:463–471

Gabagambi PN (2008) Learning ability in juvenile Atlantic cod (*Gadus morhua* L.): different CS-US relationships and reward value. Master of Science Thesis, University of Bergen
Galhardo L, Vital J, Oliveira RF (2011) The role of predictability in the stress response of a cichlid fish. Physiol Behav 102:367–372
Garcia J, Koelling RA (1966) Relation of cue to consequence in avoidance learning. Psychon Sci 4:123–124
Gibson RN (2005) The behaviour of flatfishes. In: Flatfishes: biology and exploitation. Blackwell, Oxford, pp 213–239
Girvan JR, Braithwaite VA (1998) Population differences in spatial learning in three-spined sticklebacks. Proc R Soc Lond Ser B 265:913–918
Griffin DR (1998) From cognition to consciousness. Anim Cogn 1:3–16
Grissom N, Bhatnagar S (2009) Habituation to repeated stress: get used to it. Neurobiol Learn Mem 92:215–224
Grosenick L, Clement TS, Fernald RD (2007) Fish can infer social rank by observation alone. Nature 445:429–432
Hearst E, Jenkins HM (1974) Sign-tracking: the stimulus-reinforcer relation and directed actions. Monograph of the Psychonomic Society, Austin, TX
Helfman GS, Schultz ET (1984) Social tradition of behavioural traditions in a coral reef fish. Anim Behav 32:379–384
Huneman P, Martens J (2017) The behavioural ecology of irrational behaviours. Hist Philos Life Sci 39, article number 23
Huntingford FA (2004) Implications of domestication and rearing conditions for the behaviour of cultivated fishes. J Fish Biol 65:122–142
Huntingford FA, Adams CE (2005) Behavioural syndromes in farmed fish: implications for production welfare. Behaviour 142:1207–1221
Huntingford FA, Wright PJ (1992) Inherited population differences in avoidance conditioning in three-spined sticklebacks, *Gasterosteus aculeatus*. Behaviour 122:264–273
Huntingford FA, Adams C, Braithwaite VA, Kadri S, Pottinger TG, Sandoe P, Turnbull JF (2006) Current issues in fish welfare. J Fish Biol 68:332–372
Huse G, Railsback S, Fernö A (2002) Modelling changes in migration pattern of herring: collective behaviour and numerical domination. J Fish Biol 60:571–582
Huse G, Fernö A, Holst J (2010) Establishment of novel wintering areas in herring co-occurs with peaks in the 'first time/repeat spawner' ratio. Mar Ecol Prog Ser 409:189–198
Ibrahim AA, Huntingford FA (1992) Experience of natural prey and feeding efficiency in 3-spined sticklebacks (*Gasterosteus aculeatus* L.). J Fish Biol 41:619–625
Ingraham E, Anderson ND, Hurd PL, Hamilton TJ (2016) Twelve-day reinforcement-based memory retention in African cichlids (*Labidochromis caeruleus*). Front Behav Neurosci 10:157
Ioannou CC (2017) Swarm intelligence in fish? The difficulty in demonstrating distributed and self-organised collective intelligence in (some) animal groups. Behav Process 141:141–151
Johnsson JI, Brockmark S, Näslund J (2014) Environmental effects on behavioural development consequences for fitness of captive-reared fishes in the wild. J Fish Biol 85:1946–1971
Jones A, Brown C, Gardener S (2011) Tool use in the spotted tuskfish, *Choerodon schoenleinii*. Coral Reefs 30:865
Kelley JL (2008) Assessment of predation risk by prey fishes. In: Fish behaviour. Science Publishers, Enfield, NH, pp 269–301
Kendal JR, Rendall L, Pike TW, Laland KN (2009) Nine-spined sticklebacks deploy a hill-climbing social learning strategy. Behav Ecol 20:238–244
Key B (2016) Why fish do not feel pain. Anim Sentience 1(1)
Kittilsen S (2013) Functional aspects of emotions in fish. Behav Process 100:153–159
Kittilsen S, Ellis T, Schjolden J, Braastad BO, Øverli Ø (2009) Determining stress-responsiveness in family groups of Atlantic salmon (*Salmo salar*) using non-invasive measures. Aquaculture 298:146–152

Koolhaas JM, Korte SM, de Boer SF, van der Vegt BJ, van Reenen CG, Hopster H, de Jong IC, Ruis MAW, Blokhuis HJ (1999) Coping styles in animals: current status in behavior and stress-physiology. Neurosci Biobehav Rev 23:925–935
Korte SM, Koolhaas JM, Wingfield JC, McEwen BS (2005) The Darwinian concept of stress: benefits of allostasis and costs of allostatic load and the trade-offs in health and disease. Neurosci Biobehav Rev 29:3–38
Kotrschal A, Taborsky B (2010) Environmental change enhances cognitive abilities in fish. PLoS Biol 8:e1000351
Kotrschal K, van Staaden MJ, Huber R (1998) Fish brains: evolution and environmental relationships. Rev Fish Biol Fish 8:373–408
Kraemer PJ, Golding JM (1997) Adaptive forgetting in animals. Psychon Bull Rev 4:480–491
Krause J, Ruxton GD (2002) Living in groups. Oxford University Press, Oxford
Kristiansen TS, Svåsand T (1992) Comparative analysis of stomach contents of cultured and wild cod, *Gadus morhua* L. Aquacult Fish Manag 23:661–668
Kuba MJ, Byrne RA, Burghardt GM (2010) A new method for studying problem solving and tool use in stingrays (*Potamotrygon castexi*). Anim Cogn 13:507–513
Kuzawa CW, Chugani HT, Grossman LI, Lipovich L, Muzik O, Hof PR, Wildman DE, Sherwood CC, Leonard WR, Lange N (2014) Metabolic costs and evolutionary implications of human brain development. Proc Natl Acad Sci 111:13010–13015
Laland KN, Williams K (1998) Social transmission of maladaptive information in the guppy. Behav Ecol 9:493–499
Laland KN, Atton N, Webster MM (2011) From fish to fashion: experimental and theoretical insights into the evolution of culture. Philos Trans R Soc Lond B Biol Sci 366:958–968
Langer EJ, Roth J (1975) Heads I win, tails its chance – illusion of control as a function of sequence of outcomes in a purely chance task. J Pers Soc Psychol 32:951–955
LeDoux J (1996) The emotional brain. The mysterious underpinnings of emotional life. Simon & Schuster, New York
LeDoux J (2012) Rethinking the emotional brain. Neuron 73:653–676
Leduc AOHC, Roh E, Breau C, Brown GE (2007) Learned recognition of a novel odour by wild juvenile Atlantic salmon, *Salmo salar*, under fully natural conditions. Anim Behav 73:471–477
Lieberman DA (1990) Learning, behaviour and cognition, 3rd edn. Wadsworth, Belmont, CA
Little EE (1977) Conditioned aversion to amino acid flavors in the catfish, *Ictalurus punctatus*. Physiol Behav 19:743–747
Lorenzen K, Beveridge M, Mangel M (2012) Cultured fish: integrative biology and management of domestication and interactions with wild fish. Biol Rev 87:639–660
Lucon-Xiccato T, Bisazza A (2016) Male and female guppies differ in speed but not in accuracy in visual discrimination learning. Anim Cogn 19:733–744
Macdonald JI, Logemann K, Krainski ET, Sigurdsson T, Beale CM, Huse G, Hjøxllo SS, Marteinsdóttir G (2018) Can collective memories shape fish distributions? a test, linking space-time occurrence models and population demographics. Ecography 41:938–957
MacKay B (1974) Conditioned food aversion produced by toxicosis in Atlantic cod. Behav Biol 12:347–355
Mackney PA, Hughes RN (1995) Foraging behaviour and memory window in sticklebacks. Behaviour 132:1241–1253
Madaro A, Fernö A, Kristiansen TS, Olsen RE, Gorissen M, Flik G, Nilsson J (2016) Effect of predictability on the stress response to chasing in Atlantic salmon (*Salmo salar* L.) parr. Physiol Behav 153:1–6
Mamuneas D, Spence AJ, Manica A, King AJ (2015) Bolder stickleback fish make faster decisions, but they are not less accurate. Behav Ecol 26:91–96
Manteifel YB, Karelina MA (1996) Conditioned food aversion in the goldfish, *Carassius auratus*. Comp Biochem Physiol 115A:31–35

Manuel R, Gorissen M, Roca CP, Zethof J, van de Vis H, Flik G, van den Bos R (2014) Inhibitory avoidance learning in zebrafish (Danio rerio): effects of shock intensity and unraveling differences in task performance. Zebrafish 11:341–352

Marchetti MP, Nevitt GA (2003) Effects of hatchery rearing on brain structures of rainbow trout, *Oncorhynchus mykiss*. Environ Biol Fish 66:9–14

Martins CIM, Silva PIM, Conceição LEC, Costas B, Höglund E, Øverli Ø, Schrama JW (2011) Linking fearfulness and coping styles in fish. PLoS One 6:e28084

Martins CI, Galhardo L, Noble C, Damsgard B, Spedicato MT, Zupa W, Beauchaud M, Kulczykowska E, Massabuau JC, Carter T, Planellas SR, Kristiansen T (2012) Behavioural indicators of welfare in farmed fish. Fish Physiol Biochem 38:17–41

Mas-Muñoz J, Komen H, Schneider O, Visch SW, Schrama JW (2011) Feeding behaviour, swimming activity and boldness explain variation in feed intake and growth of sole (*Solea solea*) reared in captivity. PLoS One 6:e21393

Mason GJ (2010) Species differences in responses to captivity: stress, welfare and the comparative method. Trends Ecol Evol 25:713–721

Mayer I, Meager JJ, Skjæraasen JE, Rodewald P, Sverdrup G, Fernö A (2011) Domestication causes rapid changes in heart and brain morphology in Atlantic cod (*Gadus morhua*). Environ Biol Fish 92:181–186

McEwen BS, Gianaros PJ (2011) Stress- and allostasis-induced brain plasticity. Annu Rev Med 62:431–445

McKean KA, Lazzaro BP (2011) The costs of immunity and the evolution of immunological defense mechanisms. In: Mechanisms of life history evolution. Oxford University Press, Oxford, pp 299–310

McNamara JM, Trimmer PC, Houston AI (2012) It is optimal to be optimistic about survival. Biol Lett 8:516–519

Meager JJ, Rodewald P, Domenici P, Fernö A, Järvi T, Skjæraasen JE, Sverdrup GK (2011) Behavioural responses of hatchery-reared and wild cod (*Gadus morhua* L.) to mechano-acoustic stimuli. J Fish Biol 78:1437–1450

Meager JJ, Fernö A, Skaeraasen JE, Järvi T, Rodewald P, Sverdrup G, Winberg S, Mayer I (2012) Multidimensionality of behavioural phenotypes in Atlantic cod, *Gadus morhua*. Physiol Behav 106:462–470

Meager JJ, Fernö A, Skjæraasen JE (2018) The behavioural diversity of Atlantic cod: insights into variability within and between individuals. Rev Fish Biol Fish 28:153–176

Mesoudi A, Chang L, Dall SRX, Thornton A (2016) The evolution of individual and cultural variation in social learning. Trends Ecol Evol 31:215–225

Metzinger T (2003) Being no one. the self-model theory of subjectivity. The MIT Press, Cambridge

Milinski M, Kulling D, Kettler R (1990) Tit for tat: stickleback, *Gasterosteus aculeatus*, trusting a cooperative partner. Behav Ecol 1:7–11

Miller N, Garnier S, Hartnett AT, Couzin ID (2013) Both information and social cohesion determine collective decisions in animal groups. Proc Natl Acad Sci 110:5263–5268

Millot S, Nilsson J, Fosseidengen JE, Bégout M-L, Fernö A, Braithwaite VA, Kristiansen TS (2013) Innovative behaviour in fish: Atlantic cod can learn to use an external tag to manipulate a self-feeder. Anim Cogn 17:779–785

Murren CJ, Auld JR, Callahan H, Ghalambor CK, Handelsman CA, Heskel MA, Kingsolver JG, Maclean HJ, Mase J, Maughan H, Pfennig DW, Relyea RA, Seiter S, Snell-Rood E, Steiner UK, Schlichting CD (2015) Constraints on the evolution of phenotypic plasticity: limits and costs of phenotype and plasticity. Heredity 115:293–301

Nieuwenhuys R, ten Donkelaar HJ, Nicholson C (1998) The central nervous system of vertebrates. Springer, Heidelberg

Nilsson J, Torgersen T (2010) Exploration and learning of demand-feeding in Atlantic cod (*Gadus morhua*). Aquaculture 306:384–387

Nilsson J, Kristiansen TS, Fosseidengen JE, Fernö A, van den Bos R (2008a) Learning in cod (*Gadus morhua*): long trace interval retention. Anim Cogn 11:215–222

Nilsson J, Kristiansen TS, Fosseidengen JE, Fernö A, van den Bos R (2008b) Sign- and goal-tracking in Atlantic cod (*Gadus morhua*). Anim Cogn 11:651–659
Nilsson J, Kristiansen TS, Fosseidengen JE, Stien LH, Fernö A, van den Bos R (2010) Learning and anticipatory behaviour in a "sit-and-wait" predator: The Atlantic halibut. Behav Process 83:257–266
Nilsson J, Stien LH, Fosseidengen JE, Olsen RE, Kristiansen TS (2012) From fright to anticipation: Reward conditioning versus habituation to a moving dip net in farmed Atlantic cod (*Gadus morhua*). Appl Anim Behav Sci 138:118–124
Nødtvedt M, Fernö A, Gjøsæter J, Steingrund P (1999) Anti-predator behaviour of hatchery-reared and wild juvenile Atlantic cod (*Gadus morhua* L.) and the effect of predator training. In: Stock enhancement and sea ranching. Fishing News Books, Blackwell Publishing Ltd, Oxford, pp 350–362
Odling-Smee L, Braithwaite VA (2003) The role of learning in fish orientation. Fish Fish 4:235–246
Odling-Smee LC, Boughman JW, Braithwaite VA (2008) Sympatric species of threespine stickleback differ in their performance in a spatial learning task. Behav Ecol Sociobiol 62:1935–1945
Odling-Smee L, Simpson SD, Braithwaite VA (2011) The role of learning in fish orientation. In: Fish cognition and behavior, 2nd edn. Wiley-Blackwell, Oxford, pp 166–185
Oppedal F, Dempster T, Stien LH (2011) Environmental drivers of Atlantic salmon behaviour in sea-cages: a review. Aquaculture 311:1–18
Overmier JB, Hollis KL (1990) Fish in the think tank: learning, memory and integrated behavior. In: Neurobiology of comparative cognition. Lawrence Erlbaum Associates, Hillsdale, pp 204–236
Özbilgin H, Glass CW (2004) Role of learning in mesh penetration behaviour of haddock (*Melanogrammus aeglefinus*). ICES J Mar Sci 61:1190–1194
Panksepp J (1994) Evolution constructed the potential for subjective experience within the neurodynamics of the mammalian brain. In: The nature of emotion: fundamental questions. Oxford University Press, New York, pp 396–399
Panksepp J (2005) Affective consciousness: core emotional feelings in animals and humans. Conscious Cogn 14:30–80
Panksepp J, Lane RD, Solmes M, Smith R (2017) Reconciling cognitive and affective neuroscience perspectives on the brain basis of emotional experience. Neorosci Biobehav Rev 76:187–215
Paul ES, Harding EJ, Mendl M (2005) Measuring emotional processes in animals: the utility of a cognitive approach. Neurosci Biobehav Rev 29:469–491
Petitgas P, Secor DH, McQuinn I, Huse G, Lo N (2010) Stock collapses and their recovery: mechanisms that establish and maintain life-cycle closure in space and time. ICES J Mar Sci 67:1841–1848
Pollen AA, Dobberfuhl AP, Scace J, Igulu MM, Renn SCP, Shumway CA, Hofmann HA (2007) Environmental complexity and social organization sculpt the brain in Lake Tanganyikan cichlid fish. Brain Behav Evol 70:21–39
Portavella M, Torres B, Salas C (2004) Avoidance response in goldfish: Emotional and temporal involvement of medial and lateral telencephalic pallium. J Neurosci 24:2335–2342
Rankin CH, Abrams T, Barry RJ, Bhatnagar S, Clayton DF, Colombo J, Coppola G, Geyer MA, Glanzman DL, Marsland S, McSweeney FK, Wilson DA, Wum C-F, Thompson RF (2009) Habituation revisited: an updated and revised description of the behavioral characteristics of habituation. Neurobiol Learn Mem 92:135–138
Rescorla RA (1966) Predictability and number of pairings in Pavlovian fear conditioning. Psychon Sci 4:383–384
Rodriguez F, Duran E, Vargas JP, Torres B, Salas C (1994) Performance of goldfish trained in allocentric and egocentric maze procedures suggests the presence of a cognitive mapping system in fishes. Anim Learn Behav 22:409–420
Rose JD (2002) The neurobehavioral nature of fishes and the question of awareness and pain. Rev Fish Sci 10:1–38

Rose JD, Arlinghaus R, Cooke SJ, Diggles BK, Sawynok W, Stevens ED, Wynne CDL (2014) Can fish really feel pain? Fish Fish 15:97–133

Rozin P, Kalat J (1972) Learning as a situation-specific adaption. In: Biological boundaries of learning. Appelton, New York, pp 66–97

Ruiz-Gomez ML, Huntingford FA, Øverli Ø, Thörnqvist P-O, Höglund E (2011) Response to environmental change in rainbow trout selected for divergent stress coping styles. Physiol Behav 102:317–322

Salas C, Broglio C, Durán E, Gómez A, Ocaña FM, Jiménez-Moya F, Rodríguez F (2006) Neuropsychology of learning and memory in teleost fish. Zebrafish 3:157–171

Salvanes AGV, Moberg O, Ebbesson LOE, Nilsen TO, Jensen KH, Braithwaite VA (2013) Environmental enrichment promotes neural plasticity and cognitive ability in fish. Proc R Soc Lond Ser B 280:1–7

Salwiczek LH, Pretot L, Demarta L, Proctor D, Essler J, Pinto AI, Wismer S, Stoinski T, Brosnan SF, Bshary R (2012) Adult cleaner wrasse outperform capuchin monkeys, chimpanzees and orangutans in a complex foraging task derived from cleaner – client reef fish cooperation. PLoS One 7:e49068

Schlag KH (1998) Why imitate and if so, how? A boundedly rational approach to multi-armed bandits. J Econ Theory 78:130–156

Scholtz AT, Horrall RM, Cooper JC, Hasler AD (1976) Imprinting to chemical cues: the basis for home stream selection in salmon. Science 192:1247–1249

Schreck CB, Jonsson L, Feist G, Reno P (1995) Conditioning improves performance of juvenile Chinook salmon, *Oncorhynchus tshawytscha*, to transportation stress. Aquaculture 135:99–110

Schuster S, Wöhl S, Griebsch M, Klostermeier I (2006) Animal cognition: How archer fish learn to down rapidly moving targets. Curr Biol 16:378–383

Seth AK (2013) Interoceptive inference, emotion, and the embodied self. Trends Cogn Sci 17:565–573

Sevenster P (1973) Incompatibility of response and reward. In: Constraints on learning. Academic Press, London, pp 265–283

Sheenaja KK, Thomas KJ (2011) Influence of habitat complexity on route learning among different populations of climbing perch (*Anabas testudineus* Bloch, 1792). Mar Freshw Behav Physiol 44:349–358

Shettleworth SJ (2010) Cognition, Evolution and Behaviour, 2nd edn. University Press, Oxford

Sih A, Del Giudice M (2012) Linking behavioral syndromes and cognition: a behavioral ecology perspective. Philos Trans R Soc B Biol Sci 367:2762–2772

Sih A, Bell A, Johnson JC (2004) Behavioral syndromes: an ecological and evolutionary overview. Q Rev Biol 19:372–378

Sinclair ELE, Noronha de Souza CR, Ward AJW, Seebacher F (2014) Exercise changes behaviour. Funct Ecol 28:652–659

Smithdeal M (2016) Belief in free will as an adaptive, ungrounded belief. Philos Psychol 29:1241–1252

Sneddon L (2003) The bold and the shy: individual differences in rainbow trout. J Fish Biol 62:971–975

Sneddon LU (2011) Pain perception in fish: evidence and implications for the use of fish. J Conscious Stud 18:209–229

Sneddon LU (2015) Pain in aquatic animals. J Exp Biol 218:967–976

Snell-Rood EC (2012) Selective processes in development: Implications for the costs and benefits of phenotypic plasticity. Integr Comp Biol 52:31–42

Solberg MF, Skaala Ø, Nilsen F, Glover KA (2013a) Does domestication cause changes in growth reaction norms? A study of farmed, wild and hybrid Atlantic salmon families exposed to environmental stress. PLoS One 8:e54469

Solberg MF, Zhang Z, Nilsen F, Glover KA (2013b) Growth reaction norms of domesticated, wild and hybrid Atlantic salmon families in response to differing social and physical environments. BMC Evol Biol 13:234

Spruijt BM, van den Bos R, Pijlman TA (2001) A concept of welfare based on reward evaluating mechanisms in the brain: anticipatory behaviour as an indicator for the state of reward systems. Appl Anim Behav Sci 72:145–171
Steingrund P, Fernö A (1997) Feeding behaviour of reared and wild cod and the effect of learning: two strategies of feeding on the two-spotted goby. J Fish Biol 51:334–348
Sumpter DJT, Krause J, James R, Couzin I (2008) Consensus decision making by fish. Curr Biol 18:1773–1777
Tinbergen N (1963) On aims and methods of ethology. Zeitschrift für Tierpsychologi 20:410–433
Toates F (2004) Cognition, motivation, emotion and action: a dynamic and vulnerable interdependence. Appl Anim Behav Sci 86:173–204
Torgersen T, Bracke MBM, Kristiansen TS (2011) Reply to Diggles et al. (2011): ecology and welfare of aquatic animals in wild capture fisheries. Rev Fish Biol Fish 21:767–769
Vail AL, Manica A, Bshary R (2013) Referential gestures in fish collaborative hunting. Nat Commun 4, article number 1765
van den Bos R, Flik G (2015) Editorial: decision-making under stress: the importance of cortico-limbic circuits. Front Behav Neurosci 9:203
van den Bos R, Meijer M, van Renselaar J, van der Harst J, Spruijt B (2003) Anticipation is differently expressed in rats (*Rattus norvegicus*) and domestic cats (*Felis silvestris catus*) in the same Pavlovian conditioning paradigm. Behav Brain Res 141:83–89
van den Bos R, Jolles JW, Homberg JR (2013) Social modulation of decision-making: a cross-species review. Front Hum Neurosci 7:301
van Staaden M, Huber R, Kaufman L, Liem K (1995) Brain evolution in cichlids of the African Great Lakes: brain and body size, general patterns and evolutionary trends. Zoology 98:165–178
Venkatraman A, Edlow BL, Immordino-Yang MH (2017) The brainstem in emotion: A review. Front Neuroanat 11:15
Vindas MA, Madaro A, Fraser TWK, Hoglund E, Olsen RE, Øverli Ø, Kristiansen TS (2016) Coping with a changing environment: the effects of early life stress. R Soc Open Sci 3:160382
von Uexküll J (1921) Umwelt und Innenwelt der Tiere, 2nd edn. Springer, Berlin
Ward AJW, Herbert-Read JE, Sumpter DJT, Krause J (2011) Fast and accurate decisions through collective vigilance in fish shoals. Proc Natl Acad Sci USA 108:2312–2315
Ward AJW, Krause J, Sumpter DJT (2012) Quorum decision-making in foraging fish shoals. PLoS One 7:e32411
Weary DM, Droege P, Braithwaite VA (2017) Chapter two - behavioral evidence of felt emotions: Approaches, inferences, and refinements. Adv Study Behav 49:27–48
Weinstein ND (1980) Unrealistic optimism about future life events. J Pers Soc Psychol 39:806–820
White GE, Brown C (2015) Cue choice and spatial learning ability are affected by habitat complexity in intertidal gobies. Behav Ecol 26:178–184
White SL, Wagner T, Gowan C, Braithwaite VA (2017) Can personality predict individual differences in brook trout spatial learning ability? Behav Process 141:220–228
Wingfield JC (2003) Control of behavioural strategies for capricious environments. Anim Behav 66:807–815
Wisenden BD, Harter KR (2001) Motion, not shape, facilitates association of predation risk with novel objects by fathead minnows. Ethology 107:357–364
Yoshida M, Hirano R (2010) Effects of local anesthesia of the cerebellum on classical fear conditioning in goldfish. Behav Brain Funct 6, article number 20

Chapter 8
Awareness in Fish

Ruud van den Bos

This chapter is dedicated to the late professor Alexander Cools and emeriti professors Tjard de Cock Buning and Berry Spruijt, who have been influential in my thinking on this topic.

Abstract Whether fish have awareness, and if so, of what they are aware, has been a long-standing question. Discussions on the welfare of animals, which entail whether animals suffer, have renewed the interest in this question. Here, I discuss, starting from the early work of George Romanes, different strategies that have been taken to address this question: ranging from inaccessible per se, since the private nature of mental life is impossible to study, to access, yet requiring a theoretical framework that leads to clear experimental predictions. This scientific debate will feed discussions on how to treat fish, yet should not be burdened by a desired outcome: awareness in animals is but one aspect of the moral question how to treat animals.

Keywords Mental states · Emotion · Cognition · Goal-directed · Behaviour · Animal welfare

8.1 Introduction

In 1977, a scientific meeting on welfare in humans was held in the Netherlands. One of the symposium's speakers, the Dutch ethologist Gerard Baerends, was invited to express his thoughts on the matter from the point of view of ethology. In his contribution Baerends (1978) expressed that this was a rather strange task for an ethologist: after all, welfare comprised subjective feelings, the very thing that had been set aside as being inaccessible by scientific method in the field of ethology. This view expressed the coming of age of ethology as a well-respected (natural) science

R. van den Bos (✉)
Department of Animal Ecology and Physiology, Institute for Water and Wetland Research, Radboud University, Nijmegen, The Netherlands
e-mail: ruudvdbos@science.ru.nl

T. S. Kristiansen et al. (eds.), *The Welfare of Fish*, Animal Welfare 20,
https://doi.org/10.1007/978-3-030-41675-1_8

discipline. A discipline to which people like Niko Tinbergen had contributed so strongly, for instance, by carefully describing its questions and methodology (see Tinbergen 1963). Hence Baerends' contribution (1978) focussed on abnormal behaviour of animals in captivity when these conditions would not meet the natural requirements of animals.

From the late 1960s onwards animal welfare emerged as a field of scientific study. Within this emerging field, the view expressed by Baerends was for long a dominant one. Animal welfare was defined in terms of abnormal physiology, i.e., chronic high levels of cortisol indicating a chronically activated hypothalamus–pituitary–adrenal cortex axis, and/or abnormal behavioural patterns, i.e., the occurrence of stereotypies and self-injurious behaviour (Broom and Johnson 1993; Hagen et al. 2011; Spruijt et al. 2001; Wiepkema 1985).

In contrast to this scientific approach, the political and public debate on animal welfare focussed on feelings, especially on suffering, as this founded the ethical (and societal) imperative of "doing no harm" or "promoting the good". Marian Dawkins expressed this quite clearly in her influential 1990 paper: "Let us not mince words: Animal Welfare involves the subjective feelings of animals" (Dawkins 1990, p. 1).

Romanes' reasoning by analogy, laid down in his works on comparative animal intelligence (Romanes 1882, 1884; see below), and refined later to more explicitly include physiology and anatomy of the nervous system (see Stafleu et al. 1992; Verheijen and Buwalda 1988), has been taken as a good proxy to bridge the gap between science and the ethics/public debate. When neuroanatomy and behaviour are sufficiently alike between humans and animals, say regarding behaviour and neuronal circuits related to threatening, stressful and noxious events, one may (safely) assume that the accompanying feelings in animals are similar to humans. In addition, animals were given the benefit of the doubt when called for. While in legislation on animal experimentation and welfare all vertebrate species were herewith included, especially fish have generated quite some discussions on whether they could suffer and/or experience pain (e.g., Braithwaite 2010; Braithwaite et al. 2013; Brown 2015; Rose et al. 2014; Sneddon 2015).

The tension between science and society burdened many scientific discussions on whether feelings, implying awareness, existed at all in animals, and how to study this scientifically, as the outcome of scientific discussions could have a direct impact on many activities involving animals. For instance, regarding fish species, the outcome could have direct effects on activities such as recreational and sports angling, commercial fisheries, and aquaculture.

Yet, science is not about providing nice and politically correct answers, but about asking the right questions, how uncomfortable they may seem.[1] Here, I will discuss

[1]Rephrased from professor Piet Wiepkema, one of the founding fathers of animal welfare research in the Netherlands.

the question of how to approach awareness[2] in animals in general, and in fish in particular, discussing different strategies taken by different scholars. First, I will discuss the work by George Romanes as he may be considered the first to systematically address scientifically the question whether animals have mental states. From thereon, I will discuss how different scholars have proceeded in addressing the question of mental states in animals, applying this to fish species.[3]

8.2 The Writings of George Romanes: History Is a Teacher

Inspired by Darwin, George Romanes (1848–1894) set out a comparative approach on mental faculties, i.e., a comparative psychology (Romanes 1882, 1884). While his approach has been strongly criticized by history, and dismissed as being simply (too) anthropomorphic, i.e., projecting human mental capacities to animals, this does do no justice to the way he carefully discussed the ins and outs of his comparative approach, and how well aware he was of its limitations (Romanes 1882, 1884). He carefully describes (1) a conceptual framework, defining what to observe and defining the differences between mental and non-mental activities, and (2) a data collection framework, defining how to collect the relevant data. In fact, rather than carved in stone, his writings were in pencil, leaving open adjustments on mental evolution when needed. So, even nowadays his books are more than good reading, especially as a number of issues keep returning in later writings of others. I will allude to them in the appropriate sections.

8.2.1 *Conceptual Framework*

From scratch on Romanes (1882) chose reasoning by analogy as a means to go forward. We are aware of our feelings and thoughts (two critical classes of mental states; see below), yet not of those of others. We only observe what they do, and infer from their behaviour that similar mental states accompany their activities. So, strictly speaking, we *project* or *eject* our mental states onto others. Similarly, as it is impossible to know what goes on inside animals, he continues to discuss that we could apply the same method to animals: when activities between man and animals are very much alike, so should the accompanying mental states be alike (Romanes 1882; see Box 8.1).

---

[2]I will use awareness and consciousness interchangeably in the text. In addition, I will use emotion and feeling interchangeably; emotions/feelings are the top layer of what is collectively referred to as hierarchically organized emotional systems (see text).

[3]It should be noted that in this chapter, I draw quite a few examples from mammalian species as most (conceptual) research regarding awareness thus far is done in mammals.

Next, he describes and defines a criterion of mind (see Box 8.1 for details). As a summary he writes (Romanes 1884, p. 18): "…the distinctive element of mind is consciousness, the test of consciousness is choice, and the evidence of choice is the antecedent uncertainty of adjustive action between two or more alternatives". He adds (p. 20): "agents that are able to *choose* their actions are agents that are able to *feel* the stimuli which determine the choice". Thus, critically here is that the behavioural outcome is not catered for by fixed relationships between stimuli and responses, i.e. by reflex action, and that *feeling* stimuli may critically underlie behaviour. I will return to the latter in later sections as this idea returns in works by other scholars.

While Romanes dissociated behavioural outcomes unique to the situation and based on individual experiences, from behavioural outcomes experienced by all members of a species, he endowed both behaviours as accompanied by mental capacities. The main difference is that instinct-based behaviour is brought about by evolution and selection as a general phenomenon expressed in all subjects catered to deal with recurring situations, while intelligent behaviour is related to situations, novel to the subject, leading to a unique solution.

I will return later to these distinctions in behaviour as they also form the basis of other approaches to mental life in animals. For now, it suffices to say that in fact, he preludes on what later is studied in great detail in the field of experimental psychology as stimulus–response or habit-like behaviour *versus* goal-directed behaviour in conditioning paradigms and choice behaviour in preference tests, and in the field of ethology as species-specific behaviour, i.e., behaviour brought about by natural selection expressed by all members of a species.

8.2.2 Data-Collection Framework

As no experimental work was present in his days, his database consisted of anecdotes on behaviour as observed by different people. As he realised the weakness of this approach he used a few criteria on which to accept anecdotes: (1) whether the observer is a renowned person (argument by authority), (2) when the observation would come from someone less known, whether the observation is clear and undisputed, and (3) especially in relation to point (2), whether more observers report similar findings. So the database is a mix of being built on authority and independent observations. This is not so much different from present day views on results obtained through experiments. Experimental results from renowned laboratories are taken more seriously than from less well-known laboratories (whether this is justified or not I leave open for discussion here). In addition, similar results from different laboratories (between laboratories reproducibility) are taken as stronger evidence than results from a single laboratory.

While nowadays using anecdotes may seem but naïve, in popular books anecdotes are still used to make a point forcefully to the general public, such as in Balcombe's (2016) recent book on fish emotion and cognition. In addition, anecdotes may serve as a starting point for further experimental studies: they are sources of inspiration. For instance, the work by the experimental psychologist Anthony Dickinson, discussed below, started from an incidental event happening to him on Sicily ("the Palermo protocol": see below). Finally, some behaviours may be difficult to organize in experiments in the laboratory precisely because they only occur every so often.

8.2.3 Fish

Using the methods as indicated above, Romanes asserted for fish that they possess emotions of "fear, pugnacity; social, sexual and parental feelings; anger, jealousy, play and curiosity. . .and corresponds with that which is distinctive of the psychology of a child about four months" (Romanes 1882, p. 242). He illustrates this with a series of anecdotes that he collected according to his methodology (see Box 8.2). As to mind/consciousness, he mentions on the one hand, fish migration and hunting techniques (instincts), and the way they adjust their behaviour depending on local conditions (intelligence; see Box 8.2). So Romanes (1884) grants fish elements as memory and association by similarity. Interestingly, he puts invertebrates such as ants and bees at a higher level, especially as it comes to social behaviour, while he recognizes that their brain may be less complex than of fish.

8.2.4 Critical Notes

Here I will address a few general issues regarding Romanes' work as they are still present in nowadays discussions on mental states. So, they are not meant to explicitly criticize his work, but rather discuss some general notions on studying mental states in animals.

Especially Romanes' way to study the mind of other species (by similarity), as well as his use of anecdotes, has been strongly criticised as inadequate. With the advent of ethology and experimental psychology as scientific disciplines in the last century, mental states were set aside as not being part of scientific studies of psychology and behaviour as they are private and cannot be objectively studied, i.e. it defined the boundaries of the scientific field as explicitly expressed by Baerends (1978) in his lecture. Which is not to say that people did not believe that animals may be conscious; rather it was being regarded as being inaccessible, and hence outside the realm of science. In addition, anecdotes played no role in advancing the fields, only carefully designed and controlled experiments.

Partly under influence of the way we keep and treat animals, as indicated above, and partly because of a shift in scientific paradigms, the subject returned on the table of science. Several researchers picked up on the topic of mental states using different strategies. Partly this encompassed refining Romanes' strategy, partly taking advantage of the conceptual developments in ethology, experimental psychology and neurobiology, and partly taking advantage of the developments in animal experimentation. I will return to this in the next sections.

Underlying Romanes' approach is the idea of continuity as following from Darwin's work. This argument is even nowadays often used to state that differences between humans and animals are a matter of degree and not of kind, i.e. absence and/or abrupt transition of mental states would be illogical. However, continuity is no more than an assumption, which needs to be independently proven by careful comparative research. For instance, while we have common ancestors with chimpanzees and bonobos, quite a few hominids have appeared on stage and disappeared. So, what we study now in humans, and probably also in chimpanzees and bonobos (for who says that there have not been more of these species around), is the product of a series of "natural" experiments. So, mental states could in principle have been an innovation in "our" lineage. Hence only studying similarities and dissimilarities in behaviour following from a solid theoretical framework will show which aspects were there and which may have been developed in the hominid lineage.

Related to the former argument, are the notions of a ladder-like view on evolution (a near linear accumulation of mental capacities using humans as yardstick, i.e. a kind of end product) and comparing mental states in animals with ontogenetically mental capacities in humans, e.g. the behaviour of the animal mimics the behaviour of a child of 4 years of age. Both are not productive as they assume that mental capacities are independent of the ecological niche in which species live. If we assume that the behavioural and mental repertoire of animals is tuned to the specific sensory organs and niche of the species, then it follows that we expect specialization in how, for instance, mental representations of the environment are formed (*sensu* Von Uexküll's Umwelt). While some species base this on visual information, other species may do this on other sensory information, such as in fish that sense small changes in pressure by their sideline systems. Hence, these species (members thereof) have mental representations of their surroundings, but testing them in the visual domain may erroneously suggest than one is more evolved than the other. This also applies to the comparisons regarding child ontogeny: it is not so much relevant as to know at which age it compares with a yardstick species, such as humans, as to understand how the behaviour in different domains relates to the ecological niche of the species. Finally, a phylogenetic view of evolution rooted in ecology reveals that certain traits have evolved several times in different lineages, begging the question whether similar selection pressures have lead to this (see also Bshary and Brown 2014).

A final issue that needs to be addressed is that, for instance in fish species, examples are drawn from different species of fish to make the point of mental capacities. Even in modern writing, people use examples from phylogenetically widely separated fish species to argue that fish are aware or are more than reflex machines. While the examples may be illustrative for the different mental states to be found in fish species (and only a small number of the more than 30,000 species of fish have been studied or observed), it does not follow that this is universal: there is no average fish, which can be constructed from such observations. Suppose we would do this in mammals: while mirror self-recognition can be found in some species, suggesting an element of "Theory of Mind", it does not follow that all mammals have this capacity as data also show (de Veer and van den Bos 1999). Hence, the question is more relevant to understand how in the vast number of species of fish ecological niches have been instrumental in shaping behaviour, awareness, and mental states, leading potentially to a wide variety of outcomes (see also Bshary and Brown 2014), than to construct a non-existing average fish.

8.3 Assessing Consciousness and Mental States in Animals

8.3.1 What You See, Is What You Get

As indicated above, one problem with animals, like in humans, is that we cannot observe directly what is going on inside, i.e. the mental life of the animal. We only observe the behaviour it performs. In ethology and experimental psychology, the different behavioural patterns and their interrelationships are described in neutral terms devoid of any mental connotation, as if they were no more than descriptions of rocks tumbling downhill. For ease of categorization, behaviours are labelled as belonging to classes, such as related to learning and memory, feeding, and sexual behaviour. Francoise Wemelsfelder has criticized this subject (inside)—object (outside) dualistic view and argued that behaviour, or the way the animal interacts with its environment, is always endowed with an element of how it perceives the world, i.e. behaviour is always *expressive* (Wemelsfelder 1993). This is what we essentially recognize in other human beings as we note that someone is "sad" or "happy": "sadness" or "happiness" is expressed in *the way* the subject interacts with its environment; not in a single behavioural pattern per se to which an inner unobservable state is attached. So, one could say that Wemelsfelder takes the common sense shortcut way of quick mind reading as starting point of scientific analysis of the behaviour of animals, which she has coined "Qualitative Behaviour Assessment", and which has been applied to many different species, such as pigs, sheep, horses and donkeys (see Wemelsfelder and Mullan 2014).

Her methodology has been criticized as being too anthropomorphic, i.e. it does not question shortcut reasoning, but takes it as a valid starting point. In humans, this shortcut reasoning may seem valid at first glance, yet is clearly prone to error. For instance, keeping your composure in public might hide one's feelings of anger or sadness from being noticed by others. Hence, using it for animals, while convenient in many cases, may overstate the case, as there is no independent way of assessing, whether the expression of behaviour concurs with the mental state. While close observation and knowledge of subjects of a species may lead to more precise insight into how animals conceal information from others, this remains sensitive to error. In a way, this reasoning is more or less a closed system as mental states are by definition present in the way the animals interact. It also suggests that there is no difference between online behaviour and offline processing of information generating new layers of emotions.

Yet, the method per se might be of interest as a way to quickly judge how animals fare, especially when used across time, when the behaviours of animals change, due to changes in living conditions. This method has not as yet been applied to fish species as far as I know.

8.3.2 *Romanes Refined*

Scientific discussions on suffering and pain in fish species have revolved, among others, around neuroanatomical differences between fish and mammals, in particular the prefrontal cortex. Pain involves two critical different elements: nociception and the unpleasant feeling of pain (https://www.iasp-pain.org/Taxonomy). Nociception is the mere detection of damage followed by mounting appropriate physiological and behavioural responses to the damaging (noxious) stimulus, such as withdrawal responses and other appropriate protective measures (nocifensive responses). Pain involves the unpleasant, conscious sensation (feeling) of pain: that what bothers people. Proponents of the view that fish experience pain (e.g. Sneddon 2015) argue (1) that the nociceptive and nocifensive machinery is present in fish species, (2) that fish show more than simple reflex behaviour, such as learning behaviour and optimal choice behaviour in paradigms offering conflicting situations, and (3) that functional neuroanatomy shows that homologous prefrontal/cortical structures are present, while opponents have strongly questioned these assertions, especially points 2 and 3 (e.g. Rose et al. 2014; Bermond 1999; see also Bermond 1997, 2001).

Without going into detail in the arguments pro and con, discussions partly revolve around the interpretation of data, the crude separation in reflex and complex behaviour without making clear what complex behaviour is, and the relationship between brain, mental states, and behaviour (*sensu* Romanes). In general, however, the discussions lack a strong conceptual framework leading to critical experiments for deciding for one or the other outcome. While it seems to be taken for granted that the experience of pain benefits the subjects and hence is adaptive, it is not critically articulated what, in general, the difference is between an organism which experiences pain and one that does not (what is the added value of having the experience per se), under which conditions it would evolve, and how it develops in subjects (see Braithwaite et al. 2013).

The Canadian psycho-physiologist Michel Cabanac (Cabanac 1971, 1979, 1992, 2002, 2008; Cabanac et al. 2009) has provided a conceptual framework for consciousness per se, and feelings in particular, in the organisation of behaviour. Consciousness emerged as mental space where momentary (in the here-and-now) decision-making can be optimized (Cabanac et al. 2009). Feelings are the experienced, psychological, read-outs/shortcuts of the brain's workings of motivational systems that bridge the "here-and-now" with the future, i.e. what is best (adaptive) in the long-run (Cabanac's 1971 maxim: '*pleasant is useful*'). This is akin to what

Damasio and colleagues have expressed on feelings: we are not aware of the brain's calculations or nervous system processes, only of its final result, which pops up or is presented in mental space as "this choice feels good" (see Damasio 1994). In addition, this mental space allows for comparing information coming from completely different physiological sources or motivational systems, such that momentary decisions can be optimized when conflicts arise (Cabanac's 1992 maxim: "*pleasure is the common currency*"). All motivational, physiological (including pain) systems are linked to the same emotional axis of pleasantness–unpleasantness, and hence can be momentary compared. Moreover, these emotional systems share the same neuronal machinery, in particular, limbic endorphin/dopamine-based structures (Spruijt et al. 2001).

As indicated above this choice-based feeling element was already suggested by Romanes, and echoed by Damasio (1994) in his writings: it is not "I reason, therefore I am" (as Descartes argued), but "I feel, therefore I am" (see further below).

This framework leads to several emotion-based behavioural tests indicative of consciousness (Cabanac et al. 2009): (1) emotion: handling induced changes in temperature and heart rate; (2) sensory pleasure: taste aversion learning or conditioned taste aversion, which involves a change in the value of a commodity, such as food when paired with illness (e.g. induced by lithium chloride; see below); (3) pleasure and decision-making: conflicting situations such as plain food in a cosy warm environment versus highly rewarding food in a cold environment. Applying these criteria Cabanac et al. (2009) conclude that consciousness thus defined emerged in early amniota, excluding hereby species of fish and amphibia.

Cabanac et al. (2009) underpin the behavioural observation with (comparative) neuroanatomical changes taking place between amphibia and reptiles in cortical structures and telencephalic dopaminergic projections (Cabanac et al. 2009). The nervous system has undergone several major changes from fish to amphibia to reptiles to mammals, such as repositioning and strong expansion of structures (O'Connell and Hofmann 2011; Mueller 2012). The ventral dopaminergic pathway (A10; ventral tegmental area, involved in motivation and emotion) and dorsal dopaminergic pathway (A9, substantia nigra pars compacta, involved in instrumental behaviour and motor patterning) are not separated in fish or even recognizable as such, while strongly so and clearly in mammals (see Yamamoto and Vernier 2011). Yet these changes in neuroanatomy, while present, do not follow the suggested changes in behaviour as later studies showed: (1) goldfish express conditioned taste aversion to lithium chloride dependent on the dorso-medial pallium (Martín et al. 2011), (2) zebrafish express emotion-induced behavioural fever, i.e. they seek higher temperatures following stress (Rey et al. 2015; Rey Planellas 2017), and (3) fish express optimal choice behaviour, e.g. they trade-off access to a painkiller (lidocaine) against the structure of the environment (Sneddon 2015). In fact, Cabanac also expressed doubts on his criteria as he acknowledged that behaviour has been observed in fish, such as play, that would run against fish as having no consciousness (Cabanac et al. 2009).

Hence, while it is tempting to interrelate neuroanatomy, behaviour, and mental states based on mammalian, in particular human, templates which have been done to determine, e.g. the presence of pain in fish species, this may lead to erroneous conclusions. The essential element is to understand from a bottom-up, phylogenetic view, the changes that have taken place by defining the added value of a mental space, including that in humans, by critical experiments, and by understanding how the nervous system can cater for this in different species. This will be discussed in the next section.

8.3.3 Goal-Directed Behaviour and Anticipatory Behaviour

8.3.3.1 Subjective Versus Objective

One of the issues that has been central to the discussion on awareness in animals, and as Romanes (1882) correctly pointed out also in humans, is the subjective, private, *versus* the objective, public, divide. As I have argued elsewhere, however, the relevant scientific research question is not, *what/how* human or non-human animals feel or know (i.e. a *subjective* perspective, the private contents of mental states), but *that* they feel or know (i.e. an *objective* perspective, the fact *that* they have mental states) and how this is represented in behaviour and associated circuits in the nervous system (van den Bos 1997, 2000, 2001, 2019; Braithwaite et al. 2013). So, the specific connotation I have while experiencing pain (which is private and may vary between subjects) is of less relevance than the fact *that* I experience pain (which is a property, shared by all others). In other words, emotion and cognition are not private properties just occurring by accident in only one individual, say me, but fundamental properties of the biological organization of brain and behaviour (invariants) occurring in all individuals of a species, human and non-human animals alike (van den Bos 1997, 2000, 2001, 2019; Braithwaite et al. 2013). Hence, we have to describe carefully in a brain-behaviour model encompassing awareness which behavioural patterns are dependent on this specific feature: what is the added value of awareness and experiences, and can we discriminate the "have's and the have-nots", that is can we devise critical experiments, which predict either one or the other outcome (Braithwaite et al. 2013). In addition, we have to describe how awareness is being represented by neuronal circuits, i.e. what are essential requirements. Furthermore, between species differences may emerge as to the nature of emotions and cognitions (the specific mental states), while they all share awareness as property, as the specific nature of the states may depend on their ecological niche and sensory organs as described above (van den Bos 2000).

Of course one could argue that it still is not proven that awareness exists outside myself or that it exists in animals. However, the critical point is that this approach describes a scientific, biological, model on the organization of brain and behaviour, which applies to all members of a species, of which awareness is a property of its organization, including the human species. As I happen to be a member of the human species, I have experiences critical to the human species just like my fellow human beings. The strength of any experimental scientific model is whether it describes reality more precisely than alternative models, i.e. whether the model deals adequately with the known facts and whether it generates testable hypotheses. So, it should be more than just a descriptive model. This also entails that we have to accept that current knowledge is but the best possible description of reality, and will show to what extent folk psychology or common sense notions are true or need revisions, just as progress in all fields of science occurs; hence it is written in pencil (*sensu* Romanes).

8.3.3.2 Awareness

The foregoing begs the question how to perceive or define awareness as a property inherent to the behavioural organization of the organism. As I have discussed elsewhere (van den Bos 1997, 2000, 2001, 2019; Braithwaite et al. 2013) and as also discussed by others (Cabanac 2002; Cabanac et al. 2009; LeDoux 2012; Romanes 1882) awareness may be perceived as a limited mental, working, space consisting of mental states, such as feelings or cognitions (e.g. "if…then…"), occurring serially, i.e. in succession. Whereas the brain processes incoming information through multiple hierarchically organized systems (van den Bos 1997, 2000; Cools 1985), awareness, or *that* what we perceive, is a limited working space, where information can only be processed serially, i.e. we can only fully attend to one thing at the same time. This leads to a few predictions and possibilities for experiments.

For instance, in blindsight, where the visual cortex is damaged, people do not report any conscious experience in the affected visual field (Leopold 2012). Yet, when stimuli are presented in this visual field, subjects still perform above chance level when they, for instance, have to guess how stimuli move. This is achieved by visual systems, which do not use the visual cortex. So, unconscious processing (and reaction) is intact, but conscious processing (and action) is absent. Hence, this leads to experiments with testable hypotheses on what remains of the behavioural repertoire and what disappears. For instance, people suffering from blindsight avoid obstacles as they move along a corridor, yet are not aware of these stimuli (they do not perceive them consciously, while their nervous system does) and hence cannot act on them. In general, this means that emotional and cognitive systems may consist of several layers in which external and internal information are processed and on which physiological and behavioural responses can be initiated, with only one part being dependent on awareness or consciousness (Cools 1985; LeDoux 2012).

When the limitation to awareness is that the organism can attend to only one thing at the same time, this means that those behaviours that are dependent on awareness are sensitive to distraction or to interfering information, yet those behaviours that are not dependent on awareness are insensitive to this. Indeed, as shown by Clark and Squire (1998; see in detail below), trace conditioning which is dependent on awareness is sensitive to distraction, while delay conditioning, which is independent of awareness, is not. This has also been shown in mice (Han et al. 2003).

8.3.3.3 Mental States

As to the nature of mental states, feelings or cognitions, the concept of intentional systems may be a starting point to assess the nature of behaviour (see van den Bos 1997, 2019). Three layers of intentional systems or behavioural modes may be envisioned: zero-order, first-order, and second-order, which amount, respectively, to systems or modes working on stimulus–response, action/stimulus–outcome (goal-directed behaviour) and reflections (on the previous two levels; which is often labelled "self-awareness" or "self-consciousness"). Thus, some behaviours may be zero-order, while others are first- or second-order. While some species may only have zero-order modes, others may possess all three. As subjects mature these levels may develop in succession.

Zero-order behaviour amounts to what Romanes (1882) labelled reflex action. Elsewhere I (van den Bos 2019) have argued that behaviour, which Romanes (1882) labelled instinct behaviour, is captured under zero-order behaviour and devoid of awareness. Only first- and second-order behaviours are labelled as containing awareness. Second-order behaviour entails reflection on the outcome of behavioural patterns and mental states in relation to oneself and others, i.e. it entails elements of self-awareness and mind reading in others ("Theory of Mind"). This also creates new layers of feelings. While under first-order, pleasure, anxiety, fear, and anger may be encompassed (primary emotions), under second-order, jealousy, suspicion, etc. may be encompassed, i.e. the consequence of thinking of oneself in relation to others. It should be noted that self-awareness should not be seen as a hierarchically additional layer ("an independent viewer in the cinema of life"), but as mental states with specific contents or connotations (van den Bos 1997, 2000; Dennett 1991).

A recent study in cleaner wrasses (*Labroides dimidiatus*; Kohda et al. 2019) suggested that these fish recognize themselves in a mirror. Mirror self-recognition (MSR) has been taken as one piece of evidence for "Theory of Mind" and has thus far been shown in only a few species apart from humans: elephants, dolphins, magpies, and great apes (see de Veer and van den Bos 1999; de Waal 2019). At present, the jury is still out whether the data in cleaner wrasses are convincing for MSR per se or indicative of a level of mirror understanding close to MSR (de Waal 2019). Whatever the outcome, this study clearly shows that only an open mind and careful experimentation, as well as constructive discussion, will reveal the spectrum of mental states (first-order/second-order) in fish species. For the remainder of the chapter, I will concentrate on evidence for awareness regarding first-order behaviour.

In general then, behavioural systems and their underlying neuronal circuits, can be conceived of as hierarchically organized systems, where at each level information is processed and behavioural patterns organized (Cools 1985). Hence, this entails studying which behaviours are changed by manipulations at specific levels in this hierarchy (Cools 1985). Awareness may be conceived as a working space at the highest level in the hierarchy enhancing options of processing information. In the following paragraphs I will discuss protocols, which are dependent on awareness with clear predictions as to their outcome.

Cognition: Delay *Versus* Trace Conditioning

Under both delay conditioning and trace conditioning Pavlovian procedures, subjects prepare for an upcoming event, i.e. shocks, air puffs, or food, by acquiring conditioned responses, such as freezing, closure of the eyelid or approaching the area where food will arrive when the cue comes on. A methodological difference exists between delay conditioning and trace conditioning: in delay conditioning the offset of the conditioned stimulus occurs later than the onset of the unconditioned stimulus (i.e. they overlap in time), while in trace conditioning a temporal gap exists between the offset of the conditioned stimulus and the onset of the unconditioned stimulus, i.e. they are separated in time.

Awareness as a working space allows for retaining information across a time gap for stimuli to be associated. Hence, trace conditioning would be predicted to be dependent on awareness, while delay conditioning would not, with trace conditioning, but not delay conditioning, being sensitive to distraction, as due to the serial nature of awareness, processing requires sustained attention. Studies in humans have shown (1) that trace conditioning critically depends on awareness, while delay conditioning does not, (2) that trace conditioning is susceptible to distraction, while delay conditioning is not, and (3) that trace conditioning requires more higher-order networks (prefrontal and temporal lobe structures) than delay conditioning (Carter et al. 2003; Clark and Squire 1998, 2004; Knight et al. 2004; Weike et al. 2007).

Thus, trace conditioning would qualify to show the presence of awareness in animals, that is, it is a logical expression of the presence of awareness, just as it is in humans—recall that we study awareness as a property of biological systems. Trace conditioning can be found in many different mammalian species, such as rats and cats (van den Bos et al. 2003), mice (Han et al. 2003) and rabbits (Woodruff-Pak and Disterhoft 2008). Trace conditioning, but not delay conditioning, has been shown, at least, in mice to be sensitive to distraction (Han et al. 2003). Finally, it is dependent on prefrontal—hippocampal networks, similar to humans (Han et al. 2003; Woodruff-Pak and Disterhoft 2008). Especially the anterior cingulate cortex is critical; a structure implicated, along with the insular cortex, as a hub in awareness (van den Bos 2000; Medford and Crichley 2010). As this is not the place to elaborate on the differential role of these structures in the organization of behaviour, it suffices here to say that the insular cortex may be critical in detecting bodily needs/urgencies (internal signals) whilst the anterior cingulate in organizing appropriate actions.

In fish trace conditioning has been shown, in among others, cod (Nilsson et al. 2008a, b), rainbow trout (Nordgreen et al. 2010), halibut (Nilsson et al. 2010) and goldfish (Vargas et al. 2009). As to the neuroanatomical structures it has been shown that trace, but not delay, conditioning is dependent on the lateral (Rodríguez-Expósito et al. 2017; Salas et al. 2006; see also Broglio et al. 2005) and dorsal

(Vargas et al. 2009) pallial areas, suggested to be the homologues of the hippocampus and cortical areas, respectively (see Mueller 2012; Vernier 2017; see also Woodruff 2017). No studies thus far have looked at whether trace conditioning, but not delay conditioning, is sensitive to distraction. It should be noted that lateral pallial areas are also involved in allocentric spatial learning supportive of their role in awareness (see Woodruff 2017). Finally, transitive inference, also dependent on the hippocampus in humans, can be shown in fish, i.e. in male cichlid fish; yet it has not been shown as yet that this depends on lateral pallial areas (see Woodruff 2017).

Cognition: Expectations

While animals are able to associate stimuli across a temporal window, the next question is whether subjects have an *explicit expectation* of what will happen when a cue comes on, i.e. are they *aware* of what will follow. Anticipatory behaviour in one way or the other implies an image of the future or scenario enrolling in the future, to choose an appropriate behaviour in the present based on knowledge of the upcoming event (van den Bos 2019). Thus, anticipatory behaviour assumes a form of goal-directedness. If subjects (human and nonhuman animals alike) are aware of what will happen, they are expected to scale their behaviour to the expected event, i.e. the anticipated value of the reward, or to adjust behaviour if needed; if they are not aware of the upcoming event, the behaviour is expected to be independent of the event.

Spruijt, van der Harst and co-workers (2003) have shown that rats show hyperactive behaviour, measured as transitions between behavioural patterns, in the interval between the onset of the cue and arrival of the reward, which was proportional to the (expected) rewarding properties of the stimulus, i.e. the stronger the reward the more of the behaviour was shown. The latter is akin to instrumental conditioning tasks where subjects express more lever pressing or other costly behaviours when the value of a reward is higher (Berridge 1996; Spruijt et al. 2001). The expression of this anticipatory, hyperactive, behaviour was species-specific as well as context-specific: in rats the upcoming reward elicited hyperactive behaviour, while in cats hypoactive behaviour in the laboratory, yet hyperactive behaviour in the home-setting (van den Bos et al. 2003). In fish species, several studies have shown that fish anticipate on the arrival of a food reward (cod: Nilsson et al. 2008a, b; halibut: Nilsson et al. 2010). This behaviour was species-specific, i.e. hyperactive behaviour in cod, "hypo-active" behaviour in halibut (Nilsson et al. 2008a, b, 2010) and time-interval dependent (halibut: Nilsson et al. 2010). As yet, it has not been studied, whether the expression of this behaviour is scaled to an expected outcome, say by manipulating reward values.

Thus, observations in the context of appetitive trace conditioning suggest that subjects have acquired knowledge of the emotional value of the upcoming event. While this conditioning-induced behaviour may suggest a level of knowledge of the rewarding value of the upcoming event, it would be more powerful to demonstrate that subjects change their reward-related behaviour, when the reward has, for instance, been devalued in the mean time. This would support the notion that awareness enhances flexibility in behaviour, i.e. the subject is able in its lifetime to optimize the interaction with the current features of the environment (Romanes 1882, 1884; Cabanac et al. 2009). Thus, for instance, when rats would see or hear the Pavlovian cue, announcing the upcoming reward, they should modify their behaviour when the reward was devalued prior to the experiment. For, this lower value would generate a conflict with the reward-related learned behaviour, and animals should adjust their behaviour accordingly. When this can be shown to occur, it is strong evidence that they are aware of the changed value, that they anticipate the outcome of the action, and are able to incorporate this knowledge into their ongoing behaviour. When rats do not change their behaviour, this shows that their behaviour is stimulus–response driven or habit-like rather than goal-directed. This so called "goal-directed behaviour paradigm" is up to now one of the most powerful paradigms to show that experiences may play a role in the organization of behaviour (Dickinson and Balleine 1994). Thus, the next question is whether animals are able to change behaviour when the value of biologically relevant items has changed. This will be discussed in the following sections.

The Value of Feelings

Cognitions in a general sense refer to *factual* temporal relationships, i.e. associations between stimuli like in Pavlovian conditioning, or associations between behavioural patterns and their outcome like in instrumental conditioning. As indicated above, awareness may allow to process associations across a time gap, an experienced "*if. . .then*". Feelings in a general sense refer to the *value* of a biologically relevant item, which is based on the internal, motivational, state and the properties of the specific item, and serve as momentary shortcuts to optimize interaction with the environment. Emotional systems including their underlying neuronal circuits subserve basic survival functions. For instance, Ledoux and others have characterized such a system for fear (LeDoux 2012).

Feelings as experiences (or mental states in general as said) add a new layer in the hierarchy of control of behaviour allowing new modes of processing incoming information. Cabanac has argued that intensity and duration are important determinants for emotions as experiences (Cabanac 2002). For instance, they may serve as urgency signals, i.e. related to threat (fear) or damage (pricking pain), or as highly relevant signals, i.e. related to energy (rewards, pleasures), calling for attention and subsequent processing. Thus, information may be stored more efficiently for later use or for changing subsequent behaviour, e.g. feelings enhance flexibility by re-evaluating learned behaviour and changing a course of action. Indeed, it has been suggested that memory storage is enhanced when stress increases noradrenergic signalling in the amygdala (Roozendaal et al. 2009). So, paradigms, tapping this off may show the added value of experiences, for instance, changing behaviour when the value of a reward has changed, or when conflicts arise between opposing tendencies.

Feelings-Based Changes in Behaviour

As Cabanac has shown in humans (Cabanac 1971) feelings, such as related to food (rewarding (pleasurable) or disgusting (un-pleasurable)), are based on the sensory qualities of the item (sweet, bitter) and the internal state (hunger, sated, sickness). So, feelings towards items are not fixed, but they are evaluations related to the current use of these items. For instance, when humans are hungry they report that a drop of sugar solution is pleasurable, yet when already sated with a sugar solution, as un-pleasurable. Similarly, when humans enjoy shrimp, they dislike them strongly upon the next encounter (feeling nauseous), when they have fallen ill (for whatever reason) shortly after consuming them, i.e. they express conditioned taste aversion. Conditioned taste aversion is a protective, adaptive, mechanism to avoid ingesting (rarely encountered or new) items that are (potentially) ill-making or lethal. Cabanac coined this internal state-dependent value of stimuli, alliesthesia (Cabanac 1971).

In humans, these paradigms can be used to show that subjects change learned behaviour when the value of an item has changed and that stress affects this: under normal conditions humans change learned behaviour, but not while under stress (Schwabe and Wolf 2011). Critically here, two alternatives strategies are present as will be discussed below: awareness-dependent goal-directed behaviour *versus* awareness-independent habit-like behaviour.

Dickinson used the above phenomenon, in his case an acquired dislike of watermelon after falling ill of drinking too much alcohol, as a source of inspiration to show how this affects instrumental behaviour in rats. In an elegant series of experiments Dickinson, Balleine and co-workers have shown that rats modify their learned behaviour for obtaining rewards after devaluation of the rewards, either using conditioned taste aversion or satiety-specific procedures (*the Palermo protocol*; Balleine and Dickinson 1998, 2000; Balleine and O'Doherty 2010; Dickinson and Balleine 1994, 2008). In a similar vein, in a Pavlovian conditioning paradigm Pietersen, Maes and van den Bos (unpublished data) have shown that rats adjusted their food magazine approach behaviour when rewards were devalued. The power of these experimental paradigms is that clear predictions were made as to the outcome: for instance, if rats were able to retrieve the changed value of rewards and associated this with lever pressing, they should stop lever pressing, as the outcome of this behaviour is no longer profitable (*if press lever then sugar pellet, which has changed its value, hence stop pressing*). If rats were unable to retrieve the information and associate this with lever pressing, then they should continue lever pressing. The experiments showed that rats adjusted their behaviour by combining new information (rewards have lost their initial rewarding value) with earlier acquired behaviour (pressing the lever leads to rewards) to change ongoing behaviour. Dickinson and colleagues suggested that feelings are the intermediate between motivations on the one hand (that what sets behavioural wheels in motion) and cognitions on the other (the learned factual relationships), and hence that feelings are critical in directing behaviour (*sensu* Romanes 1882, 1884; Damasio 1994; Cabanac 1971, 1992).

Of course, this experimental paradigm required many control experiments to show that the behavioural adjustments were selective, i.e. that only the operant behaviour contingent on the devalued reward was changed and that behaviour changed as result of a change in reward value (Balleine and Dickinson 1998, 2000; Balleine and O'Doherty 2010; Dickinson and Balleine 1994, 2008). Like in humans, this paradigm has been found to be sensitive to stress as (chronic) stress in rats leads to habit-like behaviour (Dias-Ferreira et al. 2009).

Thus far, in fish species only one study has attempted to run this paradigm. Nordgreen et al. (2010) trained rainbow trouts in a trace-conditioning paradigm to associate a green light with the arrival of food pellets. They subsequently devalued these pellets by associating them with a shock. Then they presented the original cue. The rainbow trouts showed a lower tendency to move towards the food magazine. While this experiment suggests that the rainbow trouts associated the cue with the changed value of the pellets as the rats above, alternative explanations are still possible as the devaluation procedure was done in the same set-up as the original training, e.g. the result may be due to the lower tendency to approach the magazine due to the expected shock. However, given that goldfish show conditioned taste aversion (Martín et al. 2011) and can easily be trained in conditioning paradigms (see, e.g. Vargas et al. 2009), similar experiments as in rats are feasible.

Decision-Making Behaviour

Goal-directed behaviour paradigms, such as discussed above, may be framed in a wider ecologically relevant context of optimizing behaviour. In human and nonhuman animals alike biologically relevant commodities, such as related to food and social interactions, are in principle not under control of the subject—although of course in humans in the case of food (and other commodities) agriculture and current infrastructure have been a tremendous step forward. Through exploration, subjects acquire information where and when to find commodities (a "cognitive" map) and what their value is (an "emotional" map; see van den Bos et al. 2002). The latter occurs through multiple interactions, in which the value of the commodity may change from moment-to-moment and of which the subject must determine a long-term average, i.e. a kind of running average. For instance, in case of foraging for an animal, a patch may contain high-quality food on one day, but low quality food on another. In case of social interactions in humans, someone may be friendly on one encounter, but grumpy on another. Hence, subjects should remain sensitive to whether the value changes over time to adjust their behaviour if needed, i.e. switch to another food patch when the quality is consistently low on encounters, or give up on a relation when someone becomes right-out nasty over time. In addition, information between alternatives may differ from moment to moment. For instance, a food patch may deliver a small amount of food on each occasion with varying quality (sometimes good, sometimes poor), yet with a long-term overall good outcome, while another may deliver a high quantity every now and then interspersed with food of very poor quality, i.e. with a long-term overall poor outcome. The same may be envisioned for social interactions. This conflicting information needs to be processed efficiently to optimize long-term choice behaviour given the limited energy available for all relevant activities (the economy of behaviour; see Spruijt et al. 2001 for discussion). The emotional system with feelings as experiences at its top and accompanying prefrontal (limbic, striatal) structures may have evolved to cater for this (van den Bos et al. 2014).

Damasio (1994) and colleagues (Bechara et al. 1994, 1997) have devised a task in humans, the Iowa Gambling Task, which essentially taps off the foregoing conflict: options with an immediate high reward with poor long-term outcome due to occasional or repeated high losses *versus* options with an immediate small reward with good long-term outcome due to occasional or repeated small losses. Experiments have shown that humans solve this task using the emotional system; in succession: subjects explore the options, without having a clue which option is better, show emotional responses (measured as skin conductance responses) to positive and negative encounters, slowly develop anticipatory emotional responses (measured as skin conductance responses) before making a choice, where at a certain moment they *feel* that one option is better than the other (they become emotionally aware of the differences), without being able to conceptualize or articulate the differences between options in detail, which is the final stage, yet which not all subjects reach (Bechara et al. 1994, 1997; Damasio 1994). Thus, emotional experiences precede insight and are sufficient to optimize choice behaviour. In addition, prefrontal areas are critical in performing this task (see Bechara 2005).

In an extensive series of experiment, we and other researchers have shown that behavioural and neuronal mechanisms are similar between rodents and humans dealing with such conflicting information in rodent versions of the Iowa Gambling Task (reviews: van den Bos et al. 2014; de Visser et al. 2011). As an anecdote, while observing the animals performing the task, when rats run into the occasional loss every now and then in the long-term good option, they "act surprised" (halting and slowly moving their head from left to right before returning to the start arm) or "annoyed" (casting the food cup aside), as if they expected something else. This is also observed in humans who perform this task behind the computer. Thus far such experiments have not been performed in fish species. Still, other paradigms have been used, which essentially tap off the same phenomenon.

As indicated above one could argue that feelings as experiences are critical in conflicting situations when no preprogrammed responses, i.e. instinctive behavioural patterns or otherwise, are present to solve the problem. Thus, imagine the following situation (van den Bos 2019). An animal is out on a foraging trip. At some point, the animal receives a cue (smell or otherwise) that a predator may be nearby. Now consider the options that expected reward levels (low-high) and cue strength (faint to strong) have been manipulated. The following possibilities may be envisioned. First, animals may respond indiscriminately: regardless of the reward level flee, or regardless of cue strength continue. In this case, we would not be able to conclude that the animal is weighing costs and benefits. Second, the outcome of the behaviour may depend on reward level and strength of the cue, which may be related to awareness, i.e. a psychological working space where costs and benefits may be assessed (van den Bos 2019). Many experiments have been run, including in fish species, using such paradigms showing that the outcome depends on specific preferences. For instance, in fish, it has been shown that fish treated with a noxious stimulus prefer to stay in a barren environment containing the pain killer, lidocaine, instead of an enriched environment which does not contain the pain killer, while they would normally prefer an enriched environment (Sneddon 2015). Similarly, depending on whether they are in an unfamiliar group or familiar group rainbow trout treated with a noxious stimulus show more or less aggressive behaviour (Ashley et al. 2009), i.e. fish express this behaviour contextually. Rainbow trout treated with a noxious stimulus did not respond to alarm pheromone by seeking shelter (Ashley et al. 2009), while zebrafish treated with a noxious stimulus did not show, e.g. freezing behaviour to alarm pheromone (Maximino 2011), suggesting that pain and anxiety interact, with pain overriding anxiety. The outcome of these different experiments seems to match the idea of Cabanac that feelings can be weighed in awareness (Cabanac 1992). Still, an alternative explanation may be that the outcome is based on motivational preferences (one is simply stronger than the other thereby determining the output) and not so much on psychological preferences (a choice determining the output). Indeed, the expression of behaviour related to the noxious stimulus was suppressed by alarm substance in zebrafish (Maximino 2011), which suggests that anxiety may also override pain, making it more than a simple and straightforward interaction. It is clear that more studies are needed to resolve this.

8.4 Concluding Remarks and Perspectives

8.4.1 General

Here, I have argued that in animal species, including in fish species, awareness is present as in humans, a global working space, allowing, for instance, the organism to adjust behaviour when stimuli have changed value; the latter implies an idea of a notion of the future. Changing behaviour based on prior experiences is a unique event as it depends on the subject's own history. You may say that all subjects possess the ability to adjust, but *what* and *how* to adjust is unique: it generates a high level of flexibility to optimize behaviour. Thus, it seems that nature's "solution" to cope with potential variation of future environments was to endow all subjects with awareness, a global working space, brought about by natural selection, allowing to process information privately in the subject's own lifetime, and without defining the optimal solution on each and single occasion.

8.4.2 Economy and Hierarchy of Behaviour

As indicated above, the essential problem every living organism faces is how to optimize behaviour in relation to environmental stimuli and energy resources available for doing so (the economy of behaviour; Spruijt et al. 2001). As discussed elsewhere (van den Bos 2019; Poli 2019) one could argue that different modes have evolved to anticipate future environments, e.g. innate behaviours to specific stimuli (fleeing for or freezing to potentially damaging stimuli), motivational priorities due to hormonal fluctuations (changes in seasonal behaviour due to changes in daytime length and/or temperature), and awareness (to allow comparisons of which activity at that particular moment in time is optimal). The former two may be called implicit anticipation, i.e. behaviour has evolved over many generations in relation to more or less predictable environmental conditions, while the latter explicit anticipation, i.e. the outcome is not determined a priori, but depends on the moment per se (van den Bos 2019; Poli 2019; Romanes 1882, 1884). Viewed from this perspective these modes are different ways of reducing uncertainty ("learning from the past") to anticipate the future (see also Romanes 1882, 1884).

As Cabanac suggested (Cabanac 1971, 1979, 1992) all motivational systems are associated with emotions [and neuroanatomically similarly organized (Spruijt et al. 2001)] allowing comparisons to be made irrespective of the specific system. Similarly, it seems that these systems use the same cognitive systems. The workings of such systems are tuned to the ecological niche of different species. The question arises whether in fish species, that inhabit so many different niches, i.e. from very stimulus-rich coral reefs to, what seems at first sight, very stimulus-poor deep ocean environments, this holds true as well. For awareness assumes that individual decisions can be made to optimize behaviour in environments in which preprogrammed behaviour might fall short, i.e. to be able to compare different emotions and/or the value of commodities to act upon this. So while they may be aware as a property of the system, *what* they may be aware of, i.e. mental states related to specific stimuli, may vary from species to species. For instance, some species may be highly tuned to social environments, while others to spatial arrangements. It is clear that an important step will be to assess the relationship between mental states and ecological niche. One mental state, which would lend itself to such an analysis, next to spatially and socially related states, is the experience of pain as damage to the subject is dependent on the ecological niche the subject lives in, which also determines the possibilities of showing appropriate behaviour. To take an example from mammals: in naked mole rats it has been shown that they are insensitive to foot-pad injections of acidic saline, which may be related to their relatively acidic environment (St John Smith and Lewin 2009).

8.4.3 Limitations

Scientific models are by definition an abstraction of reality: they entail that what can be conceptualized and measured given the current understanding and body of knowledge. They describe hence less than what may seem real as judged by folk psychology or common sense. Sometimes scientific models correct common sense (e.g. the apparent motion of the sun), sometimes it needs time before a gap is being closed—if possible. Hence current societal questions on animal welfare can only be rephrased into the particular framework to prevent overstretching the model leading to highly speculative answers. In this sense Baerends (1978) was right when the question was addressed: it was beyond the boundaries of the then prevailing framework, which by itself was cast in this way to allow ethology to mature as a scientific discipline. Hence, he could only answer it by rephrasing the question. As the field developed room was made for new venues to be explored as shown in this chapter. This has led to ways to address the question whether animals are aware: at a conceptual and experimental level. And this will continue in decades to come as there are still many conceptual and experimental hurdles to take, for instance, as it comes to insects. The question then arises how this will affect the public/ethics debate when we decide to switch to insects as a source of proteins. Hopefully, future discussions will show an evolution in our thinking on the relationship between science, ethics, and public debate to the extent that discussions on awareness are not a priori burdened by a desired outcome.

Box 8.1 Reasoning by Romanes

Early in the introductory chapter of *Animal Intelligence*, Romanes (1882) defines his reasoning by analogy (pp. 1–2): "*all our knowledge of their operations is derived, as it were, through the... activities of the organism. Hence it is evident that in our study of animal intelligence we are wholly restricted to the objective method. Starting from what I know subjectively of the operations of my own individual mind, and the activities which in my own organism they prompt, I proceed by analogy to infer from the observable activities of other organisms what are the mental operations that underlie them*". In *Mental Evolution*, Romanes (1884) states that this should be properly considered as *eject*, a projection rather than anything else. He describes the limitations of this reasoning (pp. 9–10): "*The whole organisation of such a creature is so different from that of a man that it becomes questionable how far analogy drawn from the activities of the insect is a safe guide to the inferring of mental states... so with 'inverted anthropomorphism' we must apply a similar consideration with a similar conclusion to the animal mind. The mental states of an insect may be widely different from those of a man, and yet most probably the nearest conception that we can form of their true nature is that which we form by assimilating them to the pattern of the only mental states with which we are actually acquainted. And this consideration, it is needless to point out, has a special validity to the evolutionist, inasmuch as upon his theory there must be a psychological, no less than a physiological, continuity extending throughout the length and breadth of the animal kingdom*". He stresses that it is probably more a difference in degree than in kind (pp. 12–13): "*Whether or not a neural process is accompanied by a mental process, it is in itself the same. The advent and development of consciousness,*

(continued)

Box 8.1 (continued)

although progressively converting reflex action into instinctive, and instinctive into rational, does this exclusively in the sphere of subjectivity; the nervous processes engaged are throughout the same in kind, and differ only in the relative degrees of their complexity". In addition, Romanes states that absence of evidence is not evidence of absence (p. 5): "*In other words, because a lowly organized animal does not learn by its own individual experience, we may not therefore conclude that in performing its natural or ancestral adaptations to appropriate stimuli consciousness, or the mind-element, is wholly absent; we can only say that this element, if present, reveals no evidence of the fact*". He defines his criterion of mind as (p. 4): "*It is, then, adaptive action by a living organism in cases where the inherited machinery of the nervous system does not furnish data for our prevision of what the adaptive action must necessarily be—it is only here that we recognise the objective evidence of mind.... Does the organism learn to make new adjustments, or to modify old ones, in accordance with the results of its own individual experience?*" which leads to a distinction in different classes: reflex action, instinct, reason/intelligence.

Box 8.2 Romanes on mental states in fish species

The following examples are taken from Animal Intelligence (Romanes 1882).

Affection (p. 246):

"*That adult fish are capable of feeling affection for one another would seem to be well established*: *thus Jesse relates how he once captured a female pike (Esox Lucius) during the breeding season, and that nothing could drive away the male from the spot at which he had perceived his partner slowly disappear, and whom he had followed to the edge of the water.*

Mr. Arderon gave an account of how he tamed a dace, which would lie close to the glass watching its master; and subsequently how he kept two rufis (Acerina cernua) in an aquarium, where they became very much attached to one another. He gave one away, when the other became so miserable that it would not eat, and this continued for nearly three weeks. Fearing his remaining fish might die, he sent for its former companion, and on the two meeting they became quite happy again. Jesse gives a similar account of two gold carp".

Intelligence (p. 251)

"*A morsel of food thrown into the tank fell directly in an angle formed by the glass front and the bottom. The skate, a large example, made several vain attempts to seize the food, owing to its mouth being on the underside of its head and the food being close to the glass. He lay quite still for a while as*

(continued)

Box 8.2 (continued)
though thinking, then suddenly raised himself into a slanting posture, the head inclined upwards, and the under surface of the body towards the food, when he waved his broad expanse of fins, thus creating an upward current or wave in the water, which lifted the food from its position and carried it straight to his mouth".

References

Ashley PJ, Ringrose S, Edwards KL, Wallington E, McCrohan CR, Sneddon LU (2009) Effect of noxious stimulation upon antipredator responses and dominance status in rainbow trout. Anim Behav 77:403–410

Baerends GP (1978) Welzijn–vanuit de ethologie bezien. In: Groen JJ, Groot AD (eds) Over Welzijn: Criterium, Onderzoeksobject, Beleidsdoel. Van Loghum Slaterus, Deventer, pp 83–106

Balcombe J (2016) What a fish knows; the inner lives of our underwater cousins. Scientific American/Farrar, Strauss and Giroux, New York

Balleine BW, Dickinson A (1998) Goal-directed instrumental action: contingency and incentive learning and their cortical substrates. Neuropharmacology 37:407–419

Balleine BW, Dickinson A (2000) The effect of lesions of the insular cortex on instrumental conditioning: evidence for a role in incentive memory. J Neurosci 20:8954–8964

Balleine BW, O'Doherty JP (2010) Human and rodent homologies in action control: corticostriatal determinants of goal-directed and habitual action. Neuropsychopharmacology 35:48–69

Bechara A (2005) Decision making, impulse control and loss of will power to resist drugs: a neurocognitive perspective. Nat Neurosci 8:1458–1463

Bechara A, Damasio AR, Damasio H, Anderson SW (1994) Insensitivity to future consequences following damage to human prefrontal cortex. Cognition 50:7–15

Bechara A, Damasio H, Tranel D, Damasio AR (1997) Deciding advantageously before knowing the advantageous strategy. Science 275:1293–1295

Bermond B (1997) The myth of animal suffering. In: Kasanmoentalib S, Dol M, Lijmbach S, Rivas E, van den Bos R (eds) Animal consciousness and animal ethics. Van Gorkum, Assen, pp 125–144

Bermond B (1999) Refectief bewustzijn, irreflectief bewustzijn, geen bewustzijn. In: Raat AJP, van den Bos R (eds) Welzijn van vissen. Tilburg University Press, Tilburg, pp 105–119

Bermond B (2001) A neuropsychological and evolutionary approach to animal consciousness and animal suffering. Anim Welf 10:47–62

Berridge KC (1996) Food reward: brain substrates of wanting and liking. Neurosci Biobehav Rev 20:1–25

Braithwaite V (2010) Do fish feel pain? Oxford University Press, Oxford

Braithwaite VA, Huntingford F, van den Bos R (2013) Variation in emotion and cognition among fishes. J Agric Environ Ethics 26:7–23

Broglio C, Gomez A, Duran E, Ocana FM, Jimenez-Moya F, Rodriguez SC (2005) Hallmarks of a common forebrain vertebrate plan: specialized pallial areas for spatial, temporal and emotional memory in actinopterygian fish. Brain Res Bull 66:277–281

Broom DM, Johnson KG (1993) Stress and animal welfare. Chapman & Hall, London

Brown C (2015) Fish intelligence, sentience and ethics. Anim Cogn 18(1):1–17

Bshary R, Brown C (2014) Fish cognition. Curr Biol 24(19):R947–R950

Cabanac M (1971) Physiological role of pleasure. Science 173:1103–1107

Cabanac M (1979) Sensory pleasure. Q Rev Biol 54:1–29

Cabanac M (1992) Pleasure: the common currency. J Theor Biol 155:173–200
Cabanac M (2002) What is emotion? Behav Process 60:69–83
Cabanac M (2008) The dialectics of pleasure. In: Kringelbach ML, Berridge KC (eds) Pleasures of the brain. The neural basis of taste, smell and other rewards. Oxford University Press, Oxford, pp 113–124
Cabanac M, Cabanac AJ, Parent A (2009) The emergence of consciousness in phylogeny. Behav Brain Res 198:267–272
Carter RM, Hofstotter C, Tsuchiya N, Koch C (2003) Working memory and fear conditioning. Proc Natl Acad Sci USA 100:1399–1404
Clark RE, Squire LR (1998) Classical conditioning and brain systems: the role of awareness. Science 280:77–81
Clark RE, Squire LR (2004) The importance of awareness for eyeblink conditioning is conditional: theoretical comment on Bellebaum and Daum. Behav Neurosci 118:1466–1468
Cools AR (1985) Brain and behavior: hierarchy of feedback systems and control of its input. In: Klopfer P, Bateson P (eds) Perspectives in ethology. Plenum Press, New York, pp 109–168
Damasio AR (1994) Descartes' error. Emotion, reason and the human brain. Avon Books, New York
Dawkins MS (1990) From an animal's point of view: motivation, fitness and animal welfare. Behav Brain Sci 13:1–61
de Veer MW, van den Bos R (1999) A critical review of methodology and interpretation of mirror self recognition research in nonhuman primates. Anim Behav 58:459–468
de Visser L, Homberg JR, Mitsogiannis M, Zeeb FD, Rivalan M, Fitoussi A et al (2011) Rodent versions of the Iowa gambling task: opportunities and challenges for the understanding of decision-making. Front Neurosci 5:109
de Waal FBM (2019) Fish, mirrors, and a gradualist perspective on self-awareness. PLoS Biol 17 (2):e3000112. https://doi.org/10.1371/journal.pbio.3000112
Dennett DC (1991) Consciousness explained. Penguin Books, London
Dias-Ferreira E, Sousa JC, Melo I, Morgado P, Mesquita AR, Cerqueira JJ, Costa RM, Sousa N (2009) Chronic stress causes frontostriatal reorganization and affects decision-making. Science 325:621–625
Dickinson A, Balleine B (1994) Motivational control of goal-directed action. Anim Learn Behav 22:1–18
Dickinson A, Balleine B (2008) The cognitive/motivational interface. In: Kringelbach ML, Berridge KC (eds) Pleasures of the brain. The neural basis of taste, smell and other rewards. Oxford University Press, Oxford, pp 74–84
Hagen K, van den Bos R, de Cock Buning TJ (2011) Editorial: concepts of animal welfare. Acta Biotheor 59:93–103
Han CJ, O'Tuathaigh CM, van Trigt L, Quinn JJ, Fanselow MS, Mongeau R et al (2003) Trace but not delay fear conditioning requires attention and the anterior cingulate cortex. Proc Natl Acad Sci USA 100(22):13087–13092
Knight DC, Cheng DT, Smith CN, Stein EA, Helmstetter FJ (2004) Neural substrates mediating human delay and trace fear conditioning. J Neurosci 24(1):218–228
Kohda M, Hotta T, Takeyama T, Awata S, Tanaka H, Asai J-Y et al (2019) If a fish can pass the mark test, what are the implications for consciousness and self-awareness testing in animals? PLoS Biol 17(2):e3000021. https://doi.org/10.1371/journal.pbio.3000021
LeDoux J (2012) Rethinking the emotional brain. Neuron 73:653–676
Leopold DA (2012) Primary visual cortex, awareness and blindsight. Annu Rev Neurosci 35:91–109. https://doi.org/10.1146/annurev-neuro-062111-150356
Martín I, Gómez A, Salas C, Puerto A, Rodríguez F (2011) Dorsomedial pallium lesions impair taste aversion learning in goldfish. Neurobiol Learn Mem 96:297–305
Maximino C (2011) Modulation of nociceptive-like behavior in zebrafish (*Danio rerio*) by environmental stressors. Psychol Neurosci 4:149–155

Medford N, Crichley HD (2010) Conjoint activity of anterior insular and anterior cingulate cortex: awareness and response. Brain Struct Funct 214:535–549
Mueller T (2012) What is the thalamus in zebrafish? Front Neurosci 6:64. https://doi.org/10.3389/fnins.2012.00064
Nilsson J, Kristiansen TS, Fosseidengen JE, Ferno A, van den Bos R (2008a) Learning in cod (*Gadus morhua*): long trace interval retention. Anim Cogn 11:215–222
Nilsson J, Kristiansen TS, Fosseidengen JE, Fernö A, van den Bos R (2008b) Sign and goal-tracking in Atlantic cod (*Gadus morhua*). Anim Cogn 11:651–659
Nilsson J, Kristiansen TS, Fosseidengen JE, Stien LH, Fernö A, van den Bos R (2010) Learning and anticipatory behaviour in a "sit-and-wait" predator: the Atlantic halibut. Behav Process 83:257–266
Nordgreen J, Janczak AM, Hovland AL, Ranheim B, Horsberg TE (2010) Trace classical conditioning in rainbow trout (*Oncorhynchus mykiss*): what do they learn? Anim Cogn 13:303–309
O'Connell LA, Hofmann HA (2011) The vertebrate mesolimbic reward system and social behavior network: a comparative synthesis. J Comp Neurol 519:3599–3639
Poli R (2019) Introducing anticipation. In: Poli R (ed) Handbook of anticipation. Theoretical and applied aspects of the use of future in decision making. Springer, Cham, pp 3–16
Rey Planellas S (2017) The emotional brain of fish. Anim Sentience 53(1–3)
Rey S, Huntingford F, Boltana S, Vargas R, Knowles T, Mackenzie S (2015) Fish can show emotional fever: stress-induced hyperthermia in zebrafish. Proc R Soc B Biol Sci 282(1819). https://doi.org/10.1098/rspb.2015.2266Rey
Rodríguez-Expósito B, Gómez A, Martín-Monzón I, Reiriz M, Rodríguez F, Salas C (2017) Goldfish hippocampal pallium is essential to associate temporally discontiguous events. Neurobiol Learn Mem 139:128–134
Romanes GJ (1882) Animal intelligence. Kegan Paul, Trench and Co., London
Romanes GJ (1884) Mental evolution in animals. Kegan Paul, Trench and Co., London
Roozendaal B, McEwen BS, Chattarji S (2009) Stress, memory and the amygdala. Nat Rev Neurosci 10:423–433. https://doi.org/10.1038/nrn2651
Rose JD, Arlinghaus R, Cooke SJ, Diggles BK, Sawynok W, Stevens ED, Wynne CDL (2014) Can fish really feel pain? Fish Fish 15:60–133
Salas C, Broglio C, Duran E, Gomez A, Ocana FM, Jimenez-Moya F et al (2006) Neuropsychology of learning and memory in teleost fish. Zebrafish 3:157–171
Schwabe L, Wolf OT (2011) Stress-induced modulation of instrumental behavior: from goal-directed to habitual control of action. Behav Brain Res 219:321–328
Sneddon L (2015) Pain in aquatic animals. J Exp Biol 218:967–976
Spruijt BM, van den Bos R, Pijlman F (2001) A concept of welfare based on how the brain evaluates its own activity: anticipatory behavior as an indicator for this activity. Appl Anim Behav Sci 72:145–171
St John Smith E, Lewin GR (2009) Nociceptors: a phylogenetic view. J Comp Physiol A 195:1089–1106
Stafleu FR, Rivas E, Rivas T, Vorstenbosch J, Heeger FR, Beynen AC (1992) The use of analogous reasoning for assessing discomfort in laboratory animals. Anim Welf 1(2):77–84
Tinbergen N (1963) On aims and methods of ethology. Z Tierpsychol 20:410–433
van den Bos R (1997) Reflections on the organisation of mind, brain and behavior. In: Dol M, Kasanmoentalib S, Lijmbach S, Rivas E, van den Bos R (eds) Animal consciousness and animal ethics; perspectives from the Netherlands, Animals in philosophy and science, vol 1. Van Gorcum, Assen, pp 144–166
van den Bos R (2000) General organizational principles of the brain as key to the study of animal consciousness. Psyche, 6. http://psyche.cs.monash.edu.au/v6/psyche-6-05-vandenbos.html
van den Bos R (2001) The hierarchical organization of the brain as a key to the study of consciousness in human and non-human animals: phylogenetic implications. Anim Welf 10: S246–S247

van den Bos R (2019) Animal anticipation: a perspective. In: Poli R (ed) Handbook of Anticipation. Theoretical and Applied Aspects of the Use of Future in Decision Making. pp 235–248 Springer Nature Switzerland AG

van den Bos R, Houx BB, Spruijt BM (2002) Cognition and emotion in concert in human and nonhuman animals. In: Bekoff M, Allen C, Burghardt G (eds) The cognitive animal: empirical and theoretical perspectives on animal cognition. MIT Press, Cambridge, MA, pp 97–103

van den Bos R, Meijer MK, Van Renselaar JP, Van der Harst JE, Spruijt BM (2003) Anticipation is differently expressed in rats (*Rattus norvegicus*) and domestic cats (*Felis silvestrus cattus*) in the same Pavlovian conditioning paradigm. Behav Brain Res 141:83–89

van den Bos R, Koot S, de Visser L (2014) A rodent version of the Iowa gambling task: 7 years of progress. Front Psychol 5:203

van der Harst JE, Fermont PCJ, Bilstra AE, Spruijt BM (2003) Access to enriched housing is rewarding to rats as reflected by their anticipatory behaviour. Anim Behav 66:493–504

Vargas JP, Lopez JC, Portavella M (2009) What are the functions of fish brain pallium? Brain Res Bull 79:436–440

Verheijen FJ, Buwalda RJA (1988) Report of the Department of Comparative Physiology. C.I.P. Gegevens, Utrecht

Vernier P (2017) The brains of teleost fishes. Evolution of nervous systems, vol 1, 2nd edn, pp 59–75

Weike AI, Schupp HT, Hamm AO (2007) Fear acquisition requires awareness in trace but not delay conditioning. Psychophysiology 44:170–180

Wemelsfelder F (1993) Animal boredom. Towards an empirical approach of animal subjectivity. PhD Thesis Leiden University

Wemelsfelder F, Mullan S (2014) Applying ethological and health indicators to practical animal welfare assessment. OIE Sci Tech Rev 33(1):111–120

Wiepkema PR (1985) Abnormal behaviours in farm animals: ethological implications. Netherlands J Zool 35:279–299

Woodruff ML (2017) Consciousness in teleosts: there is something it feels like to be a fish. Anim Sentience 10:1–21

Woodruff-Pak DS, Disterhoft JF (2008) Where is the trace in trace conditioning? Trends Neurosci 31(2):105–112. https://doi.org/10.1016/j.tins.2007.11.006

Yamamoto K, Vernier P (2011) The evolution of dopamine systems in chordates. Front Neuroanat 5:21. https://doi.org/10.3389/fnana.2011.00021

Chapter 9
The Predictive Brain: Perception Turned Upside Down

Tore S. Kristiansen and Anders Fernö

> *Matter (brains), thus organized, turns out to be ideally positioned to perceive, to understand, to dream, to imagine and (most importantly) to act. Perceiving, imaging, understanding, and acting are now bundled together, emerging as different aspects and manifestations of the same underlying predictive time driven, uncertainty machinery (Clark 2015).*

Abstract Accumulating evidence is turning the traditional picture of perception upside down and indicates that the brain works via the principles of predictive processing, in which the brain continuously attempts to predict its sensory input and its most probable causes, and compares these predictions to the actual sensory signals. The roots of the predictive brain paradigm stretch all the way back to Immanuel Kant's philosophy. Now new advances and insights into brain activity and function, theoretical neuroscience, and artificial intelligence have finally cleared the ground for what Kant believed would become a Copernican revolution in cognitive science. If the predictive processing theories are correct, they should also be relevant for other animals, including fish. If that is so, what implications will this have for our interpretation of observations related to perception, cognition, and learning in fish? And how can these theories help us to better understand their qualitative experience of life and improve their welfare? In this chapter, we try to explore these questions, but we also look at what challenges a fish faces and why they are dependent on a qualitative perception of the world and themselves if they are to master them. Finally, we discuss how we should reinterpret our previous

T. S. Kristiansen (✉)
Institute of Marine Research, Bergen, Norway
e-mail: torek@hi.no

A. Fernö
Department of Biology, University of Bergen, Bergen, Norway
e-mail: Anders.Ferno@uib.no

T. S. Kristiansen et al. (eds.), *The Welfare of Fish*, Animal Welfare 20,
https://doi.org/10.1007/978-3-030-41675-1_9

observations of fish behaviour and perception in the light of the predictive brain paradigm.

Keywords Fish · Allostasis · Prediction · Interoception · Exteroception · Proprioception · Living agents · Cognition · Consciousness · Anticipatory behaviour · Stress

9.1 Introduction

The past decades' enormous progress in neuroscience has provided us with a mountain of new knowledge and theories about the brain that also ought to be relevant for our understanding of fish. Innovative methods and advanced technologies have given us new insights that may lead to a fundamental change in how we think the brain works and how our perceptions of the world and ourselves are formed. One important realisation is that brains are not merely reactive, but predictive, and that all of its neurons are constantly firing, stimulating each other at various rates (Barrett 2017b). The brain processes exteroceptive sensory inputs from the external world as well as proprioceptive senses of body movements and interoceptive signals from other internal processes in the body. Accumulating evidence is turning the traditional picture of perception upside down and indicates that the brain works via the principles of predictive processing, in which the brain continuously attempts to predict its sensory input and its most probable causes, and compares these predictions to the actual sensory signals (Friston 2010; Clark 2013, 2015; Hohwy 2013; Wiese and Metzinger 2017). These top-down predictions or "prior beliefs" are based on earlier experiences of similar situations, but are continuously updated using approximate Bayesian inference to reduce (explain away) bottom-up flowing prediction errors. At any time, several competing alternative hypotheses are compared, but only the most probable prediction, i.e. the one with the smallest prediction error, is actually experienced (Clark 2015). One illustration of this is the familiar "young lady – old woman" optical illusion (Fig. 9.1), where the two hypotheses compete and flash in and out of our experienced reality, depending on our beliefs. This ability to make good predictions should be facilitated by evolutionary selection processes; those individuals who are making better predictions "intuitively" should have a higher probability to survive.

The roots of the predictive brain paradigm stretch all the way back to Immanuel Kant's philosophy (Kant 1783) and Herman von Helmholtz' theories of perception (von Helmholtz 1866), and related theories have resurfaced several times since (Gregory 1980; for review see Swanson 2016). However, new advances and insights into brain activity and function, theoretical neuroscience, and artificial intelligence have now finally cleared the ground for what Kant believed would become a Copernican revolution in cognitive science (cited in Swanson 2016):

> The situation here is the same as was that of Copernicus when he first thought of explaining the motions of celestial bodies. Having found it difficult to make progress there when he assumed that the entire host of stars revolved around the spectator, he tried to find out by

Fig. 9.1 The "young lady—old woman" optical illusion

experiment whether he might not be more successful if he had the spectator revolve and the stars remain at rest. Now, we can try a similar experiment in metaphysics, with regard to our intuition of objects. If our intuition had to conform to the character of its objects, then I do not see how we could know anything a priori about that character. But I can quite readily conceive of this possibility if the object (as object of the senses) conforms to the character of our power of intuition (Immanuel Kant, 1787, sec. B xvii (1996)).

If the predictive processing theories are correct, they should also be relevant for other animals, including fish. If that is so, what implications will this have for our interpretation of observations related to perception, cognition, and learning in fish? And how can these theories help us to better understand their qualitative experience of life and improve their welfare? In this chapter, we try to explore these questions, but first we will look at what challenges a fish faces and why they are dependent on a qualitative perception of the world and themselves if they are to master them.

9.2 Living Agents Try to Stay Alive

A difference between inanimate and living objects is that living organisms are agents—they do something to "stay alive". Living organisms are embodied within closed membranes that act as barriers towards an inside and an outside, and they can maintain a virtually steady state for hours (bacteria) to centuries (Greenland shark, *Somniosus microcephalus*) (Nielsen et al. 2016). This is achieved by consuming and spending energy to withstand the irreversibility of chemical processes and increasing entropy (second law of thermodynamics). In the course of their evolution, the millions of existing organisms have found numerous ways to stay alive, but all species face the same fundamental challenge: to persist as a species a sufficient number of individuals must survive, grow to maturity, and give birth to the next generation. Since all species can be food for others and are exposed to physical stress and chemical breakdown, this is not a simple task, so without life-supporting protective anabolic and catabolic processes and behaviours, they will soon starve to death, be killed, or just dissolve. Only species that have efficiently solved this task are still on the planet. The ray-finned fishes are the most successful of all vertebrates, both in numbers of individuals and species, and have evolved through almost all imaginable life strategies and morphologies (Chap. 3). Together they should, therefore, know all the tricks in the book to handle the challenges of life.

To be able to survive, grow, and reproduce, fish need to avoid harmful environments, escape from predators, and find enough nutritious prey for energy and growth. To achieve this, they need somehow to identify important regularities in their environment in space and time, such as where they are at risk and where to hide, where their food can be found, where and how they move, and where and in what condition they themselves are. Fish are found in almost all kinds of aquatic environments, ranging from small periodic water pools to the vast oceans (Chap. 3), so although the above tasks are challenging for all species, it may be more or less demanding for a given species, depending on the size and structure of its habitat and life strategy. As Chaps. 3 and 4 demonstrated, a fish is not just a fish, and its body and brain size and sensory abilities vary a great deal between the more than 34,000 known species of fish. We can, therefore, expect to find differences in cognitive skills and how fish species experience their lives, both between and within species and between the often profoundly different ontological states of the same individual. We know that fish have a rich toolbox of sensory organs, can learn from experience, behave in a flexible way and predict the results of future actions, all abilities that improve their survival capacity (Chaps. 3 and 7 and its references.). Even if relatively few fish species have been studied, our knowledge of the cognitive capacity of fish has greatly expanded in the course of the past few decades, and many fish species have been shown to have abilities comparable to those of some birds and mammals (Brown et al. 2011, Chap. 7).

9.3 Allostasis: The Process of Keeping the Body Alive

The brain has the core task of enabling the animal to grow, survive, and reproduce by ensuring that it has the resources needed for the multitude of physiological systems within the body (Sterling and Laughlin 2015). In this process, called allostasis (Sterling and Eyer 1988), the brain coordinates effectors to mobilize resources from modest corporal stores and enforces a system of flexible trade-offs: from each organ according to its ability, to each organ according to its need. The brain also helps regulate the internal milieu by governing anticipatory behaviour, for example by moving to a warmer location before it cools (Sterling 2012).

A well-known principle of cybernetics is that every good regulator of a system must be a model of that system (Conant and Ross Ashby 1970). This theorem has a corollary that the living brain, if it is to be successful and efficient as a regulator for survival, must learn by creating a model (or models) of its environment. All animals, therefore, run an internal model of their world and themselves for the purpose of allostasis (Sterling and Laughlin 2015). Large mobile multicellular animals are enormously complex and face tremendous challenges to coordinate the processes in their billions of specialised cells in order to function as a holistic integrated organism. To do this, they must distribute information, energy, and building materials to all cells and build new structures, refurbish old ones, and remove waste. To muster the necessary resources, they must know how to find them and need to know what is out there. Mobility enables them to explore large areas and obtain access to almost unlimited resources and to escape from predators, but mobility also increases energy costs and the risk of encountering predators and other hazards. To perform all these demanding tasks, animals need the brain and its network of advanced sensory organs.

9.4 Exteroception—Sensing the Surrounding Environment

The ability of a fish or any other organism to know something about the multitude of objects and agents in its world depends on their sensory equipment and ability to make sense of the information it supplies. Information from the environment is acquired via all our familiar senses such as vision, olfaction, taste, hearing, feeling of vibrations, touch, and various types of nociceptors (touch, heat, acid) (Von der Emde et al. 2004). Fish also possess unique sensory organs such as the lateral line system that allows them to detect even tiny water movements (Jansen 2004). Some species even have sensory organs that can both generate and sense electric fields (Keller 2004) and detect ultrasound (Higgs 2004), and receptors in their eyes that can sense UV and polarized light. Fish are found in aquatic habitats with a wide and dynamic range of physical and biological conditions, and their sensory organs and brains are specialized to collect the most information-rich qualities of their environment (Chap. 3). For example, fish species that live in muddy water have developed a sensitive olfactory sense mediated by large olfactory lobes in the brain, while fish in clear water relay more on excellent vision, large eyes, and a well-developed optic tectum (Kotrschal et al. 1998).

9.5 Proprioception—Sensing Own Movements and Actions

Proprioception is the sense through which we perceive the position and movement of our body, including our sense of equilibrium and balance. In order to act as a holistic and mobile organism, a fish needs to control and be aware of its own movements, and feedback from the proprioceptive senses is essential to keep the brain's internal representations of the environment and the body's actions up-to-date. This ability has in fact been suggested to be the origin of consciousness and a basic sense of self (Sheets-Johnstone 2007). Most fish species can move fast and precisely in a dynamic and three-dimensional environment. Organisms that live almost weightless in a fluid environment need to know if they themselves or the water are in movement. Fish can plan, sense, and adjust their movements in relation to their needs and goals by exploiting the force of their tail beats, swimming speed and direction, and positioning fins, and other parts of the body. Proprioception is the sensation of the actions and orientation of the body and the relative positions and movements of individual body parts, and is mediated by specialised sensory cells that measure internal pressures and tensions. Proprioception occurs continuously and implicitly, and body postures are constantly adjusted, most of the time unconsciously, perhaps except when a task needs special awareness, for example to protect the body against injuries.

We have identified only a few studies of proprioceptive feedback from fins and muscles in fish. Williams et al. (2013) found that the activity of fin-ray nerve fibres in bluegill sunfish (*Lepomis macrochirus*) reflected the amplitude and velocity of fin-ray bending and suggested that the pectoral fins should be considered as proprioceptive sensors. Fishes occupy an almost weightless environment and are more or less neutrally buoyant, and ought therefore to have difficulty in knowing what is up and down and how they are positioned in the water. To solve this problem, they possess a vestibular system (similar to our own), in which a fluid-filled three-dimensional system with small calcium stones called otoliths resting on sensory hair cells measures movement in the three dimensions, and controls body balance and also detects sound (Popper and Lu 2000). The lateral line can also be seen as a proprioceptive organ, and it has been shown that a functional lateral line is essential for schooling (Pitcher et al. 1976). Vision is of course also an important part of making sense of proprioceptive information and of estimating one's own movements in relation to the environment.

9.6 Interoception—Sensing the Internal State

Interoception is the representation and utilisation of sensory input to the brain from neurons and signal molecules that are continuously streamed from the internal organs and processes in the body (Barrett 2017a). This is information about internal states, such as heart rate, stomach fullness, oxygen saturation in the blood, and immune reactions. However, these signals too must be converted to concepts or signs that have meaning for the organism. For instance, certain combinations of sensory signals from the stomach and viscera should produce the feeling of "hunger". Seth and Friston (2016), Barrett and Simmons (2015), Barrett (2017a, b) and Kleckner et al. (2017) have recently argued that ascending interoceptive sensory inputs are anticipated and represented as emotional concepts. This may happen when certain combinations of interoceptive signals are associated with something of importance, for instance, if certain signals change when food is consumed, which later lead to similar patterns of signals that stimulate or inhibit food search. Kleckner et al. (2017) provide evidence of an intrinsic allostatic–interoceptive system in the human brain—comprising the salience and default mode networks—that supports not just allostasis but a wide range of psychological functions such as emotion, pain, memory, and decision-making, all of which can be explained by their reliance on allostasis. We should assume that similar allostatic–interoceptive systems also exist in fish and other animals since this is one of the major reasons why animals have a brain.

9.7 Perception Turned Upside Down?

If it is to optimise behaviour and body budget (perform allostasis), the brain needs to know what to expect, but how can it know anything about the world or what actions it should direct? The only access the brain has to this information is through the signals sent from the sensory organs and specialist cells. However, what is transferred from the sensory organs to the brain are electrochemical signals, so from this continuous stream of electromagnetic waves and chemical signals the brain must find out what is out there—or at least the parts of its environment that are necessary for its survival and reproduction.

The standard view of perceptual processing has been a bottom-up flow from the sensory receptors through the various layers of the brain (Marr 1982). As information flows forward a richer scene is built up, in which stored knowledge enriches and modulates the scene through top-down effects. The most thoroughly studied example is the visual system, which traditional neuroscience has regarded as passive and stimulus driven, in which signals from the retina are turned into a coherent percept and the scene is built up step by step: from simple intensities up to lines and edges and on to complex, meaningful shapes, accumulating structure, and complexity along the way (Clark 2015). The predictive brain or predictive processing model (PP for short) turns the view of neural processing upside down (Rao and Ballard 1999, Bubic et al. 2010)! Instead of "guessing" what the incoming signals mean, the brain makes a generative model of the world based on earlier experiences and predicts the most likely sensory input under the given circumstances (time and place). The downward flow of anticipatory prediction enables moment by moment processing to focus on the salient prediction errors, i.e. the unexplained sensory data, and it is only the unpredicted "surprises" that are fed back from the bottom up (Clark 2015). The predictive "hierarchical generative model" is continuously improved by making alternative prior hypotheses to minimise prediction errors. The brain is hypothesised to use an approximate Bayesian probabilistic model to make inferences about the most probable state, where the most probable scene under the given conditions (prior predictions and sensory signals) is the one that are experienced (Friston 2010; Hohwy 2013; Seth 2013, Clark 2013, 2015). For instance, when you go into a familiar room or meet a familiar person you have an expectation of what you will see and only have to do some crude visual checks to confirm the prediction. We are surprised when something is unexpected. The continuous process of perception and action (behaviour) is used to seek more precise information from the world and this process of "active inference" is used to support or reject the predicted scene, e.g. by moving closer or looking from a different angle (Friston et al. 2016).

The very reason for knowing something about the world and itself is, of course, to be able to perform actions that fulfil our needs efficiently and safely. In order to survive, we (and also fishes) must be able to predict accurately our behaviour and bodily movements along with their consequences (Metzinger 2009). If the PP theories are correct, PP is also causing the actions and movements. When an action is planned, the expected input from the senses when performing the action is predicted, and then by performing the predicted action, the sensory input caused by the action can be compared to the predicted one as prediction errors (Clark 2015). If the actions end as predicted, no or very little feedback is given (e.g. from under our feet when crossing the kitchen floor), but if it is not, the prediction error is registered as a perhaps unpleasant surprise (e.g. if stepping on something wet and soft).

As the neuroscientist Andy Clark (2015) puts it:

> Conceptually, this implies a striking reversal, in that the driving sensory signal is really just providing corrective feedback on the emerging top-down predictions. As ever-active prediction engines, these kinds of minds are not, fundamentally, in the business of solving puzzles given to them as inputs. Rather, they are in the business of keeping us one step ahead of the game, poised to act and actively eliciting the sensory flows that keep us viable and fulfilled. If this is on track, then just about every aspect of the passive forward-flowing model is false. We are not passive cognitive couch potatoes so much as proactive predictavores, forever trying to stay one step ahead of the incoming waves of sensory stimulation.

If these theories are true, not only are animals active constructors of their reality, but this is the only way they can experience reality at all!

9.8 Fish Must Also Be Conscious

Fish, like other organisms, are historical agents that build their "world models" based on their own experience, but the richness of the concepts (semiotic freedom, Hoffmeyer 1996) in their model (virtual reality) is constrained by their evolved sensory, neurobiological, and cognitive capacity. However, in order to experience some kind of virtual reality there must be someone to experience it—a conscious subject. As Thomas Metzinger (2005) expressed it: *If and only if a person is conscious, a world exists for her, and if and only if she is conscious, she can make the fact of actually living in a world available for herself, cognitively and as an agent.*

Consciousness or sentience is a phenomenon that we all experience and know exists, but how it occurs is still a mystery to science, and is known as the Hard Problem (Chalmers 1996; Metzinger 2005). Consciousness is often confused with awareness, which describes the narrow aspects of consciousness that are focused on at any given moment. To be unaware of something does not mean to be unconscious. Consciousness is unlimited—there is always a possibility to transfer your awareness to other parts of the environment or to part of yourself of which you were previously "unconscious". However, a conscious subject must always be aware of something that is a part of their environment or of themselves. Consciousness is not to think, but to be aware of the thoughts or other concepts created by the brain. The human experience of consciousness is clearly limited to our species, and is filled with human made concepts loaded with human values and qualia. The consciousness experienced by a fish must be very different and beyond our imagination, but it must be based on being a fish.

Fish can remember for months and years (Beukema 1970; Nilsson et al. 2008; Chap. 7), but to be conscious does not necessarily require declarative memory, even if this must be a great advantage. Conscious experience is just that something is subjectively experienced, "What it is like to be" (Nagel 1974). According to the PP theories, what is experienced is not directly "what is out there", but the self-generated guesses of what of importance for the organism in the world. Human beings have an experience of something like a continuously coherent well-focused 3D movie, with a voice-over that we call "thoughts". For example, we experience image constancy, in which the world remains still even if we move our head and we "see" a sharp image of the whole field of view even if we can only focus on a relatively narrow part of it. Based on our knowledge about fish neurobiology, advanced sensory organs, behaviour, and cognitive abilities (Bshary and Brown 2014, Brown 2015), we must assume that fish also experience their own self-generated virtual reality with meaningful objects and a similar perceptual constancy. What flatfish, with their independently moving eyes see, would be most interesting to know. Do they see two "screens" at the same time, or do they "zap" between the two in a binocular rivalry as experienced by humans, or do they fuse them? (Maier et al. 2012).

9.9 Anticipatory Behaviour and Stress

We now revisit some of our own publications on the cognitive capacities of fish and try to reanalyse them according to predictive (Bayesian) theories of brain function: how will our interpretation of the results change? The stimulus–response theory assumes that a certain stimulus (signal or sign) leads to a reflexive behavioural response. However, we now know that a response to a signal almost always depends on a range of factors (Lazarus 1991; Scherer et al. 2001). This is also true of fish: In a study by Folkedal et al. (2012), we used a light-blink that started 30s before feeding as the conditioned stimulus (CS) to condition Atlantic salmon (*Salmo salar*) to a food reward (unconditioned stimulus, US). The goal was to study the effects of recent exposure to stressors on their anticipatory behaviour before the food arrived and how it also was related to feeding motivation (hunger). A further goal was to test whether this method could be used as an indicator of well-being. We found that both stress and hunger had major effects on the anticipatory response to the signal. Repeated exposures to the CS–US procedure were necessary before the fish recovered the same anticipatory response as before stress, but the hungry fish recovered faster. If we interpret this according to a Bayesian brain and allostasis models, the "stressor" should lead to a change of the salmons' beliefs about the world and what actions and allostatic regulation should be undertaken. Should it be prepared for new stressors, or assume that the world had gone back to the previous state? The stress exposure should reduce the precision weighting (confidence) in future predictions of the state of the world, which in itself may be stressful, and lead to more agility vis-à-vis potential new stressors and therefore less appetite and feeding motivation. This was exactly what happened. After several hours and repeated exposure to the CS–US (without any new stressors occurring), the salmon gradually returned to their previous behaviour—i.e. the "prior" was again updated on the basis of new evidence of a stable environment and regular feeding.

9.10 From Fright to Anticipation

Another example is an experiment on Atlantic cod (Nilsson et al. 2012), in which we used a trace-conditioning paradigm, i.e. there was a 10s time interval between the end of signal and the reward, using an initially aversive/frightening CS and appetitive food reward as US. The aim was to find out whether fright could be turned into anticipation. As the CS we used an automatic device that suddenly pushed a dip-net down into the fish tanks. In four of eight tanks the fish was rewarded with food 10 s later. In the control groups we ran the same procedure, but without reward. The fish were of course scared, and initially showed startle responses and rises in swimming speed and oxygen consumption, but the rewarded fish rapidly learned the association between the dip net and the food and started to approach the net instead of fleeing. The unrewarded fish showed gradually less startle response and swimming speed and gradually reduced their oxygen consumption during the following exposures. During the 10 s trace period, the rewarded cod swam away from the dip-net to the feeding area, obviously expecting the food to arrive at this position, i.e. clearly showing predictive behaviour and an anticipation of future events. In a Bayesian brain terminology, we could say that the rewarded cod went from a prior belief in a stable world, in which the first entry of the dip net into the tank created a strong prediction error signal (surprise, which led to stress, fright), to a gradually stronger emerging belief whereby the net was anticipated and that it signalled food. The control group ended up with a posterior belief that the dip net was predictable and could be avoided and not very dangerous. However, these fish did not completely lose their initial startle response and maintained an elevated level of oxygen consumption, both signs of allostatic regulation, a certain increase in agility and less confidence in their predictions.

9.11 What Do Fish Feel?

According to the theory of constructed emotions, there are no "fear-centres" in the amygdala or "love-circuits" anywhere else in the brain that are turned on or off by different stimuli (Barrett 2017a,b). Depending on the context, the signals (e.g.) from a rapidly pumping heart may be due to the excitement of being in love or fear of heights. It is the probability of the emotional concept (e.g. fear or excitement) occurring in a given situation (e.g. if the heart is thumping because you are standing on a tall ladder or are going to call someone you are in love with for the first time) that makes up the totality of the signals that we experience as a specific emotion. If it is correct that emotions are constructed concepts and that this is how the fish brain also works, this must mean that the richness of a qualitative experience of well-being depends on the ability to construct emotional concepts. The resulting virtual realities must, therefore, be very different for different species according to their sensory capacity, the information flow from, and structure of, their habitat, and of course the computational capacity of the brain. A coral reef fish living in a physically and socially diverse environment needs a more advanced and complex "world model" than a blind cavefish. As we (or any animal) have no access to other models than our own, we can only use our observations (actually our own simulations) of the fishes' behaviour and physiology to model (imagine) their models. Expressed in the terminology of von Uexküll (1921), each species (and individual) is confined to its own limited Umwelt.

9.12 *Qualia* and Welfare

Our conscious experiences of specific constructed concepts (like "strawberry taste", "cold wind", "music" "fear", and "joy") are perceived in terms of emotional qualities that can be characterised on a valence scale (a sense of inner representation of value) for affects, ranging from very unpleasant to very pleasant (aversive to appetitive, etc.), and with high to low arousal measuring the "strength" of the experience (Fig. 9.2). These experiences are supported by sensory information, but are somehow constructed by the predictive brain on the basis of earlier experiences; and the experienced qualia are remarkably dependent on the predictions (priors). For instance, if you are drinking tea, and somebody without your notice switches your cup for a cup of coffee, your next sip from the cup will taste very bad and be very surprising!

We experience an almost endless number of very specific sensory qualia, like the redness or taste of an apple, the smell of a rose or the sense of touching velvet. These are all experienced with different valence and arousal, depending on their context and our prior experiences (Fig. 9.2). Every species and individual will experience their own special concepts with their set of *qualia*, depending on their sensory and brain capacity and experiences, and the range and richness of available concepts in

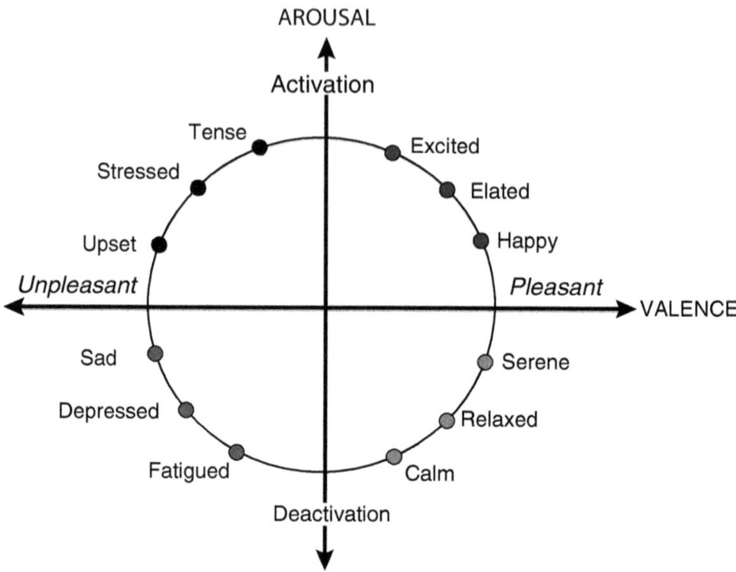

Fig. 9.2 The two dimensions of appraisal—valence and arousal—in which different combinations are transformed (by human beings) into specific emotional concepts (after Russell 1980). However, experiences of different qualia are multidimensional and can only be experienced live

their affective niche or Umwelt. The experienced qualia are signs or meanings that help to tell an animal whether something is important or not, and thus guide its actions. For instance, a smell associated with food feels "good" and guides the animal to the food, but more diversified and specific the qualia will create more precise predictions about the source (like the smell of freshly baked bread or barbecued meat) and thus produce a more efficient response (see also Kittilsen 2013). Some experiences of qualia may be "genetically" inherited, like the bad smell of rotten food or faeces, which lead animals to avoid bacterial infection or poisoning, or the sweetness of sugary food that tastes good the first time it is encountered, because consuming of sugar-rich food has been advantageous for survival. This is also the case for fish. For instance, when we gave farmed halibut juveniles boiled prawns for the first time they immediately became hyperactive when they sensed their odour (Nilsson et al. 2010).

9.13 Can the Predictive (Bayesian) Brain Paradigm Help Us to Better Understand and Improve Fish Welfare?

The concept of welfare is poorly defined, but most people agree that welfare has something to do with animals' quality of life and that we think it is ethically wrong that animals suffer. In the absence of a conscious experience of *qualia*, the welfare concept will have no meaning in this context. We should assume that acquiring resources that lead to the fulfilment of needs will be experienced as pleasant, and vice versa, that unfulfilled needs and life-threatening events are perceived as aversive. A well-functioning body may "feel good" (or perhaps in a neutral mood), whereas a poor functioning or injured body can be assumed to lead to some form of suffering that will motivate an animal to try to improve the situation and protect its body.

The Bayesian brain can be conceptualized as a probability machine that constantly makes predictions about the world and then updates them on the basis of what it receives via its senses (De Ridder et al. 2014). If we believe that PP theories are on track, this implies that what an animal experience as its reality is actually its own generative model of the world. This is the only "reality" the brain can use as a basis for allostasis—the predictive regulation of the body budget. For example, if you believe your neighbour wants to kill you, your body will feel very stressed and ready for flight or fight, irrespective of whether or not this is true. The Bayesian brain creates both "heaven and hell" based on its beliefs about the future.

When we farm fish, we have a high degree of control over their living condition throughout their life, and in principle we can construct a world for the fish in which they occupy the positive side of the valence-arousal dimensions (Fig. 9.2) and experience relatively good welfare. However, to do this we need to know how to help the fish construct this reality. First, every organism has its "preprogramed" constraints that limit the range of possible realities it can construct and of the physical environments with which it can cope (e.g. oxygen saturation and temperature). Secondly, all organisms have a certain range of plasticity that enables them to adapt to different conditions, including cognition and perception, i.e. the kind of meaning the sensory information can give them. For example, a blind cavefish has no need for better light conditions. Although there are huge variations among fish species, some are more adaptable than others and therefore better fit for farming.

According to the free energy principle (Friston 2010), fish also need to reduce, through its processes of perception, its degree of uncertainty (i.e. its prediction errors) regarding its environment. The more predictable the environment is, the better choices can be made, although some uncertainty always remains. However, living in an unpredictable world in which there is no correlation between actions and rewards/punishments will lead to helplessness and depression or anxiety (Clark 2013). The animal's prior beliefs and sensory information will be of zero accuracy (no confidence) and prediction errors will be very large. In the opposite situation, where everything is predictable, there will be no need for sensory information after the model is established, which will lead to a zombie-like state (Clark 2013). Neither of these situations occur in reality, but what degree of predictability is the ideal?

All fish behaviour is performed on the basis of beliefs and predictions about the near and more distant future (since "now" occupies no time at all), and these predictions are modulated by their state of needs (e.g. hunger) and the organism's belief in its own ability to satisfy them. The "preprogramed" goal of any organism is to satisfy its needs, the most basic of which are safety, food, and adaptable environmental conditions. The drive to obtain these goals varies with the current state of need, but at any time an inability to predict future needs can be fatal, for instance not anticipating that a predator may show up. On the other hand, to predict dangers that are not there will lead to wasted feeding opportunities and energy, but since one can die only once, unnecessarily cautious predictions may be a price worth paying.

One of the preprogrammed behavioural needs should be that of obtaining information about the hidden states of the world, in order to be able to make better predictions in the future, i.e. to continuously improve the generative model. To optimise allostasis and behaviour, a fish needs to know what to expect and acquire direct experience of a range of possible conditions. For instance, the ability to avoid and inspect predators or catch prey improves with practical experience. The rewarding experiences of improving physical and behavioural states of need are also what creates the experience of good welfare in fishes (by releasing substances such as dopamine, serotonin, and opioids/endorphins in the brain; Spruijt et al. 2001).

In accordance with this train of thought, to provide farmed fish with good welfare they should be given the following conditions:

- They should be predictable and controllable and within their environmental adaptation range.
- They should be safe (few dangers and harmful organisms) and with positive surprises (not monotonous, but rewarding experiences).
- The fish should be gradually trained to cope with stress and other environmental challenges, in order to improve their predictions and skills (controllability) and adjust their allostatic responses according to demand.

9.14 Ignorance: What We Still Do Not Know and What We Will Never Know

We will never fully know how fish experience their existence, but by studying their behaviour and neurobiology we can obtain some clues through cleverly designed experiments (Paul et al. 2005). For example, we can see the difference in behaviour between negative and positive anticipation or no anticipation at all (Folkedal et al. 2012; Nilsson et al. 2012). We still do not know if fish get trauma, i.e. if they ruminate over earlier traumatic incidents, or if memories of these simply disappear. This can be tested by well-designed experiments that enable us to check for long-lasting effects on systems ranging from behaviour to epigenetics. We do not know if the very common fin and skin injuries are painful, or if damage of proprioceptive sensors in fins and skin affect their coping ability and well-being. We still know little about how high fish density affects their neurobiology, experiences, and ability to formulate predictions. Do fish living in large shoals relax and give up their individuality and relinquish control to the school, or do they feel loss of control and stress? These and numerous other questions remain to be answered, and some will never be answered.

Predictive brain theory has important implications for how we design our experiments and how we interpret our results. Old observations will also have to be revisited and inspected from this top-down prediction perspective. For example, the prior beliefs of the experimental animals, shaped by their earlier experiences and evolved tendencies, should have large influence on individual behavioural and physiological responses. Can this explain the large individual variation we see in many experiments? As the astronomers in the fifteenth century had to turn around their way of thinking, we also have to rethink our earlier explanations of data. We look forward to seeing what this almost Copernican revolution will mean for our understanding of welfare of fish, humans, and other animals. What this will be is not easy to predict.

References

Barrett LF (2017a) The theory of constructed emotion: an active inference account of interoception and categorization. Soc Cogn Affect Neurosci 2017:1–23. https://doi.org/10.1093/scan/nsw154

Barrett LF (2017b) How emotions are made. The secret life of the brain. Macmillan, New York

Barrett LF, Simmons WK (2015) Interoceptive predictions in the brain. Nat Rev Neurosci 16:419–429

Beukema JJ (1970) Angling experiments with carp (Cyprinus carpio L.) II. Decreasing catchability through one-trial learning. Neth J Zool 20:81–92

Brown C (2015) Fish intelligence, sentience and ethics. Anim Cogn 18(1):1–17

Brown C, Krause J, Laland K (eds) (2011) Fish cognition and behaviour. Wiley, Oxford

Bshary R, Brown C (2014) Fish cognition. Curr Biol 24(19):R947–R950

Bubic A, Yves von Cramon D, Schubotz RI (2010) Prediction, cognition and the brain. Front Hum Neurosci 4:25

Chalmers D (1996) The conscious mind: In search of a fundamental theory. Oxford University Press, Oxford

Clark A (2013) Whatever next? Predictive brains, situated agents and the future of cognitive science. Behav Brain Sci 36:181–204

Clark A (2015) Surfing uncertainty: prediction, action and the embodied mind. Oxford University Press, New York, NY

Conant RC, Ross Ashby W (1970) Every good regulator of a system must be a model of that system. Int J Syst Sci 1:89–97

De Ridder D, Vanneste S, Freeman W (2014) The Bayesian brain: phantom percepts resolve sensory uncertainty. Neurosci Biobehav Rev 44:4–15

Folkedal O, Stien LH, Torgersen T, Oppedal F, Olsen RE, Fosseidengen JE, Braithwaithe VA, Kristiansen TS (2012) Food anticipatory behaviour as an indicator of stress response and recovery in Atlantic salmon post-smolt after exposure to acute temperature fluctuation. Physiol Behav 105(2):350–356

Friston K (2010) The free-energy principle: a unified brain theory? Nat Rev Neurosci 11:127–138

Friston K, FitzGerald T, Rigoli F, Schwartenbeck P, O'Doherty J, Pezzulo G (2016) Active inference and learning. Neurosci Biobehav Rev 68:862–879

Gregory RL (1980) Perceptions as hypotheses. Philos Trans R Soc Lond B 290:181–189

Higgs DM (2004) Neuroethology and sensory ecology of teleost ultrasound detection. Chapter 8. In: Von der Emde G, Mogdans J, Kapoor BG (eds) The Senses of Fish. Adaptations for the Reception of Natural Stimuli. Springer Science. Kluwer Academic, Boston

Hoffmeyer J (1996) Signs of meaning in the universe. Indiana University Press, Bloomington

Hohwy J (2013) The predictive mind. Oxford University Press, Oxford
Jansen J (2004) Lateral line sensory ecology, Chapter 11. In: Von der Emde G, Mogdans J, Kapoor BG (eds) The senses of fish. Adaptations for the reception of natural stimuli. Springer Science. Kluwer Academic Publishers, Dordrecht
Kant I (1783) Prolegomena to any future metaphysics: that will be able to come forward as science (updated edn). Hatfield G (ed). Cambridge: Cambridge University Press
Keller CH (2004) Electroreception: strategies for separation of signals from noise, Chapter 14. In: Von der Emde G, Mogdans J, Kapoor BG (eds) The senses of fish. Adaptations for the reception of natural stimuli. Springer Science. Kluwer Academic, Dordrecht
Kittilsen S (2013) Functional aspects of emotions in fish. Behav Process 100(376):153–159
Kleckner IR, Zhang J, Touroutoglou A, Chanes L, Xia C, Simmons WK et al (2017) Evidence for a large-scale brain system supporting allostasis and interoception in humans. Nat Hum Behav 1 (5):69
Kotrschal K, van Staaden MJ, Huber R (1998) Fish brains: evolution and environmental relationships. Rev Fish Biol Fish 8:373–408
Lazarus RS (1991) Progress on a cognitive-motivational-relational theory of emotion. Am Psychol 46:819–834
Maier A, Panagiotaropolous TO, Tsuchiya N, Keliris GA (eds) (2012) Binocular rivalry: a gateway to consciousness. Frontiers. Research Topics
Marr D (1982) Vision: a computational approach. W. H. Freeman, New York
Metzinger T (2005) Précis: being no one. Psyche 11:1–35
Metzinger T (2009) The Ego tunnel: the science of the mind and the myth of the self. Basic Books, New York
Nagel T (1974) What is it like to be a bat? Philos Rev 83(4):435–450
Nielsen J, Hedeholm RB, Heinemeirer J, Bushnell PG, Christiansen JS, Olsen J, Ramsey CB, Brill RW, Simon M, Steffensen KF, Steffensen JF (2016) Eye lens radiocarbon reveals centuries of longevity in the Greenland shark (Somniosus microcephalus). Science 353:702–704
Nilsson J, Kristiansen TS, Fosseidengen JE, Fernö A, van den Bos R (2008) Learning in cod (Gadus morhua): long trace interval retention. Anim Cogn 11(2):215–222
Nilsson J, Kristiansen TS, Fosseidengen JE, Stien LH, Fernö A, van den Bos R (2010) Learning and anticipatory behaviour in a "sit-and-wait" predator: the Atlantic halibut. Behav Process 83 (3):257–266
Nilsson J, Stien LH, Fosseidengen JE, Olsen RE, Kristiansen TS (2012) From fright to anticipation: reward conditioning versus habituation to a moving dip net in farmed Atlantic cod (Gadus morhua). Appl Anim Behav Sci 138(2012):118–124
Paul ES, Harding EJ, Mendl M (2005) Measuring emotional processes in animals: the utility of a cognitive approach. Neurosci Biobehav Rev 29(3):469–491
Pitcher TJ, Partridge BL, Wardle CS (1976) A blind fish can school. Science 194(4268):963–965
Popper AN, Lu Z (2000) Structure and function relationships in fish otolith organs. Fish Res 46:15–25
Rao RP, Ballard DH (1999) Predictive coding in the visual cortex: a functional interpretation of some extra-classical receptive-field effects. Nat Neurosci 2:79–87
Russell JA (1980) A circumplex model of affect. J Pers Soc Psychol 39:1161–1178
Scherer KR, Shorr A, Johnstone T (eds) (2001) Appraisal processes in emotion: theory, methods, research. Oxford University Press, Canary, NC
Seth AK (2013) Interoceptive inference, emotion, and the embodied self. Trends Cogn Sci 17 (11):565–573
Seth AK, Friston KJ (2016) Active interoceptive inference and the emotional brain. Philos Trans R Soc B 371:20160007. https://doi.org/10.1098/rstb.2016.0007
Sheets-Johnstone M (2007) Consciousness: a natural history, pp 37–41. Original Paper UDC 165.12

Spruijt BM, van den Bos R, Pijlman FTA (2001) A concept of welfare based on reward evaluating mechanisms in the brain: anticipatory behaviour as an indicator for the state of reward systems. Appl Anim Behav Sci 72(2):145–171
Sterling P (2012) Allostasis: a model of predictive regulation. Physiol Behav 106(1):5–15
Sterling P, Eyer J (1988) Allostasis: a new paradigm to explain arousal pathology, Chapter 34. In: Fisher S, Reason J (eds) Handbook of life stress, cognition and health. Wiley
Sterling P, Laughlin S (2015) Principles of neural design. MIT Press, Cambridge
Swanson LR (2016) The predictive processing paradigm has roots in Kant. Front Syst Neurosci 10 (Oct):1–13. https://doi.org/10.3389/fnsys.2016.00079
Von der Emde G, Mogdans J, Kapoor BG (eds) (2004) The senses of fish. Adaptations for the reception of natural stimuli. Springer Science. Kluwer Academic
von Helmholtz H (1866) Concerning the perceptions in general. In treatise on physiological optics, vol III, 3rd edn (translated by JPC Southall 1925 Opt Soc Am Section 26, reprinted Dover, New York, 1962)
von Uexküll J (1921) Umwelt und Innenwelt der Tiere, 2nd edn. Springer, Berlin
Wiese W, Metzinger T (2017) Vanilla PP for philosophers: A primer on predictive processing. In: Metzinger T, Wiese W (eds) Philosophy and predictive processing, vol 1. MIND Group, Frankfurt am Main. https://doi.org/10.15502/9783958573024
Williams Iv R, Neubarth N, Hale ME (2013) The function of fin rays as proprioceptive sensors in fish. Nat Commun 4(1). https://doi.org/10.1038/ncomms2751

Chapter 10
Can Fish Experience Pain?

Lynne U. Sneddon

Abstract Experiencing pain is one of the key drivers of deciding whether to protect an animal under legislation and guidelines. Over the last two decades empirical evidence for fish experiencing pain has grown and this chapter reviews the current state of our knowledge. Defining animal pain has been problematic but a definition based upon whether whole animal responses to pain differ from non-painful stimuli and whether the experience alters future behavioural decisions and motivation is adopted. Studies show that fish have a similar nociceptive system to mammals, that behaviour is adversely affected and that this is prevented by pain-relieving drugs demonstrating that fish respond to pain in a different manner to innocuous events. Further, fish are motivated to avoid areas where pain has been experienced and are consumed by the painful event such that they do not exhibit normal fear or antipredator responses. Taken together these results make a compelling case for pain in fish. However, this topic is still debated and the chapter discusses the opposing opinions. If we accept pain occurs in fish then the wider implications of the use of fish must be considered. It would be in the public's interest to keep fish healthy for a myriad of reasons including disease-free fish production, preventing zoonoses, conservation and sustainability of fish stocks and valid experimental results from laboratory studies using fish models.

Keywords Animal welfare · Aquaculture · Behaviour · Fisheries · Nociception · Rainbow trout · Zebrafish

---

L. U. Sneddon
Institute of Integrative Biology, University of Liverpool, Liverpool, UK
e-mail: lsneddon@liverpool.ac.uk

T. S. Kristiansen et al. (eds.), *The Welfare of Fish*, Animal Welfare 20,
https://doi.org/10.1007/978-3-030-41675-1_10

10.1 Introduction

This chapter sets out to review the empirical evidence for the capacity to experience pain in fish. Previous chapters have demonstrated the complexities of behaviour, learning and cognition, sensory abilities as well as emotional responses in fish, which are key to demonstrating that fish are sentient and experience adverse affective states. When considering which animals to protect, ethical guidelines and legislation usually base decisions upon whether studies have shown the ability to experience negative states such as pain as a justification for affording that animal protection. Here, the current definition of animal pain is discussed incorporating testable criteria alongside scientific results demonstrating whether fish fulfil this definition. Despite the empirical evidence, a small number of reviewers object to the concept of pain in fish thus this chapter will review opinions for and against.

Fish are used in a wide variety of contexts including as food, in recreational and sport fishing, a research model in experimentation and as a companion animal or public exhibit (Cooke and Sneddon 2007; Sneddon and Wolfenden 2012; Sneddon 2015). From a moral and ethical perspective, one could conclude that any animals under the care of humans or indeed used to benefit humanity should be treated in a manner that ensures good welfare and health. Not only this is beneficial to the animal, but also ensures the success of efforts to conserve fish species, sustainability of fishing stocks, an improved economic return in the aquaculture and the ornamental fish industries where healthy fish pose no risk to food security or public health, provide valid scientific results and promote the longevity and attractiveness of companion fish (Metcalfe 2009; Sneddon 2006, 2015; Sneddon et al. 2016). Thus, there are many positive aspects to safeguarding good health and welfare in fish. Legislation concerning fish in laboratory experiments differs across the globe with some countries affording fish protection and others not. For example, fish are protected once they become capable of independent feeding in Europe (e.g. zebrafish at 120 h post fertilisation at 28.5° C) whereas in the United States fish are not included since only warm-blooded animals are protected except rats (genus *Rattus*) or mice (genus *Mus*) (Sneddon et al. 2017). Australia and South Africa include all developmental stages of vertebrates and cephalopods whereas countries such as China and India state all animals including invertebrates are subject to experimental ethical guidelines (Sneddon et al. 2017). Europe has developed a Common Fisheries Policy (CFP, Europa 2014a) that sets quotas, determines which species are caught, which fishing method is used and proposes sustainability of fish stocks but does not directly deal with welfare issues of wild-caught fish. The CFP also extends to aquaculture and fish farming to regulate the administration, logistics and environmental impact of these practices (Europa 2014b). The Farm Animal Welfare Council of the UK reports that the fish farming industry have voluntarily embraced good welfare practices and have advised that fish should be treated as if they experience pain during rearing (FAWC 2014a) and during slaughter (FAWC 2014b) based upon the available published studies. Given the various contexts that fish are used in, it is important from a moral and ethical standpoint to understand their capacity for pain.

10.2 The Concept of Animal Pain

The terms nociception and pain are intrinsically linked and used in the literature when discussing how animals react to potentially painful stimuli. It is believed nociception is a reflex response restricted to lower brain centres and the spinal cord since it is the detection of potentially injurious stimuli and is often followed by an instantaneous reflex withdrawal response away from the noxious stimulus. Nociception occurs in all animals and is effectively a warning system that influences survival (Sneddon 2018). Nociception in humans leads to the sensation of pain, which is defined as "an unpleasant sensory and emotional experience associated with actual or potential tissue damage, or described in terms of such damage" (note there is a proposal for an updated definition: "An aversive sensory and emotional experience typically caused by, or resembling that caused by, actual or potential tissue injury", IASP 2019). Any stimulus that can possibly or does cause injury leads to the negative affective feelings associated with pain under this definition (IASP 2019). Humans can communicate their pain through spoken language, which is the typical means of assessment. This presents a problem as we do not share a common verbal language with animals thus assessing pain relies on behavioural and physiological animal responses during an event that causes pain to humans (Sneddon et al. 2014). Accepting that human-based definitions of pain are not appropriate for animals has led to new concepts being proposed that consider the differences in the evolution and life history of animals (Sneddon et al. 2014). These two main concepts state (1) any responses to pain should differ from non-painful stimuli and (2) experiencing pain should lead to long-term motivational change where future behavioural decisions are akin to exhibiting discomfort, promoting healing and avoiding future encounters with the noxious event (Sneddon et al. 2014). In order to sense pain, animals must have nociceptors, receptors that preferentially detect damaging stimuli such as extremes of temperature, high mechanical pressure and damaging chemicals. Nociceptors have been identified in non-human mammals (Lynn 1994; Rutherford 2002; Sneddon et al. 2014) and other vertebrate groups such as amphibian, reptilian and avian species (e.g. amphibians: Willenbring and Stevens 1995; Guénette et al. 2013, birds: Gentle 1992; Nasr et al. 2012) but also in invertebrates (St. John Smith and Lewin 2009; Sneddon 2015). Nociception and pain in fish have only recently been empirically studied with authors prior to 2002 (Rose 2002) denying the existence of nociceptors. In the last 15 years, research has identified and described the properties of peripheral nociceptors that preferentially detect noxious stimuli in fish (Sneddon 2002, 2003a; Ashley et al. 2006, 2007; Mettam et al. 2012); shown altered brain activity during painful treatment (Dunlop and Laming 2005; Nordgreen et al. 2007; Reilly et al. 2008a; Sneddon 2011a) and additionally described the behavioural changes to painful events (Sneddon et al. 2003a, b; Sneddon 2003b;

Dunlop et al. 2006; Reilly et al. 2008b; Roques et al. 2010; Mettam et al. 2011; Maximino 2011; Alves et al. 2013) all of which are ameliorated by effective analgesia or pain relief (Sneddon 2003b, 2012; Newby et al. 2009; Nordgreen et al. 2009; Mettam et al. 2011). Thus, the empirical evidence does support the hypothesis fish may be candidates for experiencing pain.

10.3 Concept One: Whole Animal Responses to Pain

The first concept states that whole animal responses to painful stimulation must differ from innocuous stimuli (Table 10.1; Sneddon et al. 2014). Animals must possess a nociceptive system to detect damaging stimuli. The information from peripheral nociceptors should be conveyed to the central nervous system where

Table 10.1 The two key principles and detailed criteria for pain in animals. These criteria must be fulfilled in their totality for an animal to be considered capable of pain (Adapted from Sneddon et al. (2014) with kind permission from Elsevier)

Primary principle	1. Whole animal responses to potentially painful events differ from innocuous stimulation	2. Change in motivational behaviours after a potentially painful event
	• Possession of nociceptors, pathways to central nervous system, evidence of central processing involving areas that regulate motivated behaviour (including learning and fear) • Nociceptive action responsive to endogenous modulators (e.g. Opioids) • Nociception activates physiological responses linked to stress or an elevated state over and above stress (one or a combination of the following alterations: Respiration, heart rate, or hormonal levels (e.g. Cortisol) • Evidence that responses are not just a nociceptive withdrawal reflex • Alterations in behaviour over longer term that reduce future encounters with the harmful stimulus • Protective behaviour such as wound guarding, limping, rubbing, licking or excessive grooming • All of the above reduced by analgesia or local anaesthetics	• Self-administration of analgesia • Pay a cost to access analgesia • Selective attentional mechanisms whereby the response to the noxious stimulus has high priority over other stimuli; the animal does not respond appropriately to competing events (e.g. presentation of predator; reduced performance in learning and memory tasks) • Altered behaviour after noxious stimulation where changes can be observed in conditioned place avoidance and avoidance learning paradigms • Relief learning where an animal associates a neutral stimulus with the cessation of pain • Long-lasting change in memory and behaviour, especially those relating to avoidance of repeat noxious stimulation • Avoidance of the noxious stimulus modified by other motivational requirements as in trade-offs, e.g. hungry animal will return to area where pain was given to seek food after a relevant period of time • Evidence of paying a cost to avoid the noxious stimulus

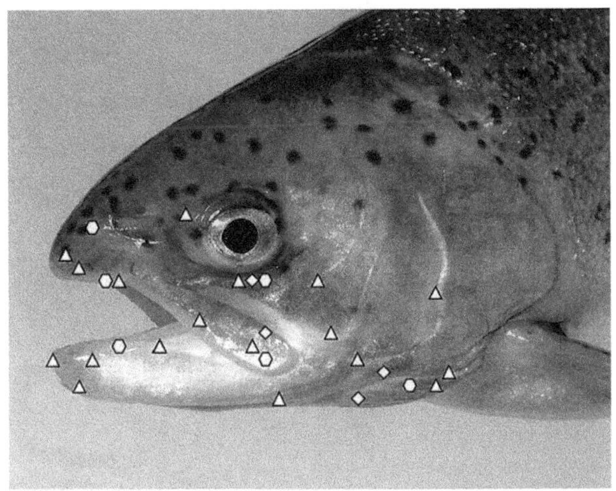

Fig. 10.1 Location of the nociceptors and chemical receptors on the head of the trout (open triangle = polymodal nociceptor, open diamond = mechanothermal nociceptor, open hexagon = mechanochemical receptor. From Sneddon et al. (2003a) by permission from the Royal Society of London

processing in key brain areas that innervate motivation, emotions and learning should occur. Parameters such as the stress response may be elicited since pain is stressful. Any behavioural responses during a painful event should not be immediate reflexes but should be prolonged and include protective or guarding behaviours and future avoidance of the painful event. Changes in behaviour and physiology should be reduced by administering effective analgesics or painkillers. These criteria are considered for fish below.

Possession of nociceptors: Using electrophysiology and neuroanatomical techniques, a number of studies have characterised nociceptors in a teleost or bony fish, the rainbow trout (*Oncorhynchus mykiss*), for the first time (Figs. 10.1, 10.2 and 10.3; Sneddon 2002, 2003a) showing they have A-delta and C fibres. These fibres act as nociceptors in mammals. Nociceptors have been found in the agnathan jawless fish, the lamprey (Matthews and Wickelgren 1978) but studies have not found nociceptors in elasmobranchs where it has been reported there is a lack of C fibres which are one type of nociceptor in mammals (e.g. Snow et al. 1996). The unmyelinated C fibres and small-diameter myelinated A-delta fibres, were found in the trout with three classes of nociceptors identified including polymodal (responsive to mechanical, thermal and chemical stimuli), mechanothermal (no response to chemicals) and mechanochemical (no response to temperature) (Sneddon 2003a; Ashley et al. 2006, 2007; Mettam et al. 2012). When compared to mammals the electrophysiological properties of these trout nociceptors are similar (Fig. 10.3; Table 10.2; Sneddon 2004, 2012). However, trout nociceptors are not responsive to cold temperatures (<4 °C) (Ashley et al. 2007); it would be maladaptive for the trout nociceptors to respond to cold since trout can live at very low temperatures. A relatively lower proportion of fish nociceptors are innervated by C fibres (4–5%, Sneddon 2002; Roques et al. 2010) in contrast to terrestrial vertebrates where some 50% of nociceptors are C fibres (Young 1977) although reptiles also have a lower

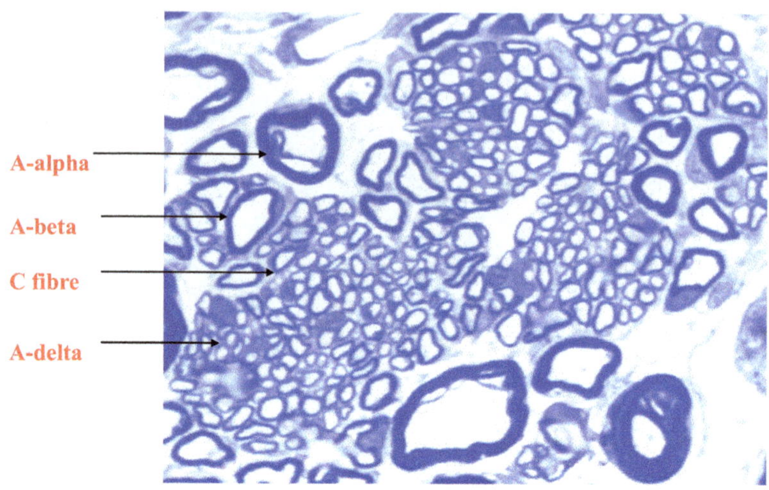

Fig. 10.2 Section of the maxillary branch of the trigeminal nerve of the rainbow trout showing the presence of A-delta and C fibres that may act as nociceptors (×1000, scale bar = 2 μm. Adapted from Sneddon, L. U. 2002. Anatomical and electrophysiological analysis of the trigeminal nerve in a teleost fish, *Oncorhynchus mykiss*. *Neurosci. Letts.*, 319, 167–171 by kind permission from Elsevier)

proportion of C fibres (Tershima and Liang 1994). C fibres in mammals contribute to dull, "thudding" pain whereas A-delta fibres are believed to signal "first" pain to the central nervous system (CNS). Sceptics have suggested that the small number of C fibres means fish cannot experience pain (Rose et al. 2014). However, A-delta fibres conduct more quickly, so perhaps the fish system is more rapid when signalling pain. Factoring in the ecological, life history and evolutionary differences that fish present, there will be a difference in how injury may occur compared with the terrestrial environment. Gravity (falling) will be counteracted by buoyancy resulting in a lower risk of damage. Noxious chemicals may be diluted in the aquatic environment and temperature changes occur less in aquatic bodies than in terrestrial environments. Therefore, damage from gravity, extremes of temperature and noxious chemicals may be experienced to a lesser degree by aquatic animals. Thus, this may explain why there is a disparity between the proportion of C fibres in trout yet it is important to note that the trout A-delta fibres respond similarly as mammalian C fibres reacting to different types of noxious stimuli and many are polymodal nociceptors (Sneddon 2002, Sneddon 2003a, b; Ashley et al. 2006, 2007; Roques et al. 2010; Mettam et al. 2012).

Most research has used teleost or bony fish with only a few on elasmobranch (cartilaginous) fish. Methodological details are lacking in one case (Leonard 1985), however, data from the long-tailed stingray (*Himantura fai)* has confirmed that there are no C fibres present, but there is a wealth of small myelinated A-delta fibres (Kitchener et al. 2010). Nevertheless, more research is required to identify nociceptors in elasmobranchs. Many sharks, skates and rays exhibit courtship biting

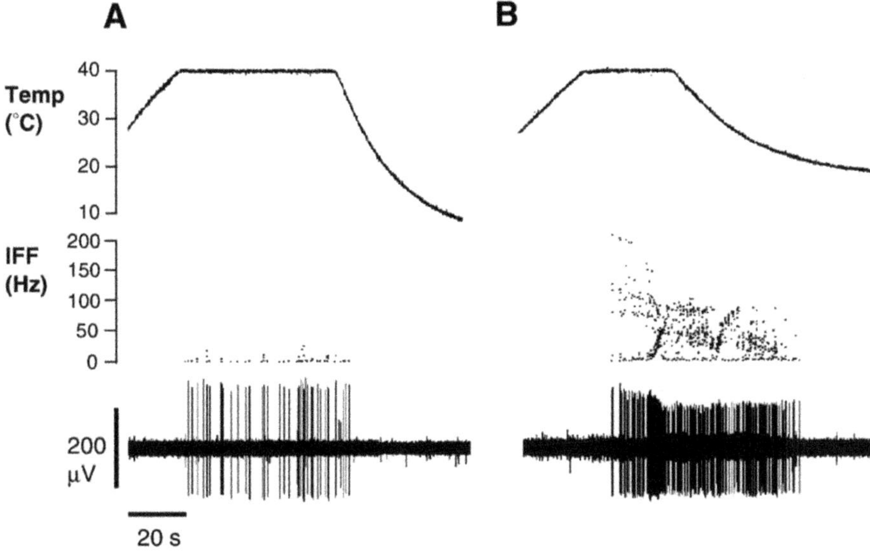

Fig. 10.3 Electrophysiological recordings from a nociceptive receptive field on the trout face showing responses of nociceptors to heat stimulation. The instantaneous firing frequency (IFF) is displayed in the centre as scatter graphs. This illustrates sensitisation of a mechanothermal receptor to heat following noxious chemical stimulation. The firing response to ramp and hold heat stimulation is shown (**a**) before and (**b**) 9 min after subcutaneous injection of 1% formalin <1 mm from the receptive field. Upper trace shows heat stimulus, middle trace plots instantaneous firing frequency (IFF) and lower trace shows extracellular single-unit recording from the trigeminal ganglion. Thermal threshold remains the same but firing frequency is greatly increased following formalin injection (Adapted from Ashley P.J., Sneddon L.U. and McCrohan C.R. 2007 Nociception in fish: stimulus–response properties of receptors on the head of trout *Oncorhynchus mykiss*. Brain Res. 1166, 47–54 by kind permission from Elsevier)

behaviour resulting in injuries thus this group may have a reduced capacity for pain (e.g. Kajiura et al. 2000; Porcher 2005), combined with slower healing (Heupel et al. 1998; Ashhurst 2004) suggesting injury may be less of a risk than for other fish groups. However, Porcher (2005) observed that courtship bite wounds in the blackfin reef shark, *Carcharhinus melanopterus*, healed rapidly within 10 days.

Using neural tract tracing the fish neuroanatomical pathways from the periphery to the brain are highly conserved when compared with mammals (Sneddon 2004) and within the teleost brain there are various connections to the thalamus and cortical areas (Rink and Wullimann 2004; see Chap. 6 for fish brain anatomy) that are involved in pain processing in mammals. Higher brain areas respond during noxious stimulation in teleost fish (e.g. gene expression in common carp, *Cyprinus carpio*, and rainbow trout, Reilly et al. 2008a; electrical activity in Atlantic salmon (*Salmo salar*), Nordgreen et al. 2007; goldfish (*Carassius auratus*) and rainbow trout, Dunlop and Laming 2005; pain-specific activity using functional magnetic resonance imaging (fMRI) in common carp, Sneddon 2011a, b) thus central activity differs from innocuous stimuli and is not restricted to the reflex centres in the

Table 10.2 Electrophysiological properties of A-delta nociceptors in a fish (Sneddon 2003a), a snake (Liang and Terashima 1993), and a mouse (Lopez de Armentia et al. 2000)

	Fish	Snake	Mouse
Conduction velocity (m/s)	0.7–5.5	3.8	0.7–5.7
AP amplitude (mV)	10–90	91	70–89
AP duration (ms)	0.8–2.4	2.4	0.7–2.8
AHP amplitude (mV)	1.8–5.5	11.9	6–12
dV/dtmax (V/s)	63–226	182	115–291

Mean values are shown for conduction velocity, action potential (AP) amplitude and duration, afterhyperpolarisation (AHP) amplitude and the maximum rate of depolarisation (dV/dtmax)
Reprinted from Neuroscience, 101, Lopez de Armentia et al., 1109–1115, (2000), with kind permission from Elsevier and from Journal of Comparative Neurology, 328, Liang & Terashima, 88–102, (1993) with kind permission from John Wiley & Sons Inc.

hindbrain and spinal cord (Rose 2002). The whole brain is involved and this may innervate the protracted changes in behaviour described below. A plethora of analgesic drugs reduce the pain-related changes in behaviour and physiology seen in noxiously stimulated fish (Fig. 10.4; Sneddon 2003b; Sneddon et al. 2003a; Mettam et al. 2011; review in Sneddon 2012). At the molecular level, responses are also conserved since opioid receptors and the action of non-steroidal anti-inflammatory drug (NSAID) on cyclooxygenase (COX2) enzyme are similar between fish and mammals (review in Malafoglia et al. 2013). Thus, the pain neural apparatus in fish is comparable to the mammalian system.

When changes in behavioural and physiological reactions during pain are observed this suggests a negative affective component to pain that may be indicative of discomfort. Common carp (*C. carpio*) withdraw from electric shock with responses reduced after anaesthesia was administered to elicit loss of consciousness yet this did not impair motor activity (Chervova and Lapshin 2011). Fish can learn to avoid electric shock which is painful to humans usually in one or a few trials (e.g. Yoshida and Hirano 2010) suggesting this is aversive to the fish. This avoidance behaviour continues for 3 days (Dunlop et al. 2006), but fish will return to the shock zone to obtain food after 3 days of food deprivation (Millsopp and Laming 2008).

Prolonged, complicated responses have been shown by a variety of fish species during a painful treatment (review in Sneddon 2009). Opercular beat rate (ventilation of the gills) increases over and above a stress response in rainbow trout and zebrafish (*Danio rerio*). Increased plasma cortisol is seen in rainbow trout (Sneddon 2003b; Ashley et al. 2009) and Mozambique tilapia (*Oreochromis niloticus*, Roques et al. 2012). Normal behaviours are often disrupted including swimming (Sneddon 2003b; Reilly et al. 2008b; Correia et al. 2011; Roques et al. 2012). Guarding behaviour (i.e. avoiding using an area into which a painful stimulus has been administered) has been recorded in trout where they suspend eating after a painful injection to the lips for up 3 h (Sneddon 2003b); sham handled (anesthetised only), saline-injected controls and acid-injected fish treated with morphine resume feeding after 80 min.

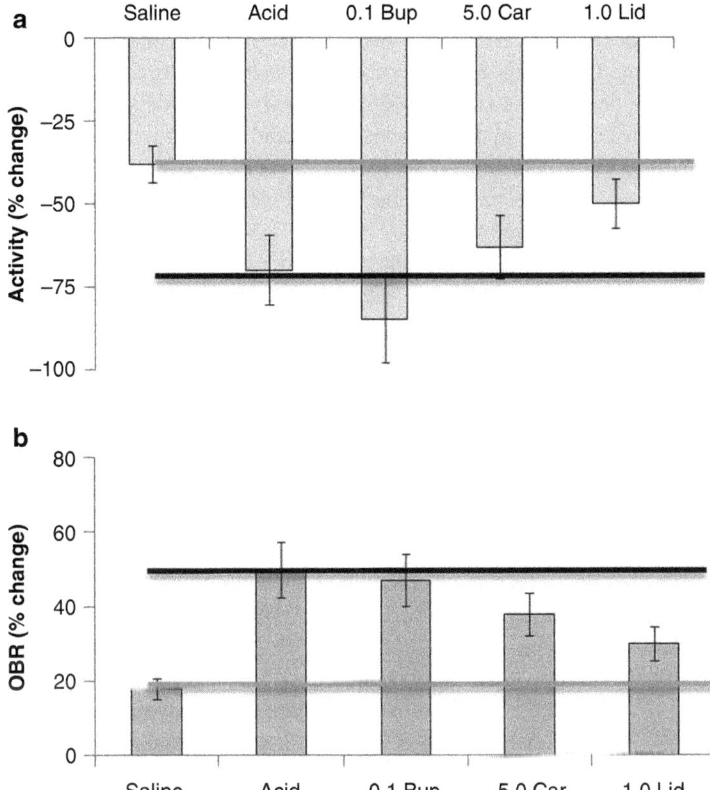

Fig. 10.4 The percentage change in (**a**) activity and (**b**) opercular beat rate (OBR) performed by rainbow trout 30 min after they were injected subcutaneously with saline or a noxious substance, 0.1% acetic acid (Acid) or acid combined with intramuscular injection of 0.1 mg/kg buprenorphine (0.1 Bup) or 5 mg/kg carprofen (5 mg/kg Car) or injected at the same site as the acid with 1 mg lidocaine (1.0 Lid). The grey line represents the impact of saline (control) treatment whereas the black line represents the impact of pain (acid injection; adapted from Mettam J.J., Oulton L.J., McCrohan C.R. and Sneddon L.U. 2011. The efficacy of three types of analgesic drugs in reducing pain in the rainbow trout, *Oncorhynchus mykiss*. Appl. Anim. Behav. Sci. 133, 265–274 by kind permission from Elsevier

Species-specific responses to pain are known to vary between mammalian species (Flecknell et al. 2007) and these species-specific behavioural responses have been recorded in fish. Piauçu (*Leporinus macrocephalus*) injected with formalin and Nile tilapia that have had a tail fin clip increase swimming (Roques et al. 2010; Alves et al. 2013). In contrast, Mozambique tilapia after electric shock and Atlantic salmon experiencing abdominal peritonitis decreased swimming (Bjørge et al. 2011; Roques et al. 2012). These contrasting responses demonstrate that behavioural indicators will have to be identified on a species-by-species basis and additionally to each type of pain. Pain-related alterations in behaviour are seen to persist from 3 h up to 2 days thus they are not simple instantaneous nociceptive reflexes (Sneddon 2003b; Bjørge et al. 2011). Pain thresholds may also differ between species. For example in rainbow trout, acetic acid, a standard mammalian pain test, elicits behavioural and physiological changes when injected subcutaneously at up to 2%, since concentrations above this destroy and silence nociceptor activity (Ashley et al. 2007; Mettam et al. 2012). In contrast, concentrations of acetic acid above 5% are needed in common carp to elicit pain-related behavioural reactions (Reilly et al. 2008b). This may indicate that cyprinids have a higher pain threshold that salmonids demonstrating species-specific differences.

Anomalous or abnormal, novel behaviours are often seen in response to painful treatment in fish. For example, tail beating in zebrafish. This is a response to injection with a known pain stimulus, acetic acid, in the caudal peduncle where zebrafish perform vigorous tail fin wafting, yet swimming and activity are both reduced (Maximino 2011; Schroeder and Sneddon 2017). Abnormal behaviours such as this has only been recorded after injection of pain-causing chemicals and include "rocking" where fish rock to and fro on the substrate, and rubbing of the injection site against the tank sides (Sneddon 2003a, b; Sneddon et al. 2003b; Reilly et al. 2008b; Newby et al. 2009). These responses have never been observed in sham-handled individuals (anaesthetised but no pain), saline-injected fish (innocuous), nor reported in toxicological research. This is compelling evidence that these novel behavioural changes are a direct result of the painful treatment, and studies have shown they are reduced by administering drugs with pain-relieving properties (Sneddon 2003a; Mettam et al. 2011; Schroeder and Sneddon 2017).

Learning to avoid a painful stimulus is evidence of the adverse and strong impact upon affective state that pain has. In goldfish, *C. auratus*, exposed to painful electric shock are able to learn to this (Portavella et al. 2002, 2004). Morphine (pain-relief) administration increased the voltage required to elicit this learnt response thus pain was diminished by morphine. Using MIF-1 and naloxone (opiate antagonists), the effects of morphine were blocked and a lower voltage elicited a learnt response. These antagonists act in a similar manner in mammals (Ehrensing et al. 1982). Additionally fish can learn through classical conditioning to associate a painful stimulus with a neutral stimulus such as a light cue resulting in fish avoiding the stimulus when the neutral stimulus is presented even without pain. Other fish species show learnt avoidance of potentially painful events (Overmier and Hollis 1983, 1990) including common carp, *C. carpio*, and pike, *Esox lucius*, avoiding hooks in angling trials (Beukema 1970a, b).

Since most research has been conducted on teleost fish, one can state that fish do appear to fulfil the criteria for the first concept of pain defined by Sneddon et al. (2014). Both agnathans and teleosts possess nociceptors to detect pain and the electrophysiological and neuroanatomical properties of these nociceptors are comparable to terrestrial vertebrates. Fish have the brain areas and pathways leading to higher brain centres that are necessary for nociception to occur and the forebrain and midbrain areas are involved rather an activity restricted to reflex centres in the hindbrain and spinal cord. Molecules known to play a role in nociception in mammals are apparent in fish. Additionally, analgesics ameliorate behavioural and physiological responses to painful stimuli. Teleost fish exhibit learned avoidance to pain. Currently, several teleost species have been tested and shown to display significant physiological and behavioural changes in response to pain over a prolonged period of time rather than instantaneous withdrawal reflex responses. This compelling evidence would suggest pain rather than a nociceptive reflex.

10.4 Concept Two: Motivational Change After a Painful Event

An experience that shapes future decision-making and motivation has obviously had a serious impact on the animal. These long-term changes in strategic decision-making can be used to infer how important a painful event was to an animal. This approach allows some interrogation of the subjective experience of the animal and studies have sought to understand the relevance and relative importance of painful treatment to fish. Self-administration paradigms to determine if animals will self-medicate with pain relief are particularly informative. In this case, food or water is dosed with an analgesic and animals can self-select this water or food to effectively reduce their pain suggesting that these animals have an internal experience or affective aspect to pain (e.g. Pham et al. 2010). Unfortunately, this approach is not feasible as fish suspend feeding during pain (Sneddon 2009). An alternative approach is to investigate whether the fish will pay a cost to accessing pain relief. If experiencing pain is a negative internal state then fish should pay a cost in either added effort or forgo access to a resource or favourable area to obtain access to pain relief. This has been explored in zebrafish where individuals were given the choice between an unfavourable barren, intensely lit chamber or a favourable less brightly-lit, enriched chamber with visual access to a shoal. The zebrafish choose to spend most of their time in the enriched chamber on six consecutive occasions. After injection with acid (pain) or saline as a non-painful treatment, their choice of the favourable enriched chamber persists. Yet when an analgesic is made available in the unfavourable chamber, zebrafish that are painfully treated lose their preference for the favoured area and are found in the unfavourable chamber for the most of the time when an analgesic, lidocaine, is added to this chamber (Sneddon 2012; Fig. 10.5). Controls (saline injected) that were given access to the lidocaine dosed unfavourable chamber did not lose their preference for the favoured chamber. These findings show that lidocaine is neither addictive nor has a sedative effect making it likely painfully treated fish are seeking pain relief and are willing to pay a cost to access analgesia in the unfavourable chamber.

Pain by its very nature dominates human attention and humans perform other tasks less well when in pain (Kuhajda et al. 2002). This idea can be used to determine the importance of pain to an animal thus if pain is imperative, an animal would be predicted to perform poorly on competing tasks or ignore them when experiencing pain. Rainbow trout do not respond to novel objects thus do not exhibit neophobia during pain, however, avoidance can be reinstated if morphine is administered

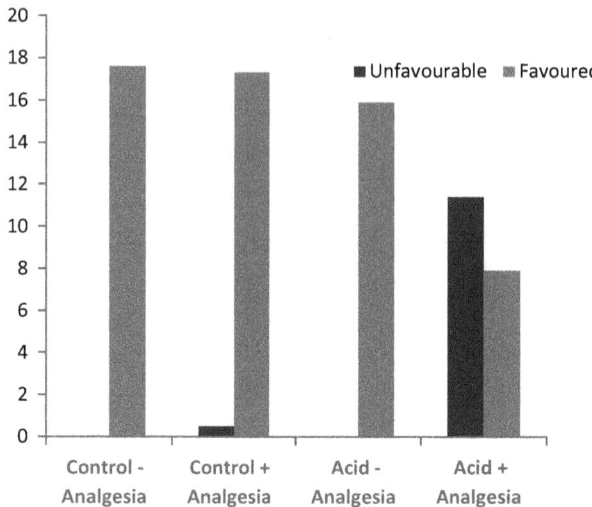

Fig. 10.5 Time spent in either a favourable or unfavourable chamber by zebrafish that were injected subcutaneously with saline (Control) or injected with 1% acetic acid (Acid) when analgesia was present (+ Analgesia) or absent (− analgesia) in the unfavourable chamber. When analgesia was present zebrafish spent more time in the unfavourable chamber ($*P < 0.001$; Sneddon, MS in prep.)

(Sneddon et al. 2003a). Painful treatment in rainbow trout also interrupts normal anti-predator behaviour such as seeking shelter and escape behaviour (Ashley et al. 2009; Fig. 10.6). Socially subordinate trout subject to chronic stress with high plasma cortisol concentrations show next to no signs of pain, possibly due to endogenous or stress-induced analgesia (Ashley et al. 2009). These findings demonstrate that pain is more important than responding to competing stimuli and that central mechanisms may be activated to reduce pain. Indeed recent studies have demonstrated that stress-induced analgesia, which is indicative of descending control of pain in mammals, occurs in piaçu fish, *Leporinus microcephalus.* When stressors are applied to piaçu they demonstrate reduced responses to pain that are blocked by the use of an antagonist naloxone (Alves et al. 2013) and this is underpinned by changes in the GABAergic system in the forebrain (Bejo Wolkers et al. 2015b) and the endocannabinoid system thus this phenomenon appears to be comparable with mammals (Bejo Wolkers et al. 2015a). The empirical evidence is clear and demonstrates that fish fulfil concept two where future behavioural decisions and motivation are altered after pain (Sneddon et al. 2014; Table 10.1).

10.5 Why Is Fish Experiencing Pain Still Debated?

The main arguments that animal pain sceptics use when suggesting animals other than non-human primates experience pain centres upon brain anatomy. Although empirical evidence is not forthcoming to argue against fish pain, some reviewers have stated that fish lack a multilayered, human-like cortex or a functional equivalent, rendering fish and other animals unconscious and not able to be aware of pain or experience the associated discomfort or suffering (Rose 2002). As described above

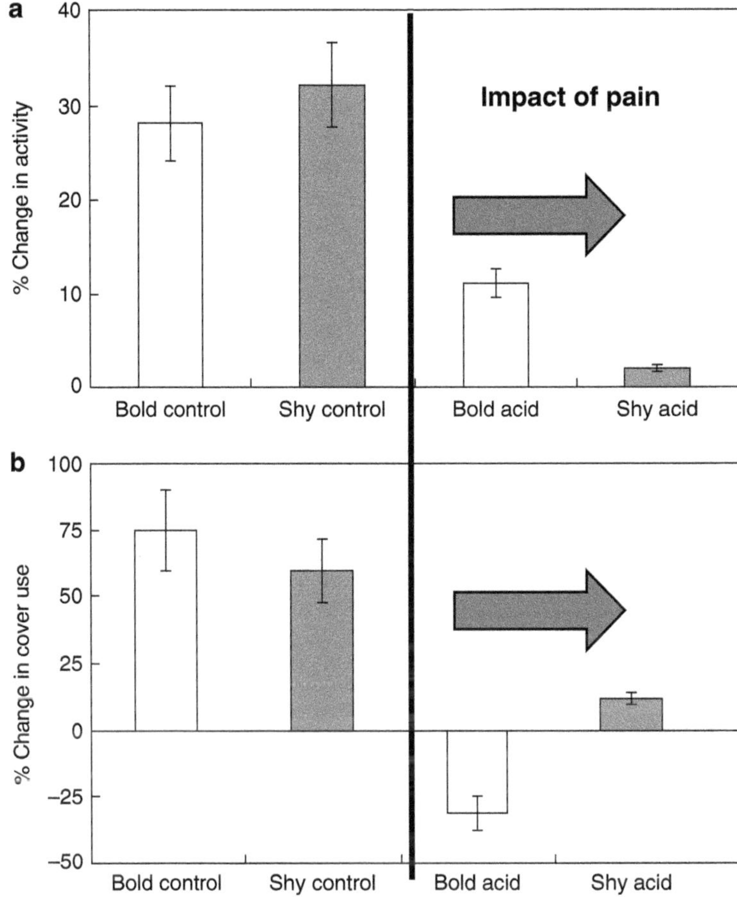

Fig. 10.6 (**a**) The median (interquartile) change in percentage time spent active in bold and shy fish injected with either saline (control) or acid (acid) from before to after the addition of alarm substance (predator cue). (**b**) The median change in the duration of time spent undercover by bold and shy fish in the control and acid groups from before to after the addition of alarm substance. The arrows indicate the impact of pain upon these behaviours ($*P < 0.01$. $N = 24$; Adapted from Ashley, P. J., Ringrose, S., Edwards, K. L., Wallington, E., Mccrohan, C. R. and Sneddon, L. U. 2009. Effect of noxious stimulation upon antipredator responses and dominance status in rainbow trout. *Animal Behaviour,* 77, 403–410 by kind permission from Elsevier)

there is a large body of empirical evidence to support the fact that pain is experienced by fish and that this is a negative state (Broom 2007, 2014; Brown 2015; Chandroo et al. 2004; Sneddon 2009, 2011a, b, 2015 Sneddon et al. 2014). The idea that a function such as pain suddenly arises in non-human primates and humans without any precursor defies the laws of evolution (Sneddon 2009). Further, this anti-fish pain opinion has been criticised as sceptics will not definitively state that birds with a singly-laminated cortex are also incapable of experiencing pain (Sneddon and Leach

2016). To address this, a recent review did indeed state birds have pain but all of the accepted evidence for birds and mammals cannot be accepted for fish (Key 2016). This led to a high number of commentaries (41) that critically evaluated this position and tellingly 73% disagreed with the proposition that fish cannot experience pain stating there was enough evidence for pain in fish; 15% declared pain in fish could not be ruled out but asked for more evidence and only 12% supported the idea of fish not able to experience pain (Animal Sentience 2016). Commentators who do not conduct research directly on pain in fish so cannot be accused of a bias argue that Key's (2016) explanation of how the brain works is flawed (Damasio and Damasio 2016; Shriver 2016) and his arguments are supported by incorrect statements and misreporting of studies (Merker 2016). To quote Bjorn Merker (2016) Key's position "looks more like a ramshackle structure gaping with holes and pieced together from imperfectly understood neuroscience and often faulty literature citations". Further, if one accepts Key's argument then fish should not be able to have other sensory functions due to having a different brain anatomy from humans. For example, fish should not be able to see or visualise since their central visual system differs from the visual cortex of the human brain (Elwood 2016). These debates are led by semantics but ultimately we cannot communicate directly with fish and as such they cannot self-report pain. Given the substantial evidence for the definition of animal pain as described above it would be prudent from an ethical and moral perspective to treat fish as if they do experience pain when subject to injury thereby safeguarding their health and welfare.

10.6 Using Fish in a Humane Manner

When considering the implications of fish experiencing some form of pain then morally and ethically we should attempt to reduce pain by avoiding practices that result in pain. Fish are a foodstuff in farming and large-scale fisheries, an experimental model in scientific research, a key species in conservation efforts and a source of leisure or entertainment in angling or individual fishing, public exhibits and scuba diving or used as pets (Sneddon 2013). The precautionary principle states that animals should be treated well and given the benefit of the doubt as to their capacity for pain. In the spirit of the precautionary principle humans should use fish in an ethical, welfare-friendly manner. This does not prevent the use of fish but that this should be done in a pain-free manner in order to improve fish health and welfare. As discussed above fish can be considered as experiencing pain (Sneddon et al. 2014) then those using fish should avoid or reduce pain where possible. The evidence for pain in fish is convincing, therefore, ethical decision-making should incorporate this when considering the welfare of fish.

There are a number of practices we subject fish to that could potentially cause injury or tissue damage and give rise to pain. Within aquaculture, fisheries, angling, experimentation, public exhibits and the ornamental animal industry research should seek to improve the treatment of fish. Humane methods for slaughter incorporate the principle that animals are killed rapidly with minimal fear and pain. In aquaculture, killing methods have been primarily developed to achieve product quality control, efficiency and processor safety (Conte 2004). Methods are variable and include electrical stunning followed by decapitation, blunt trauma to the cranium and percussive stunning using a captive bolt. Further to this damage, stress responses and mortalities occur during capture and release of fish by hook and line (Chopin and Arimoto 1995) and the use of differing nets in both commercial (Thompson et al. 1971) and recreational fishing (Steeger et al. 1994; Pottinger 1997; Cooke and Hogle 2000; Barthel et al. 2003; Cooke and Sneddon 2007) can also cause skin abrasion leaving open lesions that are readily infected (Barthel et al. 2003). Damage occurs when fish collide with commercial trawling nets or gear, resulting in injuries to jaws and vertebral column (Miyashita et al. 2000), and skin injuries are seen when fish escape from trawl cod ends and through trawl nets (Suuronen et al. 1996, Olla et al. 1997; Reviews in Metcalfe 2009; Sneddon and Wolfenden 2012). The public are willing to pay more for improved welfare in terrestrial farm animals, for example free range produce (e.g. Mulder and Zomer 2017; Frey and Pirscher 2018). This also applies to fish, where consumers are asking where the fish are caught, the method used and whether this is sustainable (MSC 2018). Inefficient management of fisheries has resulted in unsustainable fishing practices and population crashes of key species (Sneddon and Wolfenden 2012) as well as death of non-target animals or by-catch such as birds, cetaceans, turtles, etc. Aggressive behaviour is a problem in the aquaculture of some species where injuries to dorsal, pectoral and caudal fins, eyes and opercula occur and can affect feeding behaviour and growth (Abbott and Dill 1985; Turnbull 1992; Turnbull et al. 1998; Greaves and Tuene 2001; Ashley 2007). Many fish are tagged using an invasive approach, in particular fin clipping which is stressful (Sharpe et al. 1998). A recent study using zebrafish demonstrated tail fin clipping resulted in anomalous behaviour and changes in physiology that were ameliorated by aspirin and lidocaine (Schroeder and Sneddon 2017). Thus, pain relief could be provided to reduce pain during invasive tagging methods where appropriate and possible from a regulatory perspective. Improving the welfare of fish through minimising and reducing pain has obvious benefits to the aquaculture, fisheries and ornamental fish trades by yielding a better economic return.

In experimental studies, legislation of many countries protects experimental animals including fish [e.g. the European Directive (Directive 2010/63/EU)]. Although laboratory fish welfare research is not as well developed as mammals, research is specifically required on appropriate analgesic protocols for fish as well as housing since some anaesthetics may be aversive (Wong et al. 2014). When experimental procedures cause injury and the objectives of the research are not the actual study of pain, analgesia should be administered to minimise any pain. Empirical studies have revealed many analgesics can be administered by injection into relatively larger species (intramuscularly in trout, Sneddon 2003a, b; Mettam et al. 2011) but also through uptake via immersion by dissolving into tank water (e.g. zebrafish, Schroeder and Sneddon 2017; Lopez-Luna et al. 2017a, b, c). Good welfare of experimental subjects ensures the validity and reliability of the data obtained in experimental studies. Studies in laboratory rodents have demonstrated that individuals held under better welfare conditions give higher quality data with reduced intraspecific variation (Singhal et al. 2014).

10.7 Conclusions

Given the many benefits of keeping fish in good welfare and the evidence for pain in fish, it seems beneficial in the variety of contexts that we use fish in a way that ensures their well-being. There is a large and growing body of empirical evidence to suggest pain is an adverse state for a fish and this should be avoided (Sneddon 2015). Often from a legal and ethical perspective, those using fish should seek to reduce any possible suffering and discomfort by refining existing practices to make them less invasive or administer analgesia. Hypothetically, if welfare is enhanced then productivity is improved resulting in a greater economic return as well as healthy fish that pose no risk to public health, therefore, it is to our advantage to safeguard fish welfare. The ongoing debate over whether fish consciously experience pain is actually hampering progress in the field of fish health and welfare. It is clear that there is substantial scientific evidence on one hand versus personal opinions as to whether one must have a human brain for pain to occur on the other. Consciousness is an internal state so difficult to discern but studies have shown fish can recognise themselves as distinct from others (Thunken et al. 2009) and show self-directed behaviours in mirror tests (Ari and D'Agostino 2016). Fish also fulfil the criteria for sentience (Broom 2014) and show complicated behaviours indicative of higher cognitive function and intelligence (Brown 2015). Stamp Dawkins (2012) has also criticised the semantics of animal consciousness and instead suggests what is more important is whether the animal is in good health both physiologically and behaviourally and does it have what it needs in terms of many factors such as housing, diet, husbandry, environmental enrichment and so on. Here, I further suggest that pain in fish as a result of damage is an adverse state that does challenge welfare and thus it pays to ensure fish are treated in a way that results in no pain or minimal pain where possible.

References

Abbott JC, Dill LM (1985) Patterns of aggressive attack in juvenile steelhead trout (*Salmo-Gairdneri*). Can J Fish Aquat Sci 42:1702–1706

Alves FL, Barbosa Júnior A, Hoffmann A (2013) Antinociception in piauçu fish induced by exposure to the conspecific alarm substance. Physiol Behav 110–111:58–62

Animal Sentience (2016). http://animalstudiesrepository.org/animsent/vol1/iss3/

Ari C, D'Agostino DP (2016) Contingency checking and self-directed behaviors in giant manta rays: do elasmobranchs have self-awareness? J Ethol 34:167–174

Ashhurst DE (2004) The cartilaginous skeleton of an elasmobranch fish does not heal. Matrix Biol 23:15–22

Ashley PJ (2007) Fish welfare: current issues in aquaculture. Appl Anim Behav Sci 104:199–235

Ashley PJ, Sneddon LU, McCrohan CR (2006) Properties of corneal receptors in a teleost fish. Neurosci Lett 410:165–168

Ashley PJ, Sneddon LU, McCrohan CR (2007) Nociception in fish: stimulus-response properties of receptors on the head of trout *Oncorhynchus mykiss*. Brain Res 1166:47–54

Ashley PJ, Ringrose S, Edwards KL, McCrohan CR, Sneddon LU (2009) Effect of noxious stimulation upon antipredator responses and dominance status in rainbow trout. Anim Behav 77:403–410

Barthel BL, Cooke SJ, Suski CD, Philipp DP (2003) Effects of landing net mesh type on injury and mortality in a freshwater recreational fishery. Fish Res 63:275–282

Bejo Wolkers CP, Barbosa Junior A, Menescal-de-Oliveira L, Hoffmann A (2015a) Acute administration of a cannabinoid CB1 receptor antagonist impairs stress-induced antinociception in fish. Physiol Behav 142:37–41

Bejo Wolkers CP, Barbosa Junior A, Menescal-de-Oliveira L, Hoffmann A (2015b) GABA(A)-benzodiazepine receptors in the dorsomedial (Dm) telencephalon modulate restraint-induced antinociception in the fish *Leporinus macrocephalus*. Physiol Behav 147:175–182

Beukema JJ (1970a) Angling experiments with carp (*Cyprinus carpio* L.) II. Decreased catchability through one trial learning *A*. Neth J Zool 19:81–92

Beukema JJ (1970b) Acquired hook avoidance in the pike *Esox lucius* L. fished with artificial and natural baits. J Fish Biol 2:155–160

Bjørge MH, Nordgreen J, Janczak AM, Poppe T, Ranheim B, Horsberg TE (2011) Behavioural changes following intraperitonealvaccination in Atlantic salmon (Salmo salar). Appl Anim Behav Sci 133:127–135

Broom DM (2007) Cognitive ability and sentience: which aquatic animals should be protected? Dis Aquat Anim 75:99–108

Broom DM (2014) Sentience and animal welfare, CABI International, Wallingford, 185 p

Brown C (2015) Fish intelligence, sentience and ethics. Anim Cogn 18:1–17

Chandroo KP, Duncan IJH, Moccia RD (2004) Can fish suffer?: Perspectives on sentience, pain, fear and stress. Appl Anim Behav Sci 86:225–250

Chervova LS, Lapshin DN (2011) Behavioral control of the efficiency of pharmacological anesthesia in fish. J Icthyol 51:1126–1132

Chopin FS, Arimoto T (1995) The condition of fish escaping from fishing gears – a review. Fish Res 21:315–327

Conte FS (2004) Stress and the welfare of cultured fish. Appl Anim Behav Sci 86:205–223

Cooke SJ, Hogle WJ (2000) Effects of retention gear on the injury and short-term mortality of adult smallmouth bass. N Am J Fish Manag 20:1033–1039

Cooke SJ, Sneddon LU (2007) Animal welfare perspectives on recreational angling. Appl Anim Behav Sci 104:176–198

Correia AD, Cunha SR, Scholze M, Stevens ED (2011) A novel behavioral fish model of nociception for testing analgesics. Pharmaceuticals 4:665–680

Damasio A, Damasio H (2016) Pain and other feelings in humans and animals. Anim Sent 3(33). http://animalstudiesrepository.org/animsent/vol1/iss3/33/

Dunlop R, Laming P (2005) Mechanoreceptive and nociceptive responses in the central nervous system of goldfish (*Carassius auratus*) and trout (*Oncorhynchus mykiss*). J Pain 6:561–568

Dunlop R, Millsopp S, Laming P (2006) Avoidance learning in goldfish (*Carassius auratus*) and trout (*Oncorhynchus mykiss*) and implications for pain perception. Appl Anim Behav Sci 97:255–271

Ehrensing RH, Michell GF, Kastin AJ (1982) Similar antagonism of morphine analgesia by Mif-1 and naloxone in *Carassius auratus*. Pharmacol Biochem Behav 17:757–761

Elwood RW (2016) A single strand of argument with unfounded conclusion. Anim Sent 3(19). http://animalstudiesrepository.org/animsent/vol1/iss3/19/

Europa (2014a). https://ec.europa.eu/fisheries/cfp_en

Europa (2014b). https://ec.europa.eu/fisheries/cfp/aquaculture

FAWC (2014a). https://www.gov.uk/government/uploads/system/uploads/attachment_data/file/319323/Opinion_on_the_welfare_of_farmed_fish.pdf

FAWC (2014b). https://www.gov.uk/government/uploads/system/uploads/attachment_data/file/319331/Opinion_on_the_welfare_of_farmed_fish_at_the_time_of_killing.pdf

Flecknell P, Gledhill J, Richardson C (2007) Assessing animal health and welfare and recognising pain and distress. Altex-Alternativen Zu Tierexperimenten 24:82–83

Frey UJ, Pirscher F (2018) Willingness to pay and moral stance: the case of farm animal welfare in Germany. PLoS One 13:e0202193

Gentle MJ (1992) Pain in birds. Anim Welf 1:235–247

Greaves K, Tuene S (2001) The form and context of aggressive behaviour in farmed Atlantic halibut (*Hippoglossus hippoglossus* L.). Aquaculture 193:139–147

Guénette SA, Giroux M, Vachon P (2013) Pain perception and anaesthesia in research frogs. Exp Anim 62:87–92

Heupel MR, Simpfendorfer CA, Bennett MB (1998) Analysis of tissue responses to fin tagging in Australian carcharhinids. J Fish Biol 52:610–620

IASP (2019). https://www.iasp-pain.org/Education/Content.aspx?ItemNumber=1698&navItemNumber=576. Accessed 08/03/19

Kajiura SM, Sebastian AP, Tricas TC (2000) Dermal bite wounds as indicators of reproductive seasonality and behaviour in the Atlantic stingray, Dasyatis sabina. Environ Biol Fish 58:23–31

Key B (2016) Why fish do not feel pain. Anim Sent 1(1). http://animalstudiesrepository.org/animsent/vol1/iss3/1/

Kitchener PD, Fuller J, Snow PJ (2010) Central projections of primary sensory afferents to the spinal dorsal horn in the long-tailedstingray, Himantura fai. Brain Behav Evol 76:60–70

Kuhajda MC, Thorn BE, Klinger MR, Rubin NJ (2002) The effect of headache pain on attention (encoding) and memory (recognition). Pain 97:213–221

Leonard RB (1985) Primary afferent receptive field properties and neurotransmitter candidates in a vertebrate lacking unmyelinmated fibres. Prog Clin Res 176:135–145

Liang Y, Terashima S (1993) Physiological properties and morphological characteristics of cutaneous and mucosal mechanical nociceptive neurons with A-d peripheral axons in the trigeminal ganglia of crotaline snakes. J Comp Neurol 328:88–102

Lopez de Armentia ML, Cabanes C, Belmonte C (2000) Electrophysiological properties of identified trigeminal ganglion neurons innervating the cornea of the mouse. Neuroscience 101:1109–1115

Lopez-Luna J, Al-Jubouri Q, Al-Nuaimy W, Sneddon LU (2017a) Activity reduced by noxious chemical stimulation is ameliorated by immersion in analgesic drugs in zebrafish. J Exp Biol 220:1451–1458

Lopez-Luna J, Al-Jubouri Q, Al-Nuaimy W, Sneddon LU (2017b) Impact of analgesic drugs on the behavioural responses of larval zebrafish to potentially noxious temperatures. Appl Anim Behav Sci 188:97–105

Lopez-Luna J, Canty MN, Al-Jubouri Q, Al-Nuaimy W, Sneddon LU (2017c) Behavioural responses of fish larvae modulated by analgesic drugs after a stress exposure, vol 195. Appl Anim Behav Sci, p 115

Lynn B (1994) The fibre composition of cutaneous nerves and the classification and response properties of cutaneous afferents, with particular reference to nociception. Pain Rev 1:172–183

Malafoglia V, Bryant B, Raffaeli W, Giordano A, Bellipanni G (2013) The zebrafish as a model for nociception studies. J Cell Physiol 228:1956–1966

Matthews G, Wickelgren WO (1978) Trigeminal sensory neurons of the sea lamprey. J Comp Physiol A Sens Neural Behav Physiol 123:329–333

Maximino C (2011) Modulation of nociceptive-like behavior in zebrafish (*Danio rerio*) by environmental stressors. Psychol Neurosci 4:149–155

Merker BH (2016) The line drawn on pain still holds. Anim Sent 1(46). http://animalstudiesrepository.org/animsent/vol1/iss3/46/

Metcalfe JD (2009) Welfare in wild-capture marine fisheries. J Fish Biol 75:2855–2861

Mettam JM, Oulton LJ, McCrohan CR, Sneddon LU (2011) The efficacy of three types of analgesic drug in reducing pain in the rainbow trout, *Oncorhynchus mykiss*. Appl Anim Behav Sci 133:265–274

Mettam JJ, McCrohan CR, Sneddon LU (2012) Characterisation of chemosensory trigeminal receptors in the rainbow trout (*Oncorhynchus mykiss*): responses to irritants and carbon dioxide. J Exp Biol 215:685–693

Millsopp S, Laming P (2008) Trade-offs between feeding and shock avoidance in goldfish (*Carassius auratus*). Appl Anim Behav Sci 113:247–254

Miyashita S, Sawada Y, Hattori N, Nakatsukasa H, Okada T, Murata O, Kumai H (2000) Mortality of blue fin tuna *Thunnus thynnus* due to trauma caused by collision during grow out culture. J World Aquacult Soc 31:632–639

MSC (2018). https://www.msc.org/media-centre/press-releases/press-release/seafood-consumers-want-less-pollution-and-more-fish-in-the-sea

Mulder M, Zomer S (2017) Dutch consumers' willingness to pay for broiler welfare. J Appl Anim Welf Sci 20:137–154

Nasr MAF, Nicol CJ, Murrell JC (2012) Do laying hens with keel bone fractures experience pain? PLoS One 7:e42420

Newby NC, Wilkie MP, Stevens ED (2009) Morphine uptake, disposition, and analgesic efficacy in the common goldfish (*Carassius auratus*). Can J Zool 87:388–399

Nordgreen J, Horsberg TE, Ranheim B, Chen ACN (2007) Somatosensory evoked potentials in the telencephalon of Atlantic salmon (*Salmo salar*) following galvanic stimulation of the tail. J Comp Physiol A 193:1235–1242

Nordgreen J, Garner JP, Janczak AM, Ranheim B, Muir WM, Horsberg TE (2009) Thermonociception in fish: effects of two different doses of morphine on thermal threshold and post-test behaviour in goldfish (*Carassius auratus*). Appl Anim Behav Sci 119:101–107

Olla BL, Davis MW, Schreck CB (1997) Effects of simulated trawling on sablefish and walleye pollock: the role of light intensity, net velocity and towing duration. J Fish Biol 50:1181–1194

Overmier JB, Hollis KL (1983) The teleostean telencephalon in learning. In: Davis RE, Northcutt RG (eds) Fish neurobiology, vol 2: higher brain areas and functions. University of Michigan Press, Ann Arbor, MI, pp 265–283

Overmier JB, Hollis KL (1990) Fish in the think tank: learning, memory and integrated behaviour. In: Kesner RP, Olson DS (eds) Neurobiology of comparative cognition. Lawrence Erlbaum, Hillsdales, NJ, pp 205–236

Pham TM, HagmanB, Codita A, Van Loo PP, Stro�mmer, L, Baumans V (2010). Housing environment influences the need forpain relief during post-operative recovery in mice. Physiol Behav 99:663–668

Porcher IF (2005) On the gestation period of the blackfin reef shark, Carcharhinus melanopterus, in waters off Moorea, French Polynesia. MarBiol 146:1207–1211

Portavella M, Vargas JP, Torres B, Salas C (2002) The effects of telencephalic pallial lesions on spatial, temporal, and emotional learning in goldfish. Brain Res Bull 57:397–399

Portavella M, Torres B, Salas C, Papini MR (2004) Lesions of the medial pallium, but not of the lateral pallium, disrupt spaced-trial avoidance learning in goldfish (*Carassius auratus*). Neurosci Lett 362:75–78

Pottinger TG (1997) Changes in water quality within anglers' keepnets during the confinement of fish. Fish Manag Ecol 4:341–354

Reilly SC, Quinn JP, Cossins AR, Sneddon LU (2008a) Novel candidate genes identified in the brain during nociception in common carp (*Cyprinus carpio*) and rainbow trout (*Oncorhynchus mykiss*). Neurosci Lett 437:135–138

Reilly SC, Quinn JP, Cossins AR, Sneddon LU (2008b) Behavioural analysis of a nociceptive event in fish: comparisons between three species demonstrate specific responses. Appl Anim Behav Sci 114:248–259

Rink E, Wullimann MF (2004) Connections of the ventral telencephalon (subpallium) in the zebrafish (*Danio rerio*). Brain Res 1011:206–220

Roques JAC, Abbink W, Geurds F, van de Vis H, Flik G (2010) Tailfin clipping, a painful procedure: studies on Nile tilapia and common carp. Physiol Behav 101:533–540
Roques JAC, Abbink W, Chereau G, Fourneyron A, Spanings T, Burggraaf D, van de Bos R, van de Vis H, Flik G (2012) Physiological and behavioral responses to an electrical stimulus in Mozambique tilapia (*Oreochromis mossambicus*). Fish Physiol Biochem 38:1019–1028
Rose JD (2002) The neurobehavioral nature of fishes and the question of awareness and pain. Rev Fish Sci 10:1–38
Rose JD, Arlinghaus R, Cooke SJ, Diggles BK, Sawynok W, Stevens ED, Wynne CDL (2014) Can fish really feel pain? Fish Fish 15:97–133
Rutherford KMD (2002) Assessing pain in animals. Anim Welf 11:31–53
Schroeder P, Sneddon LU (2017) Exploring the efficacy of immersion analgesics in zebrafish using an integrative approach. Appl Anim Behav Sci 187:93–102
Sharpe CS, Thompson DA, Blankenship HL, Schreck CB (1998) Effects of routine handling and tagging procedures on physiological stress responses in juvenile Chinook salmon. Progress Fish Cult 60:81–87
Shriver AJ (2016) Cortex necessary for pain – but not in sense that matters. Animal Sentience 3(27)
Singhal G, Jaehne EJ, Corrigan F, Baune BT (2014) Cellular and molecular mechanisms of immunomodulation in the brain through environmental enrichment. Front Cell Neurosci 8:97. https://doi.org/10.3389/fncel.2014.00097
Sneddon LU (2002) Anatomical and electrophysiological analysis of the trigeminal nerve in a teleost fish, *Oncorhynchus mykiss*. Neurosci Lett 319:167–171
Sneddon LU (2003a) The evidence for pain in fish: the use of morphine as an analgesic. Appl Anim Behav Sci 83:153–162
Sneddon LU (2003b) Trigeminal somatosensory innervation of the head of a teleost fish with particular reference to nociception. Brain Res 972:44–52
Sneddon LU (2004) Evolution of nociception in vertebrates: comparative analysis of lower vertebrates. Brain Res Rev 46:123–130
Sneddon LU (2006) Ethics and welfare: pain perception in fish. Bull Eur Assoc Fish Pathol 26:6–10
Sneddon LU (2009) Pain perception in fish indicators and endpoints. ILAR J 50:338–342
Sneddon LU (2011a) Pain perception in fish: evidence and implications for the use of fish. J Conscious Stud 18:209–229
Sneddon LU (2011b) Cognition and welfare. In: Brown C, Laland K, Krause J (eds) Fish cognition and behavior, 2nd edn. Wiley-Blackwell, Oxford, pp 405–434
Sneddon LU (2012) Clinical anaesthesia and analgesia in fish. J Exot Pet Med 21:32–43
Sneddon LU (2013) Do painful sensations and fear exist in fish? In: van der Kemp TA, Lachance M (eds) *Animal Suffering: From Science to Law, International Symposium*. Carswell, Toronto, pp 93–112
Sneddon LU (2015) Pain in aquatic animals. J Exp Biol 218:967–976
Sneddon LU (2018) Comparative physiology of nociception and pain. Physiology 33:63–73
Sneddon LU, Leach MC (2016) Anthropomorphic denial of fish pain. Anim Sent 1(28). http://animalstudiesrepository.org/animsent/vol1/iss3/28/
Sneddon LU, Wolfenden D (2012) How are fish affected by large scale fisheries: pain perception in fish? In: Soeters K (ed) See the truth. Nicolaas G. Pierson Foundation, Amsterdam, pp 77–90
Sneddon LU, Braithwaite VA, Gentle MJ (2003a) Do fishes have nociceptors? Evidence for the evolution of a vertebrate sensory system. Proc R Soc London Ser B Biol Sci 270:1115–1121
Sneddon LU, Braithwaite VA, Gentle MJ (2003b) Novel object test: examining nociception and fear in the rainbow trout. J Pain 4:431–440
Sneddon LU, Elwood RW, Adamo S, Leach MC (2014) Defining and assessing pain in animals. Anim Behav 97:201–212
Sneddon LU, Wolfenden DCC, Thomson JT (2016) Stress management and welfare. In: Schreck CB, Tort L, Farrell A, Brauner C (eds) Biology of stress in fish – fish physiology, 1st edn. Academic Press, Cambridge, MA, pp 463–539

Sneddon LU, Halsey LG, Bury NR (2017) Considering aspects of the 3Rs principles within experimental animal biology. J Exp Biol 220:3007–3016

Snow PJ, Renshaw GMC, Hamlin KE (1996) Localization of enkephalin immunoreactivity in the spinal cord of the long-tailed ray *Himantura fai*. J Comp Neurol 367:264–273

St. John Smith E, Lewin GR (2009) Nociceptors: a phylogenetic review. J Comp Physiol A 195:1089–1106

Stamp Dawkins M (2012) Why animals matter. Animal consciousness, animal welfare, and human well-being. Oxford University Press, Oxford

Steeger TM, Grizzle JM, Weathers K, Newman M (1994) Bacterial diseases and mortality of angler-caught largemouth bass released after tournaments on Walter F. George reservoir, Alabama/Georgia. N Am J Fish Manag 14:435–441

Suuronen P, Erickson DL, Orrensalo A (1996) Mortality of herring escaping from pelagic trawl cod ends. Fish Res 25:305–321

Terashima S-i, Liang Y-F (1994) C mechanical nociceptive neurons in the crotaline trigeminal ganglia. Neurosci Lett 179(1–2):33–36

Thompson RB, Hunter CJ, Patten BG (1971) Studies of live and dead salmon that unmesh from gill nets. International North Pacific Fish Community Annual Report, pp 108–112

Thunken T, Waltschyk N, Bakker TCM, Kullmann H (2009) Olfactory self-recognition in a cichlid fish. Anim Cogn 12:717–724

Turnbull JF (1992) Studies on dorsal fin rot in farmed Atlantic salmon (Salmo salar L.) parr. Ph.D. Thesis. University of Stirling

Turnbull JF, Adams CE, Richards RH, Robertson DA (1998) Attack site and resultant damage during aggressive encounters in Atlantic salmon (Salmo salar L.) parr. Aquaculture 159:345–353

Willenbring S, Stevens CW (1995) Thermal, mechanical and chemical peripheral sensation in amphibians – opioid and adrenergic effects. Life Sci 58:125–133

Wong D, von Keyserlingk MAG, Richards JG, Weary DM (2014) Conditioned place avoidance of zebrafish (Danio rerio) to three chemicals used for euthanasia and anaesthesia. PLoS One 9: e88030

Yoshida, M. and Hirano, R. (2010). Effects of local anesthesia of the cerebellum on classical fear conditioning in goldfish. Behav. Brain Funct. 6, 20.

Young RF (1977) Fiber spectrum of the trigeminal sensory root of frog, cat and man determined by electron microscopy. In: Anderson DL, Matthews B (eds) Pain in the Trigeminal Region. Elsevier, Amsterdam, pp 137–160

Chapter 11
How Fish Cope with Stress?

Angelico Madaro, Tore S. Kristiansen, and Michail A. Pavlidis

Abstract Fish, which were considered non-conscious, non-sentient, instinct-driven animals, revealed to be affected by stress as are mammals. The goal of the chapter, thus, is to use the concept of allostasis as a conceptual framework for understanding how cultured fish are adapted or maladapted to changes in their rearing environment. The ontogeny, neuroendocrine basis and physiology of the stress response are described in relation to both acute and chronic stress conditions. Still, the chapter discusses on how cognition, appraisal and psychological factors affect the physiology of the stress response in fish. Finally, the chapter provides some of the newest insights about the long-lasting effects that early-life stress has on fish.

Keyword Fish · Stress · Allostasis · Allostatic load · Allostatic state · Allostatic overload · Cognition · Appraisal · Psychology · Neuroendocrinology · Physiology · Stress response · Acute stress · Cortisol · Chronic stress · Ontogeny · Early-life stress

11.1 Introduction

Coping styles, coping strategies, behavioural syndromes, behavioural profile, temperament, idiosyncrasy, personality. Homeostasis (Cannon 1939), homeorheusis (Signals 2006), allostasis (Sterling and Eyer 1988), predicate and reactive homeostasis (Moore-Ede 1986; Romero et al. 2009), rheostasis (Mrosovsky 1990). Eustress, stress (Selye 1936), acute stress, chronic stress, chronic mild stress (Schulkin 2004). The purpose of the chapter is not to get lost in terminologies that bear no fruit, nor does it aim to provide a thorough background of theories and

A. Madaro (✉) · T. S. Kristiansen
Institute of Marine Research, Bergen, Norway
e-mail: angelico.madaro@hi.no; torek@hi.no

M. A. Pavlidis
Department of Biology, University of Crete, Heraklion, Greece
e-mail: pavlidis@uoc.gr

T. S. Kristiansen et al. (eds.), *The Welfare of Fish*, Animal Welfare 20,
https://doi.org/10.1007/978-3-030-41675-1_11

concepts where there is "no univocal meaning of how they have been used" (Sterling 2004, 2012). Our goal is to use the concept of allostasis as a conceptual framework for understanding, at the level of the organism, how fish are adapted or maladapted to changes in their rearing environment.

The neuroendocrine basis and physiology of the stress response have been well described for several fish species. Interest in how cognition, appraisal and psychological factors affect the physiology of the stress response has also risen in the course of the past few decades (Cooper and Dewe 2004). At the opening up of this field of research, researchers mostly employed mammals as experimental models, whilst fish have only relatively recently begun to be studied (Wendelaar Bonga 1997). The main reason for this lies in the fact that most people considered (and perhaps still consider) fish to be non-conscious, non-sentient and instinct-driven animals (Laland and Hoppitt 2003). Such assumptions were supported by the lack of neocortex in fish (Rose et al. 2014; Rose 2002, 2007). However, behavioural, cognitive and neuroanatomical studies now revealed that fish perception and cognitive abilities often match or exceed those of other vertebrates (Allen 2011; Brown 2015; Brown et al. 2008; Rodriguez et al. 2002). For instance, fish learn, memorise and process information in order to gain experience, survive and exploit their immediate environment in the most efficient way (Braithwaite and Boulcott 2008; Laland and Hoppitt 2003).

This essay comprises four main sections. The first and second present the concepts of stress and the key principles of allostasis. They discuss coping styles and focus on the importance of prediction for anticipatory regulation of the allostatic mechanisms of the individual fish. The third section describes the physiology of the stress response in teleost fishes, both as an adaptive mechanism and a dysfunctional regulatory response that can lead to stress-related pathophysiology. The fourth describes the ontogenesis of the stress response and how early-life stress affects subsequent developmental stages.

11.2 The Concept of Stress

Even though it forms part of our daily vocabulary, finding a clear definition of "*stress*" is a difficult task. It is often used in a negative connotation, related to mental or physical discomfort; however, the concept is actually multifaceted. The term "biological stress" was first created by Hans Selye, as an element of the "General Adaptation Syndrome" (Selye 1936). According to the original concept "*the general adaptation syndrome is the sum of all non-specific, systemic reactions of the body which ensue upon long-continued exposure to stress*" (Selye 1936). The model comprises a stressor(s), a receiver (brain, nervous system) and a response (Seley 1950). Hence, the *non-specific responses* represent a *general adaptive* reaction of the body in response to any agent (or stressor) that threatens normal homeostasis. Stress responses are driven by the autonomic sympathetic system and the hypothalamic–pituitary–adrenal (HPA) axis (in fish called hypothalamic–pituitary–interrenal (HPI axis) that in concert modulates metabolism and behaviour ("fight or flight" response; Cannon 1932). Whilst changes in behaviour, such as an increase in vigilance, arousal and cognition enable an animal to escape or to counteract the stressor, changes in metabolic state increases energy availability (respiratory rate, cardiovascular tone, gluconeogenesis and lipolysis), with a concomitant suppression of vegetative functions that are not required for coping with the stressor (e.g. feeding, digestion, growth and reproduction) (Barton 2002; Barton and Iwama 1991). In this chapter, stress refers to processes that tune physiology and behaviour to prepare the body to handle challenges and/or extra load imposed on it, according to its needs, demands and available recourses (Sterling 2012).

"The paradox of stress lies in the simultaneity of its adaptive nature and its possible maladaptive consequences" (Korte et al. 2005). Indeed, an ability to cope efficiently implies that the body temporarily activates the stress response when necessary and shuts it down when it is no longer needed. But, if the condition persists, as during chronic and/or cumulative exposure to several stressors, this may lead to a continued activation of the stress response. As a result, a response that was initially considered to be adaptive may eventually fail in its adaptive purpose, becoming maladaptive and leading to pathologies (Korte et al. 2005; McEwen 2003; Mommsen et al. 1999; Wendelaar Bonga 1997).

Apart from the physiological and behavioural responses, the psychological dimension of stress and its influence on stress manifestation is an essential part of the stress concept [for a review see Galhardo and Oliveira (2009)]. In fact, it is hardly possible to study stress and associated consequences without considering whether or not the stressor is perceived as a serious threat, i.e. without the individual's appraisal of the predictability and/or controllability of the challenge (Koolhaas et al. 2011; Lazarus and Folkman 1984). To that end, cognition and how an individual perceives a given situation may be just as important as the actual physical challenge in determining the severity of the stress response (Von Holst 1998). The stress response thus depends on acquired information by the individual and thus their appraisal: following a perceived stress event the stimulus–response model integrates cognition and thus directly links physiology and behaviour to experiential learning (McVicar et al. 2014). The individual's previous subjective experiences become the measure by which it evaluates its own coping capacity: when these do not match with the cognitive representation of an environmental demand, the perception of danger/hazard is generated (de Kloet et al. 2005). Thus, a one-off acute stress will have two outcomes; first, a physiological and biological response, adaptive and restorative, that will help the individual to cope with the stressor, and secondly, a cognitive learning process that will influence the individual's reaction if a similar challenge occurs in the future (De Kloet et al. 2005; McVicar et al. 2014; Madaro et al. 2015). Repeated exposure to a stressor with which the animal can cope will reduce the perception of danger, resulting a waning of the response, a mechanism called *habituation* (Lieberman 2000). However, this can also be seen as an adjustment of the stress response to meet the actual demand, as described in the concept of allostasis (Sterling and Eyer 1988).

11.3 The Concept of Allostasis

Homeostasis (Cannon 1929), or preservation of the constancy of the internal milieu (Bernard 1974), was long the core model of physiological regulation. However, numerous observations have made it clear that the ability to adapt to changing environmental conditions and changing bodily needs, rather than constancy, appeared to optimise survival, growth and reproduction (Sterling and Eyer 1988). *Allostasis* has since been suggested to replace homeostasis as the model of the physiological regulation and adaptation, and the substitute concept introduces a new terminology to describe stress. It embraces the idea that the organism maintains stability through change, i.e. through a series of adjustments (shifts) of the physiological mediators according to predicted demands (Sterling 2012; Sterling and Eyer 1988).

Allostasis moves from the classical "response model" into the predictive "transactional, cognitive, appraisal model" of stress. The brain is the main coordinating organ: it constantly monitors internal and external parameters to anticipate the needs and the changes required to satisfy them, evaluate priorities, and it prepares the organism for adjustment before the changes lead to errors (Sterling 2012). The ability to prevent errors occurring is more efficient than adjusting once they have occurred (Koolhaas et al. 2011). Changes involve both physiological and behavioural alterations with a coordinated plasticity that optimises performance to meet the most likely environmental demands at minimal cost (McEwen and Lasley 2002). Part of the brain's control operations occur through the mediators of allostasis *such as* neurotransmitters and adrenal hormones (Korte et al. 2007), which target their receptors in many different tissues and organs. These mediators adjust features like metabolism, immune and cardiovascular systems to create a new dynamic state and allocate resources for the perceived challenge (McEwen 2002; McEwen and Lasley 2002; McEwen and Seeman 1999).

Animals are continuously exposed to a wide range of dynamic changes. A healthy animal possesses a large array of allostatic responses to environmental challenges that optimise performance (Korte et al. 2007; Fig. 11.1). However, when the challenges are prolonged in time, they generate an *allostatic state* in which the animal's regulatory capacity becomes reduced. The allostatic state creates a situation of chronic deviation of the regulatory system from the optimal functional level (Koob 2004; Koob and Le Moal 2001), which may lead to failure to habituate to challenges, generation of an unsuitable response (e.g. in amount), or failure to terminate the physiological response once the challenge has passed (Korte 2001). This condition is characterised by abnormal and sustained production of primary mediators, i.e. glucocorticoids, that integrate physiology and coupled behaviour following a stressor (e.g. disease or predators) or changing environmental conditions (McEwen and Wingfield 2003). The allostatic state is an "emergency" response and can thus only be maintained for a short time whilst the organism has sufficient energy (food intake or stored energy) to support the mechanisms of allostasis.

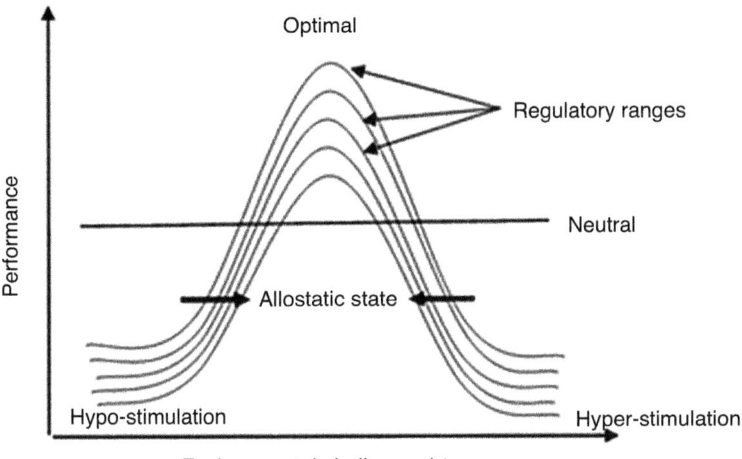

Fig. 11.1 Allostasis and animal performance in relation to environmental challenges. Excessive or insufficient environmental challenges lead to a state of chronic deviation of the regulatory system from its optimal operating level. This new equilibrium, called allostatic state, is characterised by a narrower regulatory range and by an enhanced chance of hypo-or hyper-stimulation (modified from Korte et al. 2007)

The cumulative result of an allostatic state or over activation of allostatic responses is known as *allostatic load* (McEwen and Wingfield 2003), which is the price that the body pays to adapt to the imposed adverse psychological or physical condition. In the short term, the allostatic state is an adaptive response, but if exposure is prolonged, or if exposed to additional stressors (e.g. disease, human disturbance and social interaction), the allostatic load changes to an *allostatic overload* (McEwen 2002). In such conditions of chronic stress (Fig. 11.1, right side of the curve) there will be too little resources and energy to maintain all bodily functions, and pathologies may develop due to "*wear and tear*" of the body (Korte et al. 2007; McEwen and Lasley 2002).

Allostatic load may also be very low due to *hypo-stimulation* (Fig. 11.1, left side of the curve). For instance, in mammals several diseases are related to low activation/functioning of the HPA axis, i.e. allergic reactions, inflammatory/autoimmune diseases and fatigue states (Steenbergen et al. 2011). In the mouse brain, chronic hypo-stimulation (low level of mental activity) may affect cell proliferation and neurogenesis especially in the hippocampal dental gyrus (van Praag et al. 1999). This can be explained by the concept "*use it or lose it*": the survival of neurons depends on whether they are activated by incoming signals (van Praag et al. 1999).

Individual experiences (i.e. a challenge or a predator), evolutionary history and genetic background generate variability in behavioural and physiological responses to stress (Koolhaas et al. 1999). That means that the same stressor can produce different effects amongst individuals in a population, as well as in the same individual in different contexts. This phenomenon is often termed "stress-coping style"

(Koolhaas et al. 1999; Korte et al. 2005). A coping style (also termed "behavioural syndrome" or "personality") is defined as a correlated set of individual behavioural and physiological characteristics that is consistent over time and across situations (Koolhaas et al. 1999). Essentially, individuals can be classified as being either proactive or reactive, and these two alternative coping styles have differential fitness consequences under different environmental conditions (Vindas et al. 2017). Proactive individuals are characterised by low post-stress cortisol production but high sympathetic activity; they are aggressive, exhibiting routine and high-risk taking behaviours. On the other hand, reactive individuals show high post-stress cortisol production, low sympathetic activity, low aggression level, exhibit flexible behaviour and low-risk taking (see also Chap. 12). The distribution of coping styles is often bimodal (Castanheira et al. 2015), although a continuum of behaviours between those extremes exists. Moreover, an individual's flexibility in coping with stress may become even larger when their sensitivity to psychological factors is taken into account (Feder et al. 2009).

The perception of the stressor and its predictability are the most widely discussed psychological modulators of the stress response (Galhardo et al. 2011; Galhardo and Oliveira 2009; Ursin and Eriksen 2004). As far back as 1970, Weiss described how rats repeatedly exposed to electric shock of the same intensity and duration displayed reduced harmful effects (i.e. gastric ulcers) when the stress was presented in a predictable manner (Weiss 1970). More recently, a study on cichlid fish showed that fish exposed to predictable aversive stimuli reduced stress-related behaviour such as freezing and displayed a lower cortisol response than fish that experienced unpredictable stress (Galhardo et al. 2011). In accordance with the *cognitive activation theory of stress*, predictability decreases the discrepancy between the internal expectations of the animal (set of values based on normal situations), and the reality (actual value of what is happening), thereby reducing the intensity of the stress response (Ursin and Eriksen 2004). The individual expectation are a function of the learned information about the stimuli/situations experienced and the available possibilities of coping (Galhardo and Oliveira 2009). For instance, a response to a repeated aversive stimulus can provide information on whether the stimulus is harmful or not or more harmful than initially perceived. The information gathered from previous experiences, therefore, modulates the stress response, creating a new set of values based on how it is perceived. As a result, when repeated stimuli are not found to be harmful, habituation and a consequent reduction in responsiveness will occur (Lieberman 2000). On the other hand, prediction of stimuli as being too severe or even life threatening may lead to higher activation of the stress response rather than reduction of it.

Bassett and Buchanan-Smith (Lieberman 2000) discussed how the predictability of aversive stimuli reduces stress in animals because it supplies information about safe periods, or rather when aversive stimuli are not likely to occur. The "*safety signal hypothesis*" states that if a stressor is predicted by a cue, the absence of the cue indicates that the situation is safe, and that stress episodes will not occur. Predictable stresses allow animals to remain in a state of fear only when the cue is present, just as in classical Pavlovian conditioning (Jones et al. 2014; Pavlov 1927). In fact, conditioning is often employed to study the effect of predictability on the stress response: it consists of pairing a neutral stimulus, e.g. light or sound, together with a relevant biological stimulus or unconditioned stimulus (e.g. sight of a predator) that produces a response (e.g. freezing) without any training. When the unconditioned stimulus is repeatedly announced by a neutral stimulus and the association between the two is established, the neutral stimulus alone will be able to trigger a conditioned response. Associative learning thus induces anticipatory behaviour which may offer a time advantage that is essential for animals to prepare for forthcoming events and respond in a proper manner and at an appropriate intensity (Bassett and Buchanan-Smith 2007).

Predictability also provides a perception of increased *control*, namely, the capacity to decrease exposure to, or the harmful effect of, a stressor. Previous experience becomes the key factor on the basis of which an animal evaluates its own coping capacity (Bassett and Buchanan-Smith 2007). In situations in which particular responses have been found to be successful to avoid the stressor, the animal gains a sense of control that will lower the arousal and prepare it to face new challenges. However, anticipation of the stressor may not help it to cope with the situation when nothing can be done to avoid it. Hence, if an animal learns by experience that nothing whatsoever can be done to reduce or escape the stress, it may develop *learned helplessness*, a condition associated with depression (Lovallo 2016; Ursin and Eriksen 2004). On the other hand, if an animal learns that a certain degree of control is possible, and that any of its reactions could contribute to worsening the situation, it may develop a condition of *learned hopelessness* (Ursin and Eriksen 2004). Both learned helplessness and hopelessness are associated with high cortisol activation, a condition characteristic of subordinate fish (Lovallo 2016; Ursin and Eriksen 2004).

11.4 The Physiology of the Stress Response in Teleost Fish

When a stressor is perceived, neural signals (visual, olfactory, auditory and sensory) activate the nucleus preopticus (NPO, a homologue of the mammalian paraventricular nuclei) in the hypothalamus and initiates, via the brain stem and spinal cord (Flik et al. 2006), a downstream activation of sympathetic fibres (Fig. 11.2). The chromaffin cells in the interrenal gland, the homologue of the mammals' adrenal medulla, are thereafter stimulated by the preganglionic fibres to release stored catecholamines (noradrenaline and adrenaline) into the bloodstream, as the initial stress response. In hagfish and lampreys (cyclostomes) the prevalence of one catecholamine over the other is tissue-specific, whilst in sharks and rays (elasmobranchs), noradrenaline is the predominant catecholamine (Reid et al. 1998). In teleost fish, adrenaline is the prime component of the stress response. Adrenaline, via the β-adrenoreceptors, prepares the animal for the "fight or flight" reaction by increasing ventilation rate, oxygen uptake and transport capacity, glycogenolysis, lipid degradation, etc. Secondly, following the primary stress response, the hypothalamus–pituitary–interrenal gland axis (HPI axis) which is the equivalent of

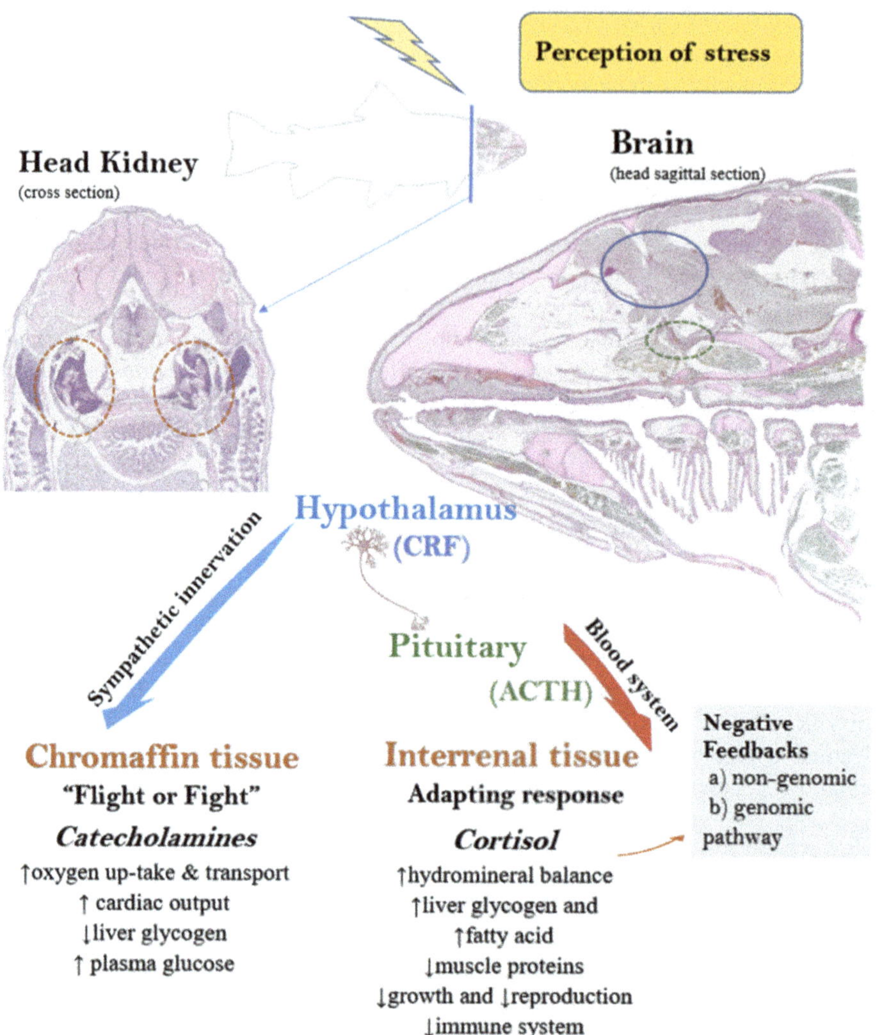

Fig. 11.2 A general fish stress response activated follow a perception of threat. As first, a fast emergency or "flight or fight" response starts through the activation of the sympathetic autonomic nervous system. Follow a more slow, adaptive response regulated by the endocrine system. Histology sections of Atlantic salmon's head (staining: Periodic acid Schiff-Orange G- Light Green) and cod's head kidney (Gadus morhua; staining: Haematoxylin Eosin) are courtesy of Tora Bardal, NTNU

the hypothalamic–pituitary–adrenal gland axis in mammals, is activated (Arends et al. 1999; Wendelaar Bonga 1997). The axis is initiated in the hypothalamic NPO, where corticotropin-releasing factor (CRF) is released to activate the pituitary corticotrophic cells (Alderman and Bernier 2007; Fryer 1989; Huising et al. 2004). In teleost fish, the NPO neurons project their fibres directly into the rostral pars distalis of the pituitary gland, close to the corticotrophs (Pepels et al. 2002). CRF then activates the CRF-R1, causing the secretion of the pro-opiomelanocortin (POMC)-derived peptide adrenocorticotropic hormone (ACTH) into the bloodstream (Sumpter et al. 1986). In the interrenal gland, ACTH induces synthesis of cortisol via the melanocortin 2 receptor (MC2R), which is expressed exclusively in cortisol producing interrenal cells (Aluru and Vijayan 2008; Wendelaar Bonga 1997). The amount of cortisol released into the blood is dependent on the type and intensity of the stressor (Doyon et al. 2006; Mommsen et al. 1999; Wendelaar Bonga 1997).

Cortisol acts through sets of intracellular ligands including mineralocorticoid (MR) and glucocorticoid (GR) receptors in the activation of *genomic pathways*. In the cell cytoplasm, cortisol binds to GR/heat shock proteins complex forming the cortisol–GR heterocomplex that moves into the nucleus. In the nucleus, this complex forms homodimers that bind the DNA in specific regions, namely glucocorticoid responsive elements (GRE), in the promoter regions of target genes, modulate transcription and protein synthesis involved in generic functions like metabolism, feed intake and/or absorption, growth, reproduction and immune function (Aluru and Vijayan 2006, 2009; Mommsen et al. 1999; Prunet et al. 2006). This wide range of the actions that make up the stress response is due to the cross-talk that occurs between the HPI axis and the other regulatory axes such as the hypothalamic–pituitary–thyroid (HPT), growth hormone/insulin-like growth factor (GH/IGF) and the hypothalamic–pituitary gonadal (HPG) axes (Chabbi and Ganesh 2012; Mommsen et al. 1999; Peter 2011).

Cortisol is also known to exert some of its effects through non-genomic pathways (Borski et al. 2002; Thomas 2012) and although knowledge of these pathway is limited in fish, this is a rapidly emerging field in other vertebrates (Bartholome et al. 2004; Falkenstein et al. 2000; Groeneweg et al. 2011; Tasker et al. 2006).

Since the second round of the whole genome duplications about 335 million years ago, teleosts have been characterised by the presence of multiple cortisol receptors. Most fish have two GR proteins, GR1 and GR2, and only one MR protein (Bury et al. 2003; Greenwood et al. 2003). However, a second MR protein was recently described in rainbow trout (*Oncorhynchus mykiss*) by Sturm (Sturm et al. 2005); the rtMRa and rtMRb. In view of the extremely high similarity of both the nucleotide and amino acid sequences (99%) the authors suggested that rtMRa and rtMRb probably represents allelic variants of the same gene. GRs and MR display different affinity for cortisol, suggesting different involvement in physiological responses. In rainbow trout, the MR receptor is 10 to 100 times more sensitive than GR1 (Sturm et al. 2005) and GR2 is more sensitive to cortisol than GR1 (Bury et al. 2003; Sturm et al. 2011).

Teleosts, unlike tetrapods, do not produce aldosterone (Mommsen et al. 1999; Wendelaar Bonga 1997), so fish use cortisol as both glucocorticoid and mineralocorticoid. In line with this, cortisol plays a major role in the restoration of the internal fluid balance in fish exposed to stress (McCormick 2001). Cortisol also promotes the ion uptake (freshwater) or secretion (saltwater) that modulate the expression of two different Na^+, K^+-ATPase isoforms, respectively, NKA $\alpha 1a$ and $\alpha 1b$; and of the cystic fibrosis transmembrane conductance regulator (CFTR) anion channel (Doyon et al. 2006; Kiilerich et al. 2007; McCormick et al. 2008; Nilsen et al. 2007).

Cortisol is also involved in negative feedback mechanisms that down-regulate the HPI axis at different levels in order to shut down the stress response. For example, in the hypothalamic NPO, cortisol downregulates the CRF gene transcription (Bernier et al. 1999; Bernier and Peter 2001; Doyon et al. 2006). In the pituitary gland, glucocorticoids inhibit basal ACTH secretion by controlling CRF-induced and POMC expression (Palermo et al. 2008) and ACTH-releasing activity (Bernier et al. 1999, 2004; Doyon et al. 2003; Fryer et al. 1984). Cortisol can also modulate the density of glucocorticoid receptors available, with patterns that depend on the species, stress condition and intensity involved (Ohl et al. 2000; Prunet et al. 2006; Sathiyaa and Vijayan 2003; Takahashi and Sakamoto 2013).

The effects of cortisol on the regulation of the HPI axis are terminated by the enzyme 11β-hydroxysteroid dehydrogenase 2 (11β-HSD2). This enzyme converts cortisol into (inactive) cortisone (Funder et al. 1988; Mommsen et al. 1999), impeding its access to the glucocorticoid receptors. CRF-binding protein (CRF-BP) also provides yet another way of modulating HPI axis activation. The protein modulates the effects of CRF and CRF-related peptides by binding them, thus reducing their bioavailability (Geven et al. 2006; Huising et al. 2008; Manuel et al. 2014; Seasholtz et al. 2002) resulting in a diminished release of ACTH.

11.4.1 Stress Response Follows an Acute Stress Episode

The stress response following a one-off, short-duration (acute) stressor, comprises a set of physiological and behavioural changes that aims to maximise its likelihood of survival. Animals that are exposed to a stressor or an environmental challenge for the first time enter a state of *emergency* (Wingfield et al. 1998), in which condition the body activates a series of adaptive processes (Sterling and Eyer 1988) that aim to conserve energy, whilst releasing what is required to deal with the immediate demands imposed by the stressor.

Most fish display a general pattern of physiological stress responses broadly grouped into primary and secondary responses. During the primary response, brain centres are activated, trigging a massive release of catecholamines via the hypothalamus–sympathetic–chromaffin cells axis, and thereafter cortisol via the HPI axis (Wendelaar Bonga 1997). The hormones released initiate a secondary response, which is characterised by the mobilisation of sources of energy, depletion of glycogen stores and thus an increase in plasma levels of glucose as readily available energy capable of sustaining a potential burst of physical activity. Depending on the physical effort required to counteract the challenge, the cardio-vascular and respiratory responses are stimulated to increase oxygen distribution and the energy substrates that are liberated in the circulation. However, strenuous muscle activity such as forced swimming lead to anaerobic glycolysis and an increase in plasma lactate. Other potential secondary responses include hydro mineral dysfunction, because adrenaline alters gill blood-flow patterns and gill permeability, both of which favour water flowing down its osmotic gradient, either in or out of the fish depending on environmental salinity. For these reasons, changes in plasma cortisol, glucose, lactate, pH and ion concentrations are frequently used as stress indicators in teleosts (Arends et al. 1999; Barton 2002). Recent findings reveal that serotonin is also released in the brain, and particularly in the brainstem, following an acute stress (Vindas et al. 2016b). It has been suggested that serotonin release increases under conditions that require the reallocation of energy resources. Serotonin thus appears to play a role similar to that of cortisol in energy regulation. It also appears that in vertebrates, serotonin plays a crucial role in other functions such as neural plasticity, and behavioural and emotional control (Winberg and Nilsson 1993).

Previous observations in rainbow trout (Pickering et al. 1991), zebrafish (Pavlidis et al. 2015) and several other Mediterranean species (Fanouraki et al. 2011), have shown that peak systemic cortisol release occurs between 30 and 60 min after stress. However, analysis of post-stress cortisol concentrations has shown that the magnitude of the response to stress differs as a function of the nature and intensity of the stressor (Madaro et al. 2015, 2016a, b). Furthermore, studies in green sturgeon (Lankford et al. 2003), sunshine bass (Davis 2004), juvenile Chinook salmon and Atlantic salmon (Madaro et al. 2018) revealed that the magnitude of a stress response can also be affected by environmental physical conditions. For example, fish acclimated to lower temperatures displayed lower and slower post-stress cortisol production than fish acclimated to higher temperatures (Fig. 11.3).

In spite of the overall similarity of responses to stress between species, their magnitude, timing and duration are species-specific (Fanouraki et al. 2007, 2011) For example, European sea bass (*Dicentrarchus labrax*) displayed an extremely high cortisol response, whilst the response of dusky grouper (*Epinephelus marginatus*) and meagre (*Argyrosomus regius*) was very low after exposure to the same acute stress protocol (5–6 min chasing and 1–1.5 min air exposure; [Fig. 11.4, as modified by Fanouraki et al. (2011)]. Another example is the measurements of the post-stress level of cortisol in two species of "cleaner fish used as biological de-lousing agents in salmon net-pens, ballan wrasse (*Labrus bergylta*) and lumpfish (*Cyclopterus lumpus*). Both species exhibit higher levels of plasma cortisol after handling, although with clear differences in peak cortisol values (two- to threefold higher in ballan wrasse), plasma lactate and ions (Jørgensen et al. 2017; Leclercq et al. 2014). Differences in their coping behaviour may help to explain these observations. Indeed, in a threat situation, wrasse, which usually tries to escape (the classical fight–flight response) expend an amount of energy much greater than that required by lumpfish, which usually freeze.

The brain initiates and coordinates the response to a stressor by integrating several factors such as experience, memories, expectations and re-evaluation of needs in anticipation of physiological requirements (Koob and Le Moal 2001). The magnitude of the response, as well as the stress tolerance may therefore also reveal significant differences between individuals of the same species and or population, depending on the animal's previous experience, e.g. exposure to a similar or other stressors (Madaro et al. 2015, 2016a, b).

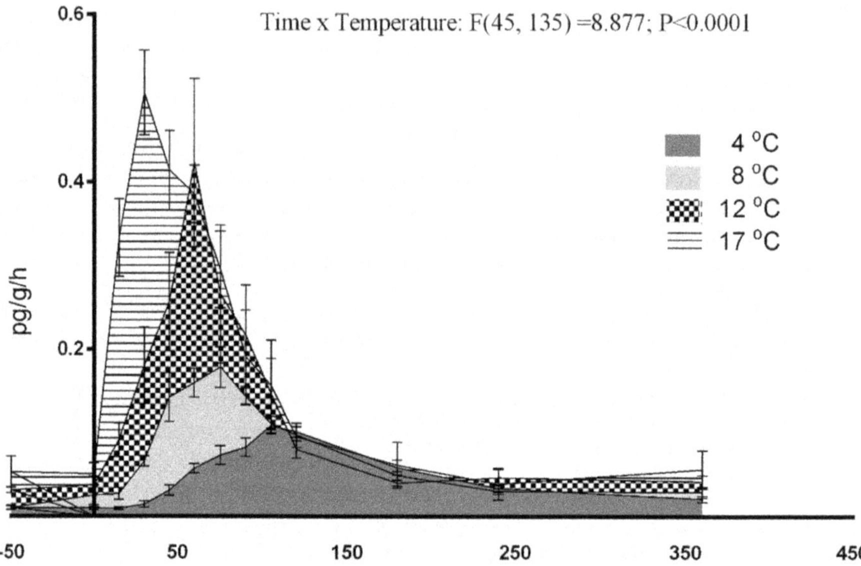

Fig. 11.3 Profile of the cortisol excreted in the water by Atlantic salmon post-smolt acclimate at four different temperatures, about 1 h before and up to 350 min after being challenged with an acute stress. Values are represented as means ± S.E.M ($n = 4$). The effects that temperature had on the fish cortisol released pattern were analysed by Repeated Measure Two-way Anova [from Madaro et al. (2018)]

As mentioned above, the adaptive purpose of the stress response is to reallocate resources to face a threat or simply an environmental challenge whilst putting on hold other processes that are not useful during the state of emergency, e.g. ovulation, copulation or digestion, until the cessation of the stress situation. A counterpart of the stress response is therefore that the metabolic reorganisation may affect the efficacy of other functions such as the immune system (Tort 2011). Indeed, some parts of the immune defence repertoire may be delayed or reduced, thus compromising immune capacity and resistance to pathogens. The result is that the stressed animal may suffer suppression of the immune system. For example, it has been shown that animals previously subjected to a stressful situation are more vulnerable to certain diseases. However, the effect of stress on the immune system may depend on the nature of the stressor and the length of exposure to it. In fact, when the exposure to a stressor is relatively brief, it may itself promote an innate response of the immune system by enhancing or activating the immune response. Neurotransmitters produced during stress by the sympathetic nervous system appear to be amongst those mainly responsible for the beneficial effect on the innate immune response (Nardocci et al. 2014). For example, short-term acute stress can enhancing adaptive responses by increasing inflammatory markers, upregulated proinflammatory cytokines IL-1β and IL-8, increased lysozyme activity, and increased complement C3 proteins as bacteriolytic mechanisms [reviewed in Yada and Tort (2016)].

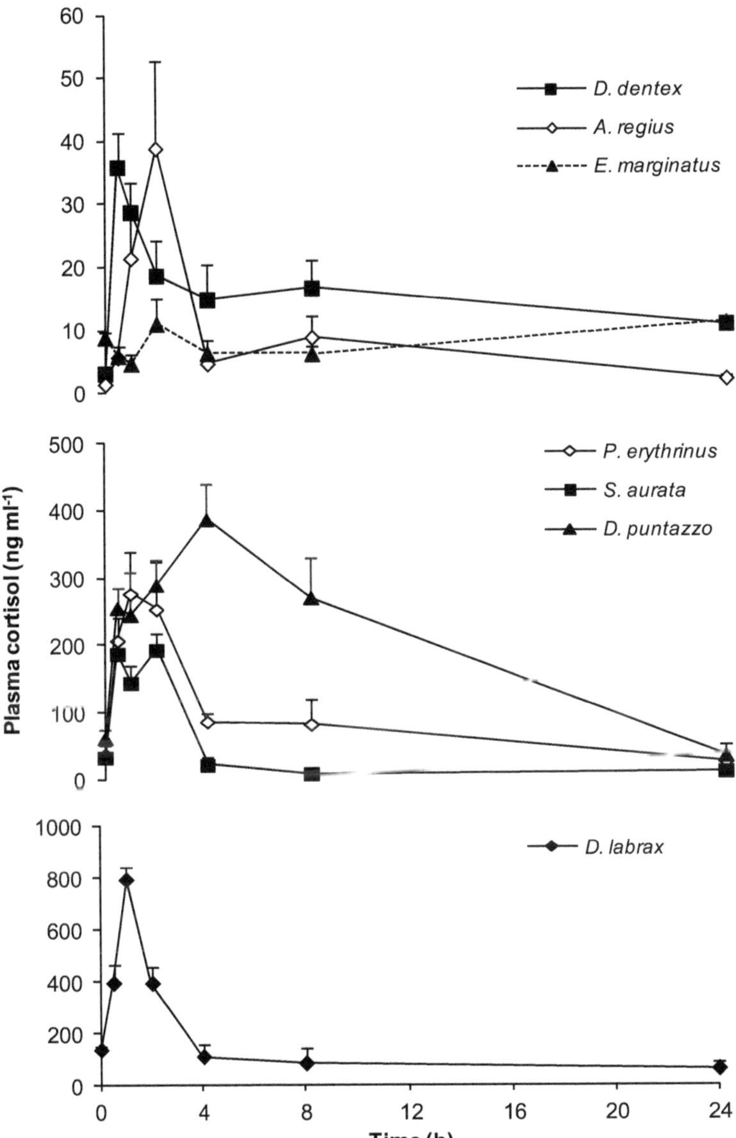

Fig. 11.4 Plasma cortisol (mean ± SEM, $n = 5$) concentrations of seven Mediterranean fish species after acute stress (5 min chasing and 1–1.5 min air exposure), during a period of 24-h after the exposure (Fanouraki et al. 2011)

11.4.2 Chronic Stress

In the course of their life, whether under natural or farmed conditions, fish may experience stressful situations that cause them discomfort. Their capacity to cope with and habituate to stressors (environmental or anthropogenic) depends on the characteristics of the stressors in terms of their severity, duration and frequency (Herman 2013). Habituation is a necessary adaptive response particularly under conditions where similar stress episodes are frequent. Animals subjected to repeated challenges (stress) reduce their response, i.e. cortisol release and oxygen consumption, over time (Grissom and Bhatnagar 2009), partly in order to avoid the negative consequences of prolonged activation of the stress response (Herman 2013; Korte et al. 2007). Indeed, cortisol and other mediators of allostasis (e.g. neurotransmitters and cytokines) promote adaptive effects only in the short term, by releasing energy to counteract the stressor. Nonetheless, when the response to stress is extended over time, it may lead to an *allostatic state*, a chronic deviation of the regulatory system from its normal operating level. In this situation, the regulatory capacity of an organism is narrowed (Korte et al. 2007) and the likelihood that an animal will be exposed to hyperstimulation (the right side of the allostatic curve; Fig. 11.1) due to new or additional stressors rises.

The cumulative effects of an allostatic state leads to an *allostatic load* when: (a) the body is subjected to repeated challenges or lasting exposure to stress; (b) the body fails to habituate to repeated challenges; (c) the body fails to turn off the stress response (constant release of glucocorticoids), and (d) the stress response is not capable of dealing with the stress episode (McEwen 2003; McEwen and Wingfield 2003). Under these conditions, the stress response loses its adaptive purpose leading the organism to an allostatic overload, that causes wear and tear of the body (Herman 2013; Korte et al. 2007). The visible effects of the allostatic overload, also known as *tertiary responses* to stress, act at the whole animal level, impairing growth, health and disease resistance, reproduction and behaviour.

Fish exposed to long (Van Weerd and Komen 1998), unpredictable (Madaro et al. 2015) or repeated mild stress (Madaro et al. 2016b) showed a significant reduction in somatic growth. Fish growth depends on a complex set of processes, starting with the foraging of nutrients in their environment until their absorption in organs and tissues. Stress affects growth in various ways. In particular, long-term cortisol and catecholamines release becomes deleterious, impairing food intake, absorption at gut level, and energy utilisation, including protein turnover. Elevated circulating cortisol levels stimulate catabolic processes and inhibit the promoters of muscle growth, i.e. growth hormone (GH) and insulin-like growth factor (IGF) (reviewed in (Sadoul and Vijayan 2016).

Loss of appetite is one of the main causes of reduced growth in fish under stressful conditions (Bernier and Craig 2005). It is reasonable to assume that in an emergency situation, the brain "commands" the body to stop eating, whilst focusing all of its attention on survival. On the other hand, side-effects appear when the stress is prolonged. Indeed, when fish stop or reduce eating, most of their available stored energy is put into maintaining the existing structure (muscle, organs, etc.) and jet to cope with the ongoing challenge, leading to a deficit in energy reserves. As in mammals, appetite and feeding behaviour in fish are regulated at the level of the hypothalamus. Unfortunately, our understanding of how stress leads to reduced appetite is far from being clear. CRF and POMC released during stress are known to act as anorexigenics, and lead to reduced food intake (Volkoff et al. 2005). During chronic stress, the persistent activation of the stress axis requires a continued supply of all the components of the HPI axis molecular cascade including CRF and POMC (Madaro et al. 2015, 2016b). This may explain why fish lose appetite for long periods. Moreover, α-MSH, a product of the cleavage of the POMC protein, also act as anorexigenic in fish [reviewed in Delgado et al. (2017)]. Cortisol may also lead to a reduced food intake and thus growth by stimulating the production of leptin in the liver, another powerful anorexigenic hormone (Madison et al. 2015).

Where fish are still capable of eating, chronic stress may also affect growth by impairing the efficient assimilation of nutrients. Acute stress can cause alteration of the gastrointestinal tract at the cellular level in Atlantic salmon (Olsen et al. 2003) and rainbow trout (Olsen et al. 2005), as well as levels and composition of the intestinal microbial population. On the other hand, information about the mechanism by which chronic stress affects the absorption of the energy substrate is larger lacking. Although the mechanism in fish still is not clear, evidence suggests that cortisol may be an important agent of impaired gut absorption during long period of stress (Sadoul and Vijayan 2016).

Chronic stress is also associated with immune system suppression. Just as for growth, immune capacity may be suppressed due to a deficit of the resources (allostatic load) needed to support the operative mechanisms of the immune system such as cell division or protein synthesis. Moreover, long-term activation of the stress axis and release of the allostasis mediators during the stress response is responsible for downregulating or reducing the intensity of the immune response (Yada and Tort 2016). For example, various studies have reported that chronic stress may suppress phagocytic and lysozyme activity, the lymphocyte activation and mobilisation, or that it can affect antibody production (reviewed by Zwollo 2017). In particular, there is evidence that catecholamine release can reduce specific immunity in fish: for example, adrenaline and noradrenaline have been shown to reduce phagocytosis in spotted murrel (*Channa punctatus*) (Roy and Rai 2008). Moreover, in vitro experiments have shown that administering adrenaline to gilthead sea bream (*Sparus aurata*) leukocytes cells lowers mRNA transcript levels of proinflammatory cytokines (Castillo et al. 2009). Is not clear whether the depressive effects of catecholamines on the immune system take place only when the hypothalamic–sympathetic–chromaffin cell axis is repeatedly activated or also after a one-off stress episode. Cortisol appears to be the main link between the stress axis and the immune system, proof of which is that most cortisol receptors (GRs) are found in all immune cells (Gorissen and Flik 2016). Cortisol can suppress the fish's immune system in several ways, affecting antibody production, leukocyte mitosis, phagocytosis and cytokine expression (reviewed in Yada and Tort 2016).

Investing energy in reproduction whilst the body is subjected to stress, or even more when safety or survival is at stake, is less prioritised in fish than in other vertebrates. In fact, stress also has an inhibitory effect on reproductive performance in male and female fish, particularly by impairing ovarian and testicular development, by delaying or inhibiting ovulation and spawning, by leading to the production of smaller eggs and larvae and/or affecting their survival. For an exhaustive presentation of the consequence of stress on reproduction and maturation, see the review by Pankhurst (2016). It is important to notice that most of the knowledge obtained to date about the effects of stress on animal reproduction are outcomes of studies performed for the most part on farmed or captive fish species, and that little is known about the extent to which these findings are valid for natural populations (Pankhurst 2011). To date, many of the mechanisms that link stress to reproduction are still unclear and often ambiguous. The activation of the HPI axis appears to be the most important factor responsible for the inhibitory effect on reproductive performance. Unfortunately, it is not always obvious which effects are directly caused by the HPI axis hormones, and which are originated by indirect effects arising from the regulatory action of the HPI axis on behaviour, metabolism and growth (Leatherland et al. 2010). Cortisol receptors have been found at all levels of the hypothalamus–pituitary–gonadal axis. High circulating levels of cortisol may thus act at different levels mediated by genomic mechanism receptors, modulating the expression of the gene that controls the endocrine cascade for the production and release of sexual hormone (reviewed by Pankhurst 2016). We stress that cortisol levels in the blood are not always the cause of deleterious effects on reproduction, and fish often maintain their reproductive activity over a wide range of cortisol levels (Pankhurst 2011).

Physiological changes due to prolonged or extreme conditions of stress also have effects on behaviour. For example, Øverli et al. (2002) induced high circulating levels of cortisol in rainbow trout by feeding them for 3 days with cortisol-enriched food. They observed that the fish with high levels of circulating cortisol displayed reduced locomotor activity and low aggressive behaviour when challenged by the introduction of a conspecific in the tank. Similarly, in salmonids, subordinate individuals in a dominance hierarchy display higher brain serotonergic activity and HPI axis activation, both of which inhibit competitive behaviours, locomotor activity and aggression (Gilmour 2005). Subordination appears to be perceived as a form of chronic stress by the subordinates, due to the aggression they receive or their exclusion by the dominant fish from resources such as food and shelter (Winberg and Nilsson 1993). Higher levels of serotonergic signalling and cortisol production are also found in growth stunted Atlantic salmon, a phenotype that frequently occurs when smolting salmon are transferred to sea cages (Vindas et al. 2016a). Interestingly, the behavioural and serotonergic profiles exhibited by the fish under these circumstances are evocative of a depressed state, similar to those described in mammals under chronic stress (Andrews et al. 2015; Blackard and Heidingsfelder 1968; Shively and Willard 2012).

The nature and frequency of the stressor seem to influence the stress response. If a stressor is repeated, especially in a predictable manner, then the animal will come to anticipate the stressor, through classical conditioning. Increased predictability seems to improve the perceived control by allowing some degree of preparation for the stressor [for review, see Galhardo and Oliveira (2009)]. However, a high degree of predictability can generate more than one outcome. Galhardo et al. (2011) showed that tilapia (*Oreochromis mossambicus*) reduce their cortisol response when negative events are predictable. When positive events are predictable (e.g. feeding), fish may display an anticipatory response related to positive emotions (i.e. exploratory behaviour). But variations in the predictability of positive events can also be frustrating and generate stress responses (Vindas et al. 2012). Madaro et al. (2016a) showed that salmon parr exposed to stress announced by a conditioning stimulus (CS, light signal) developed an "anxiety" response to the CS. Fish stopped swimming against the current and swam more randomly. In a natural environment, fish facing a perceived threat attempt to avoid or escape, whilst in a confined tank this option is not available. The anxiety observed in the conditioned fish can be the result of two learned notions. First, the CS predicts the stressor and secondly nothing can be done to avoid it. Even so, the cortisol response triggered by the stress episode was low and proportional to the stress intensity (chasing) and independent on conditioning. The parr appeared to have learned that the stressor was not life threatening, and they could look forward a period without disturbance afterwards. If this was the case, it can be speculated that predictable conditions increase the perception of control, enabling the fish to prepare themselves to face the stress (Bassett and Buchanan-Smith 2007).

Prediction is also one of the central points of the allostasis concept: when environmental conditions change and/or a challenge is presented, the physiological set points are adjusted to meet the anticipated needs of the organisms in order to optimise its performance (Sterling 2004). Obviously, unpredictable conditions will require a different regulation of allostatic mediators than predictable or habituating conditions do. For example, a comparison of Atlantic salmon parr groups exposed either to series of repeated unpredictable stressors (Madaro et al. 2015) or to a single repeated predictable stressor (Madaro et al. 2016b) showed interesting differences in the expression of genes responsible for the activation of the pituitary gland. Fish under repeated predictable stress, were able to habituate to the stressful stimuli (chasing), displaying downregulated *crfr1* receptor in the pituitary gland, thus reducing the pituitary's excitability from the hypothalamus. A possible interpretation may be that when fish have developed perception of some degree of control over a stressor/stimulus, the brain raises the threshold of the stress required to trigger a new stress response. On the contrary, in the parr subjected to unpredictable stressors, *crfr1* transcripts were similar to what was found for the naïve fish, suggesting a prompt excitability of the pituitary to unpredictable stress. These results demonstrate that under unpredictable conditions the stress axis is required to be prepared at all times to deal with a new stress episode.

11.5 Ontogeny of the Stress Response in Teleosts

The development of chromaffin and interrenal tissues in fish has been studied in few species via histological, immunohistochemical and biochemical approaches or ultra-structural studies. A study in rainbow trout showed that interrenal primordial cells are present in the head kidney of 25-days post-fertilisation (dpf) larvae, whilst chromaffin cells were first identified later on, at 27-dpf (Gallo and Civinini 2005). In the common carp, *Cyprinus carpio*, whole-body ACTH, cortisol and to a lesser extend α-MSH show a significantly elevated levels before hatching (56–72 h after fertilisation), indicating that the HPI axis is already functioning at hatching (Stouthart et al. 1998). This is further supported by the elevated whole-body cortisol concentrations observed in stressed (5 min of handling) 50 h post-fertilisation eggs (Stouthart et al. 1998). Similarly in the chum salmon, *Oncorhynhus ke*ta, and in chinook salmon, *Oncorhynchus tshawytscha,* species with relatively long developmental times in ovo, cortisol levels show a slight increase 1 week before hatching or at hatch, respectively (Feist and Schreck 2002; de Jesus and Hirano 1992).

However, in other teleosts no evidence has been found that the HPI axis is activated before or at hatching. According to Barry and colleagues (1995), the activation of the HPI axis in trout larvae occurs in the second week after hatching. In several marine teleosts, cortisol content starts to rise around first feeding, to reach a peak around flexion. In cod, *Gadhus morhua*, a measurable cortisol stress response was observed in 8-day post-hatch larvae (King and Berlinsky 2006). In European sea bass, *Dicentrarhus labrax*, an active HPI axis is first observed at first feeding (the transition from endogenous to exogenous nutritional sources; Fig. 11.5), at which time a peak in whole-body cortisol levels is observed following exposure to stressors (Pavlidis et al. 2011; Tsalafouta et al. 2014). Similar results have been reported for gilthead sea bream, *Sparus aurata*, (Szisch et al. 2005) and yellow perch, *Perca flave*scens, (Jentoft et al. 2002). In zebrafish, *Danio rerio,* physical stress raised cortisol at about 40 h post hatch (Alsop and Vijayan 2008). In addition, transcripts of genes encoding enzymes that catalyse the first and final steps in the production of cortisol have been detected before and soon after hatching in several fish species (Alsop and Vijayan 2008; Appelbaum et al. 2010; Tsalafouta et al. 2014).

Apart from the HPI axis and cortisol other systems and hormones are also involved in the manifestation of the stress response. In only a few fish species studied so far, does α-MSH also seem to be involved in the stress response (Arends et al. 1999; Sumpter et al. 1986; Tsalafouta et al. 2017)· In a recent study, the temporal patterns of whole-body α-MSH concentrations and of transcripts of melanocortin receptors during early development in response to stress were characterised in the European sea bass (Tsalafouta et al. 2017). It was shown that as development proceeds, α-MSH content gradually increases, and that at the stage of fin formation, responded with peak values at 2 h post stress. The stress challenge also resulted in elevated transcript levels of *pomc*, *mc2r* and *mc4r*, with a pattern characterised by peak values at 1 h post stress and a strong correlation with whole-body α-MSH concentrations.

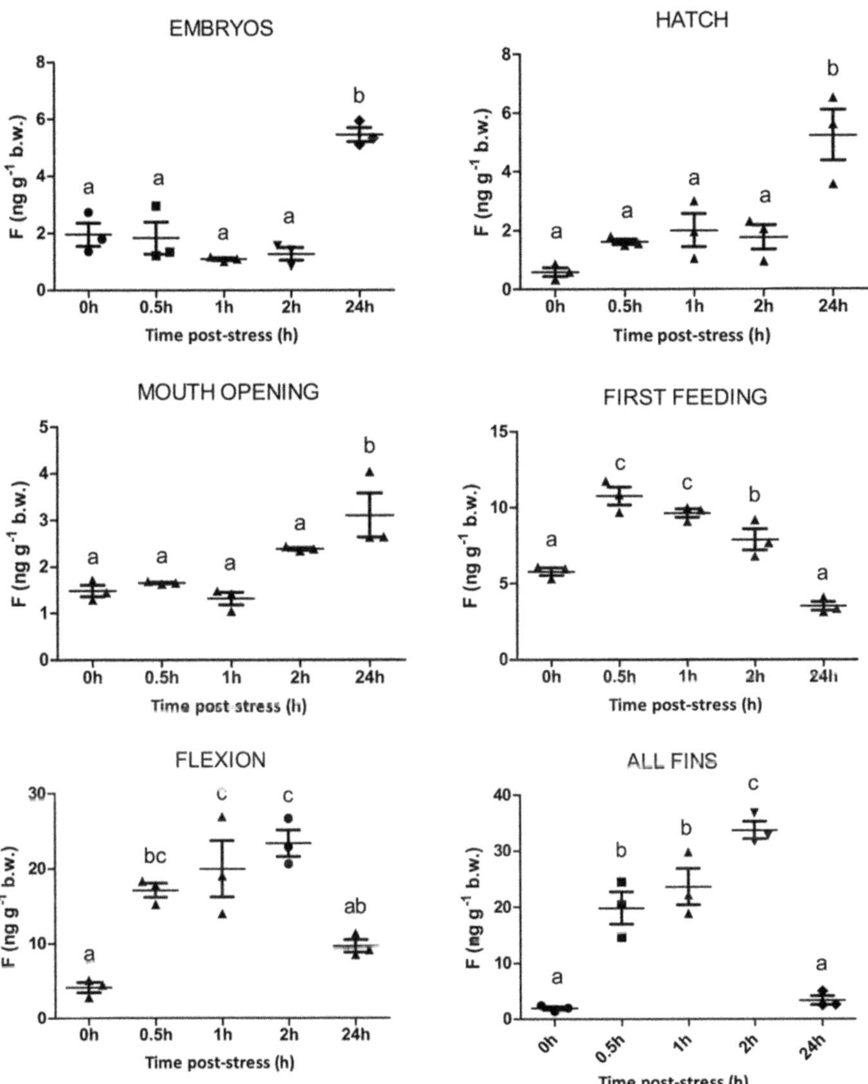

Fig. 11.5 Ontogeny of the cortisol stress response in European seabass. The cortisol response prior to (0 h) and after (0.5 h, 1 h, 2 h and 24 h) the application of the stressor during early ontogeny. Values are means ± standard error ($n = 3$). Means with different letters differ significantly from one another ($P < 0.05$; from Tsalafouta et al. 2015)

11.5.1 Early-Life Stress

How early-life stress has profound short- and long-term effects on human neurophysiology, cognition, mood and emotions is a subject of growing research interest.

It is now well accepted that early-life stress can leave lasting impacts on the brain, affecting significantly subsequent developmental stages (Bahari-Javan et al. 2017; Syed and Nemeroff 2017). Post-traumatic stress disorders, as well as depression and anxiety, appear to be sensitive to the effects of adverse early-life conditions. In mammals, laboratory early-life stress is usually induced by maternal separation or fragmented maternal care (Molet et al. 2014). Maternal stress during pregnancy also affects development of offspring and may be also associated with alterations in growth, physiology and behaviour. However, complicated genetic, environmental and social interactions present a number of challenges regarding the ultimate cause (s) of psychopathology. The use of non-parental care vertebrates could, therefore, offer an alternative approach to understanding the impact of early-life stress on subsequent phases of development. However, published data for fish are still scarce.

A recent study on the threespine stickleback (*Gasterosteus aculeatus*) reported, that even in egg-laying vertebrates, exposure to maternal steroids is mediated by both maternal and embryonic processes (Paitz et al. 2016). Specifically, it was showed that stickleback embryos could actively clear exogenous cortisol within 72 h from immersion to ^{3}H-cortisol-containing solution. Thus, not only placental mammals but also egg-laying fish can modulate foetal exposure to maternal glucocorticoids by either metabolising glucocorticoids or actively increasing their clearance rate (Paitz et al. 2016). Application of an unpredictable chronic low intensity stress protocol during different phases of early ontogeny (i.e. first feeding, flexion or development of all fins) in European sea bass larvae, resulted in increased waterborne cortisol concentrations during the treatment and in impairment of growth performance as long as 2 months after the end of the larval period (Fig. 11.6) (Tsalafouta et al. 2014). Significantly higher average resting plasma cortisol concentrations were also observed in juveniles exposed to the stress protocol at the first feeding stage and onwards to flexion and at the formation of all fins onwards to the development of melanophores, compared to fish first exposed to stress from flexion to the development of all fins and controls. The application of the same stress protocol during early ontogeny in gilthead sea bream did not evoke a cortisol stress response. However, juveniles that had experienced early-life events at the stage of all fins exhibited 2 months after the end of the larvae stage, the lowest mean total length and body weight compared to the other groups (Sarropoulou et al. 2016). On the transcriptome level distinct patterns of expression were observed both in larvae and in juveniles, with the most divergent expression pattern found to be again at the phase of the development of all fins (Sarropoulou et al. 2016). Those studies were the first to demonstrate that common husbandry practices during early life have an impact both on larvae performance and at later stages of development, affecting growth and the stress response in juveniles.

Phenotypic plasticity on stress responsiveness due to environmental changes is particularly evident at early-life stages. In a recent study, 10-month-old Atlantic salmon parr was subjected to an unpredictable chronic stress protocol for 3 weeks (Madaro et al. 2015). After 3 weeks, the fish were allowed to rest and grow for several months. Meanwhile, smoltification was induced by a light-induced signal and thereafter fish were transferred in seawater. Along this period fish performance

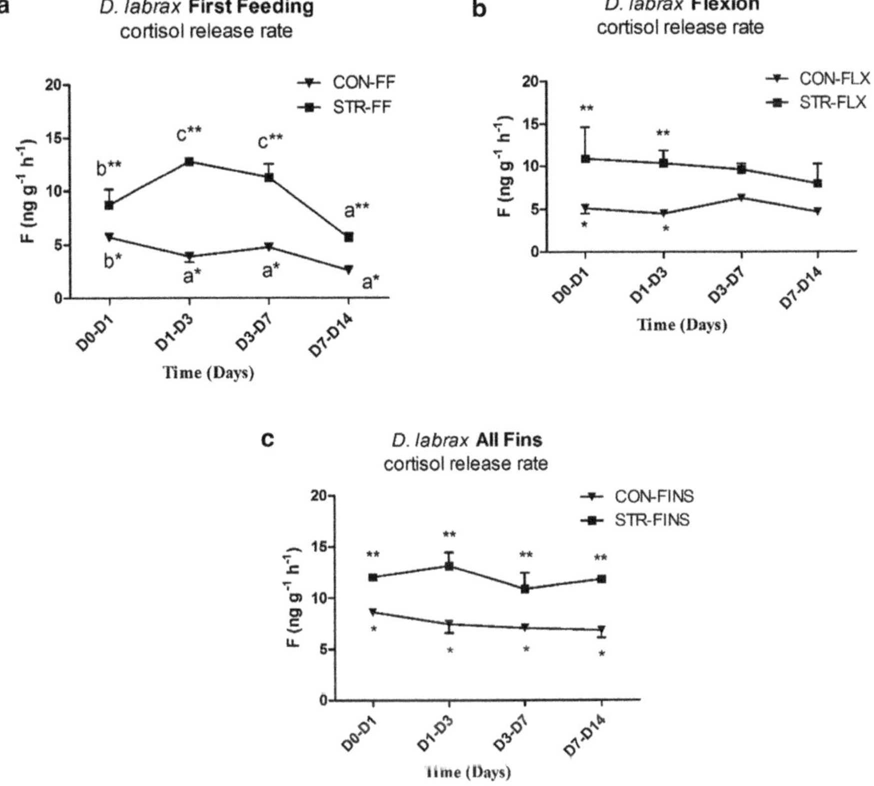

Fig. 11.6 Waterborne cortisol release rates in the holding tanks during early ontogeny of European sea bass. The unpredictable chronic low stress protocol was applied at different periods, i.e. from first feeding to flexion (group STR-FF), from flexion to the formation of all fins (group STR-FLX), and from the formation of all fins to the full cover of body with melanophores (group STR-FINS). One group remained undisturbed and served as the control (CON-FF, CON-FLX, CON-FINS). All conditions were performed in duplicate tanks. Letters indicate differences between days within the same group whereas asterisks indicate differences between the stressed and the control group (from Tsalafouta et al. 2015)

was investigated (Vindas et al. 2016b). Interestingly, the fish that had experienced early-life stress evidenced a higher rate of growth during two challenging developmental periods: during smoltification and after the transfer to seawater. Furthermore, individuals who experienced early stress responded differently to environmental stimuli later in life (up to 10 weeks after the stress regime) compared with non-stress-treated fish. Namely, after being transferred to seawater, the groups stressed in early life displayed reduced hypothalamic catecholaminergic and brainstem serotonergic responses to stress. We can, therefore, speculate that in an aquaculture environment, where stressful experiences are common, experiencing stress from an early age may help individuals to cope better with their environment later in life (Vindas et al. 2016b). These observations agree with the allostasis theory (Korte et al. 2007), which proposes that individuals that repeatedly experienced challenges are better equipped to cope with future similar stressors, for example by mounting a less intense monoaminergic stress response, which may enable them to invest more energy in other life processes such as growth.

Stress episodes, during the early stages of life, can trigger developmental paths that have lifelong impacts on the health, metabolism or behaviour of the animal (Szyf 2013). A growing amount of evidence has in fact shown that early-life experiences have a persistent impact on gene expression and behaviour, exercised through epigenetic mechanisms. Turecki and Meaney (2016) reviewed more than 40 articles (13 animal and 27 human studies). They reported that early-life adversity increased methylation on the cortisol receptor gene, namely the GR exon variant 1F in humans and on the GR17 in rats, in about 89% of human studies and 70% of animal studies. Unfortunately, to date, similar information on fish is very sparse. In a recent study, Moghadam and colleagues (Moghadam et al. 2017) investigated the effect of early-life stress (from the eye-stage until start-feeding) on Atlantic salmon embryos using cold shock and air exposure. Interestingly, 1 year after the stress trial, fish that were stressed before and after hatching displayed better growth than the other groups that had been stressed only before hatching, only after hatching or were nor stressed. The authors showed that the stress episode caused the methylation of specific area of the DNA that effected the transcriptional regulation of genes significant in developmental processes. It may be argued that low or moderate exposure (eustress) to otherwise harmful stimuli can protect or improve tolerance to the same or other stimuli/stressors during later stages of life, a concept also known as "hormesis" (Vaiserman 2010).

11.6 Summary

The neuroendocrine basis and physiology of the stress response have been well described in a number of fish species. When a noxious stimulus or stressor is perceived by a fish, neuronal signals activate the brain–sympathetic–chromaffin axis, resulting in the release of catecholamines from the chromaffin cells (equivalent to the mammalian adrenal medulla), which prepares the animal for "fight or flight" set of responses. Following this primary response, activation of the hypothalamic–pituitary–interrenal axis induces the synthesis and release of corticosteroids from the interrenal cells (equivalent to the mammalian adrenal cortex). The magnitude, timing and duration of cortisol released into the blood are all species-specific and depend on the type and intensity of the stressor.

Following exposure to an acute stressor(s), the body activates a series of adaptive processes that aim to conserve energy, but at the same time release whatever is required to deal with the immediate demands imposed by the stressor. Increases in cardiovascular tone, respiratory rate and energy availability (gluconeogenesis and lipolysis), with a concomitant suppression of functions that are not required for coping with the stressor (e.g. feeding, digestion, growth, reproduction) are typical manifestations of the acute stress response. A healthy animal has a wide array of allostatic ranges of responses that elicit optimum performance to meet internal and external challenges. However, when the challenges are prolonged, they generate a condition of *allostatic state* in which the animal's regulatory capacity becomes reduced, leading to reduced performance and impairment of health. Fish exposed to long, unpredictable or repeated mild stress show a significant reduction in somatic growth, disease resistance, reproduction, behaviour (e.g. feeding behaviour, locomotor activity, social interactions) and welfare.

It is scarcely possible to study stress and associated consequences without considering whether or not the stressor is perceived as a serious threat, i.e. without the individual's appraisal on the predictability and/or controllability of the challenge. To that end, cognition and how an individual assesses a given situation may well be just as important as the actual physical challenge in determining the severity of the stress response. The stress response therefore also depends on the individual's genetic background, acquired information, previous experiences and how they are evaluated. That means that the same stressor can produce different effects in different individuals as well as in the same individual exposed to different situations.

Phenotypic plasticity in stress responses to stress due to environmental changes is particularly evident during the early stages of life. Stressful episodes in early life can trigger developmental paths that may have lifelong impacts on the health, metabolism or behaviour of the animal, and a growing amount of evidence has in fact shown that early-life experiences have a lasting impact on gene expression and behaviour via epigenetic mechanisms. The use of non-embryo-carrying, non-parental-care fish species may provide an alternative approach to improve our understanding of the impact of early-life stress on performance and during later phases of development.

References

Alderman SL, Bernier NJ (2007) Localization of corticotropin-releasing factor, urotensin I, and CRF-binding protein gene expression in the brain of the zebrafish, *Danio rerio*. J Comp Neurol 502:783–793. https://doi.org/10.1002/cne.21332

Allen C (2011) Fish cognition and consciousness. J Agric Environ Ethics 26:25–39. https://doi.org/10.1007/s10806-011-9364-9

Alsop D, Vijayan MM (2008) Development of the corticosteroid stress axis and receptor expression in zebrafish. Am J Physiol Integr Comp Physiol 294:R711–R719. https://doi.org/10.1152/ajpregu.00671.2007

Aluru N, Vijayan MM (2006) Aryl hydrocarbon receptor activation impairs cortisol response to stress in rainbow trout by disrupting the rate-limiting steps in steroidogenesis. Endocrinology 147:1895–1903. https://doi.org/10.1210/en.2005-1143

Aluru N, Vijayan MM (2008) Molecular characterization, tissue-specific expression, and regulation of melanocortin 2 receptor in rainbow trout. Endocrinology 149:4577–4588. https://doi.org/10.1210/en.2008-0435

Aluru N, Vijayan MM (2009) Stress transcriptomics in fish: a role for genomic cortisol signaling. Gen Comp Endocrinol 164:142–150. https://doi.org/10.1016/j.ygcen.2009.03.020

Andrews PW, Bharwani A, Lee KR, Fox M, Thomson JA (2015) Is serotonin an upper or a downer? The evolution of the serotonergic system and its role in depression and the antidepressant response. Neurosci Biobehav Rev 51:164–188. https://doi.org/10.1016/j.neubiorev.2015.01.018

Appelbaum L, Wang G, Yokogawa T, Skariah GM, Smith SJ, Mourrain P, Mignot E (2010) Circadian and homeostatic regulation of structural synaptic plasticity in hypocretin neurons. Neuron 68:87–98. https://doi.org/10.1016/j.neuron.2010.09.006

Arends RJ, Mancera JM, Muñoz JL, Wendelaar Bonga SE, Flik G (1999) The stress response of the gilthead sea bream (Sparus aurata L.) to air exposure and confinement. J Endocrinol 163:149–157

Bahari-Javan S, Varbanov H, Halder R, Benito E, Kaurani L, Burkhardt S, Anderson-Schmidt H, Anghelescu I, Budde M, Stilling RM, Costa J, Medina J, Dietrich DE, Figge C, Folkerts H, Gade K, Heilbronner U, Koller M, Konrad C, Nussbeck SY, Scherk H, Spitzer C, Stierl S, Stöckel J, Thiel A, von Hagen M, Zimmermann J, Zitzelsberger A, Schulz S, Schmitt A, Delalle I, Falkai P, Schulze TG, Dityatev A, Sananbenesi F, Fischer A (2017) HDAC1 links early life stress to schizophrenia-like phenotypes. Proc Natl Acad Sci U S A 114:E4686–E4694. https://doi.org/10.1073/pnas.1613842114

Barry TP, Ochiai M, Malison JA (1995) In vitro effects of ACTH on interrenal corticosteroidogenesis during early larval development in rainbow trout. Gen Comp Endocrinol 99:382–387. https://doi.org/10.1006/GCEN.1995.1122

Bartholome B, Spies CM, Gaber T, Schuchmann S, Berki T, Kunkel D, Bienert M, Radbruch A, Burmester G-R, Lauster R, Scheffold A, Buttgereit F (2004) Membrane glucocorticoid receptors (mGCR) are expressed in normal human peripheral blood mononuclear cells and up-regulated after in vitro stimulation and in patients with rheumatoid arthritis. FASEB J 18:70–80. https://doi.org/10.1096/fj.03-0328com

Barton B (2002) Stress in fishes: a diversity of responses with particular reference to changes in circulating corticosteroids. Integr Comp Biol 42:517–525

Barton BA, Iwama GK (1991) Physiological changes in fish from stress in aquaculture with emphasis on the response and effects of corticosteroids. Annu Rev Fish Dis 1:3–26. https://doi.org/10.1016/0959-8030(91)90019-G

Bassett L, Buchanan-Smith HM (2007) Effects of predictability on the welfare of captive animals. Appl Anim Behav Sci 102:223–245. https://doi.org/10.1016/j.applanim.2006.05.029

Bernard C (1974) Lectures on the phenomena of life common to animals and plants. Charles C Thomas, Springfield

Bernier NJ, Craig PM (2005) CRF-related peptides contribute to stress response and regulation of appetite in hypoxic rainbow trout. Am J Physiol Regul Integr Comp Physiol 289:R982–R990. https://doi.org/10.1152/ajpregu.00668.2004

Bernier NJ, Peter RE (2001) The hypothalamic-pituitary-interrenal axis and the control of food intake in teleost fish. Comp Biochem Physiol B Biochem Mol Biol 129:639–644

Bernier NJ, Lin X, Peter RE (1999) Differential expression of corticotropin-releasing factor (CRF) and urotensin I precursor genes, and evidence of CRF gene expression regulated by cortisol in goldfish brain. Gen Comp Endocrinol 116:461–477. https://doi.org/10.1006/gcen.1999.7386

Bernier NJ, Bedard N, Peter RE (2004) Effects of cortisol on food intake, growth, and forebrain neuropeptide Y and corticotropin-releasing factor gene expression in goldfish. Gen Comp Endocrinol 135:230–240. https://doi.org/10.1016/j.ygcen.2003.09.016

Blackard WG, Heidingsfelder SA (1968) Adrenergic receptor control mechanism for growth hormone secretion. J Clin Invest 47:1407–1414. https://doi.org/10.1172/JCI105832

Borski RJ, Hyde GN, Fruchtman S (2002) Signal transduction mechanisms mediating rapid, nongenomic effects of cortisol on prolactin release. Steroids 67:539–548. https://doi.org/10.1016/S0039-128X(01)00197-0

Braithwaite VA, Boulcott P (2008) Can fish suffer? In: Branson E (ed) Fish welfare. Blackwell, Oxford

Brown C (2015) Fish intelligence, sentience and ethics. Anim Cogn 18:1–17. https://doi.org/10.1007/s10071-014-0761-0

Brown C, Laland K, Krause J (2008) Fish cognition and behavior. Blackwell

Bury NR, Sturm A, Le Rouzic P, Lethimonier C, Ducouret B, Guiguen Y, Robinson-Rechavi M, Laudet V, Rafestin-Oblin ME, Prunet P (2003) Evidence for two distinct functional glucocorticoid receptors in teleost fish. J Mol Endocrinol 31:141–156

Cannon WB (1929) Organization for physiological homeostasis. Physiol Rev 9:399–431

Cannon WB (1932) The wisdom of the body. W W Norton & Co., New York, NY

Cannon WB (1939) The wisdom of the body. Norton & Co., Oxford

Castanheira MF, Conceição LEC, Millot S, Rey S, Bégout M-L, Damsgård B, Kristiansen T, Höglund E, Øverli Ø, Martins CIM (2015) Coping styles in farmed fish: consequences for aquaculture. Rev Aquac 9:23. https://doi.org/10.1111/raq.12100

Castillo J, Teles M, Mackenzie S, Tort L (2009) Stress-related hormones modulate cytokine expression in the head kidney of gilthead seabream (Sparus aurata). Fish Shellfish Immunol 27:493–499. https://doi.org/10.1016/j.fsi.2009.06.021

Chabbi A, Ganesh CB (2012) Stress-induced inhibition of recruitment of ovarian follicles for vitellogenic growth and interruption of spawning cycle in the fish Oreochromis mossambicus. Fish Physiol Biochem 38:1521–1532. https://doi.org/10.1007/s10695-012-9643-z

Cooper CL, Dewe PJ (2004) Stress: a brief history. Blackwell, Oxford

Davis KB (2004) Temperature affects physiological stress responses to acute confinement in sunshine bass (Morone chrysops × Morone saxatilis). Comp Biochem Physiol Part A Mol Integr Physiol 139:433–440. https://doi.org/10.1016/j.cbpb.2004.09.012

de Jesus EGT, Hirano T (1992) Changes in whole body concentrations of cortisol, thyroid hormones, and sex steroids during early development of the chum salmon, *Oncorhynchus keta*. Gen Comp Endocrinol 85:55–61. https://doi.org/10.1016/0016-6480(92)90171-F

de Kloet ER, Joëls M, Holsboer F (2005) Stress and the brain: from adaptation to disease. Nat Rev Neurosci 6:463–475. https://doi.org/10.1038/nrn1683

Delgado MJ, Cerdá-Reverter JM, Soengas JL (2017) Hypothalamic integration of metabolic, endocrine, and circadian signals in fish: involvement in the control of food intake. Front Neurosci 11. https://doi.org/10.3389/fnins.2017.00354

Doyon C, Gilmour K, Trudeau V, Moon T (2003) Corticotropin-releasing factor and neuropeptide Y mRNA levels are elevated in the preoptic area of socially subordinate rainbow trout. Gen Comp Endocrinol 133:260–271. https://doi.org/10.1016/S0016-6480(03)00195-3

Doyon C, Leclair J, Trudeau VL, Moon TW (2006) Corticotropin-releasing factor and neuropeptide Y mRNA levels are modified by glucocorticoids in rainbow trout, *Oncorhynchus mykiss*. Gen Comp Endocrinol 146:126–135. https://doi.org/10.1016/j.ygcen.2005.10.003

Falkenstein E, Tillmann H-C, Christ M, Feuring M, Wehling M (2000) Multiple actions of steroid hormones-A focus on rapid, nongenomic effects. Pharmacol Rev 52:513–556

Fanouraki E, Divanach P, Pavlidis M (2007) Baseline values for acute and chronic stress indicators in sexually immature red porgy (Pagrus pagrus). Aquaculture 265(1–4):294–304

Fanouraki E, Mylonas CC, Papandroulakis N, Pavlidis M (2011) Species specificity in the magnitude and duration of the acute stress response in Mediterranean marine fish in culture. Gen Comp Endocrinol 173:313–322. https://doi.org/10.1016/j.ygcen.2011.06.004

Feder A, Nestler EJ, Charney DS (2009) Psychobiology and molecular genetics of resilience. Nat Rev Neurosci 10:446–457. https://doi.org/10.1038/nrn2649

Feist G, Schreck CB (2002) Ontogeny of the stress response in chinook salmon, Oncorhynchus tshawytscha*. Fish Physiol Biochem 25:31–40. https://doi.org/10.1023/a:1019709323520

Flik G, Klaren PHM, Van den Burg EH, Metz JR, Huising MO (2006) CRF and stress in fish. Gen Comp Endocrinol 146:36–44. https://doi.org/10.1016/j.ygcen.2005.11.005

Fryer JN (1989) Neuropeptides regulating the activity of goldfish corticotropes and melanotropes. Fish Physiol Biochem 7:21–27. https://doi.org/10.1007/BF00004686

Fryer J, Lederis K, Rivier J (1984) Cortisol inhibits the ACTH-releasing activity of urotensin I, CRF and sauvagine observed with superfused goldfish pituitary cells. Peptides 5:925–930

Funder J, Pearce P, Smith R, Smith A (1988) Mineralocorticoid action: target tissue specificity is enzyme, not receptor, mediated. Science 242:583–585. https://doi.org/10.1126/science.2845584

Galhardo L, Oliveira R (2009) Psychological stress and welfare in fish. ARBS Annu Rev Biomed Sci 11:1–20
Galhardo L, Vital J, Oliveira RF (2011) The role of predictability in the stress response of a cichlid fish. Physiol Behav 102:367–372. https://doi.org/10.1016/j.physbeh.2010.11.035
Gallo VP, Civinini A (2005) The development of adrenal homolog of rainbow trout *Oncorhynchus mykiss*: an immunohistochemical and ultrastructural study. Anat Embryol (Berl) 209:233–242. https://doi.org/10.1007/s00429-004-0433-y
Geven EJW, Verkaar F, Flik G, Klaren PHM (2006) Experimental hyperthyroidism and central mediators of stress axis and thyroid axis activity in common carp (Cyprinus carpio L.). J Mol Endocrinol 37:443–452. https://doi.org/10.1677/jme.1.02144
Gilmour KM (2005) Physiological causes and consequences of social status in salmonid fish. Integr Comp Biol 45:263–273. https://doi.org/10.1093/icb/45.2.263
Gorissen M, Flik G (2016) The endocrinology of the stress response in fish: an adaptation-physiological view. In: Schreck CB, Tort L, Farrell AP, Brauner CJ (eds) Biology of stress in fish: fish physiology, vol 35. Academic Press, Cambridge, MA, pp 75–111
Greenwood AK, Butler PC, White RB, DeMarco U, Pearce D, Fernald RD (2003) Multiple corticosteroid receptors in a teleost fish: distinct sequences, expression patterns, and transcriptional activities. Endocrinology 144:4226–4236. https://doi.org/10.1210/en.2003-0566
Grissom N, Bhatnagar S (2009) Habituation to repeated stress: get used to it. Neurobiol Learn Mem 92:215–224. https://doi.org/10.1016/j.nlm.2008.07.001
Groeneweg FL, Karst H, de Kloet ER, Joëls M (2011) Rapid non-genomic effects of corticosteroids and their role in the central stress response. J Endocrinol 209:153–167. https://doi.org/10.1530/JOE-10-0472
Herman JP (2013) Neural control of chronic stress adaptation. Front Behav Neurosci 7:61. https://doi.org/10.3389/fnbeh.2013.00061
Huising MO, Metz JR, van Schooten C, Taverne-Thiele AJ, Hermsen T, Verburg-van Kemenade BML, Flik G (2004) Structural characterisation of a cyprinid (*Cyprinus carpio* L.) CRH, CRH-BP and CRH-R1, and the role of these proteins in the acute stress response. J Mol Endocrinol 32:627–648
Huising MO, Vaughan JM, Shah SH, Grillot KL, Donaldson CJ, Rivier J, Flik G, Vale WW (2008) Residues of corticotropin releasing factor-binding protein (CRF-BP) that selectively abrogate binding to CRF but not to urocortin 1. J Biol Chem 283:8902–8912. https://doi.org/10.1074/jbc.M709904200
Jentoft S, Held JA, Malison JA, Barry TP (2002) Ontogeny of the cortisol stress response in yellow perch (*Perca flavescens*). Fish Physiol Biochem 26:371–378. https://doi.org/10.1023/B:FISH.0000009276.05161.8d
Jones CE, Riha PD, Gore AC, Monfils M-H (2014) Social transmission of Pavlovian fear: fear-conditioning by-proxy in related female rats. Anim Cogn 17:827–834. https://doi.org/10.1007/s10071-013-0711-2
Jørgensen EH, Haatuft A, Puvanendran V, Mortensen A (2017) Effects of reduced water exchange rate and oxygen saturation on growth and stress indicators of juvenile lumpfish (*Cyclopterus lumpus* L.) in aquaculture. Aquaculture 474:26–33. https://doi.org/10.1016/j.aquaculture.2017.03.019
Kiilerich P, Kristiansen K, Madsen SS (2007) Hormone receptors in gills of smolting Atlantic salmon, Salmo salar: expression of growth hormone, prolactin, mineralocorticoid and glucocorticoid receptors and 11beta-hydroxysteroid dehydrogenase type 2. Gen Comp Endocrinol 152:295–303. https://doi.org/10.1016/j.ygcen.2006.12.018
King W, Berlinsky DL (2006) Whole-body corticosteroid and plasma cortisol concentrations in larval and juvenile Atlantic cod Gadus morhua L following acute stress. Aquac Res 37:1282–1289. https://doi.org/10.1111/j.1365-2109.2006.01558.x
Koob G (2004) Allostatic view of motivation: implications for psychopathology. Neb Symp Motiv 50:1–18

Koob GF, Le Moal M (2001) Drug addiction, dysregulation of reward, and allostasis. Neuropsychopharmacology 24:97–129. https://doi.org/10.1016/S0893-133X(00)00195-0
Koolhaas J, Korte S, De Boer S, Van Der Vegt B, Van Reenen C, Hopster H, De Jong I, Ruis MA, Blokhuis H (1999) Coping styles in animals: current status in behavior and stress-physiology. Neurosci Biobehav Rev 23:925–935. https://doi.org/10.1016/S0149-7634(99)00026-3
Koolhaas JM, Bartolomucci A, Buwalda B, de Boer SF, Flügge G, Korte SM, Meerlo P, Murison R, Olivier B, Palanza P, Richter-Levin G, Sgoifo A, Steimer T, Stiedl O, van Dijk G, Wöhr M, Fuchs E (2011) Stress revisited: a critical evaluation of the stress concept. Neurosci Biobehav Rev 35:1291–1301. https://doi.org/10.1016/j.neubiorev.2011.02.003
Korte SM (2001) Corticosteroids in relation to fear, anxiety and psychopathology. Neurosci Biobehav Rev 25:117–142
Korte SM, Koolhaas JM, Wingfield JC, McEwen BS (2005) The Darwinian concept of stress: benefits of allostasis and costs of allostatic load and the trade-offs in health and disease. Neurosci Biobehav Rev 29:3–38. https://doi.org/10.1016/j.neubiorev.2004.08.009
Korte SM, Olivier B, Koolhaas JM (2007) A new animal welfare concept based on allostasis. Physiol Behav 92:422–428. https://doi.org/10.1016/j.physbeh.2006.10.018
Laland KN, Hoppitt W (2003) Do animals have culture? Evol Anthropol Issues News Rev 12:150–159. https://doi.org/10.1002/evan.10111
Lankford SE, Adams TE, Cech JJ Jr (2003) Time of day and water temperature modify the physiological stress response in green sturgeon, Acipenser medirostris. Comp Biochem Physiol Part A Mol Integr Physiol 135:291–302. https://doi.org/10.1016/S1095-6433(03)00075-8
Lazarus RS, Folkman S (1984) Stress, appraisal, and coping. Springer, New York
Leatherland JF, Li M, Barkataki S (2010) Stressors, glucocorticoids and ovarian function in teleosts. J Fish Biol 76:86–111. https://doi.org/10.1111/j.1095-8649.2009.02514.x
Leclercq E, Davie A, Migaud H (2014) The physiological response of farmed ballan wrasse (Labrus bergylta) exposed to an acute stressor. Aquaculture 434:1–4. https://doi.org/10.1016/j.aquacul ture.2014.07.017
Lieberman DA (2000) Learning: behavior and cognition. Wadsworth, Belmont, CA
Lovallo WR (2016) Stress and health: biological and psychological interactions, 3rd edn. Sage, Thousand Oaks, CA
Madaro A, Olsen RE, Kristiansen TS, Ebbesson LOE, Nilsen TO, Flik G, Gorissen M (2015) Stress in Atlantic salmon: response to unpredictable chronic stress. J Exp Biol 218:2538–2550. https://doi.org/10.1242/jeb.120535
Madaro A, Fernö A, Kristiansen TS, Olsen RE, Gorissen M, Flik G, Nilsson J (2016a) Effect of predictability on the stress response to chasing in Atlantic salmon (Salmo salar L.) parr. Physiol Behav 153:1–6. https://doi.org/10.1016/j.physbeh.2015.10.002
Madaro A, Olsen RE, Kristiansen TS, Ebbesson LOE, Flik G, Gorissen M (2016b) A comparative study of the response to repeated chasing stress in Atlantic salmon (Salmo salar L.) parr and post-smolts. Comp Biochem Physiol Part A Mol Integr Physiol 192:7–16. https://doi.org/10.1016/j.cbpa.2015.11.005
Madaro A, Folkedal O, Maiolo S, Alvanopoulou M, Olsen RE (2018) Effects of acclimation temperature on cortisol and oxygen consumption in Atlantic salmon (Salmo salar) post-smolt exposed to acute stress. Aquaculture 497:331–335. https://doi.org/10.1016/J.AQUACUL TURE.2018.07.056
Madison BN, Tavakoli S, Kramer S, Bernier NJ (2015) Chronic cortisol and the regulation of food intake and the endocrine growth axis in rainbow trout. J Endocrinol 226:103. https://doi.org/10.1530/JOE-15-0186
Manuel R, Metz JR, Flik G, Vale WW, Huising MO (2014) Corticotropin-releasing factor-binding protein (CRF-BP) inhibits CRF- and urotensin-I-mediated activation of CRF receptor-1 and -2 in common carp. Gen Comp Endocrinol 202:69–75. https://doi.org/10.1016/j.ygcen.2014.04.010
McCormick SD (2001) Endocrine control of osmoregulation in teleost fish. Integr Comp Biol 41:781–794. https://doi.org/10.1093/icb/41.4.781

McCormick SD, Regish A, O'Dea MF, Shrimpton JM (2008) Are we missing a mineralocorticoid in teleost fish? Effects of cortisol, deoxycorticosterone and aldosterone on osmoregulation, gill Na+, K+ -ATPase activity and isoform mRNA levels in Atlantic salmon. Gen Comp Endocrinol 157:35–40. https://doi.org/10.1016/j.ygcen.2008.03.024

McEwen BS (2002) Sex, stress and the hippocampus: allostasis, allostatic load and the aging process. Neurobiol Aging 23:921–939

McEwen BS (2003) Mood disorders and allostatic load. Biol Psychiatry 54:200–207. https://doi.org/10.1016/S0006-3223(03)00177-X

McEwen BS, Lasley EN (2002) The end of stress as we know it. Joseph Henry, Washington, DC

McEwen BS, Seeman T (1999) Protective and damaging effects of mediators of stress: elaborating and testing the concepts of allostasis and allostatic load. Ann N Y Acad Sci 896:30–47. https://doi.org/10.1111/j.1749-6632.1999.tb08103.x

McEwen BS, Wingfield JC (2003) The concept of allostasis in biology and biomedicine. Horm Behav 43:2–15. https://doi.org/10.1016/S0018-506X(02)00024-7

McVicar A, Ravalier JM, Greenwood C (2014) Biology of stress revisited: intracellular mechanisms and the conceptualization of stress. Stress Health 30:272–279. https://doi.org/10.1002/smi.2508

Moghadam HK, Johnsen H, Robinson N, Andersen ØH, Jørgensen E, Johnsen HK, Bæhr VJ, Tveiten H (2017) Impacts of early life stress on the methylome and transcriptome of atlantic salmon. Sci Rep 7:5023. https://doi.org/10.1038/s41598-017-05222-2

Molet J, Maras PM, Avishai-Eliner S, Baram TZ (2014) Naturalistic rodent models of chronic early-life stress. Dev Psychobiol 56:1675–1688. https://doi.org/10.1002/dev.21230

Mommsen TP, Vijayan MM, Moon TW (1999) Cortisol in teleosts: dynamics, mechanisms of action, and metabolic regulation. Rev Fish Biol Fish 9(3):211–268

Moore-Ede MC (1986) Physiology of the circadian timing system: predictive versus reactive homeostasis. Am J Physiol Integr Comp Physiol 250:R737–R752. https://doi.org/10.1152/ajpregu.1986.250.5.R737

Mrosovsky N (1990) Rheostasis: the physiology of change. Oxford University Press, New York, NY

Nardocci G, Navarro C, Cortés PP, Imarai M, Montoya M, Valenzuela B, Jara P, Acuña-Castillo C, Fernández R (2014) Neuroendocrine mechanisms for immune system regulation during stress in fish. Fish Shellfish Immunol 40:531–538. https://doi.org/10.1016/j.fsi.2014.08.001

Nilsen TO, Ebbesson LOE, Madsen SS, McCormick SD, Andersson E, Björnsson BT, Prunet P, Stefansson SO (2007) Differential expression of gill Na+, K+-ATPase alpha- and beta-subunits, Na+, K+,2Cl- cotransporter and CFTR anion channel in juvenile anadromous and landlocked Atlantic salmon Salmo salar. J Exp Biol 210:2885–2896. https://doi.org/10.1242/jeb.002873

Ohl F, Michaelis T, Vollmann-Honsdorf G, Kirschbaum C, Fuchs E (2000) Effect of chronic psychosocial stress and long-term cortisol treatment on hippocampus-mediated memory and hippocampal volume: a pilot-study in tree shrews. Psychoneuroendocrinology 25:357–363. https://doi.org/10.1016/S0306-4530(99)00062-1

Olsen RE, Sundell K, Hansen T, Hemre G, Myklebust R, Mayhew TM, Ringø E (2003) Acute stress alters the intestinal lining of Atlantic salmon, Salmo salar L.: An electron microscopical study. Fish Physiol Biochem 26(3):211–221

Olsen RE, Sundell K, Mayhew TM, Myklebust R, Ringø E (2005) Acute stress alters intestinal function of rainbow trout, Oncorhynchus mykiss (Walbaum). Aquaculture 250:480–495. https://doi.org/10.1016/j.aquaculture.2005.03.014

Øverli Ø, Kotzian S, Winberg S (2002) Effects of cortisol on aggression and locomotor activity in rainbow trout. Horm Behav 42:53–61. https://doi.org/10.1006/hbeh.2002.1796

Paitz RT, Bukhari SA, Bell AM (2016) Stickleback embryos use ATP-binding cassette transporters as a buffer against exposure to maternally derived cortisol. Proc R Soc B Biol Sci 283:20152838. https://doi.org/10.1098/rspb.2015.2838

Palermo F, Nabissi M, Cardinaletti G, Tibaldi E, Mosconi G, Polzonetti-Magni AM (2008) Cloning of sole proopiomelanocortin (POMC) cDNA and the effects of stocking density on POMC

mRNA and growth rate in sole, Solea solea. Gen Comp Endocrinol 155:227–233. https://doi.org/10.1016/j.ygcen.2007.05.003
Pankhurst NW (2011) The endocrinology of stress in fish: an environmental perspective. Gen Comp Endocrinol 170:265–275. https://doi.org/10.1016/j.ygcen.2010.07.017
Pankhurst NW (2016) Reproduction and development. In: Schreck CB, Tort L, Farrell AP, Brauner CJ (eds) Biology of stress in fish: fish physiology, vol 35. Academic Press, Cambridge, MA, pp 295–331
Pavlidis M, Sundvik M, Chen Y-C, Panula P (2011) Adaptive changes in zebrafish brain in dominant-subordinate behavioral context. Behav Brain Res 225(2):529–537
Pavlidis M, Theodoridi A, Tsalafouta A (2015) Neuroendocrine regulation of the stress response in adult zebrafish, Danio rerio. Prog Neuro-Psychopharmacology Biol Psychiatry 60:121–131. https://doi.org/10.1016/j.pnpbp.2015.02.014
Pavlov I (1927) Conditioned reflexes. Oxford University, Oxford
Pepels PPLM, Meek J, Wendelaar Bonga SE, Balm PHM (2002) Distribution and quantification of corticotropin-releasing hormone (CRH) in the brain of the teleost fish Oreochromis mossambicus (tilapia). J Comp Neurol 453:247–268. https://doi.org/10.1002/cne.10377
Peter MCS (2011) The role of thyroid hormones in stress response of fish. Gen Comp Endocrinol 172:198–210. https://doi.org/10.1016/j.ygcen.2011.02.023
Pickering AD, Pottinger TG, Sumpter JP, Carragher JF, Le Bail PY (1991) Effects of acute and chronic stress on the levels of circulating growth hormone in the rainbow trout, Oncorhynchus mykiss. Gen. Comp. Endocrinol. 83:86–93. https://doi.org/10.1016/0016-6480(91)90108-I
Prunet P, Sturm A, Milla S (2006) Multiple corticosteroid receptors in fish: from old ideas to new concepts. Gen Comp Endocrinol 147:17–23. https://doi.org/10.1016/j.ygcen.2006.01.015
Reid SG, Bernier NJ, Perry SF (1998) The adrenergic stress response in fish: control of catecholamine storage and release. Comp Biochem Physiol C Pharmacol Toxicol Endocrinol 120:1–27
Rodriguez F, Lopez JC, Vargas JP, Gomez Y, Broglio C, Salas C (2002) Conservation of spatial memory function in the pallial forebrain of reptiles and ray-finned fishes. J Neurosci 22:2894–2903
Romero LM, Dickens MJ, Cyr NE (2009) The reactive scope model – a new model integrating homeostasis, allostasis, and stress. Horm Behav 55:375–389. https://doi.org/10.1016/J.YHBEH.2008.12.009
Rose JD (2002) The neurobehavioral nature of fishes and the question of awareness and pain. Rev Fish Sci 10:1–38. https://doi.org/10.1080/20026491051668
Rose JD (2007) Anthropomorphism and 'mental welfare' of fishes. Dis Aquat Org 75:139–154. https://doi.org/10.3354/dao075139
Rose JD, Arlinghaus R, Cooke SJ, Diggles BK, Sawynok W, Stevens ED, Wynne CDL (2014) Can fish really feel pain? Fish Fish 15:97–133. https://doi.org/10.1111/faf.12010
Roy B, Rai U (2008) Role of adrenoceptor-coupled second messenger system in sympatho-adrenomedullary modulation of splenic macrophage functions in live fish Channa punctatus. Gen Comp Endocrinol 155:298–306. https://doi.org/10.1016/j.ygcen.2007.05.008
Sadoul B, Vijayan MM (2016) Stress and growth. In: Schreck CB, Tort L, Farrell AP, Brauner CJ (eds) Biology of stress in fish: fish physiology, vol 35. Academic Press, Cambridge, MA, pp 167–205
Sarropoulou E, Tsalafouta A, Sundaram AYM, Gilfillan GD, Kotoulas G, Papandroulakis N, Pavlidis M (2016) Transcriptomic changes in relation to early-life events in the gilthead sea bream (Sparus aurata). BMC Genomics 17:506. https://doi.org/10.1186/s12864-016-2874-0
Sathiyaa R, Vijayan M (2003) Autoregulation of glucocorticoid receptor by cortisol in rainbow trout hepatocytes. Am J Phys 284:C1508–C1515. https://doi.org/10.1152/ajpcell.00448.2002
Schulkin J (2004) Allostasis, homeostasis and the costs of physiological adaptation. Cambridge University Press
Seasholtz A, Valverde RA, Denver RJ (2002) Corticotropin-releasing hormone-binding protein: biochemistry and function from fishes to mammals. J Endocrinol 175:89–97. https://doi.org/10.1677/joe.0.1750089

Selye H (1936) A syndrome produced by diverse nocuous agents. Nature 138:32
Selye H (1950) Stress and the general adaptation syndrome. Br Med J 4667
Shively CA, Willard SL (2012) Behavioral and neurobiological characteristics of social stress versus depression in nonhuman primates. Exp Neurol 233:87–94. https://doi.org/10.1016/j.expneurol.2011.09.026
Signals C (2006) Physiologie du Comportement. Flammarion, pp 1–39
Steenbergen PJ, Richardson MK, Champagne DL (2011) The use of the zebrafish model in stress research. Prog Neuro-Psychopharmacol Biol Psychiatry 35:1432–1451. https://doi.org/10.1016/j.pnpbp.2010.10.010
Sterling P (2004) Principles of allostasis: optimal design, predictive regulation, pathophysiology and rational therapeutics. In: Sterling P (ed) Allostasis, homeostasis, and the costs of physiological adaptation. Cambridge University Press, New York, pp 17–64
Sterling P (2012) Allostasis: a model of predictive regulation. Physiol Behav 106:5–15. https://doi.org/10.1016/j.physbeh.2011.06.004
Sterling P, Eyer J (1988) Allostasis: a new paradigm to explain arousal pathology. In: Fisher S, Reason J (eds) Handbook of life stress, cognition and health. Wiley, New York, pp 629–649
Stouthart XJHX, Huijbregts MAJ, Balm PHM, Lock RAC, Bonga SEW (1998) Endocrine stress response and abnormal development in carp (Cyprinus carpio) larvae after exposure of the embryos to PCB 126. Fish Physiol Biochem 18:321–329
Sturm A, Bury N, Dengreville L, Fagart J, Flouriot G, Rafestin-Oblin ME, Prunet P (2005) 11-Deoxycorticosterone is a potent agonist of the rainbow trout (Oncorhynchus mykiss) mineralocorticoid receptor. Endocrinology 146:47–55. https://doi.org/10.1210/en.2004-0128
Sturm A, Colliar L, Leaver MJ, Bury NR (2011) Molecular determinants of hormone sensitivity in rainbow trout glucocorticoid receptors 1 and 2. Mol Cell Endocrinol 333:181–189. https://doi.org/10.1016/j.mce.2010.12.033
Sumpter JP, Dye HM, Benfey TJ (1986) The effects of stress on plasma ACTH, α-MSH, and cortisol levels in salmonid fishes. Gen Comp Endocrinol 62:377–385. https://doi.org/10.1016/0016-6480(86)90047-X
Syed SA, Nemeroff CB (2017) Early life stress, mood, and anxiety disorders. Chronic Stress 1:247054701769446. https://doi.org/10.1177/2470547017694461
Szisch V, Papandroulakis N, Fanouraki E, Pavlidis M (2005) Ontogeny of the thyroid hormones and cortisol in the gilthead sea bream, Sparus aurata. Gen Comp Endocrinol 142:186–192. https://doi.org/10.1016/J.YGCEN.2004.12.013
Szyf M (2013) DNA methylation, behavior and early life adversity. J Genet Genomics 40:331–338. https://doi.org/10.1016/j.jgg.2013.06.004
Takahashi H, Sakamoto T (2013) The role of "mineralocorticoids" in teleost fish: relative importance of glucocorticoid signaling in the osmoregulation and "central" actions of mineralocorticoid receptor. Gen Comp Endocrinol 181:223–228. https://doi.org/10.1016/j.ygcen.2012.11.016
Tasker JG, Di S, Malcher-Lopes R (2006) Minireview: rapid glucocorticoid signaling via membrane-associated receptors. Endocrinology 147:5549–5556. https://doi.org/10.1210/en.2006-0981
Thomas P (2012) Rapid steroid hormone actions initiated at the cell surface and the receptors that mediate them with an emphasis on recent progress in fish models. Gen Comp Endocrinol 175:367–383. https://doi.org/10.1016/j.ygcen.2011.11.032
Tort L (2011) Stress and immune modulation in fish. Dev Comp Immunol 35:1366–1375. https://doi.org/10.1016/j.dci.2011.07.002
Tsalafouta A, Papandroulakis N, Gorissen M, Katharios P, Flik G, Pavlidis M (2014) Ontogenesis of the HPI axis and molecular regulation of the cortisol stress response during early development in Dicentrarchus labrax. Sci Rep 4:5525. https://doi.org/10.1038/srep05525
Tsalafouta A, Papandroulakis N, Pavlidis M (2015) Early life stress and effects at subsequent stages of development in European sea bass (D. labrax). Aquaculture 436:27–33

Tsalafouta A, Gorissen M, Pelgrim TNM, Papandroulakis N, Flik G, Pavlidis M (2017) α-MSH and melanocortin receptors at early ontogeny in European sea bass (Dicentrarchus labrax, L.). Sci Rep 7:46075. https://doi.org/10.1038/srep46075

Turecki G, Meaney MJ (2016) Effects of the social environment and stress on glucocorticoid receptor gene methylation: a systematic review. Biol Psychiatry 79(2):87–96

Ursin H, Eriksen HR (2004) The cognitive activation theory of stress. Psychoneuroendocrinology 29:567–592. https://doi.org/10.1016/S0306-4530(03)00091-X

Vaiserman AM (2010) Hormesis, adaptive epigenetic reorganization, and implications for human health and longevity. Dose-Response 8:dose. https://doi.org/10.2203/dose-response.09-014. Vaiserman

van Praag H, Kempermann G, Gage FH (1999) Running increases cell proliferation and neurogenesis in the adult mouse dentate gyrus. Nat Neurosci 2:266–270. https://doi.org/10.1038/6368

Van Weerd JH, Komen J (1998) The effects of chronic stress on growth in fish : a critical appraisal. Comp Biochem Physiol Part A Mol Integr Physiol 120:107–112

Vindas MA, Folkedal O, Kristiansen TS, Stien LH, Braastad BO, Mayer I, Øverli Ø (2012) Omission of expected reward agitates Atlantic salmon (Salmo salar). Anim Cogn 15:903–911. https://doi.org/10.1007/s10071-012-0517-7

Vindas MA, Johansen IB, Folkedal O, Höglund E, Gorissen M, Flik G, Kristiansen TS, Øverli Ø (2016a) Brain serotonergic activation in growth-stunted farmed salmon: adaption versus pathology. R Soc Open Sci 3:160030. https://doi.org/10.1098/rsos.160030

Vindas MA, Madaro A, Fraser TWK, Höglund E, Olsen RE, Øverli Ø, Kristiansen TS (2016b) Coping with a changing environment: the effects of early life stress. R Soc Open Sci 3:160382. https://doi.org/10.1098/rsos.160382

Vindas MA, Gorissen M, Höglund E, Flik G, Tronci V, Damsgård B, Thörnqvist P-O, Nilsen TO, Winberg S, Øverli Ø, Ebbesson LOE (2017) How do individuals cope with stress? Behavioural, physiological and neuronal differences between proactive and reactive coping styles in fish. J Exp Biol 220(8):1524–1532

Volkoff H, Canosa LF, Unniappan S, Cerdá-Reverter JM, Bernier NJ, Kelly SP, Peter RE (2005) Neuropeptides and the control of food intake in fish. Gen Comp Endocrinol 142:3–19. https://doi.org/10.1016/j.ygcen.2004.11.001

Von Holst D (1998) The concept of stress and its relevance for animal behavior. Adv Study Behav 27:1–31

Weiss JM (1970) Somatic effects of predictable and unpredictable shock. Psychosom Med 32:397–308

Wendelaar Bonga SE (1997) The stress response in fish. Physiol Rev 77:591–625

Winberg S, Nilsson GE (1993) Roles of brain monoamine neurotransmitters in agonistic behaviour and stress reactions, with particular reference to fish. Comp Biochem Physiol Part C Pharmacol Toxicol Endocrinol 106:597–614. https://doi.org/10.1016/0742-8413(93)90216-8

Wingfield JC, Maney DL, Breuner CW, Jacobs JD, Lynn S, Ramenofsky M, Richardson RD (1998) Ecological bases of hormone—behavior interactions: the "emergency life history stage". Integr Comp Biol 38:191–206. https://doi.org/10.1093/icb/38.1.191

Yada T, Tort L (2016) Stress and disease resistance: immune system and immunoendocrine interactions. In: Schreck CB, Tort L, Farrell AP, Brauner CJ (eds) Biology of stress in fish: fish physiology, vol 35. Academic Press, San Diego, CA, pp 365–403

Zwollo P (2017) The humoral immune system of anadromous fish. Dev Comp Immunol 80:24. https://doi.org/10.1016/j.dci.2016.12.008

Chapter 12
Individual Variations and Coping Style

Ida B. Johansen, Erik Höglund, and Øyvind Øverli

Abstract By current definition, animal welfare depends on the subjective experience of cognitive and emotional processes that are engendered as individuals succeed or fail in coping with a dynamically changing environment. A functional and evolutionary approach to emotion holds that adaptive qualities such as duration, severity, controllability, and predictability of stressful stimuli determine whether a particular event or outcome is experienced as rewarding or adverse. For instance, stress-induced behavioral inhibition can be seen as an adaptive strategy during chronic, unpredictable, or uncontrollable conditions that do not merit successful active coping. In teleost fishes, such behavior can be taken to indicate a negative welfare status, since it co-occurs with reduced neural plasticity and neuroendocrine alterations akin to brain remodeling seen in depression-like states in mammals. Active responses such as aggression and active avoidance are on the other hand likely to be adaptive when stressors are mild, predictable, and of short duration. Such conditions, albeit leading to acute activation of physiological stress responses, does not necessarily impair welfare. Thus, individual variation in the threshold for when a challenge becomes inhibiting rather than stimulatory is likely key to the subjective experience of welfare in a given situation. Thresholds for employing active (proactive) and passive (reactive) responses are, however, individually variable, and complex gene–environment interactions affects the occurrence and stability of welfare relevant trait correlations. In this chapter, we will review how key components of a stress coping style (i.e., behavior, physiology, neuroendocrinology, neuronal plasticity, and immunity) are subject to great individual and heritable variation, and further how such specific trait characteristics can influence the welfare of fish. A practical outcome for applied studies in aquaculture is that if one aims to understand and measure welfare in the context of individual variation in coping

---

I. B. Johansen · Ø. Øverli (✉)
Faculty of Veterinary Medicine, Norwegian University of Life Sciences, Oslo, Norway
e-mail: ida.johansen@nmbu.no; oyvind.overli@nmbu.no

E. Höglund
Norwegian Institute of Water Research, Oslo, Norway

Centre of Coastal Research, University of Agder, Kristiansand, Norway

T. S. Kristiansen et al. (eds.), *The Welfare of Fish*, Animal Welfare 20,
https://doi.org/10.1007/978-3-030-41675-1_12

ability, relevant indicators of central nervous system function must somehow be included. The robustness of available markers of neural plasticity and their sensitivity to stress exposure is not fully resolved, but impaired welfare status is most often recognizable by characteristic behaviors such as reduced activity and feeding, and reduced ability to respond to additional stressors.

Keywords Stress · Plasticity · Neurobiology · Personality · Allostasis · Behavior · Resistance

12.1 Stress and Welfare

This chapter will focus on individual variation in physiological stress responses, neurobiology, and behavior of fish. In addition, we will address how and why such variation is relevant to fish welfare. Most current research on animal welfare confers to a few basic concepts, one being that it is the individual animal's subjective experience of its own state that determines whether it is in a condition of good or bad welfare (Broom 1991; Dawkins 1990). The subjective experience of cognitive and emotional processes in animals is an emergent phenomenon, not fully amenable to scientific scrutiny by current natural sciences. Nevertheless, these experiences are the result of measurable biological processes (communication between neurons in a network), as well as a product of natural selection (Kittilsen 2013; Nesse 1990; Panksepp 2004). Variation is the raw material for selection, and throughout the animal kingdom, in the wild as well as in captivity, there is indeed large individual variability in how animals react, should they come in harm's way (Castanheira et al. 2017; Koolhaas et al. 2007; Øverli et al. 2007). Some individuals passively withdraw from potentially harmful stimuli (i.e., passive copers), whereas others actively avoid or try to fight or remove obstacles and challenges (i.e., active copers). Such variation in behavioral responses to threat is elegantly illustrated in rodents, where some individuals will actively and vigorously bury an aversive stimulus that poses an immediate threat (e.g., an electrified shock prod) whereas others freeze and avoid the stimulus passively (De Boer and Koolhaas 2003). Similar examples of active and passive avoidance have been described also in fish (Brelin et al. 2005; Laursen et al. 2011; Martins et al. 2011b; Silva et al. 2010).

Being fundamental to survival, physiological, and behavioral responses to stress are prime examples of variable traits shaped by complex gene–environment interactions (De Kloet et al. 2005; Winberg et al. 2016). Components of the neuroendocrine stress response are also deeply involved in mood and affective processes (De Kloet et al. 2005; Graeff et al. 1996; Krishnan and Nestler 2010), and hence directly relevant to the individual experience of welfare. Furthermore, links between neuroendocrine control, adaptive behavior, and emotional states during stress appear to be conserved through the vertebrate subphylum (Broom 1998; Duman et al. 2016; Ellis et al. 2012; Silva et al. 2015; Sørensen et al. 2013; Vindas et al. 2016; Wiepkema and Koolhaas 1993).

In parallel with development of the allostasis concept (Korte et al. 2007, Chap. 11), stress research is taking on an integrative and evolutionary approach, in that physiological and behavioral stress reactions are increasingly viewed as adaptive responses that are crucial for survival in a continuously changing environment (e.g., Korte et al. 2005; Romero et al. 2009). Active responses such as aggression and active avoidance are likely to be adaptive when stressors are mild, predictable and of short duration (Wingfield 2003). Under chronic, severe, or unpredictable stress, the organism is better off by reducing risk taking and conserving energy by passive coping. In this concept, impaired welfare arises only when "allostatic overload" arises from chronic, unpredictable, or uncontrollable conditions that do not merit successful allostatic adjustment (Korte et al. 2007; McEwen and Stellar 1993).

This framework fits very well with observable and often opposing effects of acute versus long-term stress and/or exposure to stress hormones such as corticosteroids also in fishes (reviewed by Sørensen et al. 2013; Øverli and Sørensen 2016). In line with this evolutionary line of thinking, the functionalist approach to emotions (Frijda 1986; Nesse 1990) holds that emotions have evolved for a particular function, such as inducing appropriate behavioral responses to potentially dangerous stimuli (see review by Kittilsen 2013). It stands to reason, then, that cognitive changes and emotional distress are likely an essential component of passive coping in response to overwhelming stress. Therefore, individual variation in the threshold for when a challenge becomes inhibiting rather than stimulatory is likely correlated to the individual's subjective experience of welfare in a given situation.

12.2 Individual Variation and Characteristics of Coping Styles

12.2.1 Terminology

In introducing the concept of individual variation in stress responsiveness, a note on terminology is appropriate. There is a surging interest in consistent individual variation in diverse fields such as evolutionary ecology, animal husbandry, and biomedicine (Castanheira et al. 2017; Conrad et al. 2011; Korte et al. 2005; Réale et al. 2007; Rey et al. 2016; Stamps and Groothuis 2010; Øverli et al. 2007). Across these fields, there is a lack of consensus in terminology. In the following, we will adhere to the pattern defined by Castanheira et al. (2017) in that physiological–behavioral trait associations are referred to as stress coping styles (Koolhaas et al. 1999, 2010). Studies of consistency in behavior alone (animal personalities; Gosling 2001) or suites of correlated behavioral traits across situations (behavioral syndromes; Sih et al. 2004) often lack the proximate link to central nervous function, and will not be the main focus in this chapter.

12.2.2 *Characteristics of Stress Coping Style*

Regarding physiological and neurobiological correlates of consistent behavioral phenotypes (i.e., coping styles), animals are commonly classified as either "reactive" or "proactive" based on their distribution along a shy-bold or passive-aggressive continuum (for reviews, see Coppens et al. 2010; Koolhaas et al. 1999, 2010; Korte et al. 2005). Shy, reactive individuals are typically characterized by higher post-stress corticosteroid production and lower activity of the sympathetic system (e.g., low levels of circulating adrenaline and noradrenaline). They display low aggression, "freeze-and-hide" behavior, behavioral flexibility, and low-risk taking. By contrast, bold and proactive individuals are characterized by low corticosteroid responses and greater sympathetic activity. These individuals tend to employ "fight-or-flight" strategies when stressed, are aggressive, rigid, and routine forming, and generally display high-risk behavior. Individual variation indicative of divergent coping styles appear to be well conserved between fishes and mammals (Castanheira et al. 2013; Huntingford et al. 2010; Martins et al. 2011a; Silva et al. 2010; Sørensen et al. 2013; Tudorache et al. 2013; Øverli et al. 2007). In the following, we will review how key components of a stress coping style (i.e., behavior, physiology, neuroendocrinology, neuronal plasticity, and immunity) are subject to great individual and heritable variation, and further how such specific trait characteristics can influence the welfare of fish.

12.3 Coping Styles and Animal Welfare: The Role of Behavior

Behavior represents a response to the environment as fish perceive it and can, therefore, be an important indicator of welfare status. Moreover, coping style can determine how stimuli are appraised (emotionally and affectively) by an individual, emphasising the need for including stress coping style in the welfare concept of farmed fish. In accordance with this, behavioral welfare indicators, such as the effect of stress and environmental perturbations on feed intake, locomotor activity, and aggression (for references see review by Martins et al. 2012) are in fact the very same identified as contrasting behaviors between reactive and proactive stress coping styles. Moreover, if these behaviors are inappropriate in time and/or context they are likely to affect the welfare of the individual or other individuals in the same population (i.e., high aggression). One thing to keep in mind is that behaviors can be adaptive or maladaptive depending on context. A behavior that is advantageous under certain conditions can be disadvantageous in other conditions. If an animal is trapped under conditions that do not favor its inherent behavioral profile, the behavioral profile becomes maladaptive and welfare is compromised.

12.3.1 Aggression

One behavioral axis that certainly has the potential to affect the welfare of fish is aggressiveness, or the tendency to attack other individuals. This occurs in farmed fish, although aggression levels vary between species and production systems (Huntingford et al. 2010). Several farmed fish species are territorial by nature nature and carry an instinctive drive to defend a territory and form social hierarchies. Aggression is linked to several aquaculture-related problems since victims of repeated aggression suffers from reduced feed intake, chronic social stress as well as an increased risk of infectious disease due to skin and fin damage (reviewed by Damsgård and Huntingford 2012). A study by Cubitt et al. (2008) presents strong evidence that even under aquaculture conditions, Atlantic salmon (*Salmo salar*) form size hierarchies with slower growing, potentially subordinate individuals showing brain serotonergic activation indicative of chronic stress. Several authors have raised the question of how domestication affects aggressive behavior in fishes (Campbell et al. 2015; Hedenskog et al. 2002; Ruzzante 1994). Indeed, there is a distinct possibility that selection for fast growth inadvertently leads to selection for proactive and aggressive individuals. In aquaculture, aggression is often mitigated by optimizing rearing densities so that territoriality is prevented. This illustrates the complexity in predicting the welfare outcome of interactions between inherent variation in coping style and current environment: Aggressive behavior is a natural behavior, the expression of which under certain circumstances can reduce stress and increase predictability for disposed individuals (Laursen et al. 2013; Øverli et al. 2004b). Hence, the common practice to decrease the level of aggression by increasing the rearing density may actually incur negative welfare aspects for subsets of proactive individuals.

12.3.2 Behavioral Responses to Environmental Instability

Another variable that will affect proactive and reactive individuals differently is environmental stability. An underlying difference in cognition between individuals with contrasting stress coping styles is that proactive individuals show a greater tendency to base their behavior on the expected outcomes and previously learned routines. Reactive individuals, on the other hand, pay more attention and show behavioral responses to small environmental perturbations (Coppens et al. 2010; de Lourdes Ruiz-Gomez et al. 2011; Höglund et al. 2017). This suggests that intensive rearing conditions characterized by fixed and restricted feeding regimes and frequent physical or environmental stressors related to rearing routines might favor proactive individuals, prone to develop and follow routines ensuring access to resources in demand. Under the same conditions, reactive individuals are more likely to suffer from low growth rates and poorer welfare. Following this, several studies have shown that coping style affects growth performance and feed conversion in several teleost species (Huntingford et al. 2010; Millot et al. 2009). This is also reflected in feed intake, which is a commonly used indicator of welfare.

12.3.3 Feed Intake

Resumption of feed intake following a challenge is one of the behaviors that are consistently found to vary between animals with diverging stress coping styles. Proactive individuals resume feed intake quickly after stress, whereas reactive individuals are slower (Øverli et al. 2007). This has been demonstrated repeatedly in lines of rainbow trout (*Oncorhynchus mykiss*) selected for a low (LR) and high (HR) post-stress cortisol response. Using the same selection lines, Andersson et al. (2013a) showed that even time to reach first feeding at the larvae stage varies between coping styles. Obviously, resumption of feed intake following stress is an important performance trait in aquaculture together with factors such as feeding behavior and feed efficiency. In African catfish (*Clarias gariepinus*), the most successful individuals are those reacting quicker to the presence of food and quickly resume feed intake after transfer to a novel environment (Martins et al. 2005). Like LR rainbow trout, these proactive catfish respond to stress with low cortisol levels. In addition, proactive Nile tilapia (*Oreochromis niloticus*) seem to exhibit a faster recovery of feed intake after transfer to a novel environment (Martins et al. 2011a). However, in a recent growth study on farmed rainbow trout, where stress coping styles were characterized by time to first feeding, both the proactive and the reactive fraction showed lower growth than the intermediate fraction (Andersson et al. 2013b). This promotes the idea that selection criteria should perhaps favor intermediate profiles in the reactive–proactive continuum. In line with not only selecting for high competitive fast-growing proactive individuals, selection programs in the aquaculture industry have recently included "decreased within-population deviation in growth" as a desirable trait.

12.4 Coping Styles and Animal Welfare: Physiology and Organ Plasticity

Like for behavior, physiological traits used as indicators of welfare status in fish are often those identified as contrasting physiological traits in reactive and proactive stress coping styles. One of the most commonly used physiological indicators of stress in teleost fishes is hypothalamic–pituitary–interrenal (HPI) axis reactivity and output. HPI output is represented by the steroid hormone cortisol, which is essential for regulation of, for example, hydromineral balance, energy metabolism, and immune function.

12.4.1 HPI Axis Reactivity

It is fairly well established that reactive fish react to stress with higher cortisol levels than proactive conspecifics. This simple relationship, however, is associated with a range of complicating biological interactions hampering the interpretation of cortisol dynamics versus welfare outcome. For instance, coping style affects the likelihood of attaining a dominant or subordinate social position (Pottinger and Carrick 2001; Øverli et al. 2004a), which in turn may affect individual stress levels (Øverli et al. 1999a). Furthermore, prolonged elevation of plasma cortisol levels has been shown to suppress HPI axis reactivity (Jeffrey et al. 2014; Øverli et al. 1999a). Similarly, chronically stressed fish respond to an acute stressor with lower plasma cortisol levels than unstressed control fish (Moltesen et al. 2016). Moreover, the time course of the stress-induced cortisol response appears to be slower in chronically stressed fish. This effect of chronic stress on HPI axis function is obvious in socially subordinate fish. For instance, Øverli et al. (1999b) reported that socially subordinate Arctic char (*Salvelinus alpinus*) show elevated basal plasma levels of cortisol but respond to an acute netting stress with a smaller and more sluggish cortisol response than dominant char. Jeffrey et al. (2014) obtained similar results in rainbow trout. However, it is still not fully understood at what level of the HPI axis and through what mechanisms these differences originate. Still, studies suggest that alternations in the expression of glucocorticoid receptors (GRs) at the hypothalamic level (Madaro et al. 2015) and changes in the metabolism of the neurotransmitter serotonin (5-Hydroxytryptamine, 5-HT) in telencephalon (Moltesen et al. 2016) are involved in chronic stress suppression of HPI axis reactivity. In addition, Jeffrey et al. (2014) demonstrated decreased adrenocorticotropin-stimulated cortisol production in chronically stressed rainbow trout. These multilevel effects of chronic stress on HPI axis reactivity is in line with that the HPI axis is regulated by feedback mechanism on different levels. Furthermore, this accentuates that chronic stress may affect proactive and reactive individuals, with low and high HPI axis reactivity respectively, differently.

12.4.2 Cortisol-Induced Pathology

With respect to an animal welfare perspective, other than central nervous processes associated with stress, cortisol can have several directly deleterious effects on malleable organ systems if levels become high and prolonged. For example, cortisol has been shown to damage the skin (Iger et al. 1995), inhibit growth (Barton et al. 1987; McBride and van Overbeeke 1971) and reproduction (Carragher et al. 1989), and to suppress several components of the immune system (Barton and Iwama 1991; Ellis 1981; Fevolden and Røed 1993; Nardocci et al. 2014; Pickering and Pottinger 1989). Cortisol exposure has also been shown to alter the gut wall and reduce feed intake in several teleost species (Barton et al. 1987; Gregory and Wood 1999; Davis et al. 1985).

Thus, since reactive fish respond to stress with higher cortisol levels than proactive fish (Castanheira et al. 2013; Pottinger and Carrick 1999; Trenzado et al. 2003; Tudorache et al. 2013) it could be hypothesized that repeated exposure to acute stressors renders reactive fish more sensitive to cortisol-induced pathology. However, apart from factors such as feed intake and growth performance, which are clearly affected by coping style, pathological effects of high (or low) cortisol exposure have been little explored in proactive and reactive fish. Recently, however, it was shown that cortisol exerts striking effects on vital, yet highly plastic organs such as the heart and the brain. High levels of exogenous cortisol induce pathological cardiac hypertrophy and reduce brain cell proliferation in rainbow trout (Johansen et al. 2011; Sørensen et al. 2011). In line with these findings, high cortisol-responding HR rainbow trout have larger and more fibrotic hearts than LR trout (Johansen et al. 2011). The large HR ventricles also have a high expression of mammalian molecular markers of pathological cardiac hypertrophy, indicating pathology. Thus, it is likely that reactive fish are more prone to cortisol-induced cardiac disease than proactive fish. On the other hand, Korte et al. (2005) argue that the sympathetic dominance of proactive individuals may render them prone to cardiovascular problems, but this remains to be shown in proactive and reactive fish.

12.4.3 Immune System Plasticity and Disease Resistance

Another indispensable but highly plastic biological requisite in vertebrates is the immune system. Given the immune-modulating nature of the major neuroendocrine stress systems, reactive and proactive individuals likely differ in immune competence and sensitivity to infectious disease. Although the mammalian literature shows that proactive and reactive individuals typically vary in the type of diseases and health problems they acquire (Cavigelli 2005; Koolhaas 2008; Korte et al. 2005; Zozulya et al. 2008), only very few studies have looked at disease resistance or other aspects of immune competence in the context of stress coping styles in fish. Fevolden et al. (1992) showed that selecting for stress response affected disease resistance in rainbow trout and MacKenzie et al. (2009) showed that proactive and reactive common carp (*Cyprinus carpio*) differ in baseline pro-inflammatory gene expression and respond differently to inflammatory challenge for several immune genes. In our own work, we have shown that ectoparasitic sea lice develop better on HR-type Atlantic salmon (Kittilsen et al. 2012) and HR-type fish are also more variable in how many parasites they acquire (Øverli et al. 2014).

Naturally, parasites and pathogens may be detrimental to welfare, and in the case of *Lepeophtheirus* sea lice, their grazing induces not only physical damage (Johnson et al. 2004) but also a chronic increase in brain 5-HT neurotransmission (Øverli et al. 2014). Elevated brain serotonergic activity is a general response to stress and aversive experiences in all vertebrates. In fishes, increased serotonergic signaling is induced by social stress (Prunet et al. 2012; Winberg et al. 1991), exposure to toxins (Gesto et al. 2008; Weber et al. 2012), predators (Winberg et al. 1993), predator olfactory cues (Höglund et al. 2005), and confinement stress (Øverli et al. 2001). The potential influence of brain 5-HT dynamics on animal welfare is indicated by its role in mood control and emotion (Cools et al. 2008; Dayan and Huys 2009; Graeff et al. 1996) as well as in the pathophysiology of depression in mammals (Blier and El Mansari 2013; Lanfumey et al. 2008; Stockmeier 2003) and fish (Vindas et al. 2016). This conserved signaling system is also a main mediator of individual variation in stress resilience and behavior in fish (Summers and Winberg 2006; Winberg and Thörnqvist 2016; Øverli et al. 2007).

In a welfare context, the monoamine neurotransmitter/neuromodulator 5-HT is well studied for its role as a signal substance in the brain, where it controls behavioral, neuroendocrine, and autonomic responses to stress and environmental changes (see, e.g., Cubitt et al. 2008; Vindas et al. 2016) and references therein. Notably, proactive/reactive coping styles differ in 5-HT activity and responsiveness in fishes (Øverli et al. 2001) as well as mammals (Koolhaas et al. 2007). In contrast, the potential role of 5-HT as a proximate mechanism behind individual variation in immune modulation (Baganz and Blakely 2012) is much less studied in comparative models. In general, individual variation in neuroimmune interaction is poorly understood, but could have important implications for the welfare of farmed fish. After all, disease and infections are currently major challenges in aquaculture. For example, cardiac disease of a potentially immune-mediated nature is distressingly prevalent in aquaculture, and significantly contributes to morbidity, mortality, and poor welfare (Brun et al. 2003; Ferguson et al. 1990). Moreover, autoimmunity caused by vaccination is commonly observed (Haugarvoll et al. 2010; Koppang et al. 2008). Hitherto, individual variation in stress or immune reactivity has not been investigated in this context. This calls for a more fundamental and translational approach to understand individual disease vulnerability. Teleost fishes can serve as important comparative models in this context.

12.5 Coping Styles and Animal Welfare: Behavioral Flexibility and Neural Plasticity

12.5.1 Behavioral Flexibility

As mentioned above, variation in behavioral flexibility is one of the main characteristics of contrasting coping styles (reviewed by Coppens et al. 2010). Typically, proactive individuals follow learned routines and show a limited behavioral flexibility, whereas reactive individuals show a greater behavioral flexibility and will quickly respond to changing environments. Also in the HR–LR model, contrasting coping styles may encompass differences in behavior in response to subtle environmental changes and nonthreatening novelty. de Lourdes Ruiz-Gomez et al. (2011)

found that HR and LR fish took equally long to learn the location of a food reward in a T-maze tank. However, upon moving the food to a new location, the LR fish continued searching the original spot, whereas the HR fish immediately adjusted their food-seeking behavior and took food in the new location. Furthermore, Moreira et al. (2004) reported that after being trained to associate interrupted water supply with confinement stress, LR fish retained the conditioned physiological response to the cue longer than the HR fish during reversal learning (i.e., even when not followed by confinement, the signal continued to induce a stress response in the normally less responsive proactive phenotypes). More recently, we studied the link between limbic dopamine (DA) signaling and individual variation in flexibility by a reversal learning approach (Höglund et al. 2017). During interaction with a large and aggressive conspecific HR–LR trout were challenged by blocking a previously available learned escape route. LR trout performed a higher number of failed escape attempts against the transparent blockage, while HR trout were able to inhibit the now futile escape impulse. Regionally discrete changes in DA neurochemistry were observed in microdissected limbic areas of the telencephalon, supporting the view that limbic homologs control individual differences in behavioral flexibility even in nonmammalian vertebrates.

12.5.2 Neural Plasticity

In recent reviews, Sørensen et al. (2013) and Øverli and Sørensen (2016) argue that neurogenesis and neural plasticity are involved in determining the thresholds for employing contrasting coping styles. A threatening environment causes changes in monoamine neurotransmitters such as DA and 5-HT and corticosteroid hormone levels. This neuroendocrine status, in turn, directs the coping strategy adopted, directly in the short term or chronically by affecting a range of brain structural processes collectively known as neural plasticity (Øverli and Sørensen 2016; Puglisi-Allegra and Andolina 2015; Sørensen et al. 2013). During both acute and chronic stressful situations, appraisal of the situation, learning, and memory are important for shaping adaptive behavioral responses. In mammals one important function of adult neurogenesis, relevant to the expression of contrasting coping styles, seems to be to provide a substrate for cognitive flexibility required for changing between learned responses and new stimulus-evoked behavior (Opendak and Gould 2015), and neurogenesis is likely to play a role in the underlying plasticity allowing these processes to occur optimally. Johansen et al. (2012) investigated the expression of neurogenesis-related genes in response to acute and chronic stress in the HR–LR rainbow trout model. Notably, the expression of proliferating cell nuclear antigen (PCNA; a marker of actively proliferating cells; telencephalon), neurogenic differentiation factor (telencephalon and cerebellum) and doublecortin (telencephalon and hypothalamus) was generally higher in HR fish compared to LR fish. The above data are in line with results from mammals, where the flexible and perceptive behavior of reactive individuals is associated with a high hippocampal expression of cytoskeleton genes (e.g., α-tubulin, cofilin, and dynamin), suggesting greater neural plasticity. Reactive copers have morphologically better developed hippocampi and are better able to process contextual information, but with the trade-off, that they are more aware of danger signals in their environments (reviewed by Korte et al. 2005). Of note, Vindas et al. (2016) observed increased poststress expression of the neuroplasticity marker brain-derived neurotropic factor (BDNF) in both the proposed hippocampus homologue (Dorsolateral telencephalon, Dl) and the lateral septum homologue (ventral part of the ventral telencephalon, Vv) of proactive compared to reactive fish. On the other hand, PCNA (Dl) was more expressed in reactive compared to proactive fish under basal conditions.

Notably, reduced neural plasticity is associated with emotional disturbances and depressed mood in human and mammalian models of depression (see, e.g., Anacker and Hen 2017; Duman et al. 2016; Lucassen et al. 2016). In an evolutionary paradigm of emotion and mood control, there is little to suggest that a similar relationship should not exist in nonmammalian vertebrates. Reduced neural plasticity likely contributes to subjectively decreased predictability and controllability of stressors and situation outcomes. The "logical" response to such dire straits is passive behavior and associated negative mood. However, while neural plasticity in limbic regions is likely relevant for welfare also in nonmammalian vertebrates, results must be interpreted carefully due to the time—and context-dependent nature of stress-induced brain remodeling (for review, see Sørensen et al. 2013).

12.6 Summary, Conclusions, and Directions for Further Research

Trait associations indicative of contrasting coping styles in fish are summarized in Table 12.1.

Links between neuroendocrine control, adaptive behavior, and emotional states during stress are likely conserved through the vertebrate subphylum. Active responses such as aggression and active avoidance are likely to be adaptive when stressors are mild, predictable, and of short duration, while under chronic, severe, or unpredictable stress, the animals are expected to reduce risk taking and conserving energy by passive coping. Considering that mood, cognitive function, and emotional states are evolved phenomena, it follows that subjectively impaired welfare arises under conditions when "allostatic overload" arises from chronic, unpredictable, or uncontrollable conditions that do not merit successful allostatic adjustment. Such states are recognizable by characteristic behaviors, such as for instance reduced activity and feeding, and neuroendocrine changes such as reduced neural plasticity and reduced ability to respond to additional stressors. Thresholds for employing active (proactive) and passive (reactive) responses are, however, individually variable, and complex gene–environment interactions affects the occurrence and stability of welfare relevant trait correlations. A practical outcome for applied studies in

Table 12.1 Diverging trait characteristics in proactive and reactive fish

	Proactive	Reactive	References
Behavioral characteristics			
Social dominance	High	Low	Pottinger and Carrick (1999) and Øverli et al. (2004a)
Active avoidance	High	Low	Brelin et al. (2005), Laursen et al. (2011), Martins et al. (2011b), and Silva et al. (2010)
Aggressiveness	High	Low	Castanheira et al. (2013) and Øverli et al. (2004a, b)
Feeding motivation in novel environment[a]	High	Low	Kristiansen and Fernö (2007), Martins et al. (2011a), and Øverli et al. (2007)
Feed efficacy	High	Low	Martins et al. (2005), Martins et al. (2006), and van de Nieuwegiessen et al. (2008)
Risk taking and exploration	High	Low	Castanheira et al. (2013), Huntingford et al. (2010), MacKenzie et al. (2009), and Millot et al. (2009)
Behavioral flexibility	Low	High	Chapman et al. (2010), de Lourdes Ruiz-Gomez et al. (2011), Höglund et al. (2017), and Moreira et al. (2004)
Physiological characteristics			
HPI reactivity	Low	High	Pottinger and Carrick (1999), Trenzado et al. (2003), and Tudorache et al. (2013)
Sympathetic reactivity[b]	High	Low	Barreto and Volpato (2011), Schjolden et al. (2006), and Verbeek et al. (2008)
Parasympathetic reactivity	Low	High	Barreto and Volpato (2011) and Verbeek et al. (2008)
Heat shock response	Low	High	LeBlanc et al. (2012)
Myocardial hypertrophy and fibrosis	Low	High	Johansen et al. (2011)
Level of CNS neurogenesis markers	Low	High	Johansen et al. (2012)
Pro-inflammatory gene expression	High	Low	MacKenzie et al. (2009)
Parasite resistance	High	Low	Kittilsen et al. (2012)

[a]An exception was noted by de Lourdes Ruiz-Gomez et al. (2008): After transport and temporary starvation, high-cortisol HR fish had lost twice as much weight compared to similarly treated LR conspecifics, and regained feed intake faster than LR fish in a novel environment

[b]An exception was noted by LeBlanc et al. (2012). Heat chock resulted in lower plasma adrenaline in cannulated LR fish compared to HR fish. HPI hypothalamic–pituitary–interrenal axis, CNS central nervous system

aquaculture is that if one aims to understand and measure welfare in the context of individual variation in coping ability, relevant indicators of central nervous system function must somehow be included. The robustness of available markers of neural plasticity and their sensitivity to stress exposure is not fully resolved, but one emerging concept is that allostatic status, i.e., the ability to respond to novel changes, may be equally relevant to assess as routine conditions (Vindas et al. 2016). Hence, a survey to document welfare status in a production unit should include comparing the behavior, external indicators, physiology, and neurobiology of individuals under routine rearing conditions ("controls") to that of acutely stressed individuals. Ideally, information on potentially contrasting coping styles using either genetic or phenotypic markers should aid the interpretations observed versus expected responses. Despite recent efforts (see, e.g., Khan et al. 2016; Rey et al. 2013; Vindas et al. 2017; and review by Castanheira et al. 2017), the proximate and ultimate mechanisms involved in maintaining individual variation in stress coping style are still poorly understood in teleost fishes. We expect that revealing possible evolutionary roots for the emerging neuroimmune/microbiome framework for depression and related elusive syndromes (e.g., chronic fatigue; Dantzer et al. 2014) will further resolve the proximate and ultimate mechanisms behind individual variation in stress and disease resistance. In this context, fish models benefit from recently sequenced genomes and well-characterized life history biologies (Cryan and Dinan 2012; Dantzer et al. 2011; Maes et al. 2009; Miller and Raison 2016; Simopoulos 2008).

References

Anacker C, Hen R (2017) Adult hippocampal neurogenesis and cognitive flexibility; linking memory and mood. Nat Rev Neurosci 18(6):335–346

Andersson MÅ, Khan UW, Øverli Ø, Gjøen HM, Höglund E (2013a) Coupling between stress coping style and time of emergence from spawning nests in salmonid fishes: evidence from selected rainbow trout strains (Oncorhynchus mykiss). Physiol Behav 116:30–34

Andersson MÅ, Laursen DC, Silva P, Höglund E (2013b) The relationship between emergence from spawning gravel and growth in farmed rainbow trout Oncorhynchus mykiss. J Fish Biol 83:214–219

Baganz NL, Blakely RD (2012) A dialogue between the immune system and brain, spoken in the language of serotonin. ACS Chem Neurosci 4:48–63

Barreto RE, Volpato GL (2011) Ventilation rates indicate stress-coping styles in Nile tilapia. J Biosci 36:851–855

Barton BA, Iwama GK (1991) Physiological changes in fish from stress in aquaculture with emphasis on the response and effects of corticosteroids. Annu Rev Fish Dis 1:3–26

Barton BA, Schreck CB, Barton LD (1987) Effects of chronic cortisol administration and daily acute stress on growth, physiological conditions, and stress responses in juvenile rainbow trout. Dis Aquat Org 2:173–185

Blier P, El Mansari M (2013) Serotonin and beyond: therapeutics for major depression. Philos Trans R Soc Lond B Biol Sci 368:20120536

Brelin D, Petersson E, Winberg S (2005) Divergent stress coping styles in juvenile brown trout (Salmo trutta). Ann N Y Acad Sci 1040:239–245

Broom DM (1991) Animal welfare: concepts and measurement. J Anim Sci 69:4167–4175

Broom DM (1998) Welfare, stress, and the evolution of feelings. Adv Study Behav 27:371–403

Brun E, Poppe T, Skrudland A, Jarp J (2003) Cardiomyopathy syndrome in farmed Atlantic salmon Salmo salar: occurrence and direct financial losses for Norwegian aquaculture. Dis Aquat Organ 56:241–247

Campbell JM, Carter PA, Wheeler PA, Thorgaard GH (2015) Aggressive behavior, brain size and domestication in clonal rainbow trout lines. Behav Genet 45:245–254

Carragher J, Sumpter J, Pottinger T, Pickering A (1989) The deleterious effects of cortisol implantation on reproductive function in two species of trout, Salmo trutta L. and Salmo gairdneri Richardson. Gen Comp Endocrinol 76:310–321
Castanheira MF, Herrera M, Costas B, Conceição LE, Martins CI (2013) Can we predict personality in fish? Searching for consistency over time and across contexts. PLoS One 8:e62037
Castanheira MF et al (2017) Coping styles in farmed fish: consequences for aquaculture. Rev Aquac 9:23–41
Cavigelli SA (2005) Animal personality and health. Behaviour 142:1223–1244
Chapman BB, Morrell LJ, Krause J (2010) Unpredictability in food supply during early life influences boldness in fish. Behav Ecol 21:501–506
Conrad JL, Weinersmith KL, Brodin T, Saltz J, Sih A (2011) Behavioural syndromes in fishes: a review with implications for ecology and fisheries management. J Fish Biol 78:395–435
Cools R, Roberts AC, Robbins TW (2008) Serotoninergic regulation of emotional and behavioural control processes. Trends Cogn Sci 12:31–40
Coppens CM, de Boer SF, Koolhaas JM (2010) Coping styles and behavioural flexibility: towards underlying mechanisms. Philos Trans R Soc Lond B Biol Sci 365:4021–4028
Cryan JF, Dinan TG (2012) Mind-altering microorganisms: the impact of the gut microbiota on brain and behaviour. Nat Rev Neurosci 13:701–712
Cubitt KF, Winberg S, Huntingford FA, Kadri S, Crampton VO, Øverli Ø (2008) Social hierarchies, growth and brain serotonin metabolism in Atlantic salmon (Salmo salar) kept under commercial rearing conditions. Physiol Behav 94:529–535
Damsgård B, Huntingford F (2012) Fighting and aggression. In: Aquaculture and behavior. Wiley-Blackwell, West Sussex, pp 248–285
Dantzer R, O'Connor JC, Lawson MA, Kelley KW (2011) Inflammation-associated depression: from serotonin to kynurenine. Psychoneuroendocrinology 36:426–436
Dantzer R, Heijnen CJ, Kavelaars A, Laye S, Capuron L (2014) The neuroimmune basis of fatigue. Trends Neurosci 37:39–46
Davis KB, Torrance P, Parker NC, Suttle MA (1985) Growth, body composition and hepatic tyrosine aminotransferase activity in cortisol-fed channel catfish, Ictalurus punctatus Rafinesque. J Fish Biol 27:177–184. https://doi.org/10.1111/j.1095-8649.1985.tb04019.x
Dawkins MS (1990) From an animal's point of view: motivation, fitness, and animal welfare. Behav Brain Sci 13:1–9
Dayan P, Huys QJ (2009) Serotonin in affective control. Annu Rev Neurosci 32:95–126
De Boer SF, Koolhaas JM (2003) Defensive burying in rodents: ethology, neurobiology and psychopharmacology. Eur J Pharmacol 463:145–161
De Kloet ER, Joëls M, Holsboer F (2005) Stress and the brain: from adaptation to disease. Nat Rev Neurosci 6:463–475
de Lourdes Ruiz-Gomez M et al (2008) Behavioral plasticity in rainbow trout (Oncorhynchus mykiss) with divergent coping styles: when doves become hawks. Horm Behav 54:534–538
de Lourdes Ruiz-Gomez M, Huntingford FA, Øverli Ø, Thörnqvist P-O, Höglund E (2011) Response to environmental change in rainbow trout selected for divergent stress coping styles. Physiol Behav 102:317–322
Duman RS, Aghajanian GK, Sanacora G, Krystal JH (2016) Synaptic plasticity and depression: new insights from stress and rapid-acting antidepressants. Nat Med 22:238–249
Ellis A (1981) Stress and the modulation of defense mechanisms in fish. In: Pickering A (ed) Stress and fish. Academic Press, London, pp 147–170
Ellis T, Yildiz HY, López-Olmeda J, Spedicato MT, Tort L, Øverli Ø, Martins CI (2012) Cortisol and finfish welfare. Fish Pysiol Biochem 38:163–188
Ferguson H, Poppe T, Speare DJ (1990) Cardiomyopathy in farmed Norwegian salmon. Dis Aquat Organ 8:225–231
Fevolden S, Røed K (1993) Cortisol and immune characteristics in rainbow trout (Oncorhynchus mykiss) selected for high or low tolerance to stress. J Fish Biol 43:919–930

Fevolden SE, Refstie T, Røed KH (1992) Disease resistance in rainbow trout (Oncorhynchus mykiss) selected for stress response. Aquaculture 104:19–29
Frijda NH (1986) The emotions: studies in emotion and social interaction. Maison De Sciences de l'Homme, Paris
Gesto M, Soengas JL, Míguez JM (2008) Acute and prolonged stress responses of brain monoaminergic activity and plasma cortisol levels in rainbow trout are modified by PAHs (naphthalene, β-naphthoflavone and benzo (a) pyrene) treatment. Aquat Toxicol 86:341–351
Gosling SD (2001) From mice to men: what can we learn about personality from animal research? Psychol Bull 127:45
Graeff FG, Guimarães FS, De Andrade TG, Deakin JF (1996) Role of 5-HT in stress, anxiety, and depression. Pharmacol Biochem Behav 54:129–141
Gregory TR, Wood CM (1999) The effects of chronic plasma cortisol elevation on the feeding behaviour, growth, competitive ability, and swimming performance of juvenile rainbow trout. Physiol Biochem Zool 72:286–295. https://doi.org/10.1086/316673
Haugarvoll E, Bjerkås I, Szabo NJ, Satoh M, Koppang EO (2010) Manifestations of systemic autoimmunity in vaccinated salmon. Vaccine 28:4961–4969
Hedenskog M, Petersson E, Järvi T (2002) Agonistic behavior and growth in newly emerged brown trout (Salmo trutta L) of sea-ranched and wild origin. Aggress Behav 28:145–153
Höglund E, Weltzien F-A, Schjolden J, Winberg S, Ursin H, Døving KB (2005) Avoidance behavior and brain monoamines in fish. Brain Res 1032:104–110
Höglund E, Silva PI, Vindas MA, Øverli Ø (2017) Contrasting coping styles meet the wall: a dopamine driven dichotomy in behavior and cognition. Front Neurosci 11:383
Huntingford F, Andrew G, Mackenzie S, Morera D, Coyle S, Pilarczyk M, Kadri S (2010) Coping strategies in a strongly schooling fish, the common carp Cyprinus carpio. J Fish Biol 76:1576–1591
Iger Y, Balm P, Jenner H, Bonga SW (1995) Cortisol induces stress related changes in the skin of rainbow trout (Oncorhynchus mykiss). Gen Comp Endocrinol 97:188–198
Jeffrey J, Gollock M, Gilmour K (2014) Social stress modulates the cortisol response to an acute stressor in rainbow trout (Oncorhynchus mykiss). Gen Comp Endocrinol 196:8–16
Johansen IB, Lunde IG, Røsjø H, Christensen G, Nilsson GE, Bakken M, Øverli Ø (2011) Cortisol response to stress is associated with myocardial remodeling in salmonid fishes. J Exp Biol 214:1313–1321
Johansen IB, Sørensen C, Sandvik GK, Nilsson GE, Höglund E, Bakken M, Øverli Ø (2012) Neural plasticity is affected by stress and heritable variation in stress coping style. Comp Biochem Physiol Part D Genomics Proteomics 7:161–171
Johnson SC, Treasurer JW, Bravo S, Nagasawa K, Kabata Z (2004) A review of the impacts of parasitic copepods on marine aquaculture. Zool Stud 43:8–19
Khan UW et al (2016) A novel role for pigment genes in the stress response in rainbow trout (Oncorhynchus mykiss). Sci Rep 6:28969
Kittilsen S (2013) Functional aspects of emotions in fish. Behav Processes 100:153–159
Kittilsen S, Johansen IB, Braastad BO, Øverli Ø (2012) Pigments, parasites and personality: towards a unifying role for steroid hormones? PLoS One 7:e34281
Koolhaas J (2008) Coping style and immunity in animals: making sense of individual variation. Brain Behav Immun 22:662–667
Koolhaas J et al (1999) Coping styles in animals: current status in behavior and stress-physiology. Neurosci Biobehav Rev 23:925–935
Koolhaas JM, De Boer SF, Buwalda B, Van Reenen K (2007) Individual variation in coping with stress: a multidimensional approach of ultimate and proximate mechanisms. Brain Behav Evol 70:218–226
Koolhaas J, De Boer S, Coppens C, Buwalda B (2010) Neuroendocrinology of coping styles: towards understanding the biology of individual variation. Front Neuroendocrinol 31:307–321
Koppang EO et al (2008) Vaccination-induced systemic autoimmunity in farmed Atlantic salmon. J Immunol 181:4807–4814

Korte SM, Koolhaas JM, Wingfield JC, McEwen BS (2005) The Darwinian concept of stress: benefits of allostasis and costs of allostatic load and the trade-offs in health and disease. Neurosci Biobehav Rev 29:3–38

Korte SM, Olivier B, Koolhaas JM (2007) A new animal welfare concept based on allostasis. Physiol Behav 92:422–428

Krishnan V, Nestler EJ (2010) Linking molecules to mood: new insight into the biology of depression. Am J Psychiatry 167:1305–1320

Kristiansen TS, Fernö A (2007) Individual behaviour and growth of halibut (Hippoglossus hippoglossus L.) fed sinking and floating feed: evidence of different coping styles. Appl Anim Behav Sci 104:236–250

Lanfumey L, Mongeau R, Cohen-Salmon C, Hamon M (2008) Corticosteroid–serotonin interactions in the neurobiological mechanisms of stress-related disorders. Neurosci Biobehav Rev 32:1174–1184

Laursen DC, Olsén HL, de Lourdes Ruiz-Gomez M, Winberg S, Höglund E (2011) Behavioural responses to hypoxia provide a non-invasive method for distinguishing between stress coping styles in fish. Appl Anim Behav Sci 132:211–216

Laursen DC, Silva PI, Larsen BK, Höglund E (2013) High oxygen consumption rates and scale loss indicate elevated aggressive behaviour at low rearing density, while elevated brain serotonergic activity suggests chronic stress at high rearing densities in farmed rainbow trout. Physiol Behav 122:147–154

LeBlanc S, Höglund E, Gilmour KM, Currie S (2012) Hormonal modulation of the heat shock response: insights from fish with divergent cortisol stress responses. Am J Physiol Regul Integr Comp Physiol 302:R184–R192

Lucassen P, Oomen C, Schouten M, Encinas J, Fitzsimons C, Canales J (2016) Adult neurogenesis, chronic stress and depression. In: Adult neurogenesis in the hippocampus: health, psychopathol brain disease. Academic Press, Amsterdam, p 177

MacKenzie S, Ribas L, Pilarczyk M, Capdevila DM, Kadri S, Huntingford FA (2009) Screening for coping style increases the power of gene expression studies. PLoS One 4:e5314

Madaro A, Olsen RE, Kristiansen TS, Ebbesson LO, Nilsen TO, Flik G, Gorissen M (2015) Stress in Atlantic salmon: response to unpredictable chronic stress. J Exp Biol 218:2538–2550

Maes M et al (2009) The inflammatory & neurodegenerative (I&ND) hypothesis of depression: leads for future research and new drug developments in depression. Metabol Brain Dis 24:27–53

Martins CI, Schrama JW, Verreth JA (2005) The consistency of individual differences in growth, feed efficiency and feeding behaviour in African catfish Clarias gariepinus (Burchell 1822) housed individually. Aquac Res 36:1509–1516

Martins CI, Schrama JW, Verreth JA (2006) The relationship between individual differences in feed efficiency and stress response in African catfish Clarias gariepinus. Aquaculture 256:588–595

Martins CI, Conceição LE, Schrama JW (2011a) Feeding behavior and stress response explain individual differences in feed efficiency in juveniles of Nile tilapia Oreochromis niloticus. Aquaculture 312:192–197

Martins CI, Silva PI, Conceição LE, Costas B, Höglund E, Øverli Ø, Schrama JW (2011b) Linking fearfulness and coping styles in fish. PLoS One 6:e28084

Martins CI et al (2012) Behavioural indicators of welfare in farmed fish. Fish Physiol Biochem 38:17–41

McBride J, van Overbeeke A (1971) Effects of androgens, estrogens, and cortisol on the skin, stomach, liver, pancreas, and kidney in gonadectomized adult sockeye salmon (Oncorhynchus nerka). J Fish Res Board Can 28:485–490

McEwen BS, Stellar E (1993) Stress and the individual: mechanisms leading to disease. Arch Intern Med 153:2093–2101

Miller AH, Raison CL (2016) The role of inflammation in depression: from evolutionary imperative to modern treatment target. Nat Rev Immunol 16:22–34

Millot S, Bégout ML, Chatain B (2009) Risk-taking behaviour variation over time in sea bass Dicentrarchus labrax: effects of day–night alternation, fish phenotypic characteristics and selection for growth. J Fish Biol 75:1733–1749
Moltesen M, Laursen DC, Thörnqvist P-O, Andersson MÅ, Winberg S, Höglund E (2016) Effects of acute and chronic stress on telencephalic neurochemistry and gene expression in rainbow trout (Oncorhynchus mykiss). J Exp Biol 219:3907–3914
Moreira P, Pulman KG, Pottinger TG (2004) Extinction of a conditioned response in rainbow trout selected for high or low responsiveness to stress. Horm Behav 46:450–457
Nardocci G et al (2014) Neuroendocrine mechanisms for immune system regulation during stress in fish. Fish Shellfish Immunol 40:531–538
Nesse RM (1990) Evolutionary explanations of emotions. Hum Nat 1:261–289
Opendak M, Gould E (2015) Adult neurogenesis: a substrate for experience-dependent change. Trends Cogn Sci 19:151–161
Øverli Ø, Sørensen C (2016) On the role of neurogenesis and neural plasticity in the evolution of animal personalities and stress coping styles. Brain Behav Evol 87:167–174
Øverli Ø, Harris CA, Winberg S (1999a) Short-term effects of fights for social dominance and the establishment of dominant-subordinate relationships on brain monoamines and cortisol in rainbow trout. Brain Behav Evol 54:263–275
Øverli Ø, Olsen R, Løvik F, Ringø E (1999b) Dominance hierarchies in Arctic charr, Salvelinus alpinus L: differential cortisol profiles of dominant and subordinate individuals after handling stress. Aquacult Res 30:259–264
Øverli Ø, Pottinger TG, Carrick TR, Øverli E, Winberg S (2001) Brain monoaminergic activity in rainbow trout selected for high and low stress responsiveness. Brain Behav Evol 57:214–224
Øverli Ø et al (2004a) Stress coping style predicts aggression and social dominance in rainbow trout. Horm Behav 45:235–241
Øverli Ø et al (2004b) Behavioral and neuroendocrine correlates of displaced aggression in trout. Horm Behav 45:324–329
Øverli Ø, Sørensen C, Pulman KG, Pottinger TG, Korzan W, Summers CH, Nilsson GE (2007) Evolutionary background for stress coping styles: relationships between physiological, behavioral, and cognitive traits in non-mammalian vertebrates. Neurosci Biobehav Rev 31:396–412
Øverli Ø, Nordgreen J, Mejdell CM, Janczak AM, Kittilsen S, Johansen IB, Horsberg TE (2014) Ectoparasitic sea lice (Lepeophtheirus salmonis) affect behavior and brain serotonergic activity in Atlantic salmon (Salmo salar L.): perspectives on animal welfare. Physiol Behav 132:44–50
Panksepp J (2004) Affective neuroscience: the foundations of human and animal emotions. Oxford University Press, Oxford
Pickering A, Pottinger T (1989) Stress responses and disease resistance in salmonid fish: effects of chronic elevation of plasma cortisol. Fish Physiol Biochem 7:253–258
Pottinger T, Carrick T (1999) Modification of the plasma cortisol response to stress in rainbow trout by selective breeding. Gen Comp Endocrinol 116:122–132
Pottinger T, Carrick T (2001) Stress responsiveness affects dominant–subordinate relationships in rainbow trout. Horm Behav 40:419–427
Prunet P, Øverli Ø, Douxfils J, Bernardini G, Kestemont P, Baron D (2012) Fish welfare and genomics. Fish Physiol Biochem 38:43–60
Puglisi-Allegra S, Andolina D (2015) Serotonin and stress coping. Behav Brain Res 277:58–67
Réale D, Reader SM, Sol D, McDougall PT, Dingemanse NJ (2007) Integrating animal temperament within ecology and evolution. Biol Rev 82:291–318
Rey S, Boltana S, Vargas R, Roher N, MacKenzie S (2013) Combining animal personalities with transcriptomics resolves individual variation within a wild-type zebrafish population and identifies underpinning molecular differences in brain function. Mol Ecol 22:6100–6115
Rey S et al (2016) Differential responses to environmental challenge by common carp Cyprinus carpio highlight the importance of coping style in integrative physiology. J Fish Biol 88:1056–1069

Romero LM, Dickens MJ, Cyr NE (2009) The reactive scope model—a new model integrating homeostasis, allostasis, and stress. Horm Behav 55:375–389
Ruzzante DE (1994) Domestication effects on aggressive and schooling behavior in fish. Aquaculture 120:1–24
Schjolden J, Pulman KG, Pottinger TG, Tottmar O, Winberg S (2006) Serotonergic characteristics of rainbow trout divergent in stress responsiveness. Physiol Behav 87:938–947
Sih A, Bell A, Johnson JC (2004) Behavioral syndromes: an ecological and evolutionary overview. Trends Ecol Evol 19:372–378
Silva PIM, Martins CI, Engrola S, Marino G, Øverli Ø, Conceição LE (2010) Individual differences in cortisol levels and behaviour of Senegalese sole (Solea senegalensis) juveniles: evidence for coping styles. Appl Anim Behav Sci 124:75–81
Silva PI, Martins CI, Khan UW, Gjøen HM, Øverli Ø, Höglund E (2015) Stress and fear responses in the teleost pallium. Physiol Behav 141:17–22
Simopoulos AP (2008) The importance of the omega-6/omega-3 fatty acid ratio in cardiovascular disease and other chronic diseases. Exp Biol Med 233:674–688
Sørensen C, Bohlin LC, Øverli Ø, Nilsson GE (2011) Cortisol reduces cell proliferation in the telencephalon of rainbow trout (Oncorhynchus mykiss). Physiol Behav 102:518–523
Sørensen C, Johansen IB, Øverli Ø (2013) Neural plasticity and stress coping in teleost fishes. Gen Comp Endocrinol 181:25–34
Stamps JA, Groothuis TG (2010) Developmental perspectives on personality: implications for ecological and evolutionary studies of individual differences. Philos Trans R Soc Lond B Biol Sci 365:4029–4041
Stockmeier CA (2003) Involvement of serotonin in depression: evidence from postmortem and imaging studies of serotonin receptors and the serotonin transporter. J Psychiatr Res 37:357–373
Summers CH, Winberg S (2006) Interactions between the neural regulation of stress and aggression. J Exp Biol 209:4581–4589
Trenzado C, Carrick T, Pottinger T (2003) Divergence of endocrine and metabolic responses to stress in two rainbow trout lines selected for differing cortisol responsiveness to stress. Gen Comp Endocrinol 133:332–340
Tudorache C, Schaaf MJ, Slabbekoorn H (2013) Covariation between behaviour and physiology indicators of coping style in zebrafish (Danio rerio). J Endocrinol 219:251–258
van de Nieuwegiessen PG, Boerlage AS, Verreth JA, Schrama JW (2008) Assessing the effects of a chronic stressor, stocking density, on welfare indicators of juvenile African catfish, Clarias gariepinus Burchell. Appl Anim Behav Sci 115:233–243
Verbeek P, Iwamoto T, Murakami N (2008) Variable stress-responsiveness in wild type and domesticated fighting fish. Physiol Behav 93:83–88
Vindas MA et al (2016) Brain serotonergic activation in growth-stunted farmed salmon: adaption versus pathology. R Soc Open Sci 3:160030
Vindas MA et al (2017) How do individuals cope with stress? Behavioural, physiological and neuronal differences between proactive and reactive coping styles in fish. J Exp Biol 220:1524–1532
Weber R, Maceira JP, Mancebo M, Peleteiro J, Martín LG, Aldegunde M (2012) Effects of acute exposure to exogenous ammonia on cerebral monoaminergic neurotransmitters in juvenile Solea senegalensis. Ecotoxicology 21:362–369
Wiepkema P, Koolhaas J (1993) Stress and animal welfare. Anim Welf 2:195–218
Winberg S, Thörnqvist P-O (2016) Role of brain serotonin in modulating fish behavior. Curr Zool 62:317–323
Winberg S, Nilsson GE, Olsén KH (1991) Social rank and brain levels of monoamines and monoamine metabolites in Arctic charr, Salvelinus alpinus (L.). J Comp Physiol A 168:241–246
Winberg S, Myrberg AA Jr, Nilsson GE (1993) Predator exposure alters brain serotonin metabolism in bicolour damselfish. Neuroreport 4:399–402
Winberg S, Höglund E, Øverli Ø (2016) Variation in the neuroendocrine stress response. In: Fish physiology, vol 35. Elsevier, London, pp 35–74

Wingfield JC (2003) Control of behavioural strategies for capricious environments. Anim Behav 66:807–816
Zozulya AA, Gabaeva MV, Sokolov OY, Surkina ID, Kost NV (2008) Personality, coping style, and constitutional neuroimmunology. J Immunotoxicol 5:221–225

Chapter 13
Assessing Fish Welfare in Aquaculture

Lars Helge Stien, Marc Bracke, Chris Noble, and Tore S. Kristiansen

Abstract A framework for assessing the welfare of fish in aquaculture must have a suite of different welfare indicators that describe how well their welfare needs are met and thus their quality of life. The framework should utilise both input- and outcome-based welfare indicators. Input-based welfare indicators are parameters that describe the conditions the fish are subjected to, e.g. their environment. In many cases, input-based welfare indicators can give the farmer or assessor an early warning of deteriorating conditions, which can then be mitigated before they become too severe. However, it can be very challenging to have a complete overview of all the possible input parameters the fish are subjected to, at all times, and at all possible positions in the rearing facility that the fish may occupy. Further, their effects on welfare can also be subtle, delayed and also be dependent upon an array of complex interactions with other parameters and factors. It is, therefore, necessary to also include outcome-based indicators. These are parameters that are normally directly related to the animals, e.g. describing the animals themselves or their behaviour. A simple rule of thumb can be that as long as the fish look good, are doing well, are in good health, show normal behaviour and are thriving, it is not unreasonable to assume that the rearing system or operation is fulfilling, or has not markedly impacted upon, their welfare needs. If not, there is something wrong and this should be investigated further.

Keywords Welfare state · Welfare needs · Input-based welfare indicator · Outcome-based welfare indicator · Welfare assessment

---

L. H. Stien (✉) · T. S. Kristiansen
Institute of Marine Research, Bergen, Norway
e-mail: lars.helge.stien@hi.no

M. Bracke
Wageningen Livestock Research, Wageningen, The Netherlands

C. Noble
NOFIMA, Tromsø, Norway

T. S. Kristiansen et al. (eds.), *The Welfare of Fish*, Animal Welfare 20,
https://doi.org/10.1007/978-3-030-41675-1_13

13.1 Introduction

The farming of carp and tilapia in freshwater ponds has been practiced for thousands of years, and seawater farming of mullet can be traced back more than 1500 years (Costa-Pierce 1987). Despite these long traditions, fish farming has only relatively recently become a significant part of global food production; in 1950 global fish production was only 0.3 million tonnes, compared to 54 million tonnes in 2016 (FAO 2018). The late development of large-scale fish farming may be related to the fact that we (as land living animals) have an inherent lack of understanding of what fish need to survive, thrive and reproduce (Lucas and Southgate 2012). Fish live in completely different environments to ourselves, have different needs and suffer from unfamiliar diseases. It is therefore not surprising that the welfare of fish has received less public attention than the welfare of more familiar mammals and birds (Driessen 2013). However, the European food authorities and a range of different animal welfare NGOs and other bodies have now drawn attention to fish welfare in aquaculture (see Chaps. 1 and 2), and in a recent survey 30% of European consumers rated fish welfare as one of the most important aspects of sustainable aquaculture (Zander and Feucht 2018).

In response to increased public awareness, several retailers are now demanding that fish farmers meet specific welfare standards (Richards et al. 2013). Good examples of welfare certification are the RSPCA welfare standards for Atlantic salmon and rainbow trout in the United Kingdom (RSPCA 2018a, b). The salmon standards were first created in 2002 and now cover more than 70 % of the United Kingdom salmon farming industry (RSPCA 2014). Other widely used standards such as the GLOBAL G.A.P Aquaculture standard and the Aquaculture Stewardship Council (ASC) Farm Standards for sustainable aquaculture are not specific welfare standards, but focus more on sustainability and limiting environmental impact from fish farming. Nevertheless, their checklists for fish welfare provide important requirements for ensuring fish welfare and can be of great help to farmers.

Even though the term "animal welfare" is frequently used in standards and legislation, it is not always clear exactly what the term means and how animal welfare should be assessed. There are three main approaches towards defining and assessing animal welfare (Fraser 2008; Mellor et al. 2009): The most straightforward approach is to focus on biological functioning and equating a healthy animal with good growth and performance with good animal welfare. Animal rights activists and some NGOs have, however, argued for a nature-based approach, where a natural environment and the possibility to perform innate species-specific behaviours also are necessary to obtain good welfare. A third approach, often promoted by owners of companion animals and animal welfare NGOs, as well as by animal ethicists and several animal welfare scientists (e.g. Dawkins 2008; Fraser 2008; Mellor et al. 2009), is that an animals' feelings and emotional states dictate its welfare state. The feeling-based definition also reflects most people's concerns about animal suffering, which is the main reason for the public attention to animal welfare.

In 1979, the UK Farm Animal Welfare Council published five demands for good farm animal welfare conditions that they later refined and labelled "the five freedoms", each provided with a prescription as to how to achieve the "freedom" (Webster 2008; FAWC 2009).

The Five Freedoms[1]

1. Freedom from hunger and thirst—By ready access to water and a diet to maintain health and vigour.
2. Freedom from discomfort—By providing an appropriate environment.
3. Freedom from pain, injury and disease—By prevention or rapid diagnosis and treatment.
4. Freedom to express normal behaviour—By providing sufficient space, proper facilities and appropriate company of the animal's own kind.
5. Freedom from fear and distress—By ensuring conditions and treatment, which avoid mental suffering.

These five freedoms have been widely adopted as a practical checklist for the welfare assurance of terrestrial production animals (see for instance Welfare Quality, www.welfarequality.net). There has, however, been a drive to update "the five freedoms" approach as the thinking about animal welfare has moved towards a more feelings-based approach. Korte et al. (2007) argue that a certain level of stress and discomfort may be good for animals, and good animal welfare may be achieved if the animal has the capacity and resources to adapt and change in response to the demands it is subjected to. Mellor et al. (2009) suggested an alternative "five domains" model, with four physical and functional domains (nutrition, environment, physical health and behaviour) that collectively influence the fifth domain; the animals' mental state. A criticism of both the five freedoms and the "five domains" model is that they have insufficient focus on positive welfare states (Mellor 2016). In response, the "five domains" model has now been extended to include experiences animals may have that can induce positive affects (Mellor and Beausoleil 2015). Also, FAWC is moving beyond a focus on poor welfare and suffering to an approach that also incorporates welfare needs and the concept of "A life worth living" from the animals' point of view (FAWC 2009).

In many countries, land animals and farmed fish are protected by the same animal welfare legislation. For example, the Norwegian Animal Welfare Act promotes high welfare standards and respect for animals. It states: "Animals have intrinsic value independent of their usefulness for people or society. Animals must be treated well and protected from unnecessary suffering". Similarly, the UK Animal Welfare Act from 2006 also covers farmed fish and prohibits the infliction of unnecessary suffering. These laws typically demand that animals, including fish, have "good welfare" without providing specifics for what is meant by "good welfare". This

[1]Five freedoms reproduced from "FAWC, 2009. Farm animal welfare in Great Britain: Past, present and future." with permission from Crown Copyright.

makes it difficult for farmers and authorities, especially in the case of farmed fish, to determine if the animals' level of welfare is within the regulations. Both authorities and stakeholders have therefore requested scientific support for the development of science-based tools and protocols for assessing the welfare of farmed fish. The goal of this chapter is to provide a general outline or framework for assessing the welfare of farmed fish, and to give examples of operational welfare assessment.

13.2 Defining and Assessing Fish Welfare

A meaningful definition of animal welfare must consider our ethical and legal concerns for animal suffering and therefore the animal's own qualitative experiences. Assessing an animal's needs and the degree of their fulfilment can be linked directly to feelings and hence welfare (Fig. 13.1) (Mellor et al. 2009; Mellor and Beausoleil 2015; Bracke et al. 1999). We also suggest that by limiting the concept of welfare to the animal's own experience of their quality of life, we will get more conceptual clarity (Box 13.1). Therefore, in this chapter we define animal welfare as the "*quality of life as perceived by the animals themselves*".

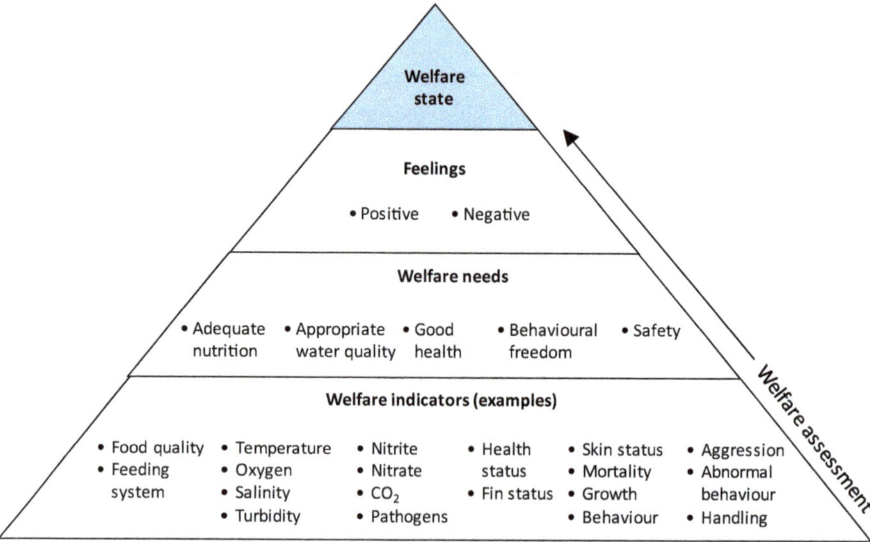

Fig. 13.1 An animal's welfare state is defined by its positive and negative feelings, which again are defined by the degree of fulfilment of the animal's welfare needs. For fish, we have defined five overarching welfare needs: Adequate nutrition, appropriate water quality, good health (fitness), behavioural freedom and safety. To assess fish welfare it is necessary to have a set of welfare indicators that describe the degree of fulfilment of the welfare needs for the given species and life stage under the prevailing farming conditions

Box 13.1 Definitions

We use the following definitions in order to have more conceptual clarity around the different concepts related to welfare assessment:

Animal Welfare—The quality of life as perceived by the animal itself.

Welfare state—An individual animal's experienced affective state ("sum" of feelings) at a given moment in time, i.e. its quality of life.

Group welfare—The distribution and variation of the individuals' welfare in a given group (which is relevant to fish farming, and other farm animals where it is difficult to monitor the welfare of all individuals on the farm).

Welfare needs—Requirements as perceived by the animals, i.e. as monitored by the animal's cognitive–emotional systems. Lack of fulfilment of welfare needs leads to a poor welfare state, and fulfilment or improvement leads to good or improved welfare.

Welfare indicators (WIs) are measurements or observations that give information about the degree of fulfilment of the animals' welfare needs, which is assumed to be correlated with the welfare state.

Input-based welfare indicators—All measurements or observations that describe factors that influence the degree of fulfilment of one or more welfare needs. Typically, indirect welfare indicators measuring the environment or describing treatments the animals are subjected to.

Outcome-based welfare indicators—All measurements or observations that describe the result or consequence of the degree of fulfilment of a welfare need; how an animal responds to impacts on its welfare state. Typically, but not necessarily, direct welfare indicators based on measurements of animal behaviour, disease symptoms, injuries, or (stress) physiology.

Operational welfare indicators (OWIs)—Welfare indicators that are practical and feasible for use on the farm.

Laboratory-based welfare indicators (LABWIs)—Welfare indicators that require access to a laboratory or other analytical facilities for evaluation.

Individual-based welfare indicators—Outcome-based welfare indicators that describe the physical appearance, health status, physiology, or behaviour of individual fish.

Group-based welfare indicators—Outcome-based welfare indicators based on observations of the fish, or from the fish (e.g. blood or scales in the water) at the group level, including schooling patterns and behaviours, group appetite and mortality levels.

Welfare assessment—An evaluation of the degree of fulfilment of all welfare needs for a selected group of animals, for a defined time period.

13.2.1 Welfare Needs

When referring to an animal's welfare needs we mean all the needs that are linked to and monitored by the animal's qualitative experiences, regulating its cognitive–emotional control mechanisms (behavioural systems), such as feeding behaviour (hunger), social contact, thermoregulation, the ability to explore the environment and move around and the need for shelter and safety. The list of possible welfare needs of animals can be extensive (e.g. Bracke et al. 1999; Stien et al. 2013; Noble et al. 2018), but for simplicity we have grouped them into five overarching needs relevant for farmed fish:

1. *Adequate nutrition*—Includes all needs related to feed and nutrition.
2. *Appropriate water quality*—Includes all needs related to the quality and constituents of the ambient rearing water that is necessary to fulfil bodily functions, such as osmoregulation, respiration and thermoregulation, and that the water has tolerable levels of metabolites and other harmful chemicals and particles that may affect fish welfare and physiology.
3. *Good health*—Includes a well-functioning physiology (brain–body) and absence or low level of malformations, diseases, parasites and injuries.
4. *Behavioural freedom*—To live in accordance with natural tendencies/inclinations, including the freedom to move as wanted, including searching for and having access to resources (foraging), social contact (in social species), migration and reproduction (at the relevant life stages) and rest.
5. *Safety*—Needs related to protecting the body against injuries, avoiding perceived dangerous conditions, shelter, hiding among conspecifics, etc.

As all welfare needs are relevant for fish welfare, a welfare assessment should account for their fulfilment (Fig. 13.1). How the different needs affect welfare and how long the fish can tolerate that a need is not fulfilled will vary. For example, the oxygen contents in the water can usually only be compromised for a very short period of time, while large well-fed fish with good energy reserves can be deprived of food for a week or more before it affects their health and welfare status. Food resources, appropriate water quality and good health are required for the fish to survive and experience good welfare over extended time periods. In contrast, full behavioural freedom may not always be necessary for survival in a farm environment. For example, Atlantic salmon post-smolts migrate over vast distances in the open ocean and into rivers. If this migration is driven by the need to search for food or simply by the need to swim, rearing the fish in enclosed cages may not compromise their welfare as farmed salmon can swim around in a large space and have ready access to food. However, if the urge to migrate is related to reproduction, then such behaviours are not possible under commercial aquaculture conditions and their behavioural welfare needs may be compromised (Huntingford et al. 2006). It should also be highlighted that although it is useful to analyse the needs one by one, the borders between welfare needs are not absolute and will in many cases overlap. Malnutrition may, for instance, result in poor health, and health and injury are closely related to safety. It should also be noted that the current lists of welfare needs for fish (Stien et al. 2013; Noble et al. 2018) are not absolute and there is still a need for consensus regarding the approach. The framework is evolving and the concept has not reached full maturity.

13.2.2 *Welfare Indicators*

As we cannot ask fish how they feel about their living conditions, we must use *welfare indicators (WIs)* to get information about their welfare state (Fig. 13.1). A welfare indicator should be scalable, meaning that the observed or measured values can be divided into two (binary—presence/absence) or more levels associated with increasing, positive or negative, welfare states. Utilising different levels is an intuitive approach and can be easily interpreted by the user. However, it can be very challenging to set the boundaries between the levels, e.g. in visual scoring of injuries or defining the thresholds of certain water quality parameters. The welfare indicator levels must also be i) valid, meaning they are well correlated with the degree of fulfilment of at least one welfare need and ii) reliable, meaning they are scored more-or-less equally by different observers and between different sampling occasions. In an ideal situation, each welfare indicator level would be linked to at least one scientific publication that supports the distinction between this level and another (preferably as a reported statistically significant difference), and it should, of course, be relevant for welfare, i.e. there should be some known correlation between the welfare indicator and the welfare state of the fish, but expert opinion can be used where the relationship is obvious (Bracke et al. 1999; Stien et al. 2013). For example, it is reasonable to assume that increasing the extent of a malformation or injury has an increasingly negative effect on welfare, even though the available data may only describe its effect when comparing normal fish and those with a profound malformation or injury.

Welfare indicators can be either input- or outcome-based (Fig. 13.2). *Input-based welfare indicators* are all measures that describe what the animals are subjected to and that influence the fulfilment of one or more welfare needs. They are typically parameters describing the resources and environment the animals have available.

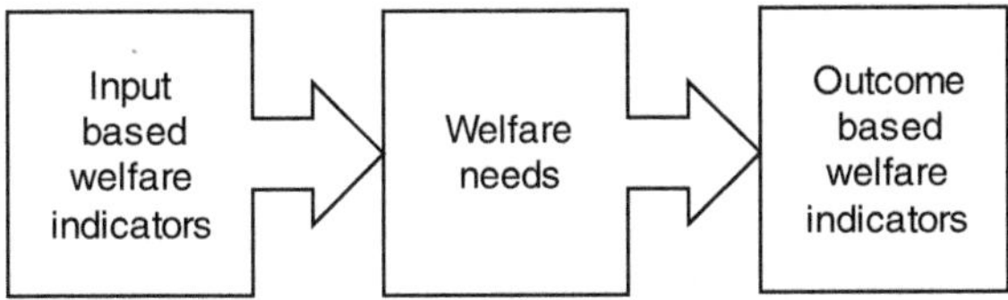

Fig. 13.2 Simplified model of the relationship between input- and outcome-based welfare indicators. Input-based welfare indicators affect the degree of fulfilment of the fish's welfare needs, while outcome-based welfare indicators describe how this fulfilment is expressed by measurable aspects of, or emanating from the fish

Outcome-based welfare indicators measure the outcome, or consequences, of how well the welfare needs of the animals are met. Outcome-based welfare indicators are typically parameters describing the animals themselves or their behaviour. In the literature, input- and outcome-based welfare indicators are often called environment- and animal-based welfare indicators, respectively. They have also been labelled indirect and direct welfare indicators, to emphasise that they measure welfare indirectly from the environment or directly from the state of the animal. This is, however, not always completely consistent and can create confusion. For example, an increase in oxygen consumption from a stressor can be estimated by comparing oxygen saturation in inlet vs. outlet water from a tank. The measurements are in other words done indirectly on the water environment, while the indicator itself is animal based and the outcome of a treatment. That it is animal based is clear if one considers that increase in oxygen consumption could have been estimated as a function of gill beat frequency or by measuring blood oxygen levels.

Input-based welfare indicators can be easy to measure, for instance, whether or not the water temperature is within the adaptable range of the species. Since welfare problems often arise in suboptimal environments, input-based welfare indicators may also warn about future welfare problems before they are visible on the fish and can, therefore, be mitigated much sooner than if the farmer were to use outcome-based welfare indicators alone. A limitation with input-based welfare indicators is, however, that their effects on the fish can be subtle and dependent on e.g. exposure time and interactions with other environmental parameters. For example, the effect of an abrupt change in water temperature upon welfare is highly dependent on other factors such as the ambient rearing temperature, oxygen levels, water current, feeding status, in addition to the physiological and health status of the fish (e.g. Folkedal et al. 2011; Hvas et al. 2017a, b). It can also be extremely challenging or impossible to make sure that absolutely all parameters that may influence the fish are monitored at all times and at all relevant positions in the rearing system.

In contrast, outcome-based welfare indicators can incorporate current and historical input-based factors that affect welfare, meaning that as long as the fish look good, behave normally and thrive, it is not unreasonable to assume that the rearing system or operation has fulfilled their welfare needs. On the other hand, if large numbers of fish show signs of abnormal behaviour, disease or if mortality is high then welfare is likely to be poor no matter how good the measured input-based welfare indicators predict the rearing environment to be. Outcome-based welfare indicators can be based on observations of individual fish (e.g. behaviour, skin and fin condition, or health status), on observations of a group of individuals (e.g. group behaviour and percentage mortality) or on the presence/absence of e.g. blood and scales in the water (Pettersen et al. 2013; Noble et al. 2018). An inherent problem in fish farming is that it is difficult to assess all of the animals within a rearing system, especially when stocking numbers or densities are high. Any assessments must, therefore, be based on a representative sample of the fish. However, there can be a large inter-individual variability on how the fish cope and if a subpopulation of the fish has their welfare needs compromised, they may be missed during sampling and remain undetected. Another weakness with outcome-based welfare indicators is that they are often visible only after welfare has been compromised for a period of time, and it may not always be clear what the underlying causes are.

The Salmon Welfare Index Model (SWIM) (Stien et al. 2013; Pettersen et al. 2013) calculates the welfare of the fish based upon several outcome-based welfare indicators describing the morphological appearance of individual fish. Based on our experience with this model (e.g. Folkedal et al. 2016; Oppedal et al. 2017; Stien et al. 2018) inexperienced assessors often state that welfare indicators which have numerous complicated levels can be difficult to score accurately. We, therefore, recommend that scoring systems for manual grading of individual fish should utilise welfare indicators that have four standardised levels: (0) good/perfect, (1) minor/slightly/suspected negative (2) evidently negative and (3) strongly negative/extreme. This makes it easy to remember and score the different indicator levels, and at the same time the resolution is high enough to get a meaningful description of welfare status. Each indicator level should also be supported by good descriptions and if relevant, also photos. This approach has been adopted in the FISHWELL handbook (Noble et al. 2018) of welfare indicators for farmed salmon (Figs. 13.3 and 13.4). If these scoring schemes were to be used at the regional or national level to e.g. score welfare between different farms or companies, we recommend providing courses where inspectors can be trained to score the indicators in the same way.

13.2.3 *Welfare Assessments*

When creating a welfare assessment scheme for a given fish species or life stage it is important to make sure that the welfare indicators cover all welfare needs. Some input-based welfare indicators such as appropriate water temperature and adequate oxygen, are easy to measure, have an immediate effect on the fish and should always be included. Monitoring all the input-based welfare indicators that are required to make sure all welfare needs are met is, however, impossible, but supplementing them with outcome-based welfare indicators can to some extent mitigate this as inappropriate living conditions will at some point result in behavioural changes, poor appetite or growth and potentially disease or mortality. Also, conditions that are apparently good do not always have positive outcomes, and conversely, relatively poor conditions may be compensated for by qualities that are difficult to measure, like good stockmanship. Including a sufficient number of outcome-based welfare indicators can, therefore, act as an insurance that no negative influences have been missed.

It is also important to choose welfare indicators that are operational and fit for the purpose, meaning that they are practical to use on-farm, whilst at the same time giving a valid indication of the fulfilment of the fish's welfare needs. Ideally, the farmer should be able to assess and interpret the welfare indicator on the farm (these WIs can be termed Operational Welfare Indicators, OWIs), but may also utilise indicators that require sending samples to a laboratory for analysis (termed Laboratory-based Welfare Indicators, LABWIs), which are acceptable if they give the farmer robust information about fish welfare within a reasonable timeframe.

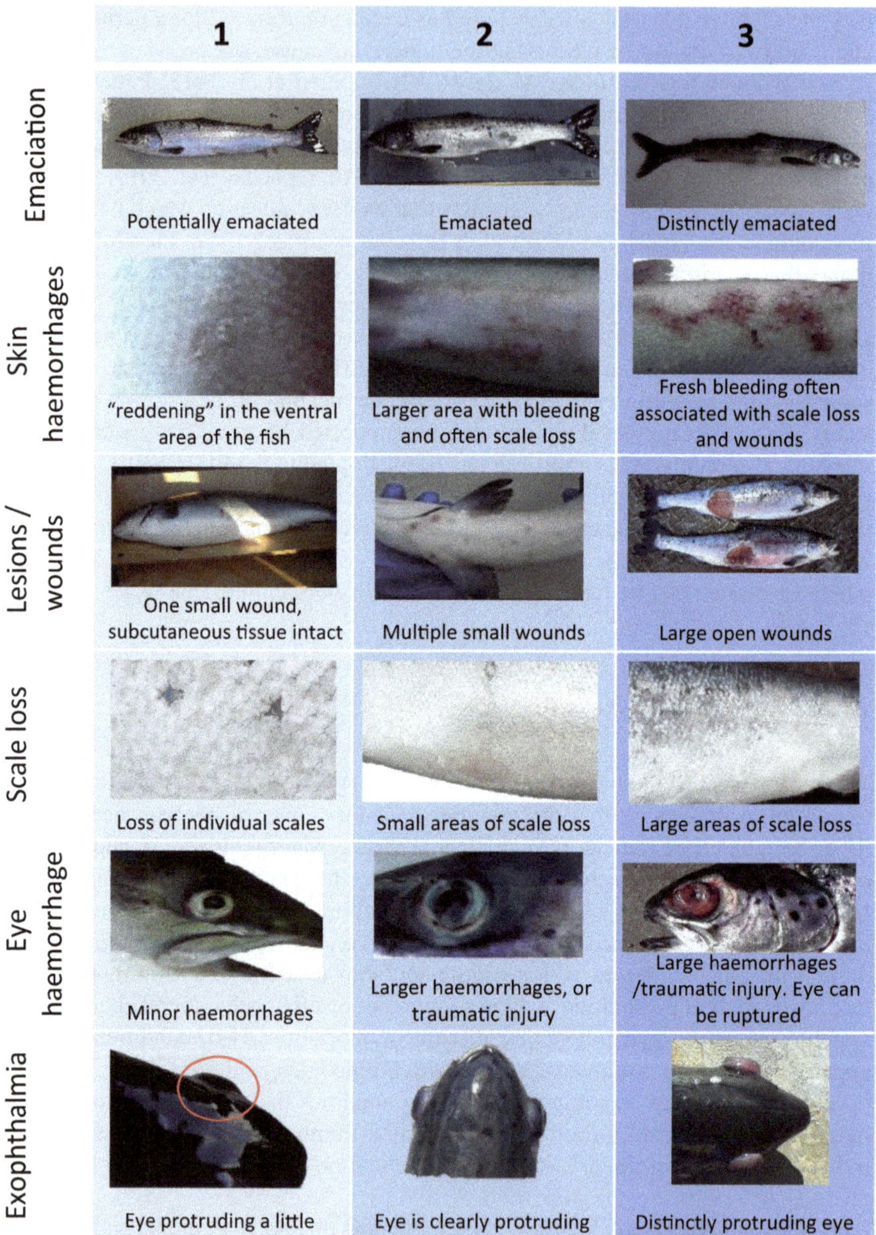

Fig. 13.3 Part 1. Morphological outcome-based welfare indicators for individual fish from the FISHWELL handbook (Noble et al. 2018)

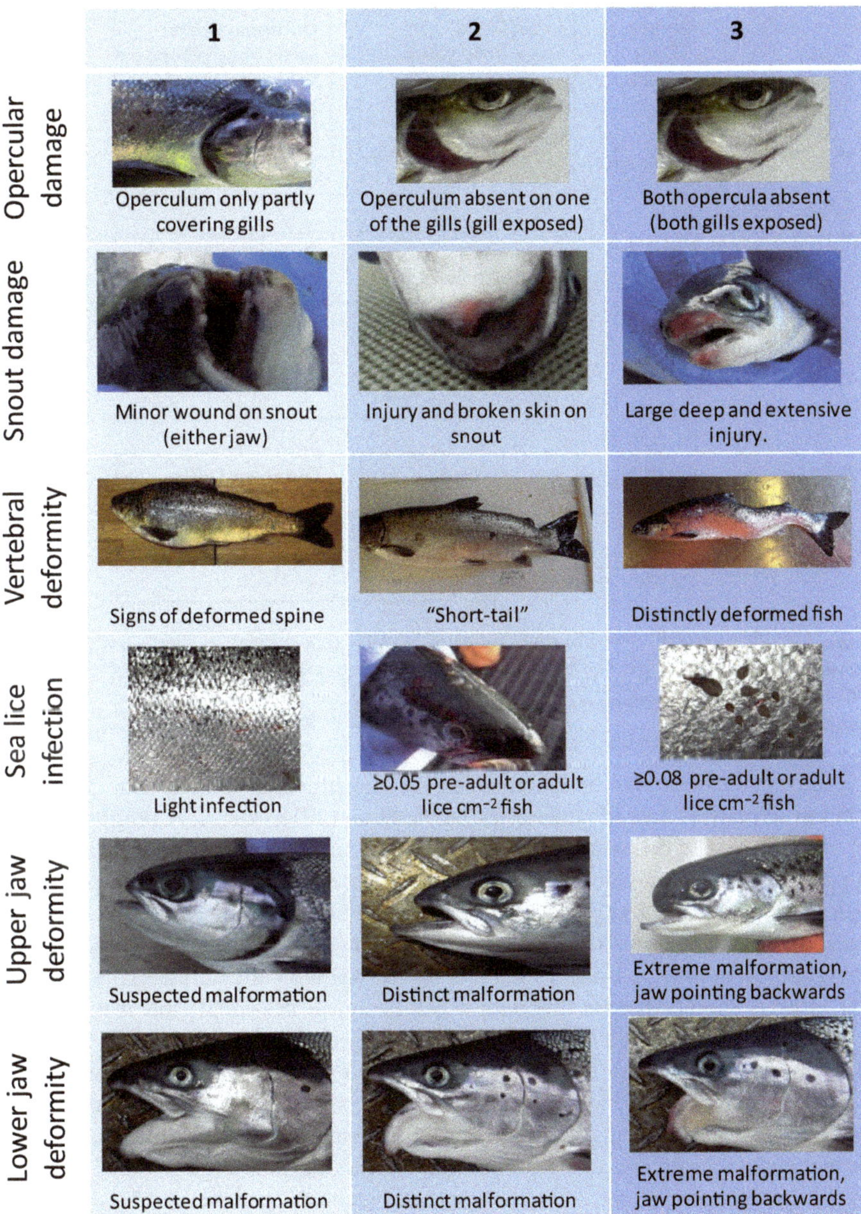

Fig. 13.4 Part 2. Morphological outcome-based welfare indicators for individual fish from the FISHWELL handbook (Noble et al. 2018)

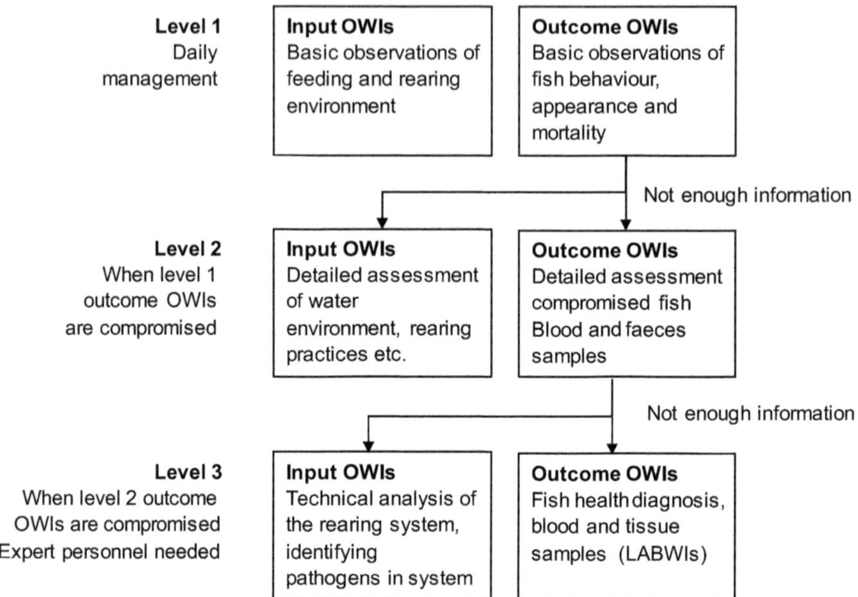

Fig. 13.5 Framework for utilising different OWIs (operational welfare indicators) and LABWIs (laboratory-based welfare indicators) in an on-farm welfare assessment (Noble et al. 2018)

Some indicators may be labour intensive and time consuming, or require some kind of expertise (trained personnel). Noble et al. (2018) therefore propose a three-level system, where Level 1 utilises simple and rapid passive OWIs, such as water quality parameters, in combination with visual observations of the fish's appearance, behaviour and mortality (Fig. 13.5). If one or more of the level 1 input-based welfare indicators are compromised, the farmers must investigate this further and perform mitigation actions if possible. If outcome-based welfare indicators are compromised, the farmers should discuss this with a fish health professional, and if there is not enough information to identify the cause and carry out a corrective action, the farmer must immediately utilise Level 2 OWIs and LABWIs. This includes sampling fish for a more accurate description of symptoms and performing a detailed investigation of the water environment, rearing practices and handling for possible causes. If the farmer still does not have enough information, then it is time to go to Level 3 and call in fish health professionals and in some cases technical personnel if it is suspected that the welfare is compromised due to some malfunction in the rearing system. This Level 3 will utilise more complex OWIs and LABWIs that require expert skills.

13.3 Salmon Welfare Index Model (SWIM)

In the salmon welfare index model (SWIM) (Stien et al. 2013; Pettersen et al. 2013), the score of the different welfare indicators is combined into an overall welfare index based on a weighting of how much each indicator affects the welfare of the fish. SWIM is based on semantic modelling (Bracke et al. 2008). It is a formalised and standardised, but also flexible model where both input- and outcome-based welfare indicators are used, and the numbers of welfare indicators can be varied and adjusted depending on the goal of the assessment (Bracke et al. 2008; Stien et al. 2013; Pettersen et al. 2013; Folkedal et al. 2016). The calculated overall welfare score ranges from 0 to 1 (bad to good). This is especially useful when comparing the welfare of different groups of fish (Oppedal et al. 2017; Stien et al. 2018). In the example below, the welfare of two different salmon populations is compared, using only outcome-based indicators, where one has a SWIM score of 0.7 and the other has a score of 0.9. The score functions as an overall welfare index, which the farmer then must break down into its different components to investigate the cause of the decreased welfare score.

Figure 13.6 shows monthly mortality for two different groups of salmon at the same sea cage facility. One of the production cages had two periods with monthly mortality (level 1 OWI) above 5%. In both cases, the farmer used the SWIM-model on 40 sampled fish as level 2 OWIs to gain more detailed information to investigate the cause of the mortality. The fish were put into the sea in November, but the increase in mortality did not occur before January. The farmer noticed a number of fish with wounds in the cage. Sampling 20 fish from each group revealed that 87% of the sampled fish in the high mortality cage had many small, but penetrating wounds (Fig. 13.7). The farmer then moved to level 3 and called in a fish health professional for help. The fish health professionals took samples of the fish that confirmed an outbreak of *Moritella viscosa*, the bacterium that causes winter ulcers. The fish in both groups performed well during the summer, but when water temperatures again started to drop in late autumn, mortally again increased to above 5% for the same group as the winter before. Sampling fish from the high mortality cage revealed that 39% of the sampled fish had wounds (sampling 2, Fig. 13.6). There were also a large percentage of fish with lower jaw deformities and snout wounds in this group. Further investigations by veterinarians revealed both *Yersiniose* and *Tenacibaculum* (Fig. 13.8). There were also many fish with signs of gill damage, but this was true for both groups (Sampling 2, Fig. 13.6). The veterinarians suggested that the gill damage was due to algae, zooplankton or jellyfish. Based on these findings the farmer decided to terminate this production in order to avoid further suffering.

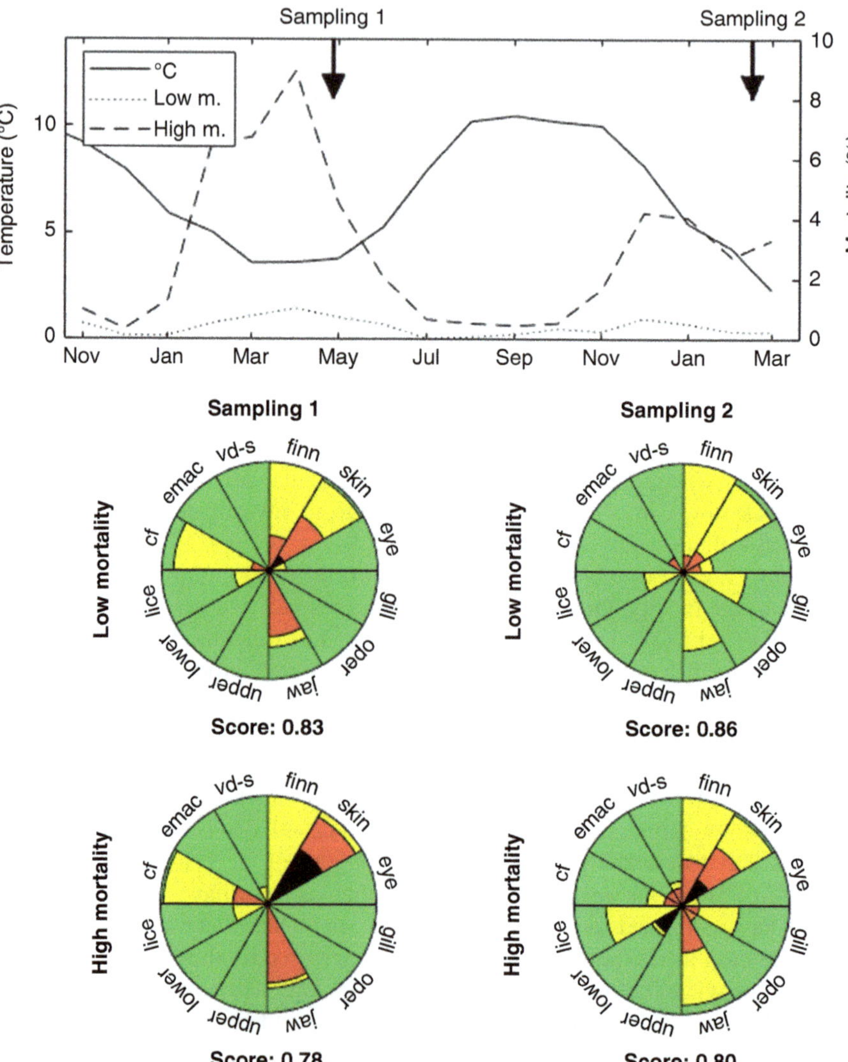

Fig. 13.6 Monitoring welfare for two different groups of salmon in separate production cages at the same farm. The upper graph shows the measured water temperature at 3 m, and the percentage monthly mortality for the two production cages. The discs show the distribution of morphological welfare indicators for 20 sampled fish from each cage at two different sampling occasions. Each sector is divided into 0: green, 1: yellow; 2: red and 3: black, where the relative area of the coloured sector indicates the proportion of the fish with different WI scores. Morphological welfare indicators: Finn condition (finn), skin condition/lesions (skin), eye condition (eye), gill condition (gill), operculum deformity/damage (oper), jaw wound/snout damage (jaw), upper jaw deformity (upper), lower jaw deformity (lower), sea lice infection (lice), condition factor (cf), emaciation state (emac), vertebral deformity (vd-s)

Fig. 13.7 An example of one of the fish from sample period 1 diagnosed with a winter ulcer. Photo: Marin Helse AS

Fig. 13.8 One of the sampled fish from sample period 2 with a cleft lower jaw, probably caused by *Tenacibaculum*. Photo: Florian Sambraus

13.4 The RSPCA Assured Standards

The RSPCA Assured standards for Atlantic salmon and rainbow trout are, to our knowledge, the most comprehensive standards for assuring fish welfare in aquaculture. They specify a range of requirements for management, health, husbandry practices, equipment, feeding, environmental quality, environmental impact, freshwater rearing for juvenile fish, seawater rearing, transport and slaughter that must be fulfilled in order to be approved by the RSPCA (2018a, b). With regard to management, the standards focus on staff training and diligent inspection and record keeping. This is to ensure that all staff members are competent in terms of stock keeping and animal welfare, and to ensure that any welfare problems are immediately discovered so that they can be dealt with appropriately. The health requirements include having site-specific veterinary health and welfare plans as well as continuously monitoring the fish for signs of disease and problems with the rearing environment or handling practices. Recurring physical damage with a common cause must be avoided and any seriously sick or injured fish found not to be recovering must be humanly killed by a blow to the head or via an overdose of a suitable anaesthetic. The requirements for husbandry practices are subdivided into handling, crowding or splitting, grading, well-boat grading, pushing or towing enclosures, protection from other animals and genetic selection and modification. For crowding, the requirements state that only healthy fish may be crowded and not for more than 2 h. The requirements for freshwater production are relatively detailed with specific limits for water quality parameters including oxygen, free ammonia, carbon dioxide, pH, alkalinity, suspended solids, nitrite and nitrate (RSPCA 2018a, b). These requirements cover different life stages and production systems. The requirements for transport are specified for a range of different transportation methods and focus on staff competence, equipment and pumping systems (to ensure they are fit for purpose and do not lead to physical trauma). In addition, specific water quality limits on oxygen, temperature and pH must be met and maximum stocking densities may not be exceeded.

13.5 The Fish Welfare Assurance System (FWAS)

Van de Vis et al. (2012) describe how Hazard Analysis and Critical Control Points (HACCP) can be used to create Fish Welfare Assurance Systems in aquaculture production. In short, firstly, the hazards in the system that may somehow impair fish welfare are determined. The hazards are then scored to assess the relative importance of each hazard (modified from EFSA 2008):

1. Probability of occurrence [extremely low (1) to high (5)]
2. Proportion of the population affected [20% (1) to >80% (5)]
3. Negative impact on fish welfare [limited (1) to very severe (4)] or mortality [20% (1) to >80% (5)]
4. Duration of the effect

Each score is normalised by dividing it by the maximum score obtained for all hazards, and the overall score for each hazard is calculated as the product of the four sub-scores. The next step is to identify Critical Control Points (CCPs). These are points in the production process at which control can be applied to prevent or reduce deterioration of fish welfare. For example, in tanks, the oxygen concentration can be measured and oxygen may be added if necessary. For each CCP operational welfare indicators with critical limits must be defined and procedures for monitoring established. For managerial factors, target levels can be more appropriate than critical limits, for instance, that fish are only removed from water when there is no other option available to do the necessary procedure. For each CCP a set of predefined actions should be taken when monitoring shows deviations outside critical limits. Finally, procedures must be in place to ensure that the FWAS works and continues to work effectively, including a record-keeping system documenting hazard analysis, a written quality assurance plan, records documenting the monitoring of CCPS, critical limits, verification activities and handling of deviations.

13.6 Concluding Remarks

In this chapter, we argue that the concept of animal welfare should be based on the quality of life as experienced by the animals themselves. We further postulate that the quality of life experienced by the animals is strongly correlated with the degree of fulfilment of their different needs necessary for biological functioning (survival, growth and reproduction), since this must be the main reason why qualitative experiences have evolved (see Chaps. 8 and 9). Attributes at the individual or group level, or from the environment, that can indicate different states of needs will, therefore, be good indicators of fish welfare.

Fish farming can involve rearing hundreds of thousands of individuals, in large tanks, ponds or sea cages, making it extremely challenging to have a complete overview of the health and welfare status of all individual fish. In sea cages, large parts of the population can avoid the surface and swim in deeper waters for long periods of time (Oppedal et al. 2011). The farmers only have a limited view of the fish if they use surface observations, and may only have a limited subsurface view if using underwater cameras that have limited mobility. If a sick or injured individual is spotted it can also often be next to impossible to catch that specific fish without, e.g. pulling up the net or chasing the fish in the crowd, stressing the individual and subjecting the rest of the group to stressful events. These practical challenges are reflected in some farming regulations. For example, the Norwegian regulation for poultry (FOR 2001-12-12-1494) states that all animals with damage or a disease that causes suffering must receive necessary treatment or be euthanised immediately, while the corresponding regulation for farmed fish (FOR-2008-06-17-822) has no such requirements. However, both poultry and fish are protected by the same Animal welfare act (LOV-2009-06-19-97). Also, point H 2.1 in the RSPCA welfare standards for Atlantic salmon (RSPCA 2018a) states “Any seriously sick or injured fish, or fish found not to be recovering, must be humanely killed without delay”. It is therefore crucial that techniques or technology is developed for conducting automatic welfare assessments of individual fish in farming systems and that it becomes possible to humanely separate sick or injured fish from the crowd for treatment or euthanasia.

To ensure future improvement of fish welfare in aquaculture it is essential that welfare assessment schemes are developed and applied for all species so that different production technology, farming practices, farm locations and handling operations can be compared. The welfare assessment schemes must cover all the overarching welfare needs described above, be based on reliable and validated welfare indicators with standardised methods for scoring the indicator levels. The fulfilment of the most important and life-sustaining welfare needs must be monitored daily, whilst frequent and more detailed checks of the fish populations and the rearing systems should also be performed regularly, in order to drive and continually stimulate future improvements of the rearing systems, routines and operations. For benchmarking and auditing purposes, the data should be stored in a common database and key statistics made publicly available for farmers and producers of new technology to benchmark their productions or handling operations against.

Acknowledgements This chapter has been written in collaboration with the Surveillance of fish welfare-project (IMR project number: 14930), the FISHWELL-project (FHF project number: 901157) and the REGFISHWELH-project (NFR project number: 267664).

References

Bracke MBM, Spruijt BM, Metz JHM (1999) Overall welfare reviewed. Part 3: welfare assessment based on needs and supported by expert opinion. Neth J Agric Sci 47:307–322

Bracke MBM, Edwards SA, JHM M, Noordhuizen JPTM, Algers B (2008) Synthesis of semantic modelling and risk analysis methodology applied to animal welfare. Animal 2:1061–1072. https://doi.org/10.1017/S1751731108002139

Costa-Pierce BA (1987) Aquaculture in ancient Hawaii. BioScience 37:320–331

Dawkins MS (2008) The science of animal suffering. Ethology 114:937–945

Driessen CPG (2013) In awe of fish? Exploring animal ethics for non-cuddly species. In: Röcklinsberg H, Sandin P (eds) The ethics of consumption: the citizen, the market and the law. Wageningen Academic Publishers, The Netherland, 537 p

EFSA (European Food Safety Authority) (2008) Scientific Opinion of the Panel on Animal Health and Welfare on a request from the European Commission on Animal welfare aspects of husbandry systems for farmed Atlantic salmon. EFSA J 736:1–31

FAO (Food and Agriculture Organization of the United Nations) (2018) Fishery and aquaculture statistics. Global aquaculture production 1950–2016 (FishstatJ) [online]. FAO Fisheries and Aquaculture Department, Rome. Updated 2018. www.fao.org/fishery/statistics/software/fishstatj/en

FAWC (Farm Animal Welfare Committee) (2009) Farm animal welfare in Great Britain: past, present and future. Farm Animal Welfare Council, London. https://www.gov.uk/government/uploads/system/uploads/attachment_data/file/319292/Farm_Animal_Welfare_in_Great_Britain_-_Past__Present_and_Future.pdf. Accessed 27 Mar 2018

Folkedal O, Stien LH, Torgersen T, Oppedal F, Olsen RE, Fosseidengen JE, Braithwaite VA, Kristiansen TS (2011) Food anticipatory behaviour as an indicator of stress response and recovery in Atlantic salmon post-smolt after exposure to acute temperature fluctuation. Physiol Behav 105:350–356. https://doi.org/10.1016/j.physbeh.2011.08.008

Folkedal O, Pettersen J, Bracke M, Stien L, Nilsson J, Martins C, Breck O, Midtlyng P, Kristiansen T (2016) On-farm evaluation of the Salmon Welfare Index Model (SWIM 1.0): theoretical and practical considerations. Anim Welf 25:135–149. https://doi.org/10.7120/09627286.25.1.135

Fraser D (2008) Understanding animal welfare. Acta Vet Scand 50(Suppl 1):S1. https://doi.org/10.1186/1751-0147-50-S1-S1

Huntingford FA, Adams C, Braithwaite VA, Kadri S, Pottinger TG, Sandøe P, Turnbull JF (2006) Current issues in fish welfare. J Fish Biol 68:332–372

Hvas M, Folkedal O, Imsland A, Oppedal F (2017a) The effect of thermal acclimation on aerobic scope and critical swimming speed in Atlantic salmon, Salmo salar. J Exp Biol 220:2757–2764. https://doi.org/10.1242/jeb.154021

Hvas M, Karlsbakk E, Mæhle S, Wright DW, Oppedal F (2017b) The gill parasite Paramoeba perurans compromises aerobic scope, swimming capacity and ion balance in Atlantic salmon. Conserv Physiol 5:1–12. https://doi.org/10.1093/conphys/cox066

Korte SM, Olivier B, Koolhaas JM (2007) A new animal welfare concept based on allostasis. Physiol Behav 92:422–428. https://doi.org/10.1016/j.physbeh.2006.10.018

Lucas JS, Southgate PC (2012) Aquaculture: farming aquatic animals and plants. Wiley, West Sussex, 648 p

Mellor DJ (2016) Updating animal welfare thinking: moving beyond the "five freedoms" towards "A lifeworth living". Animals 6(3):21. https://doi.org/10.3390/ani6030021

Mellor DJ, Beausoleil NJ (2015) Extending the 'Five Domains' model for animal welfare assessment to incorporate positive welfare states. Anim Welf 24:241–253

Mellor DJ, Patterson-Kane E, Stafford KJ (2009) The sciences of animal welfare. Wiley-Blackwell, Oxford, 212 p

Noble C, Nilsson J, Stien LH, Iversen MH, Kolarevic J, Gismervik K (2018) Velferdsindikatorer for oppdrettslaks: Hvordan vurdere og dokumentere fiskevelferd. 328 p. isbn:978-82-8296-531-6

Oppedal F, Dempster T, Stien LH (2011) Environmental drivers of Atlantic salmon behaviour in sea-cages: a review. Aquaculture 311:1–18

Oppedal F, Samsing F, Dempster T, Wright DW, Bui S, Stien LH (2017) Sea lice infestation levels decrease with deeper 'snorkel' barriers in Atlantic salmon sea-cages. Pest Manag Sci 73:1935–1943

Pettersen JM, Bracke MBM, Midtlyng PJ, Folkedal O, Stien LH, Steffenak H, Kristiansen TS (2013) Salmon welfare index model 2.0: an extended model for overall welfare assessment of caged Atlantic salmon, based on a review of selected welfare indicators and intended for fish health professionals. Rev Aquac 6:162–179. https://doi.org/10.1111/raq.12039

Richards C, Bjørkhaug H, Lawrence G, Hickman E (2013) Retailer-driven agricultural restructuring—Australia, the UK and Norway in comparison. Agric Hum Values 30:235–245

RSPCA (Royal Society for the Prevention of Cruelty to Animals) (2014) A review of farm animal welfare in the UK. Freedom Foods, Farm animal welfare: past, present and future-report, September 2014. https://www.rspcaassured.org.uk/media/1041/summary_report_aug26_low-res.pdf. Accessed 23 May 2018

RSPCA (Royal Society for the Prevention of Cruelty to Animals) (2018a) RSPCA welfare standards for Farmed Atlantic Salmon (February 2018). RSPCA, Horsham, 96 p. https://science.rspca.org.uk/sciencegroup/farmanimals/standards/salmon. Accessed May 23 2018

RSPCA (Royal Society for the Prevention of Cruelty to Animals) (2018b) RSPCA welfare standards for Farmed Atlantic Salmon (March 2018). RSPCA, Horsham, 51 p. https://science.rspca.org.uk/sciencegroup/farmanimals/standards/trout. Accessed 23 May 2018

Stien LH, Bracke MBM, Folkedal O, Nilsson J, Oppedal F, Torgersen T, Kittilsen S, Midtlyng PJ, Vindas MA, Øverli Ø, Kristiansen TS (2013) Salmon Welfare Index Model (SWIM 1.0): a semantic model for overall welfare assessment of caged Atlantic salmon: review of the selected welfare indicators and model presentation. Rev Aquac 5:33–57. https://doi.org/10.1111/j.1753-5131.2012.01083.x

Stien LH, Lind MB, Oppedal F, Wright DW, Seternes T (2018) Skirts on salmon production cages reduced salmon lice infestations without affecting fish welfare. Aquaculture 490:281–228. https://doi.org/10.1016/j.aquaculture.2018.02.045

van de Vis JW, Poelman M, Lambooij E, Bégout M-L, Pilarczyk M (2012) Fish welfare assurance system: initial steps to set up an effective tool to safeguard and monitor farmed fish welfare at a company level. Fish Physiol Biochem 38:243–257. https://doi.org/10.1007/s10695-011-9596-7

Webster J (2008) Animal Welfare: Limping Towards Eden: A Practical Approach to Redressing the Problem of Our Dominion Over the Animals. Blackwell, Oxford, 296 p

Zander K, Feucht Y (2018) Consumers' willingness to pay for sustainable seafood made in Europe. J Int Food Agribus Mark 30:251–275

Chapter 14
Welfare of Farmed Fish in Different Production Systems and Operations

Hans van de Vis, Jelena Kolarevic, Lars H. Stien, Tore S. Kristiansen, Marien Gerritzen, Karin van de Braak, Wout Abbink, Bjørn-Steinar Sæther, and Chris Noble

Abstract When fish are reared for food production in aquaculture, they can be held in different types of rearing systems and subjected to various husbandry routines and operations. Each of these systems or operations can present different welfare risks to the fish, which in turn are dependent upon both the species and its life stage. In this chapter, we address and outline potential welfare hazards the fish may encounter in a wide range of existing and emerging rearing systems used for on-growing. These systems include: (1) pond-based aquaculture, (2) flow-through systems, (3) semi-closed containment systems, (4) RAS, (5) net cages and (6) farming offshore using sea cages in exposed conditions. We also outline potential welfare hazards for two key farming operations: transport and slaughter. We present the tools the farmer can use to assess fish welfare during on-growing and also outline relevant welfare actions that can be taken to militate against welfare hazards.

Keywords Farmed fish · Welfare · Rearing system · Pond · Flow-through · RAS · Semi-closed containment system · Cage · Welfare indicator · Transport · Stunning · Slaughter

H. van de Vis (✉) · M. Gerritzen · W. Abbink
IMARES, Wageningen University & Research Centre, Yerseke, The Netherlands
e-mail: hans.vandevis@wur.nl

J. Kolarevic · B.-S. Sæther · C. Noble
Nofima, Tromsø, Norway

L. H. Stien · T. S. Kristiansen
IMR, Bergen, Norway

K. van de Braak
Sustainable Aquaculture Solutions, Horst, The Netherlands

T. S. Kristiansen et al. (eds.), *The Welfare of Fish*, Animal Welfare 20,
https://doi.org/10.1007/978-3-030-41675-1_14

14.1 Introduction

In the previous chapter of this book, "Chapter 13: Assessing Welfare in Farmed Fish", *animal welfare* was defined as the "quality of life as perceived by the animals themselves" (Stien et al. 2013), and *welfare needs* were defined as all the requirements that animals have that influence their qualitative experience of life. We will follow and expand upon this approach in the current chapter, addressing some of the differing welfare challenges farmed fish face when reared under different production systems and being subjected to various operations, with a specific focus on transport and slaughter. We will give an overview of system- and operation-specific threats with examples from commonly farmed species in each rearing system and operation, and we will also summarise some potential operational mitigation strategies.

To get information on how a fish's welfare needs are affected by different rearing systems or handling operations, we must rely on welfare indicators (WIs). Since a fish's experienced welfare is assumed to be directly related to the fulfilment of its welfare needs, all measurements or observations that give information about the degree of fulfilment of welfare needs qualify as welfare indicators.

Welfare indicators can be broken down into various types, for example input- and outcome-based welfare indicators. *Outcome-based welfare indicators* are all the measures that describe the result of welfare needs being fulfilled or compromised, typically measures based upon the fish themselves, such as health status and behaviour. *Input-based welfare indicators* are all measures that describe what the animals are subjected to and that influence the fulfilment of one or more welfare needs. WIs that are practical and suitable for use on the farm are termed operational welfare indicators (OWIs). WIs that require access to a laboratory or require further analysis in a lab are termed laboratory-based welfare indicators (LABWIs); see Chap. 13 or Noble et al. (2018) for further information.

Farmed fish can be raised in a wide array of rearing systems. However, on a global scale, aquaculture is dominated by a relatively small number, which can be classified as (1) land-based, such as natural and artificial ponds, various flow-through (FT) tanks and raceways, and recirculating aquaculture systems (RAS), and (2) water-based, such as freshwater/inshore/offshore floating net cages (Funge-Smith and Phillips 2001). In this chapter, we will therefore address potential fish welfare issues in ponds, flow-through systems, RAS and net cages to give the reader an introduction to some of the welfare challenges that affect current, widely used production systems (see Table 14.1 for a summary of rearing systems). We will also briefly discuss some emergent systems such as semi-closed containment systems (S-CCS) and the challenges presented by offshore aquaculture. The welfare of fish in other rearing systems such as snorkel cages or submersible cages is covered elsewhere (see, e.g. Kolarevic et al. 2018; Noble et al. 2018).

Farmed fish can also be subjected to a range of different handling operations during a production cycle. These include operations such as grading (separating smaller and larger fish), changing the rearing unit, crowding, netting/pumping, vaccination, medicinal treatments, parasite treatments, transport and slaughter.

Table 14.1 Brief description of different production systems covered in this chapter

Production system	Definition
Pond	The use of natural or artificial ponds with the surface area from a few square metre (intensive) to several hundred hectares (extensive). Gravity fed systems with a low-tech approach to water treatment; inlet and outlet. Utilise natural conditions for aquaculture production
Flow-through	Tanks and raceways with water supplied from gravity pipes from lakes or rivers, or pumped from the sea or lakes. No reuse of water. Oxygenation and CO_2 degassing systems are commonly used to reduce water demand
RAS	RASs are systems in which water is (partially) reused after biological and mechanical treatment and where limiting waste compounds are removed from the water
Flow-through system: S-CCS	A floating production unit where the fish are confined within a watertight or semi-permeable structure and where new water must be actively transported in and out of the unit (typically pumped from deeper in the water column)
Net cages	Floating production units where fish are enclosed within a net or mesh cage and are subjected to the water current–driven open exchange of water into and out of the rearing system (e.g. Beveridge 2004; Belle and Nash 2008)
Offshore cages	Net cages exposed to high wave, and often also strong currents in the open ocean unprotected by islands

RAS recirculating aquaculture system, *S-CCS* semi-closed containment system

These operations can stress the fish and can also potentially physically harm the fish. It is not within the scope of this chapter to address the welfare challenges of each of these routines, and we will therefore focus on the potential welfare challenges of two widespread handling operations: transport and slaughter. Other important routines, such as crowding (which can certainly be stressful for fish), are addressed in other publications such as the fish welfare book edited by Branson (Branson 2008), in the Royal Society for the Prevention of Cruelty to Animals (RSPCA) welfare standards (RSPCA 2018a, b) and, e.g. in the FISHWELL welfare indicator handbook (Noble et al. 2018). Fish welfare is also addressed in standards for organic aquaculture. The EU has developed standards for organic aquaculture for a coherent and consistent aquaculture policy. However, Council Regulation (EC) No 889/2008 (2008) does not specify which methods should be used to protect fish at slaughter.

14.2 Farming of Fish Worldwide: Some Key Facts and Figures

Global aquaculture has rapidly expanded over the last few decades. In 1996 the reported production of farmed fish was only 17 million tonnes and this tripled to more than 54 million tonnes 20 years later in 2016 (FAO 2017). This production was dominated by freshwater species (ca. 46 million tonnes), followed by diadromous

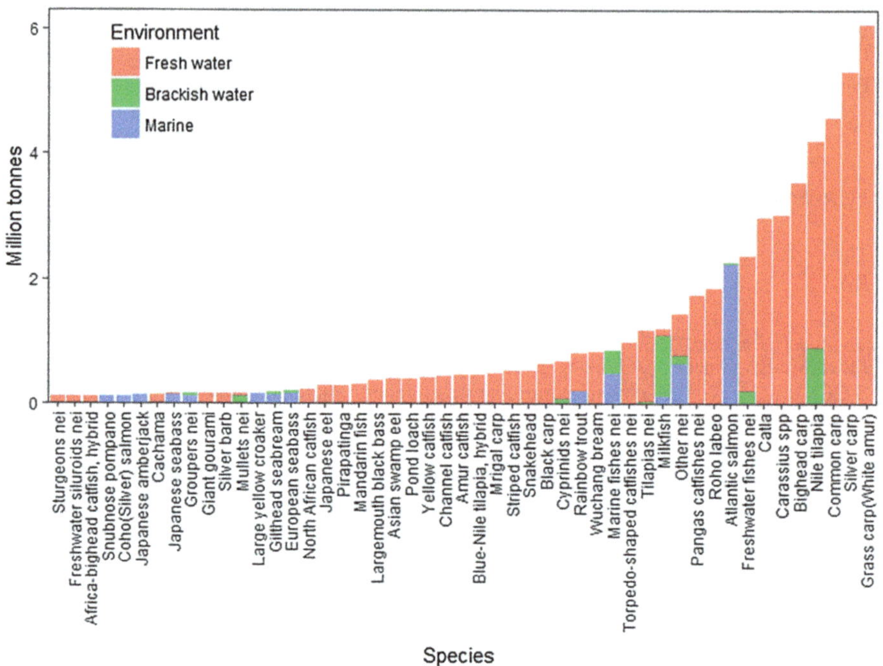

Fig. 14.1 The production tonnage of farmed fish broken down into species and their rearing environment (freshwater, brackish or marine) as reported to the FAO in 2016 (FAO 2017). 'nei' = registration not elsewhere included

species (ca. 5 million tonnes) and marine species (ca. 3 million tonnes, see Fig. 14.1). Asia accounted for 88.2% of the global aquaculture production of fish by volume in 2016, whilst Europe accounted for 4.3%, the Americas 3.7%, Africa 3.6% and Oceania 0.2%.

Freshwater species such as grass carp, silver carp and common carp dominate the global production of farmed fish. Nile tilapia, which can be farmed in both freshwater and brackish water, is the fourth largest in terms of production volume, while the first diadromous farmed species, Atlantic salmon, is only the ninth largest. The first marine fish is the European seabass in 35th place in terms of reported production volume (Fig. 14.1).

The wide variety of species (362 species FAO 2018a) that are reared in aquaculture can be explained to a large extent by their biology and their ease of rearing and domestication but also by historical and socio-economic factors. Cyprinids, such as grass carp, silver carp and common carp, are herbivores or omnivores living on cheap and easily available food resources. They are also robust and adaptable and have evolved to tolerate still, turbid and oxygen-poor water and a wide range of temperatures. This made them fairly easy to domesticate, and they have been grown in ponds for millennia. On the other hand, the farming of pelagic ocean ranging predatory fish as Atlantic salmon did not become a significant part of global production until much later, when the technology for rearing high numbers of fish in tanks on land and in cages at sea became available.

Fish have a huge diversity with respect to phylogeny, habits and habitats. Due to this diversity, their welfare needs can differ markedly, e.g. in relation to the water quality requirements. An example of this is related to, e.g. the threshold concentrations for unionized ammonia (NH_3) in rearing water in different species. For African catfish, it is recommended that the NH_3 concentration should not exceed 24 μM during the on-growing phase (Schram et al. 2010), whereas for rainbow trout ammonia concentrations should be 0.9–1.4 μM (MacIntyre et al. 2008). When the level of ammonia in the water is too high, it causes neurotoxicity in fish. Water ammonia should therefore be kept well below species-specific threshold concentrations.

Regardless of the specific criteria for ensuring appropriate water quality in order to meet the species-specific welfare needs of the fish, there is also a large variety in technologies and procedures used for fish rearing, transportation and slaughtering in the aquaculture sector worldwide. This variety is related to the specific fish species, the availability/affordability of equipment, levels of staff training, cultural aspects or standards set by e.g. NGOs, supermarkets or other purchasers (amongst many other factors); all these factors may vary within and between countries and even within a specific company.

All in all, the great variety of fish species, aquaculture systems and production chains makes it impossible to provide a comprehensive overview that takes into account all welfare-related issues in aquaculture within a single book chapter. We will therefore here limit ourselves to a general overview with some more specific examples to illustrate the main points. Furthermore, when contemplating fish welfare in different production systems and handling operations, and deciding which welfare indicators to use to assess fish welfare, it is also important to consider how welfare needs can differ with different life stages, especially in diadromous fish such as salmon, where differing parts of the life cycle are borne out under markedly different environmental conditions. In comparison to birds and mammals, fish hatch as very small and poorly developed larvae, often only a few millimetres in length, and are very vulnerable, in this first period, to starvation, malnutrition, injuries and stress. However, due to space constraints, life stage-specific challenges will not be considered here and we refer the reader to other work on this topic in other chapters in the book, e.g. Huntingford et al. (2006), Ashley (2007), Branson (2008), EFSA (2008a, b, c, d, e) and Van de Vis et al. (2012).

14.3 Welfare of Farmed Fish

14.3.1 A Brief Overview of Rearing Systems

As stated in the introduction, although there are a wide range of existing and emerging systems for rearing the diverse range of farmed fish species, global aquaculture production is dominated by a relatively small number of farming systems. In this section, we will look at some of the potential welfare challenges and mitigation strategies associated with farming fish in (1) ponds, (2) flow-through systems, (3) RAS, (4) net cages and (5) semi-closed containment systems (S-CCS) and how the challenges of (6) farming offshore are met.

In many countries, especially Europe and North America, most of the production systems listed in Table 14.1 are not integrated into agricultural systems. In China, however, integrated systems used to be common practice, especially in inland aquaculture (Edwards 2008). Fish farming can be integrated in various ways with the farming of crops (mainly rice), vegetables and livestock. Nevertheless, fish is the major commodity (Jena et al. 2017).

A major characteristic of integrated fish farming is that by-products/wastes from one system become the inputs to another system. In the past, carp polyculture systems in China were often integrated with the production of poultry, pigs and/or ducks. Manure produced by livestock can also be used as pond fertilizer to produce organisms as feed for fish. Due to the possible impacts and concerns of such farming systems with regard to food safety, this type of integration is rapidly disappearing in China (Edwards 2008). With regard to pond farming, polyculture is the most common approach utilised in China, which is the largest aquaculture producer in the world, contributing to nearly 60% of global production of farmed finfish (Anonymous 2017). In other integrated aquaculture systems, by-products that originate from the production of horticulture or agriculture (e.g. rice bran) are used as a source for fish feed. In China approx. 1 million tonnes of fish was produced in rice fields (Weimin 2010).

A further integrated approach is aquaponics, which is a combination of aquaculture and hydroponics (the growing of plants without soil, Jena et al. 2017). In aquaponics, nutrient-rich water, which is a by-product or waste product from rearing fish, is used as fertilizer for hydroponic production beds. At present, aquaponics is only applied on a small scale in various countries.

Another integrative approach is IMTA (Integrated multi-trophic aquaculture). IMTA is a concept where species of different trophic levels in fresh water or seawater are farmed in the same system at the same time (Jena et al. 2017; Chopin et al. 2012; Troell et al. 2009). These levels are linked in such a way that by-products/wastes resulting from finfish or shrimps (i.e. fed species) or their uneaten feed become resources for species at lower trophic levels (i.e. extractive species). Extractive species are, for instance, invertebrates or plants. A special case of IMTA is the use of cleaner fish in salmon aquaculture. Here wrasses and lumpfish are kept together with salmon, not for future human consumption, but to pick and eat sea lice of the salmon as a parasite management measure.

14.3.2 Welfare of Farmed Fish in Different Production Systems

In this section, we will focus on growing life stages of aquaculture production. As stated earlier, the sheer diversity of farmed fish species means we cannot go into great detail with regard to their species-specific welfare requirements under farming conditions but we can give an overview of:

1. The general welfare challenges that fish can be subjected to when reared in different production systems, with examples of some of the most commonly farmed species in each system
2. The framework and approach that can be followed to identify and deliver potential operational mitigation strategies

An overview of the appropriate operational (OWI) and laboratory-based (LABWI) welfare indicators and tools (Chap. 13) that can be used for assessing welfare in each rearing system can be a good starting point for identifying welfare threats and also for developing operational mitigation strategies. The European Food Safety Authority have published a series of comprehensive reviews on the welfare requirements and challenges of rearing Atlantic salmon, rainbow trout, European sea bass, gilthead seabream, European eel and the common carp in a selection of different rearing systems in Europe (EFSA 2008a, b, c, d, e). A recent handbook has also been published outlining which OWIs and LABWIs are suitable for Atlantic salmon aquaculture (Noble et al. 2018), and this has a large section on fit-for-purpose OWIs and LABWIs for salmon in different rearing systems (Kolarevic et al. 2018).

There are a number of welfare challenges that are applicable to all rearing systems and their relevant species and life stage-specific OWIs and LABWIs can generally be applicable across all rearing systems, especially with regard to input-based WIs and morphological outcome-based WIs. We have highlighted some major production system OWIs in Table 14.2, and whilst this table does highlight that many WIs are general, other WIs are very specific for a given rearing system or a group of rearing systems. A good example of this is related to the potential water quality challenges presented by farming in open or closed systems (due to external water quality challenges in open production systems and system-driven challenges when water is recirculated). Some water quality parameters should also be continuously monitored in all systems such as temperature and oxygen, which are two of the most important water quality welfare indicators (EFSA 2008b, c).

14.3.2.1 Pond-Based Aquaculture

Inland fish farming is the most dominant type of fish farming operation on a global scale, and the culture of fish in earthen ponds (Fig. 14.2) is the largest aquacultural contributor to food production, especially in economically less developed countries (FAO 2018a). Farming of fish in ponds is a low-tech approach to farming, using

Table 14.2 Overview of the major OWIs and LABWIs[a] that should be monitored in different rearing systems, covering input- and outcome-based group and individual WIs

		Some major operational welfare indicators				
		Land-based systems			Water-based systems	
	All systems	Ponds	Flow-through	RAS	S-CCS	Sea cages
Input-based WIs	Oxygen levels, temperature, salinity, water velocity, light conditions, algae, stocking density, *total suspended solids, turbidity, pathogens*	Algae, predators, temperature, water velocity, pH	Water velocity, oxygen, ammonia, feed load	CO_2, pH, ammonia, nitrate, feed load	Water velocity, oxygen, ammonia, feed load	Water velocity, oxygen, temperature, algae
Outcome-based WIs at the group level	Appetite, growth, mortality, deviations from normal behaviour, emaciated fish, *health and disease*	Deviations from normal behaviour, appetite mortality, *tissue sampling and off-site analyses*	Deviations from normal behaviour, appetite, mortality, *tissue sampling and off-site analyses*	Deviations from normal behaviour, appetite, mortality, *tissue sampling and off-site analyses*	Deviations from normal behaviour, appetite, mortality, *tissue sampling and off-site analyses*	Deviations from normal behaviour, appetite, mortality, *tissue sampling and off-site analyses*
Outcome based WIs at the individual level	Emaciation state, scale loss and skin condition, eye status, deformities, fin damage, opercula damage, mouth/jaw damage, HSI, CSI, condition factor, gill status, sexual maturation, precocious maturation, feed in intestine	Emaciation state, scale loss and skin condition, eye status, deformities, fin damage, opercula damage, mouth/jaw damage	Emaciation state, scale loss and skin condition, eye status, deformities, fin damage, opercula damage, mouth/jaw damage	Emaciation state, scale loss and skin condition, eye status, deformities, fin damage, opercula damage, mouth/jaw damage	Emaciation state, scale loss and skin condition, eye status, deformities, fin damage, opercula damage, mouth/jaw damage	Emaciation state, scale loss and skin condition, eye status, deformities, fin damage, opercula damage, mouth/jaw damage

WIs for flow-through, RAS, S-CCS and sea cages are reproduced with permission from Kolarevic et al. (2018)
[a]LABWIs are indicated with italics

Fig. 14.2 Farming of fish in earthen ponds

systems that vary in production intensity, depending on stocking density and external feeding (Stickney 2000). With regard to feeding, there are a whole range of options, from unfertilised ponds where the fish graze on natural plants to fertilised ponds with supplemental feeding. Fertilisation increases primary plant production, and these plants serve as a base for the food webs that support the nutritional needs of the fish. These techniques have a long history, and production strategies developed more than a thousand years ago are still applicable for producers today. The long history of these systems tells us that they have some advantages when compared to more technologically advanced farms, primarily by being the most cost-efficient approach to aquaculture production and, in practice, the only viable option for producing low-value species (Boyd and Tucker 1998). As the pond water supply is based on gravity and limited, with water otherwise untreated, there is very little that can be done to avoid unfavourable conditions or even militate against them if they were to occur.

There are also ponds that are operated with full feeding, supplemental water flow and aeration. Such intensive pond production is similar to flow-through tanks. These usually have little or no water treatment systems and variable water quality dependent on its natural source. Temperature can be affected by season and to some degree also during the day. Soluble gases, pH and particle density (suspended solids) may also vary, with low oxygen and particulate matter in the water being key welfare risks. Production in small intensive systems is a low-tech variant of farming in tanks, but with less control of the fish and the production environment.

Extensive pond cultures are very different to intensive production. The water production volumes are typically large and fish density is low; e.g. a typical Polish carp pond is 100–300 ha with a production of only 600–1500 kilo per hectare of water surface (Horváth and Urbányi 2000). The fish have a good opportunity to express natural behaviour, to seek refuge and to avoid aggressive conspecifics. These ponds are managed as closed ecosystems with the farmer rearing several species, which also means that the welfare of each species has to be ensured.

The dominant fish produced in ponds are various species of carp (grass carp, silver carp, common carp and others) and tilapia (Nile tilapia and others), with the production of grass carp being the dominant aquaculture species worldwide (Bostock et al. 2010). As extensive ponds are large-scale open systems, they are often difficult to control and monitor. The production system is exposed to and dependent on natural conditions. There are few opportunities to improve water quality parameters if conditions are unfavourable. This is challenging for health control, and Koi Herpes Virus (KHV) has been spreading in Asia and in Europe since the early 2000s (Haenen et al. 2004). This virus causes severe disease and mortality in common carp and koi carp at all life stages.

14.3.2.2 Flow-Through Systems

Raceways, tanks or ponds where the water only flows through the system once are called flow-through systems. A raceway is an artificial channel, which usually consists of rectangular basins or canals constructed of concrete and equipped with an inlet and outlet (Fig. 14.3).

Fig. 14.3 Raceway as flow-through system for the on-growing of trout

The flow of water through the rearing system provides fish with oxygen and carries dissolved and suspended wastes out of the system (Heller 2017). However, a part of the suspended wastes settles and accumulates at the bottom of raceways or tanks, and these systems need to be cleaned periodically. In Europe, effluents are usually treated before being released back to the aquatic environment (FEAP 2018).

In general, there are two types of flow-through systems, conventional and intensive (Anonymous 2015; FEAP 2018):

- In conventional systems, an abundant flow of water ensures enough oxygen for the fish. When the temperature of the inflowing water is, for instance, 20 °C, the required specific flow rate is estimated at 2.4 m^3/(day × kg fish) to ensure an adequate supply of oxygen (FAO 1984). Fish species like Nile tilapia are farmed in these systems.
- In intensive flow-through systems, an aerator or pure oxygen is used to supply additional oxygen for the fish. In an intensive system, the flow rate that is needed for the flushing of the metabolic wastes also becomes a critical factor. In an intensive system, the required specific flow rate is estimated at 0.248 m^3/(day × kg fish) (FAO 1984). Trout, Atlantic salmon, European sea bass, sea bream, eel and turbot are reared in intensive systems (EFSA 2008a, b, c, d).

A major advantage of flow-through systems is that investment costs for this technology are low, compared to, e.g. RASs. Flow-through systems can be applied in many locations, provided that a large amount of water is available at appropriate temperatures for the species concerned. However, flow-through systems are, in principle, open systems and therefore vulnerable to pathogens and sudden changes in the quality of the inlet water.

Water oxygen content may constitute one of the most important welfare risks in flow-through systems. If the oxygen supply (aeration or added oxygen) is not balanced with the biomass, hypoxic conditions may affect the fish's growth and welfare. A rapid reduction in the available dissolved oxygen can result in metabolic alkalosis, rapid changes in blood pH and, in severe cases, mortality. Adding oxygen to avoid hypoxic conditions can also result in supersaturation (>100% O_2 saturation) if this supplementation is not properly balanced. Oxygen supersaturation leads to a decreased ventilation rate and respiratory acidosis in fish, e.g. Espmark and Baeverfjord (2009). To control oxygen levels in the water and ensure an adequate removal of metabolic wastes from a tank, monitoring systems and a back-up system in case of pump failure should be in place. In addition, trained staff should operate flow-through systems in accordance with good manufacturing practices.

14.3.2.3 Semi-closed Containment Systems

Semi-closed containment systems (S-CCS) in the sea are flow-through production systems that provide a dense or relatively dense physical barrier between the water environment in which the fish is reared and the surrounding environment (Rosten et al. 2011). In S-CCS, water from the deeper water column is pumped into the system to avoid contamination from the surface water and organic waste can be collected and removed from the system. In addition, the physical barrier reduces the risk of escapees and provides more stable temperature conditions within the rearing system that can be beneficial in the early seawater grow-out phase (Handeland 2016).

Existing S-CCS are constructed from different materials (GRP panels, tarpaulin), have different shapes (circular tanks, raceways, "bag-shaped" systems) and can also be in a range of sizes (from 1000 m^3 to 21,000 m^3). Different pilot S-CCS are currently being tested for the production of Atlantic salmon post-smolt up to 1 kg, although the potential of S-CCS for full sea water grow-out phase is also being considered (Kolarevic et al. 2018). It has been reported that S-CCS for Atlantic salmon have no adverse effects on growth and survival rates (Handeland et al. 2015; Kolarevic et al. 2016; Nilsen et al. 2017).

S-CCS are designed to allow for efficient self-cleaning and optimization of water quality; however, technical issues can lead to reduced flow in the system and the accumulation of particles. This can have negative effects on water quality in the system. The monitoring of carbon dioxide, oxygen and ammonia should be in place to avoid any adverse effects of suboptimal water quality on fish welfare and health. Emergency oxygenation and back-up systems in case of pump failures should also be considered. The majority of existing S-CCS do not treat intake water prior to entering the system, which means pathogens from marine sediments can potentially enter the system (Rud et al. 2016). In addition, the fish are exposed to the periodical occurrence of pathogens due to, e.g. bacterial or algal blooms and the occurrence of jellyfish (Kolarevic et al. 2018) that can negatively affect the health and welfare of fish. The treatment of the intake water in floating S-CCS is currently being considered as it could potentially lead to better control of the water quality, increased biosecurity, reduced mortality and improved fish health and welfare.

14.3.2.4 RAS

Recirculating aquaculture systems are systems in which water is (partially) reused after limiting waste compounds are removed (Kolarevic et al. 2018). Water treatment in RAS is mainly dependent on the degree of water reuse, economical considerations and water quality requirements based upon the fish species and fish size (Blancheton et al. 2007). A classical RAS loop consists of solid removal (mechanical filtration, decantation), gas control (oxygenation and degassing of carbon dioxide) and biological treatment (ammonia removal by biofilters), UV and ozone treatment, e.g. Roque d'Orbcastel et al. (2009a). In addition, automatic pH and alkalinity regulation, heat exchange and denitrification systems are components that can be added to increase the efficiency of the system. RAS can be used outdoors, as in the case of the Danish Model farm recirculating systems described in Jokumsen and Svendsen (2010). RAS technology can reduce effluent loads to, e.g. meet the demands of increased environmental regulations (Martins et al. 2010), can lead to a 100-fold reduction in make-up water requirements (Roque d'Orbcastel et al.

2009b) and more effective economies of scale (Timmons and Ebeling 2007). Further developments of the technology and the arrival of cost-efficient indoor RAS allows production to be carried out in a controlled environment that can mitigate the risks of outdoor aquaculture, e.g. natural disasters, escapees, exposure to pollution, disease and predators, which can mean that growth can be optimised on a year-round basis close to the market (Timmons and Ebeling 2007).

Although RAS technology allows for the production of a wide range of aquatic organisms, the choice of species is largely restricted to higher value species or more sensitive and faster growing early life stages that can secure a high economic return, due to the high investment costs and more complex operations involved (Bostock et al. 2010; Bregnballe 2015). Some of the species with good biological performance and market conditions for production in RAS are arctic char, Atlantic salmon smolts, eel, grouper, rainbow trout, European sea bass, gilthead seabream, yellowtail kingfish (yellowtail amberjack) and sturgeons. Species like perch, tilapia, African catfish, barramundi, pangasius, carps and whitefish can also be produced in RAS, but they may not be particularly profitable due to low market prices (Bregnballe 2015). Pikeperch and Atlantic salmon post-smolt are species/life stages that are emerging in current markets and still need production optimisation.

Apart from a number of advantages, RAS technology also faces a number of challenges such as increased investment and operational costs, a need for more skilled personnel, the intensification of production with the use of higher temperatures, potentially higher stocking densities and elevated inlet dissolved oxygen saturations that can affect fish welfare (Kolarevic et al. 2018). The inadequate design and dimensioning of RAS that does not take into consideration species specific needs for water quality can pose a risk for fish welfare. The use of RAS can lead to the accumulation of carbon dioxide, the depletion of oxygen and an increase in nitrogen compounds (ammonia, nitrate and nitrite) if the production exceeds the carrying capacity of the system. Therefore, the monitoring of the above-mentioned compounds is essential in RAS. Long-term chronic exposure to suboptimal water quality could have subclinical and clinical effects on numerous fish species in RAS, making them more susceptible to diseases (e.g. Hjeltnes et al. 2017). It has been shown that when water quality is maintained at an appropriately safe level, it is possible to produce rainbow trout in RAS at high stocking densities without a negative effect on fish performance or certain morphological OWIs such as dorsal and pectoral fin damage, but caudal fin damage can be adversely affected (Roque d'Orbcastel et al. 2009b). Atlantic salmon smolts produced in RAS can exhibit similar production performance to those produced in flow-through systems, with significantly less fin damage (whilst using 98% less water) (Kolarevic et al. 2014).

An adequate alkalinity is necessary for the nitrification of ammonia to nitrate in biofilters, and alkalinity loss is compensated for by supplementation with a base, such as NaOH or $NaHCO_3$ (Summerfelt et al. 2015). It has been documented that the method by which alkalinity is managed in the RAS for Nile tilapia can lead to nephrocalcinosis, e.g. by the use of agricultural-grade lime, calcium carbonate (Chen et al. 2001). Nephrocalcinosis problems were mitigated once this substance was substituted with sodium bicarbonate (Chen et al. 2001). In low- and near-zero-exchange RAS, the accumulation of nitrate and potassium has been indicated as the cause of increased swimming speed and abnormal side swimming behaviour in fish (Davidson et al. 2011).

Good biosecurity is a prerequisite for the successful operation of RAS, and it is one of its advantages over traditional outdoor production systems. The potential sources of disease in RAS are biological material (eggs and fish), fish feed and new (make-up) water. The eradication of introduced diseases are more difficult in RAS due to the disturbance the treatment might have on biofilter and its function and the segregation of different life stages. An all-in and all-out procedure coupled with the disinfection of the production system is essential for fish health management (Hjeltnes et al. 2017).

Visibility in low-exchange RAS with high feed loads and a lack of ozonation can be significantly reduced, which makes it difficult to observe the fish both manually and automatically with cameras. Acoustic telemetry has been shown to be a promising technology for the real-time monitoring of fish in RAS (Kolarevic et al. 2016). With regard to the monitoring of water quality, a number of in-line systems have been adapted from the water treatment industry or created specifically for use in aquaculture (Kolarevic et al. 2015). In order to ensure the accurate monitoring of water quality, it is important to prevent biofouling on the sensor surfaces (Kolarevic et al. 2018).

14.3.2.5 Net Cages

Net cages are floating production units where fish are enclosed within a net or mesh cage and are subject to the open exchange of water into and out of the rearing system (see Table 14.1 for more information). Cages utilise existing water resources, be they in the ocean (e.g. Fig. 14.4), estuary, lake or river. One of the advantages of this open production approach is that the currents that drive new water into the cages both replenish oxygen and provide the fish with a flowing medium whilst removing feed waste, faecal matter and dissolved wastes (Kolarevic et al. 2018).

Net cage aquaculture is dominated by salmonids, specifically Atlantic salmon (51%). Another quarter of the production is dominated by rainbow trout, coho salmon, Japanese amberjack (yellowtail) and Pangasius spp. (Tacon and Halwart 2007). This section will therefore use Atlantic salmon as a case study species, although the approach and challenges regarding the production system may be just as applicable to other species.

The benefits of readily replenishing water in net cages from the surrounding environment can also present challenges to fish welfare. In terms of the environment, when compared to RAS and S-CCS farmers have very little control of the water quality the fish are subjected to and it can be difficult to improve the rearing situation if the fish are exposed to suboptimal water conditions. Fish can be exposed to a wide range of daily and seasonal changes in water quality parameters that can also change markedly in relation to water depth (Oppedal et al. 2011). In the case of Atlantic salmon farmed in marine net cages, this variability can also be affected by tidal

Fig. 14.4 Sea cage for the on-growing of Atlantic salmon

currents, storms or freshwater run-off (Oppedal et al. 2011; Kolarevic et al. 2018). In cases where some water parameters, such as oxygen, drop below optimal levels and become a welfare threat to the fish, farmers can militate against this to a certain extent by either aerating or oxygenating the cages, or by ceasing feeding to reduce metabolic demand and avoiding various handling operations or routines that may be stressful for the fish. They can also manage biofouling to ensure adequate water flows in and out of the cages and stocking densities to reduce the risks of low oxygen conditions (EFSA 2008a). However, for other factors such as undesirably high or low seasonal temperatures, very little can be done to alleviate the problem, aside from ceasing feeding, until the threat has passed. However, large and deep cages can circumvent these challenges to some extent as they can give the fish the opportunity to gather and school, e.g. deeper in the cage or in other areas in the water column where oxygen and temperature gradients may be less of a welfare threat (e.g. EFSA 2008a; Oppedal et al. 2011).

In terms of biosecurity, open water transfer can introduce pathogens and other harmful organisms into the rearing system. These can be bacteria or viruses, parasites, stinging organisms or blooms of phyto- and zooplankton (e.g. EFSA 2008a, b). The consequences for fish health and welfare in relation to bacterial and viral infections can be very large and is a key welfare issue. These are covered in other reviews and articles (EFSA 2008a, b, c; Segner et al. 2012; Hjeltnes et al. 2017) and so will not be covered here. However, the impacts of parasitic infections and harmful algal blooms (HABs) and zooplankton blooms can also be a problem. With regard to Atlantic salmon, sea lice can be a major problem (EFSA 2008a), both in terms for the fish themselves if levels get high enough (as >0.12 lice cm^{-2} of salmon epidermis can be lethal, Stien et al. 2013) and also due to the numerous handling and treatments that may be required to keep lice numbers low (Nilsson et al. 2018). A way to mitigate against the risks of diseases in fish is the coordinated fallowing of large regions until there are practically no infective stages or parasite broodstock left. Fallowing means that a site is emptied and not restocked for a period of time. Atlantic salmon sites are fallowed for 2–6 months after harvest (Marine Harvest 2018; Werkman et al. 2011). Fallowing should be applied in combination with the systematic surveillance of farms.

Phytoplankton and zooplankton blooms are rare and usually unpredictable (EFSA 2008a, b) and can be detrimental to the state of the water that the fish are reared in, by depleting oxygen levels (e.g. Hallegraeff 2003). They can also be directly detrimental to the fish by, e.g. producing toxins (EFSA 2008b) and damaging the gills (Hallegraeff 1993) or digestive tract (Roberts et al. 1983). They can also lead to mass mortalities, as evidenced by the blooms of the flagellate *Chattonella* spp., which have impacted the cage aquaculture of Japanese amberjack in Japan for many years (Cho et al. 2016). Mitigating against algal and zooplankton blooms can be difficult in cage farming, but some contingency actions are available, such as ceasing feeding, pumping, e.g. deeper water into the cage to dilute the bloom, oxygenating or aerating the water, towing the cages away from the outbreak if the blooms are actively monitored (and if there are logistics in place for the mobilisation of resources to carry this out), water treatment and the applications of therapeutics (see Hallegraeff 2003; Rensel and Whyte 2003 for more information). However, not all of these approaches are effective and slaughtering or euthanizing fish on ethical grounds may be required (EFSA 2008b).

14.3.2.6 Farming Offshore Using Sea Cages in Exposed Conditions

The main limitations for the use of sea cages in exposed conditions are high waves and strong currents. New technologies for more exposed farming are now being developed as the need for new aquaculture areas grows, and these can utilise, e.g. platform- or ship-like structures with inflexible net walls. Submersible cages can also be used to avoid suboptimal surface conditions and high waves/storms (e.g. Dempster et al. 2009). The two main fish welfare challenges with regard to exposed farming are strong currents and waves. When the current velocity through the sea cage increases, the schooling structure of Atlantic salmon changes from circular swimming at a voluntary cruising speed to standing on the current swimming at speeds dictated by the environment (Johansson et al. 2014; Hvas et al. 2017a). If the current velocity exceeds the swimming capacity of the fish, they will reach physiological fatigue and can get pushed onto the rear wall of the sea cage. Physiological fatigue induces a state of maximum stress with dramatic endocrine, osmotic and respiratory disturbances that may cause mortality (Wood 1991; Wendelaar Bonga 1997). Wounds from contact with net wall and other equipment can also be infected by opportunistic bacteria, inhibiting the healing process and meaning the wounds can develop into severe ulcers (Karlsen et al. 2012). For responsible and ethical farming at exposed locations, it is therefore necessary to quantify the swimming capabilities of farmed fish and also how they are affected by biological and environmental factors. Generally, larger salmons are able to swim faster, which makes them more robust to exposed environments (Remen et al. 2016). Hence, producing larger smolts prior to transfer to exposed sea cages is a way to adapt production to exposed locations.

Temperature is the primary environmental factor for determining the swimming performance of fish in sea cages. At the either extreme of the thermal niche, swimming capabilities are reduced (Brett 1964; Hvas et al. 2017b). For exposed farming, the main concern will be in winter where very low temperatures may coincide with stormy weather. Diseases and parasites can also reduce the swimming ability (Hvas et al. 2017c) and may therefore have further negative welfare impacts in exposed settings. When assessing the suitability of new exposed locations from a welfare perspective, a detailed documentation of water current and wave conditions should be obtained. Specifically, it is important to consider both the magnitude and duration of strong water current events and how it relates to the sustained swimming speed of farmed fish (Hvas et al. 2017a). With regard to how the fish tolerate high wave action, knowledge is relatively limited as this is difficult to test experimentally. Anecdotal evidence suggests the fish simply go deeper into the cage where waves and currents are less strong, but to the authors' knowledge, this remains to be documented. Cages for exposed farming conditions should therefore be designed to be deep enough so the fish can avoid strong waves.

14.4 Welfare of Farmed Fish Under Different Farming Operations

Fish are subjected to a range of different husbandry routines and operations throughout the production cycle, and these will also have a variation in impacts upon fish welfare. The stress caused by these operations cannot be avoided. Farmers need robust guidance on the issues fish can potentially face when subjected to different routines and daily operations. The European Food Safety Authority has written comprehensive reviews on the welfare risks of numerous routines and operations for numerous European species, including Atlantic salmon, rainbow trout, European sea bass, gilthead seabream, European eel and common carp (see EFSA 2008a, b, c, d, e); we refer the reader to these reviews for further information on the welfare risks of different routines such as vaccination, handling, crowding, feeding, and grading. In this section, we will present an overview of two key operations farmed fish are subjected to, irrespective of species and rearing system. These are transport and stunning/slaughter.

14.4.1 Transport

Farmed fish are often transported numerous times during their life cycle. This transport can be carried out by truck, boat or aircraft (Dalla Villa et al. 2009) and can take place between companies or sites. The transportation of live fish can mean they are exposed to various stressors during the procedure such as crowding, the loading of the transport vehicle, exposure to vibrations that may occur during transport, unloading upon arrival at the facility and exposure to a potentially novel environment for on-growing or storage for slaughter. In general, the process of transporting live fish involves the following steps:

Pre-transport

1. Establishing that fish are fit for travel
2. Subjecting the fish to a period of fasting, which may range from a few days to approximately 2 weeks

Transport Phase

3. Crowding of the fish
4. Loading a transport vehicle
5. Transport in a tank, plastic bag or vessel hold
6. Unloading

Post-transport

7. Fish are released into a new environment or subjected to a slaughter procedure.

Exposure to potential stressors either simultaneously or in rapid succession during and around the transport process may induce severe physiological stress (Dalla Villa et al. 2009; Koolhaas et al. 2011; McEwen and Wingfield 2003; Sampaio and Freire 2016).

It is obvious that stress in fish should be minimised, as transportation results in a higher metabolic rate in fish and can cause shedding of fish mucus, leading to a deterioration of water quality and also making the fish less robust and resilient for further on-growing (if they are being transported between production sites). The interaction between the stockperson and animal is also very important. Poorly controlled crowding, loading or unloading of fish may result in severe injuries, an increase in aggression among fish or even mortality. Injuries, aggression and mortality are important output-based OWIs for transportation. As a consequence, it is essential to ensure that trained staff are present (appropriate staff training is also an OWI) to ensure that loading and unloading are smooth and quick operations. With respect to the density of fish during transport, the deterioration of water quality is the dominant factor, but other factors such as the length of the journey, road conditions during overland transport and weather conditions during transport by boat or aircraft, fish size, life stage and fish species and whether the system is open or closed will all determine appropriate densities. In a closed system, carbon dioxide (King 2009) and ammonia—which are both excreted by fish—will increase during transport. Oxygen needs to be administered to avoid hypoxia, and CO_2-degassers should be used to control the levels of CO_2. Oxygen, ammonia and CO_2 levels in the water during transport are important input-based OWIs. As fish can be susceptible to motion sickness (Hilbig et al. 2002), road conditions, weather conditions and the skill levels of the transport staff are also important OWIs.

Practical experience with, e.g. Atlantic salmon shows that a well-controlled transport procedure is possible and thus the impact of the process on welfare can be substantially reduced. In fact, a smooth and incident-free transport phase may itself be important for helping the fish recover from the stress of the crowding and loading aspects of the procedure, if it is long enough (Iversen et al. 2005; Nomura et al. 2009). Key recommendations for water quality and management criteria for a well-controlled transport are available for, e.g. Atlantic salmon and rainbow trout (RSPCA 2018a, b), and we refer the reader to these comprehensive guidelines for full details.

It should be noted that there is great variation in oxygen, pH, salinity and temperature requirements between species, and species-specific requirements are needed to ensure the welfare needs of each species are met during the transport process. For example, Nile tilapia can cope with rather low levels of oxygen in the presence of suspended solids, whereas these conditions would be very stressful for salmonids. As a consequence, it is not possible to set optimal conditions that apply to all species in the very diverse group of farmed fish (EFSA 2004).

The type of vehicle used for transport depends on whether the fish need to be moved between companies or sites that are, e.g. both land-based or when one site is land-based and the other company or site is at sea. With regard to the transport of salmon smolts to cages for on-growing, a helicopter may be used. In practice, the use of a helicopter is exceptional (its market share in the UK is less than 1%, Schrijver et al. 2017) and this method will not be covered here. An overview of transport methods used in Europe for fry, fingerlings or smolts to a facility for on-growing and for transporting market-sized fish to a facility for slaughter are presented in Tables 14.3 and 14.4, respectively. Some of the advantages and disadvantages of the transport methods are also outlined in these tables.

14.4.1.1 Road Transport

Road transport is frequently used for transferring fish from a hatchery to an on-growing facility. Road transport is also used to transport market-sized fish to live markets, as is common practice, for instance, in Poland for 70% of the common carp produced (Lambooij et al. 2007; Schrijver et al. 2017), crucian carp (FAO 2018c), bighead carp (FAO 2018d), silver carp (FAO 2018e) and also Nile tilapia (FAO 2018f), e.g. in Asia. It can also be used to transport live fish to a slaughter facility, e.g. for market-sized rainbow trout (Schrijver et al. 2017), European eel (Boerrigter et al. 2015) and African catfish (Manuel et al. 2014).

Table 14.3 Transport methods used for the transfer of Atlantic salmon, European sea bass, gilthead seabream, carp species, Nile tilapia and European eel to on-growing facilities (Dalla Villa et al. 2009; Da Silva et al. 2009; Sampaio and Freire 2016; Schrijver et al. 2017)

Fish species	Transport method	Advantages	Disadvantages
Atlantic salmon, Pangasius	Well-boat	Fish may recover from loading stressors during transportation	Loading and unloading causes stress. Water quality may deteriorate during closed transports
European sea bass/gilthead seabream	Transport by truck to a ferry or well-boat	Flexibility with regard to the planning of the transport process	Water quality may deteriorate. Loading and unloading are stressful
	Well-boat	Fish may recover from loading stressors during transportation	Loading and unloading are stressful
Carp species Nile tilapia glass eel	Truck	Flexibility with regard to the planning of the transport process	Loading and unloading of the tanks is stressful. Water quality may deteriorate
Glass eels, yellowtail kingfish, other species	Aircraft	Suitable to cover long distances within a country or across continents	Loading and unloading of the transport bags/tanks can be stressful. Water quality may deteriorate

Table 14.4 Transport methods used for the transfer of Atlantic salmon, common carp, European eel, African catfish, rainbow trout, European sea bass and gilthead seabream to an abattoir for slaughter (Boerrigter et al. 2015; Manuel et al. 2014; Schrijver et al. 2017)

Fish species	Transport	Advantages	Disadvantages
Atlantic salmon Pangasius	Well-boat	Fish may recover from loading stressors during transportation	Loading and unloading are stressful. Water quality may deteriorate during closed transports
Salmon, blue fin tuna	Cage towed by boat	Stress caused by loading a vehicle is avoided	No data available
Grass carp, silver carp, common carp, bighead carp crucian carp Nile tilapia European eel African catfish Freshwater rainbow trout	Transport by truck	Flexibility with regard to the planning of the transport process	Water quality may deteriorate Loading and unloading are stressful
European sea bass/ gilthead Seabream	NA[a]	NA[a]	NA[a]

[a]*NA* not applicable, European sea bass/gilthead sea bream are not transported to a facility of slaughter

Purpose-built tanks on flatbed trucks are used to transport live fish. In China, live market-sized common carps are transported in tanks by truck or other vehicles (Anonymous 2006a). At restaurants and markets, these animals are kept alive for a few days in aquaria and at markets the fish are also sold as a fresh product (Schrijver et al. 2017). Although the literature on the current practices for transport of grass carp, silver carp, bighead carp and crucian carp is scarce, substantial numbers of these species are transported live to markets and restaurants in European and Asian countries. We assume that road transport is the dominant method.

For fry or fingerlings, two methods for road transport are available. These animals can be transported in insulated tanks at stocking densities that are appropriate for the species, with controlled temperature and oxygen levels (Dalla Villa et al. 2009). Fry or fingerlings are also transported in plastic bags, or in large tanks or vats (Sampaio and Freire 2016). Transport of fry in polyethylene bags with oxygen is widespread, and vehicles such as any motorized land vehicle or animal cart can be used for the live transport of fry or fingerlings in plastic bags (Anonymous 2006a). For instance, in China, common carp fingerlings are mainly transported in plastic bags with oxygen (Schrijver et al. 2017).

When using tanks to transport the fish, the tank water is aerated with compressed air and oxygen to prevent the build-up of carbon dioxide and avoid hypoxia, respectively. A combination of compressed air and pure oxygen is preferred, as this combination avoids the supersaturation of the water with oxygen. Supersaturation should not occur, as fish can be at risk of gas bubble disease. Probes should be used to monitor oxygen levels in the water during transport.

Transport in tanks or bags is a closed system, and carbon dioxide and TAN (total ammonia nitrogen, i.e. the sum of NH_3 and NH_4^+ concentrations) can accumulate in the water (Dalla Villa et al. 2009), which may cause stress in fish. Species-specific threshold values for TAN, which are also pH dependent, and carbon dioxide should not be exceeded. It is not possible to give an exact limit for the maximum transportation time in a closed system. This will depend on fish species, density, and temperature and water treatment in the transport unit.

With regard to fish welfare, attention should also be given to loading and unloading procedures. In general, fish are netted or pumped during loading and unloading. This can mean fish are subjected to air exposure or mechanical injury due to collisions with other fish, the netting materials, or the constituents of the pump and holding tanks, especially if the process is not carried out correctly. With regard to common carp, this fish is able to cope with a well-controlled exposure to air, whereas for Atlantic salmon risk thresholds of air exposure are currently not known and the RSPCA welfare standards (RSPCA 2018a) offer precautionary advice that air exposure should not exceed 15 s. When pumps are used, exposure to air can be avoided. It should be noted that the use of a fish pump should not result in high total gas pressure (TGP) and/or supersaturation of the water with nitrogen gas. This hazard can be avoided/minimised by degassing water in a tank on a truck on in a well-boat before loading fish (Rosten and Kristensen 2011).

14.4.1.2 Transport by Boat

Water quality in a well-boat is controlled by continuously pumping in new water or recirculating existing holding water. Well-boats are commonly used to transport market-sized Atlantic salmon to a facility for slaughter (Iversen et al. 2005). They are also used for the transfer of salmon smolts to sea cages for on-growing (Dalla Villa et al. 2009) or to transport European sea bass or gilthead seabream to sea cages for on-growing (Schrijver et al. 2017). Transport by well-boat can be performed with an open or closed haul system. As stated earlier, it should be noted that the use of a closed system may result in elevated levels of CO_2 (King 2009) and TAN in the water. In an open haul system, which is a flow-through system, the accumulation of these substances is avoided. The use of an open or a closed system depends upon whether there are biosecurity issues associated with the fish itself, local regulations or the risks of contamination with fish pathogens along the transport route (Rosten and Kristensen 2011). As stated earlier, two studies on the transport of Atlantic salmon smolts showed no negative impact on their welfare when a well-boat with an open system was used (Iversen et al. 2005; Nomura et al. 2009). Iversen et al. (2005) showed that the cortisol levels of Atlantic salmon smolts increased during the loading stage of transport but then decreased and returned to normal during the actual transport phase. However, if transport took place during rough weather conditions, plasma cortisol levels remained elevated (Iversen et al. 2005). A lowering of the water temperature during the transport of Atlantic salmon can reduce the impact of several factors on fish physiology and welfare (Lines and Spence 2012). The reduction in temperature must be carried out with great care and should not exceed 1.5 °C/h (RSPCA 2018a). For Atlantic salmon and rainbow trout, the temperature during transport should not be below 6 °C (Schrijver et al. 2017).

A well-boat is also used for the transport of pangasius to ponds for on-growing and also for the transport of market-sized pangasius to a land-based facility for processing. Another type of transport is towing sea cages with fish by boat. Some market-sized fish species can be transported in cages themselves, as reported for, e.g. Atlantic bluefin tuna (Reglero et al. 2013). In Newfoundland in Canada, salmon cages are towed to a land-based processing plant where the fish are pumped out, stunned by percussion and slaughtered (Schrijver et al. 2017).

14.4.1.3 Air Transport

When fish are traded between countries or have to be transported within large countries/across continents, air transport is used for fry or fingerlings. Fish are placed in plastic bags and these are inflated by oxygen from a cylinder. Subsequently, the bags are packaged in insulated polystyrene boxes for air transport (Dalla Villa et al. 2009; Sampaio and Freire 2016). Air transport is used for, e.g. glass eels (Dalla Villa et al. 2009) and yellowtail kingfish (pers. comm. Hans van de Vis).

14.4.2 Stunning and Killing

Slaughter is a common process used for the killing of food animals intended for human consumption. The term "slaughter" is also used to depict the killing of animals by bleeding. With regard to slaughter or killing, an important issue is whether the methods can be stressful or painful for the fish. The number of studies which indicate that fish are able to perceive pain and fear is increasing (see, e.g. Braithwaite et al. 2013; Braithwaite and Ebbesson 2014). Accordingly, to protect farmed fish at slaughter, these animals should be rendered unconscious and insensible by stunning to avoid pain, fear or distress prior to the procedure, which is a general provision in the EU legislation to protect animals at slaughter (Council Regulation (EC) No. 1099/2009). This general provision in the EU legislation for warm-blooded slaughter animals can be used as a general term of reference for the slaughter of farmed fish. The general term of reference is met when stunning induces an immediate loss of consciousness and sensibility in fish, which lasts until death or, when an instantaneous induction is not possible, the animal should be rendered unconscious and insensible, without causing avoidable pain, fear or distress.

All steps in the slaughter process should be examined when considering fish welfare. However, within the scope of this chapter, we will only focus on stunning and killing methods used for a wide range of farmed species (grass carp, silver carp, common carp, Nile tilapia, bighead carp, crucian carp, catla, Atlantic salmon, striped catfish—also known as striped pangasius or pangasius—, rainbow trout, gilthead seabream, European sea bass, tuna spp., African catfish, turbot and the European eel).

Specific Definitions Used in This Section

We use the following definitions in relation to stunning and slaughter:

Insensible—fish is unable to perceive (and as a consequence respond to) stimuli (Van de Vis et al. 2014).

Slaughter—the killing of animals, especially farmed ones, for the production of food (Van de Vis et al. 2014).

Killing—defined in council regulation (EC) no. 1099/2009 as any intentionally induced process which causes the death of an animal.

Stunning—any intentionally induced process which causes the loss of consciousness and sensibility without pain, including any process resulting in instantaneous death (council regulation (EC) no. 1099/2009, 2009). When stunning is reversible it should be followed by the application of a killing method. In case of an irreversible stun, the application of a method also induces death.

Unconsciousness—a state of unawareness (loss of consciousness) in which the brain is unable to process sensory input, e.g. during (deep) sleep, anaesthesia or due to temporary or permanent damage to brain function (Van de Vis et al. 2014).

Various studies have reported that slaughter can be stressful for fish, as not all farmed fish species are stunned effectively prior to killing or not at all (see for instance Van de Vis and Lambooij 2016). A brief overview of stunning and killing methods is given in Table 14.5. It is known that behavioural measures alone are insufficient to assess the level of brain function of fish unequivocally (Van de Vis et al. 2014). Hence, an integrated approach is needed to assess whether fish are stunned. A two-step approach, as recommended by EFSA (2018), is needed to establish specifications to protect fish at slaughter (i.e. by stunning them prior to killing) and to subsequently put these results into practice. EFSA's two-step approach is: (1) to establish specifications in a laboratory setting to protect fish at slaughter and (2) to evaluate the subsequent implementation of the results in practice or under similar conditions. It should be noted that live chilling of, e.g. Atlantic salmon can be used to calm fish prior to stunning (EFSA 2009a). As it is not a stunning method for this fish species, live chilling of Atlantic salmon is not be presented in Table 14.5.

To establish whether the general term of reference is met after the application of a stunning method, the onset and duration of unconsciousness and insensibility in fish have to be assessed in a laboratory setting (step 1). Behavioural measures alone may be insufficient to assess the level of brain function of fish unequivocally. By using an EEG, a trained observer can monitor the electrical activity in the brain. In addition, stimuli are administered to determine whether the stunned fish can be roused, both on the EEG and behaviourally. When the fish do not respond, this implies that the fish remain unconscious and are insensible until death occurs. The electrical activity of the heart (as determined using an ECG) provides additional information for the assessment of stunning and killing methods, both EEG and ECG and LABWIs. It should be noted that the levels of stress hormones in fish or product quality parameters cannot serve as robust readout parameters for the loss of consciousness and sensibility in fish.

Step 2 is performed to assess whether the equipment for the stunning and killing of fish in practice meets the established criteria in step 1 for stunning and killing. In this step, the following OWIs are relevant: physical measurements in combination with behavioural observations and visual inspection of product quality parameters. For electrical stunning, the strength of the electrical current (in water it is the height of the current density), its waveform, the applied voltage (in water it is the field strength) and duration of exposure of fish to the electricity need to be established, as well as the time interval between fish leaving the stunner and the application of a killing method. For electrical stunning, a killing method needs to be applied to prevent the recovery of the stunned fish. When percussion is applied, the air pressure, which drives the bolt, should be checked to make sure it is sufficiently high and that the blow is given correctly to the head of each fish. For both electrical and percussive stunning carcass damage should be prevented or minimized. Well-trained staff should be available at the slaughter facility to control the process of stunning and killing.

Table 14.5 Overview of methods used for stunning or stunning and killing (Anonymous 1997, 2006b; EFSA 2009a, b, e; Erikson 2011; Lines and Spence 2012, 2014; Roth et al. 2007; Sattari et al. 2010; Schrijver et al. 2017; Van de Vis and Lambooij 2016)

Stunning or stunning and killing	Fish species	Advantage	Disadvantage
Electrical stunning (reversible stun)	Atlantic salmon, pangasius, rainbow trout, common carp, European eel, African catfish	An immediate stun can be achieved. Allows pre-rigor filleting in salmon For pangasius: Very few data available	An effective killing method is needed to avoid recovery Product quality can be affected; mis-stuns[a] may occur due to varying resistance among fish
Percussion (irreversible stun)	Atlantic salmon	An immediate stun can be achieved When applied correctly, no recovery; the method stuns and kills Allows pre-rigor filleting	Mis-stuns due to variation in size Damage to the head may occur
	Common carp	When applied correctly, no recovery; the method stuns and kills	Manual application may lead to mis-stuns Damage to the head may occur
	Pangasius	Very few data available	
	Rainbow trout	When applied correctly, no recovery; the method stuns and kills	Manual application may lead to mis-stuns Damage to the head may occur
Spiking/coring (Iki jime) or lupara (underwater shooting) (both irreversible stun)	Tuna spp.	Spiking/coring for small tuna and underwater and lupara for large tuna is preferred to spiking/coring on board and shooting from the surface	These methods have not been assessed with EEGs When lupara is used, a back-up diver is required in case a second shot is needed
CO_2 stunning	Rainbow trout Atlantic salmon		Distress. Fish exhibit vigorous avoidance and panic behaviour. Its use is banned in Norway
Clove oil and Aqui-S™		EEG registration shows that Atlantic cod is stunned effectively	Not allowed in the European Union and Norway for the slaughtering of fish Its use may lead to an off-odour in fish

[a]A mis-stun occurs when the application of a stunning method is not effective. For electrical and percussive stunning, this implies that consciousness is not lost immediately

14.4.2.1 Electrical Stunning

For the electrical stunning of fish, two methods are available: (1) in water or (2) after dewatering. The key factor is that a sufficient current should be passed through the brain of the fish and to achieve this an adequate voltage across electrodes after dewatering or an adequate field strength in water needs to be applied. It should be noted that the waveform, the orientation of fish in a stunner and the density of fish (kg/l) impact the effectiveness of an electrical stun. Mis-stuns may occur due to varying resistance among fish in a stunner. Nevertheless, the introduction of electrical stunning significantly reduces stress in salmon during the slaughter process, compared to CO_2 stunning.

Brain stimulation by the application of electricity should induce a generalised epileptiform activity or an immediate onset of a quiescent EEG (as judged from EEG recordings), and these patterns are indicative of unconsciousness and insensibility. The evoked electrical activity (somatosensory or visual) in the brain is also abolished during the manifestation of epileptiform activity and quiescent EEG (EFSA 2018).

Electrocution is stunning and killing by the use of electricity. Electrocution induces death by cardiac arrest. However, existing data for, e.g. Atlantic salmon, European sea bass, European eel, turbot, and common carp show that these animals cannot be killed by the use of electricity alone, as the fibrillation of the heart is not permanent (see for instance Van de Vis et al. 2014). Hence, electrical stunning of fish needs to be followed by a further killing method.

For Atlantic salmon, freshwater rainbow trout, European sea bass, gilthead seabream and European eel both methods for electrical stunning are used at facilities for slaughtering fish, whereas for common carp, electricity is only applied in water. For African catfish and pangasius, electricity is only used after dewatering. The electrical stunning of turbot is still experimental. Regarding pangasius, published data on electrical stunning after dewatering are lacking; however, some practical experience is available (Lines and Spence 2012).

14.4.2.2 Percussion

Percussion is a manually or automatically applied blow to the head of a fish. At commercial slaughterhouses for Atlantic salmon, an automatic device is commonly used for percussive stunning. To the authors' knowledge, percussion is applied manually to common carp and rainbow trout (Table 14.5). In general, fish are removed from water prior to being subjected to percussion. The main hazard for automated percussive stunning is related to variation in the size of fish within the population, causing a mis-stun in some fish. It should be noted that the implementation of percussive stunning at slaughterhouses has significantly improved the welfare of salmon at slaughter, compared to, e.g. CO_2 stunning.

On the EEGs, the appearance of theta, delta waves and spikes, followed by an iso-electric EEG, indicates unconsciousness and insensibility (Lambooij et al. 2007).

The recovery of percussed Atlantic salmon is prevented due to cerebral haemorrhage (Lambooij et al. 2010). Very little information is available on percussive stunning of pangasius. Only one paper reports that some practical experience is available (Lines and Spence 2012).

14.4.2.3 Spiking/Coring/Lupara

For tuna, spiking/coring under water and underwater shooting (lupara) are preferred. Shooting from the surface and spiking or coring on board, which requires crowding, hosting or gaffing, can both have a severe effect on the welfare of these fish (EFSA 2009a). In a laboratory study, the method used for the coring of tuna was applied for Atlantic salmon. For this study, a hollow bolt was constructed and the registration of EEGs showed that consciousness was not lost immediately in Atlantic salmon (Robb et al. 2000). EEGs have not been recorded in tuna.

14.4.2.4 Clove Oil/Aqui-S™

Clove oil/Aqui-S™ is not allowed for the stunning/slaughtering of fish in the EU and Norway. The costs of overcoming the legislative requirements form a barrier for its application at slaughter in the EU and Norway (Van de Vis et al. 2014). In a study on Atlantic cod, EEG recordings showed that Aqui-S™ renders this species unconscious, as judged from the appearance of theta and delta waves on the EEG (Erikson et al. 2012).

14.4.2.5 Live Chilling with Low to Moderate Levels of CO_2

Live chilling in combination with added carbon dioxide is no longer used for the slaughter of Atlantic salmon in Norway. When it was used previously, the fish were exposed to a temperature of −0.5–3 °C and carbon dioxide was added to achieve low and moderate levels (65–257 mg/l) whilst maintaining oxygen at levels of 70–100% saturation (Erikson 2011). In a commercial setting, the water was re-used. This method was followed by gill-cutting and bleeding in chilled seawater. An assessment of live chilling with added carbon dioxide showed that this method is stressful and does not stun the fish (Erikson 2011).

14.4.2.6 CO_2 Stunning

Carbon dioxide stunning has been banned in Norway (Anonymous 2006b). It is known that the method is highly stressful for Atlantic salmon (Robb et al. 2000) and other fish species. Nevertheless, it is still used in some other countries for the killing of Atlantic salmon and rainbow trout (Schrijver et al. 2017). When this method is applied, carbon dioxide is bubbled into a tank with seawater until a pH of approximately 5.5–6.0 is obtained. This corresponds to CO_2 levels of 200–450 mg/l. Subsequently, the fish are placed in the water and after 2–4 min the struggling stops (EFSA 2009e). In the next step, the fish are taken from the tank and bled.

14.4.2.7 Ice or Ice Water Slurry

The transfer of fish into ice flakes or crushed ice leads to asphyxia in rainbow trout, European sea bass and gilthead seabream (EFSA 2009c). For the carp species listed in Table 14.6 and turbot (EFSA 2009b), the steep drop in temperature may have a bigger impact on these species than the lack of oxygen. It is known that carp species are more tolerant to low levels of oxygen than sea bass and seabream. EFSA reported that it may take 4 h for turbot to die in ice (EFSA 2009d). Crucian carp is exceptional, as it can survive long periods of anoxia. This carp species is able to cope with the acid–base consequences of anaerobiosis, as it is able to convert anaerobically produced lactic acid into ethanol, which can freely diffuse across the gills into the ambient water (Shoubridge and Hochachka 1980). No data could be

Table 14.6 An overview of methods used for killing without stunning (Erikson 2011; EFSA 2009d, f; FAO 2018b, f; Lines and Spence 2012, 2014; Robb and Kestin 2002; Roth et al. 2007; Sattari et al. 2010; Schrijver et al. 2017; Van de Vis et al. 2003; Van de Vis and Lambooij 2016)

Killing without stunning	Fish species	Advantage	Disadvantage
Live chilling with low to moderate levels of CO_2 followed by a gill-cut	Atlantic salmon	Food quality and safety	Fish are not stunned prior to killing by gill-cutting. Fish behavioural indicators suggest the method is probably stressful
Ice or ice/water slurry	European sea bass Gilthead seabream Rainbow trout Turbot Catla Grass carp Nile tilapia	Easy to use Food quality and safety	Distress in fish due to steep drop in temperature. The occurrence of asphyxia is fish species dependent
Exsanguination in ice water	Turbot, pangasius	Easy to use Food quality and safety	Distress in fish due to steep drop in temperature. Likely to result in asphyxia
Asphyxia in air	Abandoned for gilthead seabream and rainbow trout	Easy to use	Distress in fish. Attempts to escape from a dry tub or tank may lead to carcass damage
Salt or ammonia	European eel		Distress. It has been banned in Germany since 1997 and in the Netherlands as of 1 July 2018

found whether crucian carp and bighead are slaughtered by transferring the species into ice. For catla and grass carp, ice is used (FAO 2018b; Scherer et al. 2005).

An ice water slurry is a mixture of ice and water and is another common method for the killing of some farmed fish species (see Table 14.6). Depending on the fish species, a transfer into ice water may also lead to asphyxia. The reason is that the density of fish is substantially increased without any aeration or addition of oxygen to the slurry. Live European sea bass and gilthead seabream can be put into tubs in a slurry/fish ratio of 2:1 (EFSA 2009c). The ratio of flake ice: water in the slurry ranges from 1:2 to 3:1. This implies that at the start of the transfer of fish into ice water the density ranges from 430 to 660 kg of fish/1000 l water. It is known that both fish species exhibit a vigorous behavioural response to the steep drop in temperature. Due to the transfer, the density also increases from 20 kg fish/1000 l in a sea cage (EFSA 2008f) to 430–660 kg/1000 l for both species. This increase in density by a factor of 22–33 without any supplemented oxygen will likely lead to asphyxia in both sea bass and seabream that make vigorous attempts to escape. It should be noted that even after a complete transition from flake ice into water, the density increases by a factor of 17, compared to the density in a sea cage. In the view of EFSA, this method poses a severe welfare risk (EFSA 2009c). EFSA reported that the transfer of rainbow trout into a mixture of ice and water also results in asphyxia (EFSA 2009d).

Nile tilapia and carp species are known for their capacity to survive in an environment with lower levels of oxygen than sea bass and sea bream. Hence, it is likely that the steep drop in temperature leads to distress in Nile tilapia and carp species, instead of low levels of oxygen per se. No reported data could be found whether the ice slurry is used for the slaughtering of crucian carp and bighead carp. The response of the fish brain to a cold shock has been studied in common carp. For this species, exposure to a temperature drop of 10 °C was studied by functional magnetic resonance imaging. Van den Burg et al. (2005) found that after 90 s the brain blood volume was reduced. Limiting the entry of cold oxygen-rich blood from the gills leads to a slower decrease in brain temperature. This slows the rate of cooling of the brain due to the cold shock and thereby prolongs the period of consciousness. In the view of EFSA, adverse effects are apparent (EFSA 2009g).

14.4.2.8 Exsanguination in Ice Water

Turbot are exsanguinated in a slurry of ice and seawater. EFSA (2009b) reported that these animals showed escape behaviour and other responses to physical handling. A study by Morzel et al. (2002) revealed that behavioural responses in turbot were lost within 15–30 min after bleeding in an ice slurry. EFSA (2009b) concluded that exsanguination combined with exposure to an ice slurry constitutes a considerable welfare risk. Most pangasius are killed by bleeding and chilling in ice water (Lines and Spence 2012).

14.4.2.9 Asphyxia in Air

Asphyxia in air is traditionally used for captured fish (Poli et al. 2005). Prolonged air exposure collapses the gill tissue, hindering respiration. However, fish do not lose consciousness rapidly. The reason is that a variety of strategies, which are species dependent, occur in fish to cope with asphyxia to some extent. A common strategy is decreasing metabolic rate, hypometabolism (Jackson 2004). The time to lose consciousness in air is affected by the ambient temperature; for example, rainbow trout loses consciousness after 2.6 min exposure in air at 20 °C, whereas it took 9.6 min at 2 °C (Kestin et al. 1991). Asphyxia in air has earlier been used for the killing of rainbow trout (Robb and Kestin 2002), but this method is no longer used for rainbow trout in European aquaculture (Schrijver et al. 2017) due to welfare concerns. For gilthead seabream, it can take 5.5 min before consciousness is lost after exposure to air at 22 °C. The method has now been abandoned for the slaughtering of gilthead seabream in the Mediterranean (Van de Vis et al. 2003).

14.4.2.10 Salt Bath or Ammonia to De-slime Conscious European Eel

Van de Vis and Lambooij (2016) give a brief overview of the use of ammonia or salt for the killing of conscious European eels. It is known that these animals make extremely vigorous attempts to escape from a salt bath or ammonia. EEG recordings showed that it may take longer than 10 min to lose consciousness and sensibility, based on the time to induce unconsciousness in eels that are exposed to salt. The process of de-sliming by salt or ammonia causes an osmotic shock that kills these animals. It is, however, possible that the de-slimed eels are eviscerated while still conscious. In the view of EFSA (2009f), this method results in severe stress and pain. As of the first of July 2018, the use of salt to kill conscious eels has been banned in the Netherlands.

14.4.2.11 Live Chilling

Live chilling of Atlantic salmon is not included in Tables 14.5 or 14.6 as its use it not used for killing. Live chilling of Atlantic salmon in water whilst the fish is supplied with sufficient oxygen is used to calm fish prior to stunning (EFSA 2009a). It has previously been reported that a controlled decrease in temperature from 16 to 4 °C in 1 h and 16 to 0 °C in 5 h did not result in a significant increase in plasma cortisol levels in Atlantic salmon (Foss et al. 2012).

14.5 Summary, Conclusion, Future Directions

The aim of this chapter is to give the reader a general overview of both the mutual and system- specific welfare challenges that farmed fish face in different rearing systems, in addition to potential mitigation strategies. We have also included an overview of the tools the farmer can use to assess these threats and evaluate the efficacy of any welfare actions. We have also utilised this approach with two key operations that are common across all aquaculture production sectors: transport and slaughter practices.

In terms of differing rearing systems, the largest difference in risk factors is between land-based recirculating aquaculture systems with limited water supply and open rearing systems where water passes through them. In open systems, dissolved and particulate wastes are removed when the system is replenished with new water. In systems where the water is recirculated, dissolved or particulate wastes require some mechanical or biological treatments. However, closed or semi-closed systems offer numerous benefits with regard to the level of control the farmer has and also in relation to biosecurity, waste management and treatment and, e.g. escape prevention. The recent development of S-CCS systems for deployment in existing water bodies also introduces some control of the water environment within the rearing system, whilst also introducing some challenges that are seen in land-based flow through or RAS associated with the water supply. For example, when water supply is recirculated or limited, there is an increased possibility for the accumulation of dissolved waste products from the fish, e.g. CO_2 and ammonia. And if the water supply is drawn from soft water with low alkalinity, the accumulation of CO_2 can lead to a reduction of water pH, which increases the risk of metal toxicity (e.g. aluminium toxicity). This can then lead to a decrease in blood oxygen-carrying capacity and reduced growth (Kolarevic et al. 2018). The uncontrolled flow of water through open systems, such as net cages, also creates risks. For example, the largest limiting factor for the continued growth of salmon aquaculture in Norway is, for instance, the various pathogens that are brought by the currents, such as sea lice and various gill-infecting amoeba and viruses (Palić et al. 2017). As described previously, when fallowing is used in combination with the systematic surveillance of Atlantic salmon sea cages, the prevalence of diseases can be reduced.

The section on the transport of fry/fingerlings and also on market-sized farmed fish shows that detailed protocols to safeguard and monitor fish during transport are lacking for numerous species, with some notable exceptions such as the comprehensive RSCPA welfare standards for both farmed rainbow trout and Atlantic salmon in the UK (RSPCA 2018a, b). For other species, such protocols need to consider (1) the species and their needs, which also depend on their life stage, with respect to, e.g. water quality, (2) handling of fish during loading and unloading of a transport vehicle, (3) transport duration, especially for fish in a closed system, in combination with road conditions and at sea weather conditions and (4) the training of staff responsible for handling and transportation of live fish. With regard to stunning and killing, an integral approach should be applied to protect farmed fish at slaughter. The first step is to establish specifications for effective stunning without recovery. In the next step, the implementation of the results obtained in a laboratory setting should be evaluated under commercial conditions (as recommended by EFSA 2018). Thirdly, adequately trained staff should be available to control the process of stunning and killing.

Better monitoring and auditing of both the fish and the environment they are reared in, or the operation they are subjected to, will also help the farmer and stakeholder to circumvent some welfare challenges by providing an early warning of potential threats and give useful learning opportunities and experience for future decisions that may impact fish welfare. It should be noted that in the aquaculture production chain many preventive measures are already taken to reduce welfare threats, e.g. providing guidance on species-specific water quality requirements and also guidance for handling and other operations (e.g. EFSA 2008a, b, c, d, e, 2009a, b, c, d, e, f, g; RSPCA 2018a, b; Noble et al. 2018).

As with all animal production systems, there are still processes throughout the aquaculture chain that can be improved with regard to fish welfare. During the last decades, the aquaculture industry and other stakeholders such as NGOs have made initiatives to communicate their achievements with respect to the sustainability of production of farmed fish to a wide audience in a clear manner (Gamborg and Sandoe 2005) by, e.g. the labelling of fish products for consumers. However, with the notable exception of the RSPCA Assured welfare label for farmed Atlantic salmon and rainbow trout, to the authors' knowledge, there are no other consumer label initiatives or a government label such as the EU label for organic aquaculture products that specify, in detail, how the welfare of other farmed species are assured throughout the aquaculture production chain.

References

Anonymous (1997) Verordnung zum Schutz von Tieren in Zusammenhang mit der Schlachtung oder Tötung – TierSchlV (Tierschutz-Schlachtverordnung), vom 3. März 1997, Bundesgesetzblatt Jahrgang 1997 Teil I S. 405, zuletzt geändert am 13. April 2008 durch Bundesgesetzblatt Jahrgang 2008 Teil I Nr. 18, S. 855, Art. 19 vom 24. April 2006

Anonymous (2006a) Transporting fish. https://thefishsite.com/articles/transporting-fish

Anonymous (2006b) Forskrift om slakterier og tilvirkingsanlegg for akvakulturdyr Kapittel 4. In: kystdepartementet F-O (ed) Nasjonale tilleggsbestemmelser om fiskevelferd. Oslo, pp 13–14

Anonymous (2015). http://worldwideaquaculture.com/quick-easy-fish-farming-the-raceway-aquaculture-system

Anonymous (2017) China fishery statistical yearbook 2017. China Agriculture Press. ISBN: 9787109229419

Ashley PJ (2007) Fish welfare: current issues in aquaculture. Appl Anim Behav Sci 104:199–235

Belle SM, Nash CE (2008) Better management practices for net-pen aquaculture. In: Tucker CS, Hargreaves JA (eds) Environmental best management practices for aquaculture. Blackwell, Ames, pp 261–330

Beveridge MCM (2004) Cage aquaculture, 3rd edn. Blackwell, Oxford

Blancheton JP, Piedrahita R, Eding EH, Roque D'orbcastel E, Lemarie G, Bergheim A, Fivelstad S (2007) Intensification of landbased aquaculture production in single pass and reuse systems. In: Aquaculture engineering and environment (Chapter 2)
Boerrigter JG, Manuel R, Bos R, Roques JA, Spanings T, Flik G, Vis HW (2015) Recovery from transportation by road of farmed European eel (*Anguilla anguilla*). Aquac Res 46:1248–1260
Bostock J, McAndrew B, Richards R, Jauncey K, Telfer T, Lorenzen K, Little D, Ross L, Handisyde N, Gatward I, Corner R (2010) Aquaculture: global status and trends. Philos Trans R Soc B 365:2897–2912
Boyd CE, Tucker CS (1998) Pond aquaculture water quality management. Kluwer Academic, Springer Science
Braithwaite V, Ebbesson LO (2014) Pain and stress responses in farmed fish. Rev Sci Tech 33:245–253
Braithwaite V, Huntingford F, Van den Bos R (2013) Variation in emotion and cognition among fishes. J Agric Environ Ethics 26:7–23
Branson EJ (2008) Fish welfare. Blackwell, Oxford, 300 p
Bregnballe J (2015) A guide to recirculation aquaculture. Copenhagen, Eurofish, p 96
Brett JR (1964) The respiratory metabolism and swimming performance of young sockeye salmon. J Fish Res Board Can 21:1183–1226
Chen C-Y, Wooster GA, Getchell RG, Bower PR, Timmons MB (2001) Nephrocalcinosis in Nile Tilapia from a recirculation aquaculture system: a case report. J Aquat Anim Health 134:368–372
Cho K, Sakamoto J, Noda T, Nishiguchi T, Ueno M, Yamasaki Y, Yagi M, Kim D, Oda T (2016) Comparative studies on the fish-killing activities of *Chattonella marina* isolated in 1985 and *Chattonella antiqua* isolated in 2010, and their possible toxic factors. Biosci Biotechnol Biochem 80:811–817
Chopin T, Cooper JA, Reid G, Cross S, Moore C (2012) Open-water integrated multi-trophic aquaculture: environmental biomitigation and economic diversification of fed aquaculture by extractive aquaculture. Rev Aquac 4:209–220
Commission Regulation (EC) No 889/2008 (2008) Laying down detailed rules for implementation of Council Regulation (EC) No 834/2007 on organic production and labelling of organic products with detailed rules on production, labelling and control. Off J Eur Union, L 250:1–84
Council Regulation (EC) No 1099/2009 (2009) On the protection of animals at the time of killing. Off J Eur Communities, L 303:1–30
Da Silva JM, Coimbra J, Wilson JM (2009) Ammonia sensitivity of the glass eel (*Anguilla anguilla* L.): salinity dependence and the role of branchial sodium/potassium adenosine triphosphatase. Environ Toxicol Chem 28:141–147
Dalla Villa P, Marahrens M, Velarde A, Calvo A, Di Nardo A, Kleinschmidt N, Fuentes Alvarez C, Truar A, Di Fede E, Otero JL Müller-Graf C (2009) Final report on project to develop animal welfare risk assessment guidelines on transport-project developed on the proposal CFP/EFSA/AHAW/2008/02, 127 pp
Davidson J, Good C, Welsh C, Summerfelt S (2011) Abnormal swimming behavior and increased deformities in rainbow trout *Oncorhynchus mykiss* cultured in low exchange water recirculating aquaculture systems. Aquac Eng 45:109–117
Dempster T, Korsøen O, Folkedal O, Juell JE, Oppedal F (2009) Submergence of Atlantic salmon (*Salmo salar* L.) in commercial scale sea-cages: a potential short-term solution to poor surface conditions. Aquaculture 288:254–263
Edwards P (2008) The changing face of pond aquaculture in China. Glob Aquac Advocate 77–80. http://pdf.gaalliance.org/pdf/GAA-Edwards-Sept08.pdf
EFSA (2004) Opinion of the scientific panel on animal health and welfare (AHAW) on a request from the commission related to the welfare of animals during transport. Question N° EFSA-Q-2003-094. EFSA J 44, 181 pp

EFSA (2008a) Scientific opinion of the panel on animal health and welfare on a request from the European Commission on animal welfare aspects of husbandry systems for farmed Atlantic salmon. EFSA J 736:1–31

EFSA (2008b) Scientific opinion of the panel on animal health and animal welfare on a request from the European Commission on the animal welfare aspects of husbandry systems for farmed trout. EFSA J 796:1–22

EFSA (2008c) Scientific opinion of the panel on animal health and welfare on a request from the European Commission on animal welfare aspects of husbandry systems for farmed European seabass and gilthead seabream. EFSA J 844:1–21

EFSA (2008d) Scientific opinion of the panel on animal health and welfare on a request from the European Commission on animal welfare aspects of husbandry systems for farmed European eel. EFSA J 809:1–18

EFSA (2008e) Scientific opinion of the panel on animal health and welfare on a request from the European Commission on animal welfare aspects of husbandry systems for farmed fish: carp. EFSA J 843:1–28

EFSA (2008f) Welfare aspects of husbandry systems for farmed European seabass and gilthead seabream. EFSA J 844:1–89

EFSA (2009a) Species-specific welfare aspects of the main systems of stunning and killing of farmed tuna. EFSA J 1072:1–53

EFSA (2009b) Species-specific welfare aspects of the main systems of stunning and killing of farmed turbot. EFSA J 1073:1–34

EFSA (2009c) Species-specific welfare aspects of the main systems of stunning and killing of farmed sea bass and sea bream. EFSA J 1010:1–52

EFSA (2009d) Species-specific welfare aspects of the main systems of stunning and killing of farmed rainbow trout. EFSA J 1013:1–55

EFSA (2009e) Species-specific welfare aspects of the main systems of stunning and killing of farmed Atlantic salmon. EFSA J 1012:1–77

EFSA (2009f) Species-specific welfare aspects of the main systems of stunning and killing of farmed eel (*Anguilla anguilla*). EFSA J 1014:1–42

EFSA (2009g) Species-specific welfare aspects of the main systems of stunning and killing of farmed carp. EFSA J 1013:1–37

EFSA (2018) Guidance on the assessment criteria for applications for new or modified stunning methods regarding animal protection at the time of killing. EFSA J 16(7):5343, 35 pp

Erikson U (2011) Assessment of different stunning methods and recovery of farmed Atlantic salmon (*Salmo salar*): isoeugenol, nitrogen and three levels of carbon dioxide. Anim Welf 20:365–375

Erikson U, Lambooij B, Digre H, Reimert HGM, Bondø, Van de Vis H (2012) Conditions for instant electrical stunning of farmed Atlantic cod after de-watering, maintenance of unconsciousness, effects of stress, and fillet quality – a comparison with Aqui-S™. Aquaculture 324–325:135–144

Espmark ÅM, Baeverfjord G (2009) Effects of hyperoxia on behavioural and physiological variables in farmed Atlantic salmon (*Salmo salar*) parr. Aquac Int 17:341–353

FAO (1984) Inland aquaculture engineering. FAO, Rome. ISBN: 92-5-102168-6. http://www.fao.org/docrep/X5744E/x5744e0e.htm

FAO (2017) Fisheries and aquaculture software. FishStat Plus – Universal software for fishery statistical time series. In: FAO Fisheries and Aquaculture Department [online]. Rome. Updated 14 September 2017

FAO (2018a) FAO yearbook. Fishery and aquaculture statistics 2016. FAO, Rome, p 108. ISBN: 9789250099873. http://www.fao.org/3/i9942t/I9942T.pdf

FAO (2018b) Cultured aquatic species information programme – *Catla catla* (Hamilton, 1822), 11 pp. http://www.fao.org/fishery/culturedspecies/Catla_catla/en

FAO (2018c) Cultured aquatic species information programme *Carassius carassius* (Linnaeus, 1758), 9 pp. http://www.fao.org/fishery/culturedspecies/Carassius_carassius/en

FAO (2018d) Cultured aquatic species information programme *Hypophthalmichthys nobilis* (Richardson, 1845), 10 pp. http://www.fao.org/fishery/culturedspecies/Hypophthalmichthys_nobilis/en

FAO (2018e) Cultured aquatic species information programme – *Hypophthalmichthys molitrix* (Valenciennes, 1844), 9 pp. http://www.fao.org/fishery/culturedspecies/Hypophthalmichthys_molitrix/en

FAO (2018f) Cultured aquatic species information programme, 12 pp. *Oreochromis niloticus* (Linnaeus, 1758). http://www.fao.org/fishery/culturedspecies/Oreochromis_niloticus/en

FEAP (2018). www.feap.info/Default.asp?CAT2=0&CAT1=0&CAT0=0&SHORTCUT=590, visited May 2018

Foss A, Grimsbo E, Vikingstad E, Nortvedt R, Slinde E, Roth B (2012) Live chilling of Atlantic salmon: physiological response to handling and temperature decrease on welfare. Fish Physiol Biochem 38:565–571

Funge-Smith S, Phillips MJ (2001) Aquaculture systems and species. In: Aquaculture in the third millennium. In: Subasinghe P, Bueno MJ, Phillips C, Hough SE, McGladdery, Arthur JR (eds) Technical proceedings of the conference on aquaculture in the third millennium, Bangkok, 20–25 Feb 2000, pp 129–135. NACA, FAO, Bangkok, Rome

Gamborg C, Sandoe P (2005) Sustainability in farm breeding: a review. Livest Prod Syst 92:221–231

Haenen OLM, Way K, Bergmann SM, Ariel E (2004) The emergence of koi herpesvirus and its significance to European aquaculture. Bull Eur Assoc Fish Pathol 24:293–307

Hallegraeff GM (1993) A review of harmful algal blooms and their apparent global increase. Phycologia 32:79–99

Hallegraeff GM (2003) Harmful algal blooms: a global overview. Manual on harmful marine microalgae. Monogr Oceanogr Methodol 11:25–49

Handeland SO (2016) Postsmoltproduksjon I semi-lukkede anlegg; Resultat fra en komparativ feltstudie. Fjerde konferanse om resirkulering av vann i akvakultur på Sunndalsøra, 25–26 oktober 2016

Handeland S Vindas M, Nilsen T, Ebbesson L, Sveier H, Tangen S, Nylund A (2015) Documentation of post smolt welfare and performance in large-scale Preline semi-containment system (CCS). CtrlAQUA annual report, pp 60–64

Heller M (2017) Food product environmental footprint literature summary: land-based aquaculture. Center for Sustainable Systems, University of Michigan, 17 pp

Hilbig R, Anken RH, Bauerle A, Rahmann H (2002) Susceptibility to motion sickness in fish: a parabolic aircraft flight study. J Gravit Physiol 9:29–30

Hjeltnes B, Bornø G, Jansen MD, Haukaas A, Walde C (eds) (2017) Fiskehelserapporten 2016. Oslo, Veterinærinstituttet, p 121

Horváth L, Urbányi B (2000) Fish species bred in Hungary. In: Horváth L (ed) Fish biology and fish breeding. Mezőgazda Kiadó, Budapest, pp 229–343 (in Hungarian)

Huntingford FA, Adams C, Braithwaite VA, Kadri S, Pottinger TG, Sandøe P, Turnbull JF (2006) Current issues in fish welfare. J Fish Biol 68:332–372

Hvas M, Folkedal O, Imsland A, Oppedal F (2017a) The effect of thermal acclimation on aerobic scope and critical swimming speed in Atlantic salmon, *Salmo salar*. J Exp Biol 220:2757–2764

Hvas M, Folkedal O, Solstorm D, Vågseth T, Gansel LA, Oppedal F (2017b) Assessing swimming capacity and schooling behaviour in farmed Atlantic salmon *Salmo salar* with experimental push-cages. Aquaculture 473:423–429

Hvas M, Karlsbakk E, Mæhle S, Wright DW, Oppedal F (2017c) The gill parasite *Paramoeba perurans* compromises aerobic scope, swimming capacity and ion balance in Atlantic salmon. Conserv Physiol 5, cox066

Iversen M, Finstad B, McKinley RS, Eliassen RA, Carlsen KT, Evjen T (2005) Stress responses in Atlantic salmon (*Salmo salar* L.) smolts during commercial well boat transports, and effects on survival after transfer to sea. Aquaculture 243:373–382

Jackson DC (2004) Acid-base balance during hypoxic hypometabolism: selected vertebrate strategies. Respir Physiol Neurobiol 141:273–283

Jena AK, Biswas P, Saha H (2017) Advanced farming systems in aquaculture: strategies to enhance the production. Innov Farming 2:84–89

Johansson D, Laursen F, Fernö A, Fosseidengen JE, Klebert P, Stien LH, Vågseth T, Oppedal F (2014) The interaction between water currents and salmon swimming behaviour in sea cages. PLoS One 9:e97635

Jokumsen A, Svendsen LM (2010) Farming of freshwater rainbow trout in Denmark. Charlottenlund: DTU aqua. Institut for Akvatiske Ressourcer. DTU Aqua-rapport; no. 219-2010

Karlsen C, Sørum H, Willasses NP, Åsbakk K (2012) Moritella viscosa bypasses Atlantic salmon epidermal keratocyte clearing activity and might use skin surfaces as a port of infection. Vet Microbiol 154:353–362

Kestin SC, Wotton SB, Gregory NG (1991) Effect of slaughter by removal from water on visual evoked activity in the brain and reflex movement of rainbow trout (*Oncorhynchus mykiss*). Vet Rec 128:443–446

King HR (2009) Fish transport in the aquaculture sector: an overview of the road transport of Atlantic salmon in Tasmania. J Vet Behav 4:163–168

Kolarevic J, Bæverfjord G, Takle H, Ytteborg E, Megård Reiten BK, Nergård S, Terjesen BF (2014) Performance and welfare of Atlantic salmon smolt reared in recirculating or flow through aquaculture systems. Aquaculture 432:15–25

Kolarevic J, Espmark AM, Aas-Hansen Ø, Terjesen BF, Saether BS (2015) Real time monitoring of water quality and fish welfare in recirculation aquaculture systems (RAS). Aquaculture Europe 2015, Rotterdam, 20–23 October 2015

Kolarevic J, Aas-Hansen Ø, Espmark ÅM, Baeverfjord G, Terjesen BF, Damsgård B (2016) The use of acoustic acceleration transmitter tags for monitoring of Atlantic salmon swimming activity in recirculating aquaculture systems (RAS). Aquacult Eng 72–73:30–39

Kolarevic J, Stien LH, Espmark ÅM, Izquierdo-Gomez D, Sæther B-S, Nilsson J, Oppedal F, Wright DW, Nielsen KV, Gismervik K, Iversen MH, Noble C (2018) Velferdsindikatorer for oppdrettslaks: Hvordan vurdere og dokumentere fiskevelferd – Del B. Bruk av operative velferdsindikatorer for ulike produksjonssystem. In: Noble C, Nilsson J, Stien LH, Iversen MH, Kolarevic J, Gismervik K (eds) Velferdsindikatorer for oppdrettslaks: Hvordan vurdere og dokumentere fiskevelferd, pp 142–223. ISBN: 978–82–8296-531-6

Koolhaas JM, Bartolomucci A, Buwalda BD, De Boer SF, Flügge G, Korte SM, Meerlo P, Murison R, Olivier B, Palanza P, Richter-Levin G (2011) Stress revisited: a critical evaluation of the stress concept. Neurosci Biobehav Rev 35:1291–1301

Lambooij E, Pilarczyk M, Bialowas H, Van den Boogaart JGM, Van de Vis JW (2007) Electrical and percussive stunning of the common carp (*Cyprinus carpio* L.): neurological and behavioural assessment. Aquac Eng 37:171–179

Lambooij E, Grimsbø E, Van de Vis JW, Reimert HGM, Nortvedt R, Roth B (2010) Percussion and electrical stunning of Atlantic salmon (*Salmo salar*) after dewatering and subsequent effect on brain and heart activities. Aquaculture 300:107–112

Lines JA, Spence J (2012) Safeguarding the welfare of farmed fish at harvest. Fish Physiol Biochem 38:153–162

Lines JA, Spence J (2014) Humane harvesting and slaughter of farmed fish. Rev Sci Tech 33:255–264

MacIntyre C, Ellis T, North BP, Turnbull JF (2008) The influences of water quality on the welfare of farmed trout: a review. In: Branson E (ed) Fish welfare. Blackwells Scientific, London, pp 150–178

Manuel R, Boerrigter J, Roques J, van der Heul J, van den Bos R, Flik G, Van de Vis H (2014) Stress in African catfish (*Clarias gariepinus*) following overland transportation. Fish Physiol Biochem 40:33–44

Marine Harvest (2018) Salmon farming industry handbook, 113 pp

Martins CIM, Eding EH, Verdegem MCJ, Heinsbroek LTN, Schneider O, Blancheton JP, d'Orbcastel ER, Verreth JAJ (2010) New developments in recirculating aquaculture systems in Europe: a perspective on environmental sustainability. Aquac Eng 43:83–93

McEwen BS, Wingfield JC (2003) The concept of allostasis in biology and biomedicine. Horm Behav 43:2–15

Morzel M, Sohier S, Van de Vis JW (2002) Evaluation of slaughtering methods of turbots with respect to animal protection and flesh quality. J Sci Food Agric 82:19–28

Nilsen A, Nielsen KV, Biering E, Bergheim A (2017) Effective protection against sea lice during the production of Atlantic salmon in floating enclosures. Aquaculture 466:41–50

Nilsson J, Stien LH, Iversen MH, Kristiansen TS, Torgersen T, Oppedal F, Folkedal O, Hvas M, Gismervik K, Ellingsen K, Nielsen KV, Mejdell CM, Kolarevic J, Izquierdo-Gomez D, Sæther B-S, Espmark ÅM, Midling KØ, Roth B, Turnbull JF, Noble C (2018) Velferdsindikatorer for oppdrettslaks: Hvordan vurdere og dokumentere fiskevelferd – Del A. Fiskevelferd og oppdrettslaks, kunnskap og teoretisk bakgrunn. In: Noble C, Nilsson J, Stien LH, Iversen MH, Kolarevic J, Gismervik K (eds) Velferdsindikatorer for oppdrettslaks: Hvordan vurdere og dokumentere fiskevelferd, pp 10–141. ISBN: 978-82-8296-531-6

Noble C, Gismervik K, Iversen, MH, Kolarevic J, Nilsson J, Stien LH, Turnbull JF (eds) (2018) Welfare indicators for farmed Atlantic salmon: tools for assessing fish. 351 pp. ISBN 978-82-8296-556-9

Nomura M, Sloman KA, Von Keyserlingk MAG, Farrell AP (2009) Physiology and behaviour of Atlantic salmon (*Salmo salar*) smolts during commercial land and sea transport. Physiol Behav 96:233–243

Oppedal F, Dempster T, Stien LH (2011) Environmental drivers of Atlantic salmon behaviour in sea-cages: A review. Aquaculture 311:1–18

Palić D, Norheim K, De Briyne N (2017) Fish diseases lacking treatment- gap analysis outcome. Report prepared for 15 pp. http://www.fve.org/uploads/publications/docs/fishmed_plus_gap_analysis_outcome_final.pdf

Poli BM, Parisi G, Scappini F, Zampacavallo G (2005) Fish welfare and quality as affected by pre-slaughter and slaughter management. Aquac Int 13:29–49

Reglero P, Balbın R, Ortega A, Alvarez-Berastegui D, Gordoa A, Torres AP, Moltó V, Pascual A, De la Gándara F, Alemany F (2013) First attempt to assess the viability of bluefin tuna spawning events in offshore cages located in an a priori favourable larval habitat. Sci Mar 77:585–596

Remen M, Solstorm F, Bui S, Klebert P, Vågseth T, Solstorm D, Hvas M, Oppedal F (2016) Critical swimming speed in groups of Atlantic salmon *Salmo salar*. Aquac Environ Interact 8:659–664

Rensel JE, Whyte JNC (2003) Finfish mariculture and harmful algal blooms. Manual on harmful marine microalgae. Monogr Oceanogr Methodol 11:693–722

Robb DFH, Kestin SC (2002) Methods used to kill fish: field observations and literature reviewed. Anim Welf 11:269–282

Robb DHF, Wotton SB, McKinstry JL, Sorensen NK, Kestin SC (2000) Commercial slaughter methods used on Atlantic salmon: determination of the onset of brain failure by electroencephalography. Vet Rec 147:298–303

Roberts RJ, Bullock AM, Turners M, Jones K, Tett P (1983) Mortalities of *Salmo gairdneri* exposed to cultures of *Gyrodinium aureolum*. J Mar Biol Assoc UK 63:741–743

Roque d'Orbcastel E, Blancheton J-P, Belaud A (2009a) Water quality and rainbow trout performance in a Danish model farm recirculating system: comparison with a flow through system. Aquac Eng 40:135–143

Roque d'Orbcastel E, Person-Le Ruyet J, Le Bayon N, Blancheton J-P (2009b) Comparative growth and welfare in rainbow trout reared in recirculating and flow through rearing systems. Aquac Eng 40:79–86

Rosten TW, Kristensen T (2011) Best practice in live fish transport. NIVA REPORT SNO 6102-2011, 25 p

Rosten TW, Ulgenes Y, Henriksen K, Terjesen BF, Biering E, Winther U (2011) Oppdrett av laks og ørret i lukkede anlegg – forprosjekt. SINTEF, Trondheim, 76 pp

Roth B, Imsland A, Gunnarsson S, Foss A, Schelvis-Smit R (2007) Slaughter quality and rigor contraction in fanned turbot (*Scophthalmus maximus*); a comparison between different stunning methods. Aquaculture 272:754–761

RSPCA (2018a) RSPCA welfare standards for farmed Atlantic salmon. RSPCA, Horsham, 96 p. https://science.rspca.org.uk/sciencegroup/farmanimals/standards/salmon. Accessed 25 May 2018

RSPCA (2018b) RSPCA welfare standards for farmed rainbow trout. RSPCA, Horsham, 51 p. https://science.rspca.org.uk/sciencegroup/farmanimals/standards/trout. Accessed 25 May 2018

Rud I, Kolarevic J, Holan AB, Berget I, Calabrese S, Terjesen BF (2016) Deep-sequencing of the microbiota in commercial-scale recirculating and semi-closed aquaculture systems for Atlantic salmon post-smolt production. Aquac Eng 78:50–62

Sampaio FD, Freire CA (2016) An overview of stress physiology of fish transport: changes in water quality as a function of transport duration. Fish Fish 17:1055–1072

Sattari A, Lambooij E, Sharifi H, Abbink W, Reimert H, Van de Vis JW (2010) Industrial dry electro-stunning followed by chilling and decapitation as a slaughter method in Claresse® (*Heteroclarias* sp.) and African catfish (*Clarias gariepinus*). Aquaculture 302:100–105

Scherer R, Augusti PR, Steffens C, Bochi VC, Hecktheuer LH, Lazzari R, Radünz-Neto J, Pomblum SCG, Emanuelli T (2005) Effect of slaughter method on postmortem changes of grass carp (Ctenopharyngodon idella) stored in ice. J Food Sci 70:348–353

Schram E, Abbink W, Roques J, Spanings T, De Vries P, Bierman S, Van de Vis H, Flik G (2010) The impact of elevated exogenous ammonia levels on growth, feed intake and physiology of African catfish (*Clarias gariepinus*). Aquaculture 306:108–115

Schrijver R, Van de Vis H, Bergevoet R, Stokkers R, Dewar D, Van de Braak K, Witkamp S (2017) Welfare of farmed fish: common practices during transport and at slaughter. Final report written for the European Commission Directorate Health and Food Safety (SANTE), reference SANTE/2016/G2/009, Contract SANTE/2016/G2/SI2.736160, 186 p. https://publications.europa.eu/en/publication-detail/-/publication/59cfd558-cda5-11e7-5d5-01aa75ed71a1/language-en. ISBN: 978-92-79-75336-7

Segner H, Sundh H, Buchmann K, Douxfils J, Sundell KS, Mathieu C, Ruane N, Jutfelt F, Toften H, Vaughan L (2012) Health of farmed fish: its relation to fish welfare and its utility as welfare indicator. Fish Physiol Biochem 38:85–105

Shoubridge EA, Hochachka PW (1980) Ethanol: novel end-product in vertebrate anaerobic metabolism. Science 209:308–309

Stickney RR (ed) (2000) Encyclopedia of aquaculture. Wiley-Interscience, New York

Stien LH, Bracke M, Folkedal O, Nilsson J, Oppedal F, Torgersen T, Kittilsen S, Midtlyng PJ, Vindas MA, Øverli Ø, Kristiansen TS (2013) Salmon welfare index model (SWIM 1.0): a semantic model for overall welfare assessment of caged Atlantic salmon: review of the selected welfare indicators and model presentation. Rev Aquac 5:33–57

Summerfelt ST, Zühlke A, Kolarevic J, Reiten BKM, Selset R, Gutierrez X, Terjesen BF (2015) Effects of alkalinity on ammonia removal, carbon dioxide stripping, and system pH, in semi-commercial scale WRAS operated with moving bed bioreactors. Aquacult Eng 65:46–54

Tacon AGJ, Halwart M (2007) Cage aquaculture: a global overview. In: Halwart M, Soto D, Arthur JR (eds) Cage aquaculture – regional reviews and global overview. FAO Fisheries Technical Paper, No. 498. FAO, Rome 2007, pp 1–16, 241 p

Timmons M, Ebeling J (2007) Recirculating aquaculture, 2nd edn. NRAC publication no 01-007, Cayuga aqua ventures, Ithaca, 769 pp

Troell M, Joyce A, Chopin T, Neori A, Buschmann AH, Fang J-G (2009) Ecological engineering in aquaculture – potential for integrated multi-trophic aquaculture (IMTA) in marine offshore systems. Aquaculture 297:1–9

Van de Vis H, Lambooij B (2016) Fish stunning and killing. In: Velarde A, Raj M (eds) Animal welfare at slaughter. 5M Publishing, Sheffield, pp 152–176

Van de Vis H, Kestin S, Robb D, Oehlenschlager J, Lambooij B, Munkner W, Kuhlmann H, Kloosterboer K, Tejada M, Huidobro A, Ottera H, Roth B, Sørensen NK, Akse L, Byrne H, Nesvadba P (2003) Is humane slaughter of fish possible for industry? Aquac Res 34:211–220
Van de Vis H, Kiessling A, Flik G, Mackenzie S (eds) (2012) Welfare of farmed fish in present and future production systems. Springer, Heidelberg, 312 pp
Van de Vis H, Abbink W, Lambooij B, Bracke M (2014) Stunning and killing of farmed fish: how to put it into practice? In: Devine C, Dikeman M (eds) Encyclopedia of meat sciences 2e, vol 3. Elsevier, Oxford, pp 421–426
Van den Burg EH, Peeters RR, Verhoye M, Meek J, Flik G, Van der Linden A (2005) Brain responses to ambient temperature fluctuations in fish: reduction of blood volume and initiation of a whole-body stress response. J Neurophysiol 93:2849–2855
Weimin M (2010) Recent developments in rice-fish culture in China: a holistic approach for livelihood improvement in rural areas. In: De Silva SS, Davy FB (eds) Success stories in Asian aquaculture. International Development Research Centre, Ottawa, ON, pp 15–40
Wendelaar Bonga SE (1997) The stress response of fish. Physiol Rev 77:591–625
Werkman M, Green DM, Murray AG, Turnbull JF (2011) The effectiveness of fallowing strategies in disease control in salmon aquaculture assessed with an SIS model. Prev Vet Med 98:64–73
Wood CM (1991) Acid-base and ion balance, metabolism, and their interactions, after exhaustive exercise in fish. J Exp Biol 160:285–308

Chapter 15
Ornamental Fish and Aquaria

Thomas Torgersen

Abstract Ornamental fish share only one feature: they are kept as display or hobby animals. Apart from that, they are as diverse a group of fish as can be. The typical ornamental fish, irrespective of species, will go through capture (or being bred and raised in captivity), transport to a capture station, a wholesale facility and a retailer's display tanks, and being dip-netted into a bag for transport to the customer's home aquarium, where it is released and typically kept until it dies. In all these stages of the life of the fish, a range of welfare problems can and do arise: the fish are subject to various stressful events, they may experience water of poor or deteriorating quality, or with chemistry (pH, ionic composition, temperature) with which they are not compatible, they may be placed in aquaria that are too small or in other ways physically not suited to their needs and lifestyle, they may be housed together with the fish with which they are not compatible and they may not be offered appropriate nutrition.

On one side, the topic of ornamental fish welfare includes a lot more than what can be included in a book chapter; on the other side, the published relevant literature has serious knowledge gaps. This chapter provides an overview of the welfare problems of ornamental fish and attempts to identify the areas of major importance.

Keywords Aquaria · Garden ponds · Capture · Transport · Holding environment · Welfare · Acclimation · Motion · Acceptance

---

Thomas Torgersen died suddenly in May 2018. This was a shock and great loss for all who knew him. We are missing his intelligent writings and comments, humour and loud laughs. Thomas made the world a more beautiful, exciting and interesting place. He is deeply missed.

T. Torgersen (✉)
Institute of Marine Research, Bergen, Norway
e-mail: torek@imr.no

T. S. Kristiansen et al. (eds.), *The Welfare of Fish*, Animal Welfare 20,
https://doi.org/10.1007/978-3-030-41675-1_15

15.1 Introduction

Ornamental fish share only one feature: they are kept as display or hobby animals (Fig. 15.1). Taxonomy, size and environmental requirements of ornamental fish cover most fish families and aquatic environments. They span everything from small, robust and adaptive to large, fragile and with small capacity to adapt to different feeds, water qualities, tank layouts and stressors. Further, ornamentals are subject to welfare problems in different parts of the ornamental fish trade and hobby. From its origin millennia ago, when a few species were kept and eventually domesticated as display animals in ponds, the practice of keeping ornamental fish has expanded enormously in volume and diversity (Teletchea 2016). According to the ornamental fish trade association OFI (http://www.ofish.org/ornamental-fish-industry-data), more than 120 countries are involved in the collection, breeding, import and export of ornamental fish. FAO data suggest that the export value of ornamental fish was approximately $330 million USD in 2011. The estimated number of fish traded annually was approximately 1.5 billion. OFI argues that the FAO figures, although generally believed to be the most accurate, are probably significantly lower than the numbers actually involved. The great majority of ornamental fish are freshwater species, although about 20 million marine fish are traded annually. The number of species traded is reported to be about 6000–7000, about two-thirds of which are freshwater species. Together, the species described in

Fig. 15.1 Two mobile phone photos from the public aquarium in Busan in 2017. The requirements of the common sea dragon and the sand tiger shark include overlapping water quality tolerances, but possibly little else. Photo: Runar Torgersen

the atlases of freshwater and marine aquarium fish of Axelrod et al. (2007) and Burgess et al. (2000), respectively, exceed 10,000. The freshwater aquarium trade and hobbyists rely mainly on captive-bred fish (>90%), whereas the marine trade exhibits the opposite pattern, with >90% of the traded fish being wild-caught. Therefore, for the marine aquarium industry, capture-related issues probably account for the majority welfare problems. In freshwater aquaristics, the capture stage probably accounts for a smaller proportion of welfare problems.

There are a number of different phases in the life of a fish in the ornamental and hobby fish trade: For wild-captured fish, it starts with capture, the first in a series of stressful events and transient states—capture, holding, transfer, transport, purchase and introduction. The captive-bred fish starts its life in the trade very differently: it is spawned, hatched (unless it is a live-bearing species) and raised in tank environments that are, at the very least, sufficient for growth and survival until it reaches the marketable size. From that point, it experiences similar events and conditions as the wild-caught fish (unless it is to be kept as a brood-fish, in which case it will presumably continue to live in a fairly stable environment that not only supports survival and growth, but also reproduction), and the typical ornamental fish destined for the home aquarium, irrespective of species, will be transported to a retailer, kept in a display tank and be dip-netted into a bag for transport to the customer's home aquarium where it is released. The last and usually longest phase is that in the tank of the hobbyist who purchases the fish. It might be useful to divide welfare problems associated with the ornamental and hobby fish trade into those that arise from inherently stressful and potentially damaging transient processes before the fish ends up in the hobbyist's tank and those that arise from the mismatch between the requirements of the fish and the conditions offered by the hobbyist. In all the stages of the trade and the life of the fish, different welfare problems can occur. The fish are subject to various stressful events, they may experience poor-quality water, with chemistry (pH, ionic composition, temperature) with which they are not compatible, they may be placed in aquaria that are too small or in other ways unsuitable to their needs and lifestyle, they may be housed together with fish with which they are not compatible, they may not be offered appropriate food and they may suffer from diseases and physiological impairments. The aquaculture industry was quite early to consider animal welfare, and the trade organization OATA's code of conduct addresses various welfare aspects of the hobby and trade (http://www.ornamentalfish.org/wp-content/uploads/2015/10/CODE-OF-CONDUCT-FINAL-OCT-2015.pdf).

15.2 Scope of the Chapter

This chapter discusses welfare issues in the ornamental fish trade and hobby on a broad and aggregated level: what welfare issues are involved, and where in the trade chain and at what phase in the life of the fish do they occur. It also suggests some complementary approaches to the topic of ornamental fish welfare to that provided by Walster (2008) and the recent review of Stevens et al. (2017). The chapter does not address specific welfare issues of individual species. Regarding which conditions cause welfare problems for different species of fish, and to what extent ornamental fish meet such conditions in the hobby and trade, our knowledge is fragmentary. Only a few reports of some welfare proxies or aspects of individual species exist in the literature, and since the diversity of ornamental fish is so enormous, extrapolating knowledge about requirements is of limited value.

A number of definitions and understandings of animal welfare exist, but the term “welfare”, however it is measured, should be restricted to addressing the quality of the lives of animals that are able to experience it. Assessing the experienced quality of life of an animal directly is inherently difficult, but there are sound reasons for using different proxies as measures of welfare: Function-based and nature-based animal welfare approaches make welfare measurable under the assumptions that animals that do not function properly or are unable to express their innate behaviours are not experiencing good welfare. Behavioural responses such as preferences and aversive responses can sometimes be used as indicators of the fish’s experience, but in general we are left with the option of assuming that the fish’s experienced quality of life is a function of and correlated with measurable proxies like disease, nutrition, physiological function, stress and the ability to display innate behaviour and to cope with challenges. One highly measurable proxy—mortality—usually implies poor antemortem welfare, as well as poor welfare of the surviving fish living under the same conditions.

15.3 Welfare in Hobbyist Aquaria

The basis for poor welfare in hobbyist aquaria can generally be attributed to discrepancies between the requirements of the kept fish and the environment provided by the fish keeper. Certain conditions are detrimental to all fish, such as water qualities that do not support their physiological function, e.g. temperatures that are far removed from preferred ranges, high nitrite and low chloride levels, or high ammonia and high pH. The critical values of such parameters, however, vary widely from species to species, and guideline values for aquarium fish in general are of limited value. The same is true of generic guidelines regarding stocking densities, and minimum aquarium size as a function of fish length. By default, larger fish or fish held in greater numbers require larger aquaria, but beyond that, the specific requirements of the species must be understood and met.

Good welfare is achieved by offering the fish an appropriate environment in all relevant parameters, i.e. water quality, aquarium size and layout, fish community and feed. This is more easily achieved for some species than for others. An easier fix for general welfare problems in home aquaria would obviously be to have retailers marketing fish and aquarium owners buying fish whose environmental requirements are more easily met, rather than aiming at aquarium owners becoming more competent and spending more on meeting the demands of inherently difficult species.

For which fish is it easy to offer an environment that provides good welfare? As pointed out earlier, one approach is to look at which species are generally able to grow and even reproduce, staying alive and healthy, and not showing signs of behaviours interpreted as signs of sustained and severe stress in different tank set-ups.

Social interactions between fish include aggression, both within and between species, that may lead to physical injury or death. Also, dominant fish may suppress other individuals to such an extent that they become chronically stressed, as manifested by continuous hiding and reduced feeding. On the other hand, many ornamental species are schooling or in other ways social, and fish living in large schools and in multi-species assemblages often enjoy better welfare than when kept solitary, in small groups or in single species set-ups. Saxby et al. (2010) and Sloman et al. (2011) have reported examples of such relationships. Further, in many cases the presence of aggression and territoriality in aquaria does probably not represent major welfare problems. A large aquarium may allow several territorial fish to be able to defend their territories, and coping with an aggressive fish in an adjacent territory is different from being caught within the territory of a dominant fish. Since relationships between the welfare of fish of different species and their social environment, i.e. group size and presence of fish of other species, are highly dependent on tank size and layout and fish states (stage of maturity, dominance), generic operational rules are of limited value.

As Sales and Janssens (2003) pointed out, the feeds available for ornamental fish are to a large degree based on the extrapolation of results derived from food for fish kept under intensive farming conditions. The studies of the dietary requirements of individual species are few and do not cover most of the species available in the trade. Furthermore, in community aquaria, the different species will generally have different dietary requirements and not only in terms of nutritional quality: different fish have different diel feeding patterns and require feed of different particle size; some feed near the surface, others at the bottom, some tackle their feed with high intensity, while others are slow. Furthermore, fish kept in tanks may be very reluctant to eat processed feeds even if they have the right nutritional composition and particle size and are available in the parts of the tank where they would normally feed. The failure of fish that feed on only selected prey items and of timid and subordinate fish to feed properly is probably the cause of most welfare problems related to feeding.

In most cases, animals probably suffer from poor welfare for some time before death. This time may be short if they end their life as prey, or longer when they eventually succumb to disease, starvation or respiration failure. Environmental conditions and diseases that in extreme cases lead to death, e.g. infections, environmental hypoxia, high nitrite levels, will cause sublethal welfare problems when the poor conditions are less severe. The conditions that lead to mortality in weakened or sensitive individuals are also likely to reduce welfare among surviving fish. The death of some fish in a population is therefore often a proxy for poor welfare among survivors. Since a large proportion of aquarium fish undoubtedly die at an earlier age than their potential lifespan, it can be argued that the causes of death in aquarium fish are also important causes of poor welfare that do not result in mortality. However, aquarium fish die from a number of causes, and Engelhardt (1992) found that in more than a thousand dead fish of different species collected from home aquaria and retailers, various non-infectious causes were detected in 45% of cases, while different infectious causes were found in 38%. The rest of the sample displayed no pathological symptoms.

Although it is clear that poor welfare is the result of keeping a fish in an environment with which it is not compatible, the opposite relationship may not be true: fish that are capable of surviving and growing in a given environment may still experience poor welfare. Fish are evolved organisms and so are their abilities to adapt and the rates at which they acclimate, move and habituate, and these reflect their evolutionary histories. The patterns of change in the environments in which an animal evolves, combined with physiological and behavioural constraints of the fish, determine the adaptation capacities of the fish. Physical, chemical, biological and social environments differ and change in space and time. Such variability can have very small or very large amplitudes and occur on everything from small (centimetres, seconds) to very large (annual, cross-ocean) scales. Fish cannot position themselves in time, but they can choose, to a greater or lesser extent, their position in space, and thus do not have to deal with all the spatial environmental variability of their ecosystems. Furthermore, migrating out of temporarily unfavourable locations into favourable ones can reduce the experienced temporal environmental variability. A fish that is not satisfied with its current environment has three possible ways of coping: first, it can adapt its physiology to its current suboptimal environment (acclimation) and thereby, given sufficient time, overcome the mismatch between the organism and the environment. This will effectively remove, or at least reduce, the reason for dissatisfaction with prevailing conditions. Second, it can move to a place more suited to its needs and preferences. Third, it can simply accept its current situation and rely on its tolerance of suboptimal environments. Acclimation, motion and acceptance occur on different scales and have different constraints and positive and negative consequences, and many fish undoubtedly employ all three strategies.

Physiological acclimation is useful, yet slow and typically takes days to accomplish (Allen and Strawn 1971; Brett 1946; Chung 2001; Evans 1990; Jobling 1994; Jones and Sidell 1982; Peterson and Anderson 1969). If the change to which the animal acclimates turns out to be a transient one, the acclimation process may leave it worse off than before. The cost of acclimation, like synthesis of new protein, is not easily distinguishable from the increased metabolic costs induced by environmental stress, but if significant it is another reason for animals not to be too hasty with acclimation if their environment changes often or their behaviour tends to take them into different environments.

An animal moves only if motivated to do so, and the motivation to move is derived from dissatisfaction with the current environment and the need for something different. Migratory behaviour ends when the animal is exhausted, dies or accepts its current position. The question then is how rapidly dissatisfaction should attenuate: when should the animal stop searching and accept its suboptimal environment? The benefits of accepting suboptimal conditions are obvious; removing the motivation to move or make an effort saves energy and may reduce the risk of mortality, and it allows the animal to utilise its "hedonic scope", i.e. its capacity to respond positively and negatively to fitness-related events and environmental conditions on something else. However, giving up too early, i.e. in a situation where continued effort could be successful, will leave the animal in unnecessarily poor conditions and at risk of death if the environment deteriorates further.

The spatial and temporal environmental variability patterns and the ability of fish to exploit spatial variability vary greatly between environments and species. Instead of swimming away from unattractive environments, fish can acclimate to different water quality parameters, adapt their behaviour according to social hierarchies, learn to exploit less-preferred food resources, etc. The potential for behavioural control over its environment, and therefore the need for acclimation or physiological tolerance for suboptimality, differs between systems. Fish that have evolved in ponds or streams where conditions are occasionally uniformly poor, for example, should not waste neither their hedonic nor their metabolic scope on frustration-driven relentless swimming. At the other end of the scale are large, migratory, oceanic species. Oceans are spatially variable. Many marine fish species exploit this variability by migration in order to overcome temporal fluctuations and avoid the areas of low profitability (Haugland et al. 2006; Holm et al. 2000; Itoh et al. 2003; Polovina 1996). These fish have both the necessary swimming performance and access to vast areas to be able to emigrate out of a poor-quality environment.

It might be argued that the survival and growth of oceanic migrators like salmon and tuna in aquaculture show that these animals cope well with their captive, confined environments, and that this implies that current culture practices ensure reasonable levels of welfare. However, animal welfare is not a mere function of performance and a widely accepted definition is that of Broom (1988), that "the welfare of an individual is its state as regards its attempts to cope with its environment". Although health and physiological functioning is important for animal welfare, the most important aspect is arguably the subjective experience the animal has of its quality of life; i.e. is it getting what it wants? (Dawkins 2004). A more relevant question regarding welfare than whether the animal can perform in the environment it is offered is whether it has evolved to accept it. Animals that are capable of having a subjective experience of the quality of life do not necessarily have good welfare just because their survival probability and growth rate support population persistence. Selection acts on fitness differentials between competing strategies. If the unwillingness to accept a small environmental deterioration is beneficial because available spatial environmental variability will generally lead a searching animal to find a better place, selection may produce a population of individuals that will never settle for anything less. From this line of reasoning, it follows that restricting the access to spatial environmental variability relative to what the animals have evolved to explore may have serious effects on their welfare, and especially so when temporal variability is imposed on the animal, such as when a naturally vertically migrating fish is both denied access to deep water and is exposed to diel or seasonal temperature fluctuations outside its range of preference.

We may conclude from this that the ability of a given fish to accept an environment different from its optimum can be inferred from the temporal and spatial environmental variability patterns in which it has evolved, and that fish that have evolved in systems that offer small possibilities of finding better conditions through sustained effort should experience less dissatisfaction in suboptimal aquarium conditions. Extreme examples of such fish are annual killifish, which include members of the genus *Notobranchius*. Eggs hatch after rehydration when drought seasons end, and the fish live their lives in small ephemeral pools of water of unpredictable size and longevity, which offer changeable conditions (Lucas-Sánchez et al. 2014).

15.4 Welfare in Garden Ponds

The number of species that are kept in garden ponds is much smaller. In temperate areas, various cyprinid species and sturgeons dominate, and the two dominant species are goldfish (*Carassius auratus*) and koi carp (*Cyprinus carpio*). These fish are hardy and tolerate a wide range of different water qualities. As shown by Ford and Beitinger (2005), once acclimatised, the goldfish survives temperatures within the range of 0.3–43.6 °C. Due to their large potential size, an adequate size of pond, and sufficient aeration, filter capacity and water exchange rate are critical for keeping these species. In addition, the concept of keeping fish in relatively small water bodies in fluctuating climates (dielly and seasonally) offers some additional challenges: shallow ponds in particular may provide fish with very rapid and extreme rises in temperature on sunny summer days. At the other end of the temperature scale, in areas with subzero winter temperatures, ice cover introduces a further challenge: even though the fish survive well in the deep, 4 °C water under the ice (provided that the pond is sufficiently deep to offer such an environment), if gas exchange through an open-water area in the pond is not provided, environmental hypoxia and the build-up of metabolites from the fish's respiration and the breakdown of detritus may gradually deteriorate their environment to such a degree that they succumb in the course of the winter.

15.5 Welfare in Capture, Transport, and Holding Tanks

Fish capture inevitably results in stresses. In addition to the stress imposed on fish by being chased and constrained, various capture methods and postcapture treatment may injure or kill them. The literature has been reviewed by Rubec and Cruz (2005). The combination of invasive capture methods, including cyanide, poor water quality, crowding, infections and insufficient time for fish to recover between stressful procedures results in high levels of mortality. According to Rubec and Soundararajan (1991), the cumulative mortality rates of marine ornamental fish from capture to retailer may exceed 90%. Undoubtedly, the surviving 10% will also have experienced poor welfare on their way to the retailer. Moreover, a large proportion of the survivors that are delivered to buyers by capturers and capture stations are not exported, but are discarded for various reasons (Militz et al. 2016). In their study, the majority of the discarded fish were not discarded for belonging to a species or being of a size that made them unmarketable, but because of fin damage, wounds and emaciation, presumably caused by capture or post-capture treatment. The size of holding tanks, tank layout and stocking density at capture stations, breeders, wholesalers and retailers represent trade-offs between costs and reasonable survival rates and product quality. The same can be said about packaging systems and standards for ornamental fish transport, and Lim et al. (2003) argued that survival rates could be greatly improved. Without doubt, shipping is a major source of poor welfare of aquarium fish.

15.6 Welfare Costs of Downstream Mortality and Low Value

I will argue that high rates of mortality late in the value chain cause an increased volume of welfare problems earlier in the value chain (sic). A fish that dies in a home aquarium will usually be replaced by a new one, while fish that die in transit must be compensated for by shipping more fish from wholesalers, etc. Furthermore, fish that are discarded due to low marketability will also have lived through stressful periods of low welfare, even if the culling process itself is quick and painless. Mortality can be substantial at each stage in the value chain, so a single fish in a home aquarium is the end-point of a much larger number of fish that have passed through repeated decimations on their way from capture or breeding. Since most operations and stages until the fish reaches the home aquarium are inherently stressful, there is inevitably a high level of poor welfare in the wake of a well-acclimatised fish in a well set-up home aquarium.

The enthusiastic home aquarium owner can spend amounts of time and money on caring for his fish that far exceed its potential sales value, but until the fish reaches the enthusiast aquarium, it represents a sale value and a set of holding and transport costs. By default, the money spent by the supply chain on holding and transporting the fish cannot exceed its sales value. The prices of aquarium fish vary enormously between species, strains (e.g. discus) and individuals (especially koi carp) at the retailer. The relative price difference decreases throughout the value chain, since even the cheapest fish require handling, storage, shipping, medication and feeding. Therefore, a fish that retails at a low price has been of little value for the breeder, fisherman or wholesaler. The relative cost of losing fish because of less efficient transport and handling, or crowding and poor water quality in holding tanks and transport bags is smaller for a cheap fish than for a more expensive one, and poor welfare and mortality probably increase among cheaper fish. A good rule of thumb for improving ornamental fish welfare could thus be to keep few but expensive fish.

Acknowledgements The editors offered valuable comments on earlier drafts of the manuscript.

References

Allen KO, Strawn K (1971) Rate of acclimation of juvenile channel catfish, *Ictalurus punctatus*, to high temperatures. Trans Am Fish Soc 100:665–671

Axelrod GS, Warren FZS, Burgess E, Pronek N, Axelrod HR, Wall JG (2007) Dr. Axelrod's atlas of freshwater aquarium fishes, 11th edn. T.F.H, Neptune City

Brett JR (1946) Rate of gain of heat-tolerance in goldfish (*Carassius auratus*). Univ Tor Stud Biol Ser 53:8–28

Broom DM (1988) The scientific assessment of animal welfare. Appl Anim Behav Sci 20:5–19

Burgess WE, Axelrod HR, Hunziker RE (2000) Dr. Burgess' atlas of marine aquarium fishes, 3rd edn. T.F.H, Neptune City

Chung KS (2001) Critical thermal maxima and acclimation rate of the tropical guppy *Poecilla reticulata*. Hydrobiologia 462:253–257

Dawkins MS (2004) Using behaviour to assess animal welfare. Anim Welf 13:3–7

Engelhardt A (1992) Causes of disease and death in ornamental fish – frequency and importance. Berl Munch Tierarztl Wochenschr 105:187–192

Evans DO (1990) Metabolic thermal compensation by rainbow trout – effects on standard metabolic rate and potential usable power. Trans Am Fish Soc 119:585–600

Ford T, Beitinger TL (2005) Temperature tolerance in the goldfish, *Carassius auratus*. J Therm Biol 30:147–152

Haugland M, Holst JC, Holm M, Hansen LP (2006) Feeding of Atlantic salmon (*Salmo salar* L.) post-smolts in the Northeast Atlantic. ICES J Mar Sci 63:1488–1500

Holm M, Holst JC, Hansen LP (2000) Spatial and temporal distribution of post-smolts of Atlantic salmon (*Salmo salar* L.) in the Norwegian Sea and adjacent areas. ICES J Mar Sci 57:955–964

Itoh T, Tsuji S, Nitta A (2003) Migration patterns of young Pacific bluefin tuna (*Thunnus orientalis*) determined with archival tags. Fish Bull 101:514–534

Jobling M (1994) Fish bioenergetics. Chapman and Hall, London

Jones PL, Sidell BD (1982) Metabolic responses of striped bass (*Morone Saxatilis*) to temperature acclimation. 2. Alterations in metabolic carbon sources and distributions of fiber types in locomotory muscle. J Exp Zool 219:163–171

Lim LC, Dhert P, Sorgeloos P (2003) Recent developments and improvements in ornamental fish packaging systems for air transport. Aquac Res 34:923–935

Lucas-Sánchez A, Almaida-Pagán PF, Mendiola P, de Costa J (2014) *Nothobranchius* as a model for aging studies. A review. Aging Dis 5:281–291

Militz TA, Kinch J, Foale S, Southgate PC (2016) Fish rejections in the marine aquarium trade: an initial case study raises concern for village-based fisheries. PLoS One 11(3):e0151624. https://doi.org/10.1371/journal.pone.0151624

Peterson RH, Anderson JM (1969) Influence of temperature change on spontaneous locomotor activity and oxygen consumption of Atlantic salmon *Salmo salar* acclimated to two temperatures. J Fish Res Board Can 26:93–109

Polovina JJ (1996) Decadal variation in the trans-Pacific migration of northern bluefin tuna (*Thunnus thynnus*) coherent with climate induced change in prey abundance. Fish Oceanogr 5:114–119

Rubec PJ, Cruz FP (2005) Monitoring the chain of custody to reduce delayed mortality of net-caught fish in the aquarium trade. SPC Live Reef Fish Inf Bull 13:13–23

Rubec PJ, Soundararajan R (1991) Chronic toxic effects of cyanide on tropical marine fish. In: Chapman P et al (eds) Proceedings of the seventeenth annual toxicity workshop, November 5–7, 1990, Vancouver (Can Tech Rep Fish Aquat Sci 1774(1):243–251)

Sales J, Janssens GPJ (2003) Nutrient requirements of ornamental fish. Aquat Living Resour 16:533–540

Saxby A, Adams L, Snellgrove D, Wilson RW, Sloman KA (2010) The effect of group size on the behaviour and welfare of four fish species commonly kept in home aquaria. Appl Anim Behav Sci 125:195–205

Sloman KA, Baldwin L, McMahon S, Snellgrove D (2011) The effects of mixed-species assemblage on the behaviour and welfare of fish held in home aquaria. Appl Anim Behav Sci 135:160–168

Stevens CH, Croft DP, Paull GC, Tyler CR (2017) Stress and welfare in ornamental fishes: what can be learned from aquaculture? J Fish Biol 91:409–428

Teletchea F (2016) Domestication level of the most popular aquarium fish species: is the aquarium trade dependent on wild populations? Cybium: Int J Ichthyol 40(1):21–29

Walster C (2008) The welfare of ornamental fish. In: Branson EJ (ed) Fish welfare. Blackwell, Oxford

Chapter 16
Fish as Laboratory Animals

Anne Christine Utne-Palm and Adrian Smith

Abstract In this chapter, we aim at giving an overview of the extent to which fish are used in laboratory studies. We look at the most commonly used species (zebrafish, salmonids, goldfish, medaka and three-spined stickleback) and give some reasons why these species—and fish in general—are so popular experimental animals. Further, we give an overview of some of the legislation governing the use of fish as experimental animals and the areas in which they are used. We describe general and specific efforts to improve laboratory fish welfare and the quality of the research being performed on them. Given that fish is such a diverse group of species with species-specific needs and adaptations, it is difficult to make general guidelines, but we argue that we can achieve a great deal by building a knowledge database and welfare guidelines for the most commonly used species. Guidelines for planning and conducting fish experiments are mentioned, as well as advice on the assessment of welfare in fish species.

Keywords Fish · Laboratory · Research · Zebrafish · Salmonid · Goldfish · Medaka · Stickleback

16.1 Introduction

Millions of fish are used annually in research worldwide. More than 1.2 million were reported to be used in EU countries in 2017 (European Union 2020). Much of present-day legislation regulating animal research offers fish the same protection as other vertebrates, but it often refers to "fish" if they were one species. More than

A. C. Utne-Palm (✉)
Institute of Marine Research, Bergen, Norway
e-mail: annecu@hi.no

A. Smith
Norecopa, Oslo, Norway
e-mail: adrian.smith@norecopa.no

T. S. Kristiansen et al. (eds.), *The Welfare of Fish*, Animal Welfare 20,
https://doi.org/10.1007/978-3-030-41675-1_16

34,000 fish species have been described (FishBase 2019), and they exhibit greater species diversity than any other vertebrate group (IUCN Red List 2014). This huge diversity poses ethical and scientific challenges, since we cannot adequately address species-specific needs in general legislation on the use of fish in research. There is an urgent need for more specific guidelines, at least for the species which are commonly used in research.

Fish are used in research for many reasons. Traditionally, fish are an important food source for humans, and increasing pressure on wild populations has resulted in a thriving aquaculture industry. Thus, much research is carried out on aquaculture species to improve the health of the species, by the development of vaccines and improvements in environmental conditions. This research is, however, largely for human benefit, based upon the desire to increase production efficiency under intensive husbandry conditions.

However, fish are now being increasingly used in basic and applied research within areas such as physiology, genetics, behaviour and ecology. Much of this work is being performed on zebrafish, which have proven to be robust animals with a conveniently short generation interval and a number of other advantages, described later.

In this chapter, we discuss the use of fish in laboratory research, including a review of the commonly used species and the studies for which they are used. We will provide some relevant questions to ask before obtaining fish and using them in experiments. We also look at legislation regarding the use of fish in research today, as well as means of improving fish research and animal welfare. Valid scientific results are dependent upon research animals exhibiting normal behaviour representative for their species, without physical or behavioural changes caused by the environment or other stressors. We make some suggestions for how the validity and quality of research can be improved by improving the conditions for laboratory fish.

16.2 The Extent to Which Fish Are Used in Laboratory Studies

The total number of fish used worldwide in research today is not easy to estimate, since many countries do not report their use of research animals. Even in the European Union reports are not issued annually, and the methods used to collect these statistics are changing with the implementation of the latest Directive, 2010/63 (European Union 2010, 2019). The EU publishes compiled data for all Member States). Templates for reporting annual statistics have now been developed, and statistics from the individual Member States are published on the Commission website. The latest compiled report was published in early 2020 (European Union 2020).

The most recent compiled report from the EU covers the years 2015–2017 (European Union 2020). A total of nearly 9.4 million animals were used for the first time in 2017, of which mice and rats accounted for 61% and 12%, respectively. The total number of research animals declined by nearly 430,000 in the EU compared to 2016, and the number of fish decreased from 1.3 million to 1.2 million in 2017.

The countries outside the EU with a large aquaculture industry use large numbers of fish in research for, among other things, the development and testing of fish vaccines. Norway, for instance, used 1.6 million fish in research in 2018, compared to a total of 93,000 animals of all other species (Mattilsynet 2019). In 2016 11.5 million fish were used, of which two experiments alone accounted for 10.6 million animals, related to the treatment of salmon lice in the aquaculture industry.

Fish are used for research into fundamental biology, genetics, cancer research, vaccine production, physiopathology and diagnostic work. Fish are also used for testing biocides and for telemetric monitoring in the wild, but this group represents only a small percentage of the total use of experimental fish.

In Canada, 4.3 million research animals were used in 2016, of which 1.6 million were fish. Here again the trend is towards increasing numbers (Speaking of Research 2019).

In the USA, mice, rats, fish and birds are not covered by the Animal Welfare Act, and it is therefore not possible to estimate the numbers used, since they are not part of the annual statistics. The organisation Speaking of Research has estimated that somewhere between 12 and 27 million animals are used in total (Speaking of Research 2019). In Australia, 9.9 million animals were used in research in 2015, of which 1.2 million were fish (Humane Research Australia 2016).

As mentioned earlier, global statistics for research animal use are not available. One way to gain an impression of their use worldwide is to look at the number of published studies on fish compared to the number on other typical research animals such as rodents and birds. According to the ISI Web of Science, approximately 350,000 studies on fish were published between 1945 and 2017. These data indicate that fish comprise the group of species whose use, together with birds and mice, is increasing most. Since 1990, the number of published fish studies has increased by 63%, compared to 62% for birds and 46% for mice. The ISI statistics correspond well with the numbers reported from EU countries and strengthen the picture of fish becoming increasingly popular as a laboratory model animal worldwide.

Fish are frequently tagged in field studies which do not have to be reported. According to NOAA Fisheries (2014), 35 million fish have been tagged in the Columbia River Basin alone during the last 25 years. The number of tagged fish worldwide must be many times higher.

16.3 The Most Commonly Used Species and the Reasons for Their Popularity

Research on fish covers many areas, but some are more common than others. One of these is the use of model species to study biological mechanisms, for example, within physiology, genetics, toxicology and evolution. These model species are typically robust, small, easily bred fish with a short generation interval, such as the zebrafish, medaka, guppy, stickleback and goldfish. Another common category is research into species-specific traits and requirements. This category, which includes the four most common aquaculture species, typically consists of salmonids, carp, telapia, sea bream, rainbow trout (*Oncorhynchus mykiss*), Atlantic salmon (*Salmo salar*), tilapiine cichlids, zebrafish (*Danio rerio*), sea bass (*Dicentrarchus labrax*), Atlantic halibut (*Hippoglossus hippoglossus*), Atlantic cod (*Gadus morhua*), turbot (*Scophthalmus maximus*) and African catfish (*Clarias gariepenus*). Furthermore, there is a great deal of research related to fisheries and ecosystems, where a large range of species are used, including salmonids, cod fish, clupeoids, mackerel, perch, bass and bream.

Salmonids (salmon and trout) are the most studied fish group, while zebrafish (*Danio rerio*) are the most studied fish species, with 97,000 and 36,000 publications, respectively, in the period 1945–2017 according to the ISI Web of Science. Zebrafish and salmonids together represented 38% of all fish publications in this period.

After these, the most common species are goldfish (*Carassius auratus*), medaka (*Oryzias latipes*) and the three-spined stickleback (*Gasterosteus aculeatus*) with 13,300, 6300 and 3700 registered publications, respectively (Table 16.1). Medaka and stickleback have become increasingly popular in recent decades, while the goldfish has a longer history as a laboratory species.

Traditionally, most fish studies have been on the aspects of general biology, related to fisheries or ecology. However, in recent decades there have been an increasing number of studies related to the environment and to both veterinary and human medicine. Fish are now used extensively in toxicological studies and research into gene expression (Fig. 16.1). Studies on gene function have become feasible for many following the introduction of the CRISPR-Cas9 methodology, which can specifically target and mutate genes in any organism, thus also allowing studies on the genetics of key traits (Ledford 2015). For this type of research, smaller fish such

Table 16.1 Total number of publications (grouped in 15-year periods) from 1945 until 2017

	1945–1960	1961–1975	1976–1990	1991–2005	2002–2017
Goldfish	95	1004	2169	5214	6216
Medaka	26	181	513	2081	5517
Salmonids	383	3030	9967	37,668	58,953
Stickleback	145	128	292	1271	3731
Zebrafish	5	37	176	7925	32,391

Data are based on registrations in the ISI Web of Science

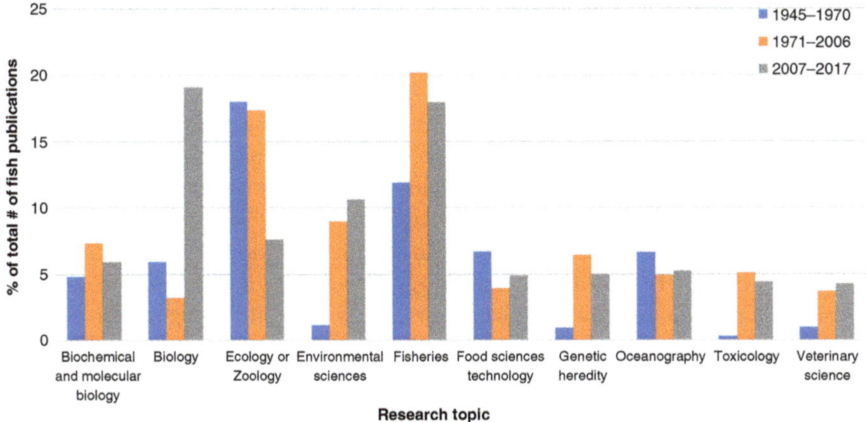

Fig. 16.1 Percentage of fish publications by research topic from 1945 until 2017. Based on data from the ISI Web of Science

as zebrafish and medaka are ideal, as they are easy to keep and breed readily in captivity. Their egg and larvae stages are transparent and easy to manipulate. Also, their short generation interval provides opportunities to study multiple generations and to look for evolutionary effects. The use of zebrafish and medaka as research animals is likely to increase even faster in the near future.

*16.3.1 Zebrafish (*Danio reiro*)*

Until the 1990s, there were not many studies on zebrafish (Table 16.1), and these were within topics such as basic zoology, neuroscience, anatomy, developmental biology, environmental science and toxicology. However, since 1990 there has been an enormous increase in their use within research on molecular biology and genetics. These new research areas account for 66% of all zebrafish publications since 1991.

Mice and rats are evolutionarily more like us because they are mammals. So, what is the advantage of zebrafish compared to their furry mammalian relatives?

Zebrafish do not require much space, as they are small and prefer to be in large groups or shoals. They breed readily, have a short generation interval (3–4 months) and high productivity, and lay several hundred eggs every 2–3 days (Spence et al. 2008). Their eggs and larvae are transparent and relatively large, which allows microscopic observation of embryos as they grow into fully formed larvae. The larvae grow rapidly, with all major organs developing within 36 h (Kimmel et al. 1995). Recent research has shown that 70% of human genes are present in zebrafish (Howe et al. 2013), which makes them a useful model organism for studies of vertebrate development and gene function. Another advantage is that the embryo develops outside the mother, unlike mice and rats, so the zebrafish egg and larva can easily be manipulated. Furthermore, their transparency enables the visualisation of fluorescent-labelled tissues. Zebrafish can also be used as indicators of pollution (see Example 16.1).

Example 16.1 Chinese researchers have cloned oestrogen-sensitive genes and injected them into fertile eggs of zebrafish. These genetically manipulated fish turn green when exposed to oestrogen-polluted water. Oestrogen is linked to male infertility (Chen et al. 2010). Zebrafish have the ability to regenerate their fins, skin, heart, lateral line hair cells and brain during their larval stages (Wade 2010; Goldshmit et al. 2012; Kishimoto et al. 2012). This ability, in combination with their close genetic relations to humans, has made zebrafish an interesting model organism for many types of medical research.

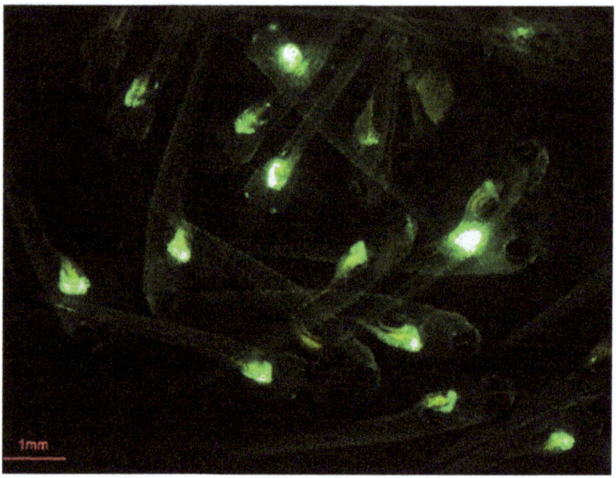

Zebrafish larvae cloned with an oestrogen-sensitive gene, exposed to oestrogen-polluted water. The figure is from Chen et al. (2010).

16.3.2 Salmonids

Long before fish became scientifically interesting as research objects, they were an important food source. Fish have been kept in ponds for thousands of years. Pacific trout, also called rainbow trout (*Oncorhynchus mykiss*), and European trout (*Salmon trutta*) have been farmed since around 1850 (FAO 2019a) to restock and enhance natural populations. However, the commercial farming of Atlantic salmon (*Salmo salar* L.), and its Pacific relatives the Chinook salmon (*Oncorhynchus tshawytscha*), rainbow trout (*O. mykiss*) and Coho salmon (*O. kisutch*), started in the 1970s. By 1990, salmon farming was a great success and now dominates the aquaculture industry (FAO 2019b).

Fish represent the greatest proportion of the world's annual production from aquaculture (44 of 74 million tonnes), and salmon and trout are the largest single fish commodity by value in world trade (FAO 2016). Due to the huge economic importance of salmonids, any study that can improve the efficiency or productivity of the farming industry will have high priority. Thus, farmed salmonids have, in recent decades, become the most commonly used experimental fish. It is impossible to get a full overview of all published papers on fish. Although ISI Web of Science does not cover all publications, we believe that it is the source that gives us the best indication on which species and topics are dominating fish research. According to this source, more than 25% of all published fish papers during the last 10 years have been on salmonids. From 1945 until 1990, approximately 13,000 studies on salmonids were published, but since then there have been over 84,000 papers.

Aquaculture's struggle with salmon disease has also influenced the research focus. Since 1991, there has been an increase in the number of studies within veterinary science, vaccine development and testing, genetics, food science and technology, environmental science and toxicology. Salmon lice have represented the greatest problem since the turn of the century. Close to 90% of all publications (1197 papers in total) on salmon lice (*Lepeophtheirus salmonis*) have been published since the year 2000, and 70% of these have been published in the last 10 years.

Another challenge created by the salmon industry has been the escape of farmed fish, polluting wild salmon gene pools (Glover et al. 2017). The use of sterile triploid salmon (with three sets of chromosomes) has been suggested as a solution, but triploid salmon are less tolerant to low oxygen levels or higher temperatures (SALMOTRIP 2013). These fish have also an increased incidence of malformations and exhibit higher mortality under commercial production conditions (Sambraus 2016).

Recently, CRISPR-Cas9 methodology has been used to knock out genes controlling the development and survival of germ cells in salmon to produce sterile salmon. This ongoing research aims at developing a vaccine that can sterilise the fish (Wargelius et al. 2016). Most probably, farmed salmon will be sterile in the near future. This may save what is left of the genes in wild salmon populations but may be at a cost of poorer fish welfare. Farmed salmon and trout may undergo stressful handling in relation to, among other things, vaccination and delousing (see Stien et al. 2013 and references within). Furthermore, farmed salmonids are kept at densities which are much higher than they would experience in their natural environment.

*16.3.3 Goldfish (***Carassius auratus***)*

The use of goldfish as a research animal increased from the 1960s and through the 1980s (Blanco et al. 2017). Only 95 of the 13,300 papers on this species were published before 1960 (Table 16.1). Most of these papers are on topics within neuroscience and endocrinology. Important areas include the regulation of growth, appetite, metabolism, reproduction, gonadal physiology and stress responses (for a review, see Blanco et al. 2017). Goldfish have been described as excellent model organisms for neuroendocrine signalling and the regulation of reproduction in vertebrates (Popesku et al. 2008). They have also been used for many years as a model organism for vision research (Martinez-Conde and Macknik 2008).

16.3.4 Medaka or Japanese Killifish (Oryzias latipes)

Medaka come from Asia, where they have been used for heritability studies for more than 100 years (see a review by Wittbrodt et al. 2002). As a model species they have many of the same advantages as zebrafish, being small with a short generation interval and high productivity, needing little space and living in shoals. They are hardier, thriving in a temperature range of 6–40 °C, and they are less susceptible to disease (Wittbrodt et al. 2002). Furthermore, they have a relatively small genome for a vertebrate, half the size of that of the zebrafish (Lamatsch et al. 2000). In contrast to zebrafish, medaka have sex chromosomes, in which there is large variability. Medaka are therefore a frequently used model species for studying mechanisms related to sex determination (Myosho et al. 2015; Wittbrodt et al. 2002; Woods et al. 2000). Medaka are also an important test system for environmental research. They are widely used for carcinogenetic studies and for testing endocrine disruptors in ecotoxicology (Hawkins et al. 2003). Salt water species of medaka have been used for marine research (Koyama et al. 2008; Dong et al. 2014).

Medaka have even been reared on board the International Space Station to study the effects of microgravity on osteoclast activity and the vertebrate system for sensing gravity (Ljiri 2003).

16.3.5 Three-Spined Stickleback (Gasterosteus aculeatus)

Three-spined stickleback are found in most inland and coastal waters above 30° North. The marine population is anadromous, breeding in fresh or brackish water. They are, therefore, highly tolerant to changes in salinity and temperature, which makes them a good model species for physiological studies. That they come from a variable environment, are easy to find in nature, and are easy to breed and keep in aquaria make them popular as research animals. They show great morphological variation throughout their area of distribution, making them an ideal model species for the studies of evolution and population genetics (McKinnon and Rundle 2002). Every fifth publication on three-spined stickleback is on evolutionary biology. Furthermore, sticklebacks display elaborate breeding behaviour. The male defends a territory, builds a nest and takes care of the eggs and fry. These intricate reproductive behaviours are all controlled hormonally, making stickleback popular subjects for endocrinological, ethological and behavioural studies (Katsiadaki et al.

2007; Ostlund-Nilsson et al. 2007). They have also been used to study the host–parasite relationship between the tapeworm *Schistocephalus solidus* and the three-spined stickleback. This has become a classic model within parasitology to study interactions, evolution and virulence (Barber and Scharsack 2010; Barber et al. 2000).

16.4 Types of Research Where Fish Are Used as Experimental Animals

A closer look at the approximately 350,000 studies performed on fish from 1945 to 2017, using the ISI Web of Science, gives us an idea of how the choice of fish species and research topics have changed over the last 50 years. In general, the focus has changed from a dominance by more general ecological, zoological and oceanographic studies towards laboratory studies within genetics, molecular biology, toxicology, veterinary science and fundamental biology. However, research related to fisheries and environmental science has also increased, which means that there are still a lot of fish studies being conducted in the field (Fig. 16.1).

16.5 Legislation on the Use of Fish for Experimental Purposes

Requirements for Education and Training of Fish Researchers

Current EU legislation (Directive 2010/63) states that each breeder, supplier and user must have sufficient staff on site, and that they must be adequately educated and trained before they perform any of the following functions:

(a) Carrying out procedures on animals
(b) Designing procedures and projects
(c) Taking care of animals
(d) Killing animals

Member States are required to publish, on the basis of topics in Annex V of the Directive, minimum requirements for education and training and details of how these people are to obtain, maintain and demonstrate competence for these functions. Topics mentioned in Annex V include legislation and ethics, biology and behaviour, anaesthesia, analgesia and humane killing and experimental design. It is also expected that staff are supervised until they have obtained and demonstrated the necessary competence. The European Commission has issued extensive guidance on the operation of the Directive and, specifically, a guidance document on an *education and training framework* with a modular learning outcome-based training structure, principles and criteria for supervision, competence assessment, continued professional development and mutual approval/accreditation of courses (European Union 2019).

There is, however, still a shortage of species-specific educational and training material for use on courses directed towards fish researchers.

Legislative Requirements and Guidelines for Fish Research

Modern legislation regulating animal research offers fish broadly the same protection as for other vertebrates, but they are often referred to as if they were one species. There is, therefore, a large and as yet poorly satisfied need for species-specific and situation-specific guidelines for, at the very least, those species of fish which are used in large numbers in research.

For example, zebrafish are the only fish species specifically referred to in the EU Directive, under Annex I which lists those research species that must be purpose-bred. Annex III, which covers care and accommodation, contains only a very short and general section on fish, in contrast to the detail given for other species. This section covers water supply and quality, oxygen, nitrogen compounds, pH and salinity, temperature, lighting and noise, stocking density and environmental complexity, and feeding and handling. The descriptions are, however, very brief and no species are mentioned by name. This Annex even states that 'appropriate environmental enrichment' (such as hiding places or bottom substrate) does not have to be provided if 'behavioural traits suggest none is required'. Annex IV, which describes acceptable methods of killing, has a separate column for fish, but no species are mentioned by name.

The same can be said of the Council of Europe's Convention for the Protection of Vertebrate Animals used for Experimental and other Scientific Purposes, ETS 123 (Council of Europe 1986). Appendix A of the Convention gives guidelines for the accommodation and care of animals. The Appendix includes general guidance on the environment and its control, health, housing and enrichment, care, humane killing and transport. When this Appendix was revised, expert working groups were convened for the most widely used species, including a group for fish. The background documents produced by these working groups, which contain a large number of references to the scientific literature, are publicly available in the online FELASA library (FELASA 2019a), except for the document on fish. This is particularly unfortunate, since, as it states in the Appendix, 'species-specific guidance on rainbow trout (*Oncorhynchus mykiss*), Atlantic salmon (*Salmo salar*), tilapiine cichlids, zebra fish (*Danio rerio*), sea bass (*Dicentrarchus labrax*), Atlantic halibut (*Hippoglossus hippoglossus*), Atlantic cod (*Gadus morhua*), turbot (*Scophthalmus maximus*), African catfish (*Clarias gariepenus*) is available in the background document elaborated by the Group of Experts'.

In the absence of more details on fish, other parts of the EU Directive must be used to increase fish welfare. These include the use of the local Animal Welfare Body, which is expected to advise on matters relating to welfare, including housing and care. These bodies are expected to receive advice from the National Committees, which in turn shall share knowledge between themselves. There is also a requirement in the Directive for establishments to have a person who is responsible for ensuring that staff have access to species-specific knowledge about the animals used there. This, of course, applies to fish research as well.

16.6 The Three Rs and Other Concepts to Improve Fish Research

In the 1950s, the British organisation Universities Federation for Animal Welfare (UFAW) commissioned a survey among researchers on animal experimentation. The work was carried out by William Russell and Rex Burch. The science of alternatives to animal experiments was still in its infancy at that time, not least due to technological barriers. Russell and Burch focused, therefore, on methods to reduce inhumanity, which indirectly would lead to an increase in the humanity of animal research. They identified two main types of inhumanity:

1. Direct inhumanity: the direct infliction of distress caused by a procedure
2. Contingent inhumanity: distress caused by other factors such as transport, housing conditions, social groups and poor husbandry

Contingent inhumanity, nowadays more often referred to as contingent suffering, is today an important concept in the field of laboratory animal science, as it is easily overlooked by scientists as they plan their experiments. It should be discussed at all stages of the acquisition, care and use of research animals, to reduce contingent suffering to a minimum, not only on welfare grounds but because it will reduce the validity of the data collected.

Russell and Burch presented the results of their survey in a book entitled *The Principles of Humane Experimental Technique* which was published in 1959 (Russell and Burch 1959). The book describes in detail a concept which they developed during the work, which is now known worldwide as *the Three Rs: Replacement, Reduction, Refinement.*

By Replacement they meant any scientific method employing non-sentient material which may replace methods that use conscious living vertebrates.

By Reduction they meant any decrease in the number of animals used to obtain a given amount of information.

By Refinement they referred to developments leading to a decrease in the severity of procedures which still had to be performed on animals.

Russell and Burch distinguished between two types of Replacement: *absolute replacement*, where no animals are used at all, and *relative replacement*, which includes non-recovery (terminal) experiments on anaesthetised animals and the humane killing of animals to obtain material for research.

Michael Balls (2008) has suggested that it can be helpful to split replacement techniques into two more categories: *direct replacement*, in which the alternative method yields roughly the same information as an animal study, and *indirect replacement*, in which the information achieved is different but can be used for a similar purpose.

The 3Rs are now mentioned specifically or indirectly in legislation worldwide, including the EU Directive 2010/63. Many institutions have written their own interpretations of Russell and Burch's concept (Tannenbaum and Bennett 2015). Nowadays, many would not consider the use of animals in terminal anaesthesia as Replacement, even though they are non-sentient. Many people today reserve the use of Replacement for non-animal methods.

Article 13 of the EU Directive requires Member States to ensure that a procedure is not carried out if another method or testing strategy for obtaining the result, not entailing the use of a live animal, is recognised under the legislation of the Union. It is therefore important that fish researchers are made aware of the development of alternative methods (see Example 16.2).

Example 16.2 Fish may be used instead of mice and rats in toxicity tests (Hodson 1985). They are also used to test vaccines for the aquaculture industry. The LD_{50} method has traditionally been used for toxicity testing (Paget 1983). This test aims at estimating the concentration at which 50% of the test animals die, and it is therefore associated with pain and suffering in many animals (see Botham (2004) for a history of the work to refine the methods used in toxicity tests). To avoid death as an endpoint, a method called Fixed Dose Procedure (FDP) was developed (van den Heuvel et al. 1990). In this procedure, the test substance is given at four fixed-dose levels, aiming at identifying a dose that produces clear signs of toxicity but no mortality (Stallard et al. 2002).

These types of testing using the whole animal are called in vivo testing. However, in many cases the test fish can be replaced by using microorganisms, cell cultures or biological molecules (so-called in vitro methods) (Castaño et al. 2003). Cell cultures have several advantages, and in vitro testing is likely to replace much of the in vivo testing in the future (Erkekoglu et al. 2011).

Likewise, the concept of Refinement, which in Russell and Burch's book is used for methods to reduce inhumanity, has evolved to include techniques which actively improve animal welfare by, for example, the use of environmental enrichment (Chap. 5).

Many other "Rs" have been proposed to supplement those of Russell and Burch (Rowan and Goldberg 1995). Within fish research, the Rs of Relevance and

Reproducibility are particularly constructive concepts. Example 16.3 gives an example of why this is the case.

Example 16.3 Research has shown that the Reduction of the number of fish in a tank is not always best for Atlantic salmon under laboratory conditions. A careful balance between Reduction and Refinement has to be maintained, depending on the life stage of the fish. Recent studies have shown more fighting, higher cortisol levels and larger individual variations in salmon smolts stored in small groups (<50 individuals) compared to larger groups (Nilsson 2017).

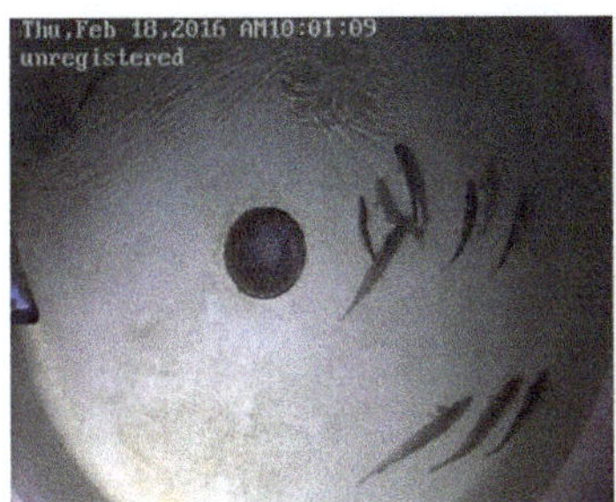

16.6.1 *The Three S's*

Although the 3Rs have dominated discussions on the ethics of animal research, another tenet, proposed in the 1970s before the 3Rs attracted much attention, deserves a mention. This concept is the 3S's of Carol Newton (1925–2014). Unlike the 3Rs, the 3S's were never published, and they are known only from a remark in the proceedings of a symposium held in Washington, D.C. in 1975, when they were mentioned by one of the other participants, Harry Rowsell. Newton's 3S's were, according to Rowsell, *Good Science, Good Sense and Good Sensibilities*. The concept has been brought into the limelight again in the form of a publication by Smith and Hawkins, who offer their interpretation (Smith and Hawkins 2016).

16.6.2 *A Culture of Care*

Recital 31 of EU Directive 2010/63 states that the local animal welfare body at a research institution is expected to 'follow the development and outcome of projects at establishment level, foster a climate of care and provide tools for the practical application and timely implementation of recent technical and scientific developments in relation to the principles of replacement, reduction and refinement, in order to enhance the life-time experience of the animals'. Fostering a climate of care has, during the last few years, emerged as yet another means of improving welfare, in this case not just for the animals but also for all those involved in the care and use of animals, under the name of 'A Culture of Care' (Klein and Bayne 2007). Some have taken this one step further and urge users also to adopt a *Culture of Challenge*, not necessarily accepting current practice but searching for better (i.e. less harmful and more scientifically valid) means of conducting animal research (Louhimies 2015). An international Culture of Care Network was established in 2016 (Norecopa 2019a).

16.7 How Can We Increase Research Quality and Better the Conditions for Laboratory Fish?

The consideration of animal welfare and the concept of the three Rs, with the ultimate aim of replacing sentient animals with alternatives, is embedded in all modern legislation regulating animal research. Improved animal welfare has also scientific advantages, since animals that cope best with their surroundings, experiencing a minimum of stress, deliver the most valid data.

16.7.1 Classification of the Severity of Fish Experiments

The EU Directive requires the severity of each procedure to be classified on the basis of the '*degree of pain, suffering, distress or lasting harm expected to be experienced by an individual animal during the course of the procedure*', with the aim of enhancing transparency, facilitating the project authorisation process and providing tools for monitoring compliance. Member States have to ensure that all procedures are classified as 'non-recovery', 'mild', 'moderate' or 'severe' on a case-by-case basis. Criteria for assignment to these categories are described in a European Commission Working Group report on severity classification. This focuses heavily on procedures that are relevant to the 'traditional', terrestrial laboratory animal species (Expert Working Group Report 2009).

As 'fish' are an extremely diverse vertebrate class, the effect of a procedure is likely to vary markedly between species. Criteria for the assignment of severity classification to fish procedures should not only pay attention to species differences, but also to the fact that many procedures are performed out of water, which in itself involves the stress of capture, handling and immobilisation (Example 16.4). In addition, many species of fish undergo large physiological changes during their natural part of their life cycle. The same procedure may therefore affect different age groups in different ways. For these reasons, Norecopa commissioned a Working Group which produced a set of guidelines for severity classification of procedures used on fish, using a similar format to the EU Commission report (Hawkins et al. 2011). More resources are available on Norecopa's website (Norecopa 2019b).

Example 16.4 Routine handling, such as moving fish from one tank to another using a dip net (a), has shown to affect the outcome in behavioural studies performed afterwards (Brydges et al. 2009; Thompson et al. 2016). However, better nets are available (b) where fish are moved together with some of their tank water, inside a dark bag (see the discussion in Pelletier et al. 2007).

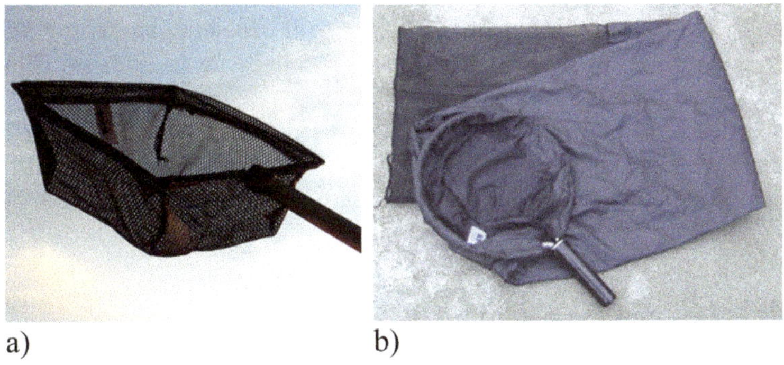

a) b)

16.7.2 *Detection and Alleviation of Pain, Suffering and Distress in Fish*

Techniques to manage pain in fish are in their infancy and in many cases are based on extrapolation from experience with mammalian species. There is an urgent need for reliable methods to detect and alleviate suffering for each of the commonly used species. This would also make predictions of severity and retrospective reporting more reliable. Currently available indicators include clinical signs such as respiratory rate, food consumption and health status, together with the assessment of stress levels in groups of fish (Sneddon 2009; Bert et al. 2016).

Traditionally there has been greater tolerance of stress, disease and mortality in fish when used in research compared to mammals, which probably reflects general attitudes to fish in society (Byrd et al. 2017). Such attitudes should be challenged. Mortality rates can, however, be difficult to assess in some instances, if the species in question has a naturally high mortality rate. There is a great need for more species and situation-specific guidance on anaesthesia and analgesia in fish (see Example 16.5).

Example 16.5 Tricaine methanesulfonate (also called MS-222) is probably the most widely used anaesthetic worldwide, as it is the only anaesthetic drug approved by the US Food and Drug Administration for use in fish that are to be used for human consumption. It is registered for veterinary use in fish by Health Canada, and in the

UK, Italy, Spain and Norway (see review on MS-222 by Popovic et al. 2012). However, it is well known that many fish species show strong aversive behaviour when introduced to water containing MS-222 (Readman et al. 2013). Although MS-222 anaesthesia benefits the fish by minimizing the impact of a more severe stressor, it is nonetheless stressful and causes physiological effects such as an increase in cortisol levels (Gressler et al. 2014; Readman et al. 2013). Other anaesthetics such as clove oil, eugenol (Atlantic salmon, Iversen et al. 2003), metomidate (catfish, Small 2003), ethomidate and tribromoethanol (zebrafish, Readman et al. 2013) have been shown to have fewer negative effects (lower cortisol responses). The choice of optimal anaesthetic will depend upon the species, exposure time and dose (Sneddon 2012). Thus, guidelines for anaesthesia of the most commonly used species are urgently needed.

Analgesics have until recently only been used occasionally in fish. There is a comparable need to increase our knowledge of suitable analgesics for fish species and to develop guidelines for their use (Mettam et al. 2011; Sneddon 2012).

16.7.3 Humane Endpoints

In research involving mammals, a lot of attention has been applied to the concept of humane endpoints, in which animals are humanely killed (or removed from the experiment and treated) as soon as the objective has been reached, or if the animals reach a level of suffering which has been designated in advance as the maximum they are to experience. This avoids animals suffering unnecessarily. Article 13 in the EU Directive describes this concept in detail, requiring where possible the use of early humane endpoints. Where death as the endpoint is unavoidable, the procedure shall be designed so as to:

(a) Result in the deaths of as few animals as possible
(b) Reduce the duration and intensity of suffering to the animal to the minimum possible and, as far as possible, ensure a painless death

There is a great potential to refine the endpoints of fish experiments. Many claim that fish often show no clear clinical signs of distress until just before they die, making it difficult to apply humane endpoints in a study. No doubt, as our knowledge of the behavioural physiology of fish increases, scientists and technicians will become better at detecting earlier signs of distress, which will reduce the use of death as an endpoint.

16.7.4 Guidelines for Fish Research

Guidelines for the care and use of fish in research can be difficult to track down, as they are typically produced in one country by one scientific organisation. To improve the situation, a global list of guidelines has been made available in the Norecopa database (Norecopa 2019c). These include general guidelines for health and welfare monitoring of fish in research (Johansen et al. 2006). It is the authors' hope that more species-specific guidelines will be produced within this area, as has been the case with health monitoring guidelines for mammalian species (FELASA 2019b).

Research such as the ENRICH Fish project is also playing a role in helping to disseminate information on the refinement of housing for both laboratory and farmed Atlantic salmon (Norecopa 2019d). There is a need for many more projects of this type.

An overview of guidelines for handling, bleeding and administration techniques in fish (Hawkins 2009) was presented at a consensus meeting entitled *Harmonisation of the Care and Use of Fish in Research* at Gardermoen, Norway in 2009 (Norecopa 2019e). Not much progress has been made since then in producing new guidelines. An overview of guidelines and 3R resources for zebrafish is also available (Norecopa 2019f).

16.7.5 Validity, Reproducibility and Translatability of Fish Experiments

Analyses of scientific papers reporting animal experiments have revealed alarming omissions (see references in Smith et al. 2017), even after the development and widespread journal endorsement of reporting guidelines such as ARRIVE (Kilkenny et al. 2010). Reporting guidelines designed specifically for fish experiments have been produced (Brattelid and Smith 2000), based upon an existing set which were more suitable for studies on mammals (Ellery 1985), but there is clearly an additional, more fundamental, need to improve the quality of the research itself. Reporting guidelines are valuable drivers for improving the quality of scientific publications, but they are not in themselves sufficient to improve the quality of the research leading up to the publications.

There is in fact widespread concern about the lack of reproducibility and translatability of laboratory animal research (Chalmers et al. 2014; Macleod et al. 2014; Munafò et al. 2017), and a new term has been coined: therioepistemology, the study of how knowledge is gained from animal research (Garner et al. 2017). This concern comes in addition to more general concern in science about publication bias, in which the reporting of positive results is favoured (Nissen et al. 2016).

Guidelines for reporting animal experiments cannot solve problems associated with poor design. If experiments on animals are to meet current ethical, welfare and scientific standards, they must be assessed for quality at all stages, including planning, performance, interpretation of results and publication. Experiments on fish bring extra challenges, due among other things to the fact that they live in a medium with which we are unfamiliar, to our incomplete knowledge of welfare requirements, pain perception, and the fact that 'fish' tend to be treated as one species. Systematic and thorough planning increases the likelihood of success and is an important step in the implementation of the three R's. To this end, advice specifically for *planning* animal experiments has also been produced. The PREPARE guidelines (Smith et al. 2017) combine a checklist (currently available in 21 languages) with a comprehensive website (https://norecopa.no/PREPARE) where there are links to a large number of species- and situation-specific guidelines for each of the 15 topics in the checklist. New guidelines are added to this website as they become available. The English version of the checklist is shown in Fig. 16.2.

Besides the ethical reasons for focusing on the welfare of experimental animals, there is also a great scientific advantage in doing so. To obtain scientifically valid results, the animals being studied must be in good condition and not stressed. Good welfare is therefore also good science.

16.8 Concluding Remarks

The vast amount of research that has been conducted on fish has contributed greatly to our understanding of fish behaviour, cognition and sentience. Through these studies, we have gained insights into fish biology which in turn have fuelled public and regulatory concerns for fish welfare. This insight creates a positive feedback loop which improves the welfare of fish in new research projects (Fig. 16.3).

It is not always easy to identify advances in fish welfare in the scientific literature, as these are often part of papers on other subjects. There is a need for more work specifically on the welfare of fish in research.

As described earlier, an additional challenge with research on fish is that they comprise over 30,000 individual species, each with their own needs. It is therefore not possible to make detailed general guidelines for fish research. However, since most fish studies are done on a small number of species (in particular salmonids, zebrafish, medaka, guppies and sticklebacks), by focusing on the construction of detailed guidelines and regulations for these species, we can relatively rapidly increase research quality and conditions for laboratory fish.

Welfare studies require the ability to detect the welfare state of the animal. Fish have, by mammalian standards, very limited facial expression and body language, as far as we can tell today. Other good indicators of welfare must be sought after. Welfare indicators are quantifiable internal or external parameters which are associated with predictable variation in the fish's well-being. These indicators can be:

- External pathological features, e.g. skin infections and wounds, scale loss or fin erosion
- Internal parameters, e.g. levels of cortisol, lactate or toxins
- Behavioural indicators, e.g. changes in activity patterns, cessation of feeding, loss of natural reflexes or surface swimming
- Morphological indicators, e.g. pigmentation of the skin or the presence of gas in the eye

PREPARE

The PREPARE Guidelines Checklist

Planning Research and Experimental Procedures on Animals: Recommendations for Excellence

Adrian J. Smith[a], R. Eddie Clutton[b], Elliot Lilley[c], Kristine E. Aa. Hansen[d] & Trond Brattelid[e]

[a]Norecopa, c/o Norwegian Veterinary Institute, P.O. Box 750 Sentrum, 0106 Oslo, Norway; [b]Royal (Dick) School of Veterinary Studies, Easter Bush, Midlothian, EH25 9RG, U.K.; [c]Research Animals Department, Science Group, RSPCA, Wilberforce Way, Southwater, Horsham, West Sussex, RH13 9RS, U.K.; [d]Section of Experimental Biomedicine, Department of Production Animal Clinical Sciences, Faculty of Veterinary Medicine, Norwegian University of Life Sciences, P.O. Box 8146 Dep., 0033 Oslo, Norway; [e]Division for Research Management and External Funding, Western Norway University of Applied Sciences, 5020 Bergen, Norway.

PREPARE[1] consists of planning guidelines which are complementary to reporting guidelines such as ARRIVE[2].
PREPARE covers the three broad areas which determine the quality of the preparation for animal studies:

1. **Formulation of the study**
2. **Dialogue between scientists and the animal facility**
3. **Quality control of the components in the study**

The topics will not always be addressed in the order in which they are presented here, and some topics overlap. The PREPARE checklist can be adapted to meet special needs, such as field studies. PREPARE includes guidance on the management of animal facilities, since in-house experiments are dependent upon their quality. The full version of the guidelines is available on the Norecopa website, with links to global resources, at **https://norecopa.no/PREPARE**.
The PREPARE guidelines are a dynamic set which will evolve as more species- and situation-specific guidelines are produced, and as best practice within Laboratory Animal Science progresses.

Topic	Recommendation
(A) Formulation of the study	
1. Literature searches	☐ Form a clear hypothesis, with primary and secondary outcomes. ☐ Consider the use of systematic reviews. ☐ Decide upon databases and information specialists to be consulted, and construct search terms. ☐ Assess the relevance of the species to be used, its biology and suitability to answer the experimental questions with the least suffering, and its welfare needs. ☐ Assess the reproducibility and translatability of the project.
2. Legal issues	☐ Consider how the research is affected by relevant legislation for animal research and other areas, e.g. animal transport, occupational health and safety. ☐ Locate relevant guidance documents (e.g. EU guidance on project evaluation).
3. Ethical issues, Harm-Benefit Assessment and humane endpoints	☐ Construct a lay summary. ☐ In dialogue with ethics committees, consider whether statements about this type of research have already been produced. ☐ Address the 3Rs (Replacement, Reduction, Refinement) and the 3Ss (Good Science, Good Sense, Good Sensibilities). ☐ Consider pre-registration and the publication of negative results. ☐ Perform a Harm-Benefit Assessment and justify any likely animal harm. ☐ Discuss the learning objectives, if the animal use is for educational or training purposes. ☐ Allocate a severity classification to the project. ☐ Define objective, easily measurable and unequivocal humane endpoints. ☐ Discuss the justification, if any, for death as an end-point.
4. Experimental design and statistical analysis	☐ Consider pilot studies, statistical power and significance levels. ☐ Define the experimental unit and decide upon animal numbers. ☐ Choose methods of randomisation, prevent observer bias, and decide upon inclusion and exclusion criteria.

Fig. 16.2 The PREPARE checklist for planning research and experimental procedures on animals (Smith et al. 2017)

Topic	Recommendation
	(B) Dialogue between scientists and the animal facility
5. Objectives and timescale, funding and division of labour	☐ Arrange meetings with all relevant staff when early plans for the project exist. ☐ Construct an approximate timescale for the project, indicating the need for assistance with preparation, animal care, procedures and waste disposal/decontamination. ☐ Discuss and disclose all expected and potential costs. ☐ Construct a detailed plan for division of labour and expenses at all stages of the study.
6. Facility evaluation	☐ Conduct a physical inspection of the facilities, to evaluate building and equipment standards and needs. ☐ Discuss staffing levels at times of extra risk.
7. Education and training	☐ Assess the current competence of staff members and the need for further education or training prior to the study.
8. Health risks, waste disposal and decontamination	☐ Perform a risk assessment, in collaboration with the animal facility, for all persons and animals affected directly or indirectly by the study. ☐ Assess, and if necessary produce, specific guidance for all stages of the project. ☐ Discuss means for containment, decontamination, and disposal of all items in the study.
	(C) Quality control of the components in the study
9. Test substances and procedures	☐ Provide as much information as possible about test substances. ☐ Consider the feasibility and validity of test procedures and the skills needed to perform them.
10. Experimental animals	☐ Decide upon the characteristics of the animals that are essential for the study and for reporting. ☐ Avoid generation of surplus animals.
11. Quarantine and health monitoring	☐ Discuss the animals' likely health status, any needs for transport, quarantine and isolation, health monitoring and consequences for the personnel.
12. Housing and husbandry	☐ Attend to the animals' specific instincts and needs, in collaboration with expert staff. ☐ Discuss acclimation, optimal housing conditions and procedures, environmental factors and any experimental limitations on these (e.g. food deprivation, solitary housing).
13. Experimental procedures	☐ Develop refined procedures for capture, immobilisation, marking, and release or re-homing. ☐ Develop refined procedures for substance administration, sampling, sedation and anaesthesia, surgery and other techniques.
14. Humane killing, release, re-use or re-homing	☐ Consult relevant legislation and guidelines well in advance of the study. ☐ Define primary and emergency methods for humane killing. ☐ Assess the competence of those who may have to perform these tasks.
15. Necropsy	☐ Construct a systematic plan for all stages of necropsy, including location, and identification of all animals and samples.

References

1. Smith AJ, Clutton RE, Lilley E, Hansen KEA & Brattelid T. PREPARE:Guidelines for Planning Animal Research and Testing. *Laboratory Animals,* 2017, DOI: 10.1177/0023677217724823.
2. Kilkenny C, Browne WJ, Cuthill IC *et al.* Improving Bioscience Research Reporting: The ARRIVE Guidelines for Reporting Animal Research. *PloS Biology,* 2010; DOI: 10.1371/journal.pbio.1000412.

Further information
https://norecopa.no/PREPARE / *post@norecopa.no* / *@norecopa*

Fig. 16.2 (continued)

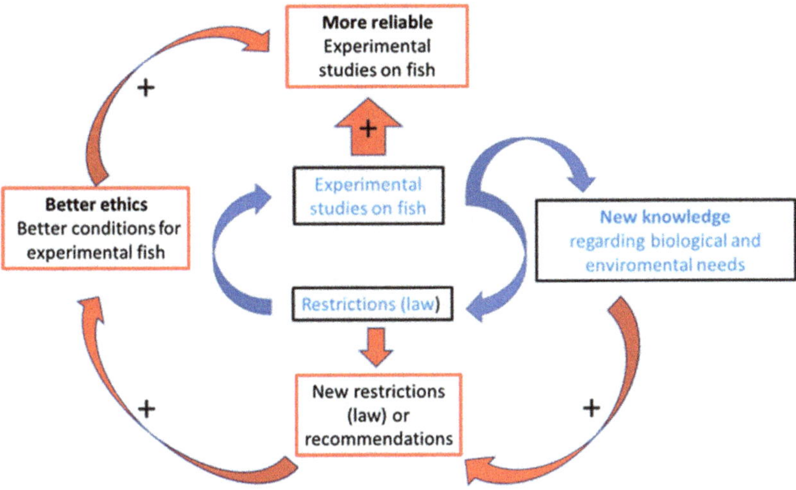

Fig. 16.3 The ethical and scientific positive feedback loop. It is important that we take advantage of the new knowledge achieved through experimental studies to improve legislation

To be able to utilize these welfare indicators, we need basic knowledge about normal variations so that we can validate the most commonly used welfare indicators. From this we will be able to build a knowledge database and guidelines to score the welfare of the most commonly used species. If we start with salmonids and zebrafish, we will better the life of approximately 30% of laboratory fish. This work has indeed started. Stien and co-workers recently made an overall welfare assessment model for Atlantic salmon (SWIM 1.0) (Stien et al. 2013). The model was designed to help in the assessment of welfare status by fish farmers, but it is of course also applicable to research, and the results should be integrated in future regulations and guidelines.

Acknowledgement We thank Gunvor Kristin Knudsen for assistance with statistical data regarding the use of research animals in non-EU countries.

References

Balls M (2008) Professor W.M.S. Russell (1925–2006): Doyen of the three Rs. ALTEX 14, Special Issue, 1–7. In: Proceedings of the 6th World Congress on Alternatives & Animal Use in the Life Sciences, August 21–25, 2007, Tokyo. http://www.asas.or.jp/jsaae_old/zasshi/WC6_PC/paper1pdf. Accessed 13 Mar 2020

Barber I, Scharsack JP (2010) The three-spined stickleback-Schistocephalus solidus system: an experimental model for investigating host-parasite interactions in fish. Parasitology 137:38

Barber I, Hoare D, Krause J (2000) Effects of parasites on fish behaviour: a review and evolutionary perspective. Rev Fish Biol Fish 10:131–165

Bert B, Chmielewska J, Bergmann S, Busch M, Driever W, Finger-Baier K, Hößler J, Köhler A, Leich N, Misgeld T, Nöldner T, Reiher A, Schartl M, Seebach-Sproedt A, Thumberger T, Schönfelder G, Grune B (2016) Considerations for a European animal welfare standard to evaluate adverse phenotypes in teleost fish. EMBO J 35:1151–1154. https://doi.org/10.15252/embj.201694448. Accessed 13 Mar 2020

Blanco AM, Sundarrajan L, Bertucci JI, Unniappan S (2017) Why goldfish? Merits and challenges in employing goldfish as a model organism in comparative endocrinology research. Gen Comp Endocrinol 257:13. https://doi.org/10.1016/j.ygcen.2017.02.001. Accessed 13 Mar 2020

Botham PA (2004) Acute systemic toxicity – prospects for tiered testing strategies. Toxicol In Vitro 18(2):227–230

Brattelid T, Smith AJ (2000) Guidelines for reporting the results of experiments on fish. Lab Anim 34:131–135. https://journals.sagepub.com/doi/abs/10.1258/002367700780457590. Accessed 13 Mar 2020

Brydges NM, Boulcott P, Ellis T, Braithwaite VA (2009) Quantifying stress responses induced by different handling methods in three species of fish. Appl Anim Behav Sci 116:295–301

Byrd E, Widmar NO, Fulton J (2017) Of fur, feather and fin: human's use and concern for non-human species. Animals 7(3):22. https://doi.org/10.3390/ani7030022. Accessed 13 Mar 2020

Castaño A, Bols N, Braunbeck T, Dierickx P, Halder M, Isomaa B, Kawahara K, Lee LEJ, Mothersill C, Pärt P, Repetto G, Sintes JR, Rufli H, Smith R, Wood C, Segner H (2003) The use of fish cells in ecotoxicology the report and recommendations of ECVAM workshop 47. ATLA 31:317–351. https://journals.sagepub.com/doi/abs/10.1177/026119290303100314. Accessed 13 Mar 2020

Chalmers I, Bracken MB, Djulbegovic B, Garattini S, Grant J, Gülmezoglu AM, Howells DW, Ioannidis JPA, Oliver S (2014) How to increase value and reduce waste when research priorities are set. Lancet 383:156–165. https://doi.org/10.1016/s0140-6736(13)62229-1. Accessed 13 Mar 2020

Chen H, Hu J, Yang J, Wang Y, Xu H, Jiang Q, Gong Y, Gu Y, Song H (2010) Generation of a fluorescent transgenic zebrafish for detection of environmental estrogens. Aquat Toxicol 96:53–61

Council of Europe (1986) European convention for the protection of vertebrate animals used for experimental and other scientific purposes (ETS 123). http://www.coe.int/en/web/conventions/full-list/-/conventions/treaty/123. Accessed 13 Mar 2020

Dong S, Kang M, Wu X, Ye T (2014) Development of a promising fish model (*Oryzias melastigma*) for assessing multiple responses to stresses in the marine environment. BioMed Research International. ID 563131, 17 p. https://doi.org/10.1155/2014/563131. Accessed 13 Mar 2020

Ellery AW (1985) Guidelines for specification of animals and husbandry methods when reporting the results of animal experiments. Report of the Working Committee for the Biological Characterization of Laboratory Animals/GV-SOLAS, Laboratory Animals 19, pp 106–108. https://journals.sagepub.com/doi/abs/10.1258/002367785780942714. Accessed 13 Mar 2020

Erkekoglu P, Giray BK, Başaran N (2011) 3R principle and alternative toxicity testing methods. FABAD J Pharm Sci 36(2):101–117

European Union (2010) Directive 2010/63/EU of the European Parliament and of the Council of 22 September 2010 on the Protection of Animals Used for Scientific Purposes. https://eur-lex.europa.eu/legal-content/EN/TXT/?uri=celex%3A32010L0063. Accessed 13 Mar 2020

European Union (2019) Implementation, interpretation and terminology of Directive 2010/63/EU. https://ec.europa.eu/environment/chemicals/lab_animals/interpretation_en.htm. Accessed 13 Mar 2020

European Union (2020) 2019 report on the statistics on the use of animals for scientific purposes in the Member States of the European Union in 2015–2017. https://eur-lex.europa.eu/legal-content/EN/TXT/?qid=1581689520921&uri=CELEX:52020DC0016. Accessed 13 Mar 2020

Expert Working Group Report (2009) On severity classification of scientific procedures performed on animals. Brussels, 2009. http://ec.europa.eu/environment/chemicals/lab_animals/pdf/report_ewg.pdf. Accessed 13 Mar 2020
FAO (2016) The state of world fisheries and aquaculture 2016. Contributing to food security and nutrition for all Rome, 190 pages. http://www.fao.org/3/a-i5555e.pdf. Accessed 13 Mar 2020
FAO (2019a) *Salmo trutta* (Berg, 1908). http://www.fao.org/fishery/culturedspecies/Salmo_trutta/en. Accessed 13 Mar 2020
FAO (2019b) *Salmo salar* (Linnaeus, 1758). http://www.fao.org/fishery/culturedspecies/Salmo_salar/en. Accessed 13 Mar 2020
FELASA (2019a) Library. http://www.felasa.eu/about-us/library. Accessed 13 Mar 2020
FELASA (2019b) Recommendations. http://www.felasa.eu/working-groups/recommendation. Accessed 13 Mar 2020
FishBase (2019). https://www.fishbase.se. Accessed 13 Mar 2020
Garner JP, Gaskill BN, Weber EM, Ahloy-Dallaire J, Pritchett-Corning KR (2017) Introducing Therioepistemology: the study of how knowledge is gained from animal research. Lab Anim 46:103–113
Glover KA, Solberg MF, McGinnity P, Hindar K, Verspoor E, Coulson MW, Hansen MM, Araki H, Skaala Ø, Svåsand T (2017) Half a century of genetic interaction between farmed and wild Atlantic salmon: status of knowledge and unanswered questions. Fish Fish 1–38. https://doi.org/10.1111/faf.12214. Accessed 13 Mar 2020
Goldshmit Y, Sztal T, Jusuf PR, Hall TE, Nguyen-Chi M, Currie PD (2012) Fgf-dependent glial cell bridges facilitate spinal cord regeneration in zebrafish. J Neurosci 32(22):7477–7492. https://doi.org/10.1523/JNEUROSCI.0758-12.2012. Accessed 13 Mar 2020
Gressler LT, Riffel APK, Parodi TV, Saccol EMH, Koakoski G, Costa ST, Pavanato MA, Heinzmann BM, Caron B, Schmidt D, Llesuy SF, Barcellos LJG, Baldisserotto B (2014) Silver catfish *Rhamdia quelen* immersion anaesthesia with essential oil of *Aloysia triphylla* (L'Hérit) Britton or tricaine methanesulfonate: effect on stress response and antioxidant status. Aquac Res 45(6):1061–1072
Hawkins P (2009) An overview of existing guidelines for handling, bleeding, administration and identification techniques. https://norecopa.no/media/6342/fish-guidelines.pdf. Accessed 13 Mar 2020
Hawkins WE, Walker WW, Fournie JW, Manning JS, Krol RM (2003) Use of the Japanese medaka (*Oryzias latipes*) and guppy (*Poecilia reticulata*) in carcinogenesis testing under national toxicology program protocols. Toxicol Pathol 31:88–91. https://journals.sagepub.com/doi/10.1080/01926230390174968. Accessed 13 Mar 2020
Hawkins P, Dennison N, Goodman G, Hetherington S, Llywelyn-Jones S, Ryder K, Smith AJ (2011) Guidance on the severity classification of scientific procedures involving fish: report of a working group appointed by the Norwegian consensus-platform for the replacement, reduction and refinement of animal experiments (Norecopa). Lab Anim 45:219–224. https://doi.org/10.1258/la.2011.010181. Accessed 13 Mar 2020
Hodson PV (1985) A comparison of the acute toxicity of chemicals to fish, rats and mice. J Appl Toxicol 5:2220–2226
Howe K, Clark MD, Torroja CF et al (2013) The zebrafish reference genome sequence and its relationship to the human genome. Nature 496:498–503. https://doi.org/10.1038/nature12111. Accessed 13 Mar 2020
Humane Research Australia (2016) 2015 Australian Statistics of Animal Use in Research and Teaching. http://www.humaneresearch.org.au/statistics/statistics_2015. Accessed 13 Mar 2020
IUCN Red List (2014) Table 1. Numbers of threatened species by major groups of organisms (1996–2014). http://cmsdocs.s3.amazonaws.com/summarystats/2014_3_Summary_Stats_Page_Documents/2014_3_RL_Stats_Table_1.pdf. Accessed 13 Mar 2020
Iversen I, Finstad B, McKinley RS, Eliassen RA (2003) The efficacy of metomidate, clove oil, Aqui-Sk and Benzoak® as anaesthetics in Atlantic salmon (*Salmo salar* L.) smolts, and their potential stress-reducing capacity. Aquaculture 221:549–566

Johansen R, Needham JR, Colquhoun DJ, Poppe TT, Smith AJ (2006) Guidelines for health and welfare monitoring of fish used in research. Lab Anim 40:323–340. https://journals.sagepub.com/doi/abs/10.1258/002367706778476451. Accessed 13 Mar 2020

Katsiadaki I, Sanders M, Sebire M, Nagae M, Soyano K, Scott AP (2007) Three-spined stickleback: an emerging model in environmental endocrine disruption. Environ Sci 14:263–283

Kilkenny C, Browne WJ, Cuthill IC, Emerson M, Altman DG (2010) Improving bioscience research reporting: the ARRIVE guidelines for reporting animal research. PLoS Biol 8: e1000412. https://doi.org/10.1371/journal.pbio.1000412. Accessed 13 Mar 2020

Kimmel CB, Ballard WW, Kimmel SR, Ullmann B, Schilling TF (1995) Stages of embryonic development of the zebrafish. Dev Dyn 203:253–310

Kishimoto N, Shimizu K, Sawamoto K (2012) Neuronal regeneration in a zebrafish model of adult brain injury. Dis Model Mech 5:200–209. https://doi.org/10.1242/dmm.007336. Accessed 13 Mar 2020

Klein HJ, Bayne KA (2007) Establishing a culture of care, conscience, and responsibility: addressing the improvement of scientific discovery and animal welfare through science-based performance standards. ILAR J 43(1):3–11

Koyama J, Kawamata M, Imai S, Fukunaga M, Uno S, Kakuno A (2008) Java medaka: a proposed new marine test fish for ecotoxicology. Environ Toxicol 23(4):487–491. https://doi.org/10.1002/tox.20367. Accessed 13 Mar 2020

Lamatsch DK, Steinlein C, Schmid M, Schartl M (2000) Noninvasive determination of genome size and ploidy level in fishes by flow cytometry: detection of triploid *Poecilia Formosa*. Cytometry 39:91–95

Ledford H (2015) CRISPR, the disruptor. Nature 522(7554):20–24

Ljiri K (2003) Life-cycle experiments of medaka fish aboard the international space station. Adv Space Biol Med 9:201–216

Louhimies S (2015) Refinement facilitated by the culture of care. In: Proceedings of the EUSAAT congress, 20–23 Sept 2015, Linz. ALTEX, vol 4(2), p 154. Abstract available at http://www.altex.ch/resources/ALTEX_Linz_proceedings_2015_full.pdf. Accessed 13 Mar 2020

Macleod MR, Michie S, Roberts I, Dirnagl U, Chalmers I, Ioannidis JPA, Al-Shahi Salman R, Chan A-W, Glasziou P (2014) Biomedical research: increasing value, reducing waste. Lancet 383:101–104. https://doi.org/10.1016/s0140-6736(13)62329-6. Accessed 13 Mar 2020

Martinez-Conde S, Macknik SL (2008) Fixational eye movements across vertebrates: comparative dynamics, physiology, and perception. J Vis 8:28. https://doi.org/10.1167/8.14.28. Accessed 13 Mar 2020

Mattilsynet (2019) Forsøksdyr. https://www.mattilsynet.no/dyr_og_dyrehold/dyrevelferd/forsoksdyr/. Accessed 13 Mar 2020

McKinnon JS, Rundle HD (2002) Speciation in nature: the threespine stickleback model systems. Trends Ecol Evol 17:480–488

Mettam JJ, Oulton LJ, McCrohan CR, Sneddon LU (2011) The efficacy of three types of analgesic drugs in reducing pain in the rainbow trout, *Oncorhynchus mykiss*. Appl Anim Behav Sci 133 (3):265–274

Munafò MR, Nosek BA, Bishop DVM, Button KS, Chambers CD, du Sert NP, Simonsohn U, Wagenmakers E-J, Ware JJ, Ioannidis JPA (2017) A manifesto for reproducible science. Nat Hum Behav 1:0021. https://doi.org/10.1038/s41562-016-0021. Accessed 13 Mar 2020

Myosho T, Takehana Y, Hamaguchi S, Sakaizumi M (2015) Turnover of sex chromosomes in *Celebensis* group medaka fishes. Genes Genomes Genet 5:2685–2691

Nilsson (2017) Social enrichment and requirements for the tank rearing of Atlantic salmon. https://norecopa.no/media/7276/nilsson.pdf. Accessed 13 Mar 2020

Nissen SB, Magidson T, Gross K, Bergstrom CT (2016) Publication bias and the canonization of false facts. eLife 5:e21451. https://doi.org/10.7554/eLife.21451. Accessed 13 Mar 2020

NOAA Fisheries (2014) Salmon restoration and PIT tags: big data from a small device. https://www.fisheries.noaa.gov/feature-story/salmon-restoration-and-pit-tags-big-data-small-device. Accessed 13 Mar 2020

Norecopa (2019a) The International Culture of Care Network. https://norecopa.no/coc. Accessed 13 Mar 2020
Norecopa (2019b) Severity classification. http://norecopa.no/categories. Accessed 13 Mar 2020
Norecopa (2019c) The 3R Guide database. https://norecopa.no/3r-guide-database. Accessed 13 Mar 2020
Norecopa (2019d) ENRICH fish. http://www.enrich-fish.net. Accessed 13 Mar 2020
Norecopa (2019e) Harmonisation of the care and use of fish in research. https://norecopa.no/meetings/fish-2009. Accessed 13 Mar 2020
Norecopa (2019f) 3rs resources and guidelines for zebrafish. http://norecopa.no/media/7724/zebrafish-resources.pdf. Accessed 13 Mar 2020
Ostlund-Nilsson S, Maier I, Huntingford F (2007) Biology of the three-spined stickleback. Taylor & Francis Group, London, p 392
Paget E (1983) The LD_{50} test. Acta Pharmacol Toxicol 52:1–14
Pelletier C, Hanson KC, Cooke SJ (2007) Do catch-and-release guidelines from state and provincial fisheries agencies in North America conform to scientifically based best practices? Environ Manag 39:760–773. https://doi.org/10.1007/s00267-006-0173-2. Accessed 13 Mar 2020
Popesku JT, Martyniuk CJ, Mennigen J, Xiong H, Zhang D, Xia X, Cossins AR, Trudeau VL (2008) The goldfish (*Carassius auratus*) as a model for neuroendocrine signaling. Mol Cell Endocrinol 293:43–56
Popovic NT, Strunjak-Perovic I, Coz-Rakovac R, Barisic J, Jadan M, Berakovic AP, Klobucar RS (2012) Tricaine methane-sulfonate (MS-222) application in fish anaesthesia. J Appl Ichthyol 28:553–564
Readman GD, Owen SF, Murrell JC, Knowles TG (2013) Do fish perceive anaesthetics as aversive? PLoS One 8(9):e73773. https://doi.org/10.1371/journal.pone.0073773. Accessed 13 Mar 2020
Rowan A, Goldberg A (1995) Responsible animal research: a riff of Rs. ATLA 23:306–311
Russell WMR, Burch RL (1959) The principles of humane experimental technique. Universities Federation for Animal Welfare, Wheathampstead. https://caat.jhsph.edu/principles/the-principles-of-humane-experimental-technique. Accessed 13 Mar 2020
SALMOTRIP (2013) Feasibility study of triploid salmon production. http://cordis.europa.eu/result/rcn/60437_en.html. Accessed 13 Mar 2020
Sambraus F (2016) Solving bottlenecks in triploide Atlantic salmon production. Temperature, hypoxia and dietary effects on performance, cataracts and metabolism. PhD thesis University of Bergen, p 69. http://bora.uib.no/handle/1956/15352. Accessed 13 Mar 2020
Small BC (2003) Anesthetic efficacy of metomidate and comparison of plasma cortisol responses to tricaine methanesulfonate, quinaldine and clove oil anesthetized channel catfish *Ictalurus punctatus*. Aquaculture 218:177–185
Smith AJ, Hawkins P (2016) Good science, good sense and good sensibilities: the three Ss of Carol Newton. Animals 6:70. https://doi.org/10.3390/ani6110070. Accessed 13 Mar 2020
Smith AJ, Clutton RE, Lilley E, Hansen KEA, Brattelid T (2017) PREPARE: guidelines for planning animal research and testing. Lab Anim 52:135–141. https://journals.sagepub.com/doi/full/10.1177/0023677217724823. Accessed 13 Mar 2020
Sneddon LU (2009) Pain perception in fish: indicators and endpoints. ILAR J 50(4):338–342. https://doi.org/10.1093/ilar.50.4.338. Accessed 13 Mar 2020
Sneddon LU (2012) Clinical anesthesia and analgesia in fish. J Exotic Pet Med 21:32–43
Speaking of Research (2019) Worldwide animal research statistics. https://speakingofresearch.com/2017/09/27/canada-sees-rise-in-animal-research-numbers-in-2016/. Accessed 13 Mar 2020
Spence R, Gerlach G, Lawrence C, Smith C (2008) The behaviour and ecology of the zebrafish, *Danio rerio*. Biol Rev 83:13–34. https://doi.org/10.1111/j.1469-185X.2007.00030.x. Accessed 13 Mar 2020
Stallard N, Whitehead A, Ridgway P (2002) Statistical evaluation of the revised fixed-dose procedure. Hum Exp Toxicol 21:183–196
Stien LH, Bracke MBM, Folkedal O, Nilsson J, Oppedal F, Torgersen T, Kittilsen S, Midtlyng PJ, Vindas MA, Øverli Ø, Kristiansen TS (2013) Salmon welfare index model (SWIM 1.0): a

semantic model for overall welfare assessment of caged Atlantic salmon: review of the selected welfare indicators and model presentation. Rev Aquac 5:33–57. https://doi.org/10.1111/j.1753-5131.2012.01083.x. Accessed 13 Mar 2020
Tannenbaum J, Bennett BT (2015) Russell and Burch's 3Rs then and now: the need for clarity in definition and purpose. J Am Assoc Lab Anim Sci 54:120–132. https://www.ingentaconnect.com/content/aalas/jaalas/2015/00000054/00000002/art00002. Accessed 13 Mar 2020
Thompson RRJ, Paul ES, Radford AN, Purser J, Mendl M (2016) Routine handling methods affect behaviour of three-spined sticklebacks in a novel test of anxiety. Behav Brain Res 306:26–35
Van den Heuvel MJ, Clark DG, Fielder RJ, Koundakjian PP, Oliver GJA, Pelling D, Tomlinson NJ, Walker AP (1990) The international validation of a fixed-dose procedure as an alternative to the classical LD50 test. Food Chem Toxicol 28:469–482
Wade N (2010) Research offers clue into how hearts can regenerate in some species. NY Times, March 24. https://www.nytimes.com/2010/03/25/science/25heart.html. Accessed 13 Mar 2020
Wargelius A, Leininger S, Skaftnesmo KO, Kleppe L, Andersson E, Taranger GL, Schulz RW, Edvardsen RB (2016) Dnd knockout ablates germ cells and demonstrates germ cell independent sex differentiation in *Atlantic salmon*. Sci Rep 6:21284. https://doi.org/10.1038/srep21284. Accessed 13 Mar 2020
Wittbrodt J, Shima A, Schartl M (2002) Medaka – a model organism from the far east. Nat Rev 3:53–64
Woods IG, Kelly PD, Chu F, Ngo-Hazelett P, Yan Y-L, Huang H, Postlethwait JH, Talbot WS (2000) A comparative map of the zebrafish genome. Genome Res 10(12):1903–1914

Chapter 17
Catch Welfare in Commercial Fisheries

Mike Breen, Neil Anders, Odd-Børre Humborstad, Jonatan Nilsson, Maria Tenningen, and Aud Vold

Abstract The introduction of catch welfare to commercial wild-capture fisheries will be challenging. In this chapter, we discuss how taking a science-based approach to understanding catch welfare in commercial fisheries could lead to practical solutions to improving welfare that will not only have ethical benefits, but may also have tangible benefits for the fishery, including improved sustainability, product quality and shelf life, and hence profitability. There has been little research to date specifically directed at the development of catch welfare in commercial fisheries. However, there is a substantial and growing body of literature on the fate and vitality of released animals from commercial fisheries—most recently catalysed by the introduction of the Landing Obligation in the EU. Furthermore, there is much to be learned from the aquaculture industry with regard to good welfare practices and product quality, particularly regarding catch handling and slaughter. This chapter utilises this available knowledge to develop a risk assessment-based framework for identifying capture-related stressors and suggests ways of mitigating their impact on the welfare of the catch, as well as on product quality. This framework is developed in context with four contrasting case study capture methods: trawl, purse seine, gill/trammel nets and pots. Finally, it concludes with a summary of the current research priorities and significant strategic challenges for developing welfare-conscious practices in commercial fisheries.

Keywords Catch welfare · Commercial fisheries · Science-based approach · Sustainability

M. Breen (✉) · O.-B. Humborstad · J. Nilsson · M. Tenningen · A. Vold
Fish Capture Division, Institute of Marine Research (IMR), Bergen, Norway
e-mail: michael.breen@hi.no

N. Anders
Fish Capture Division, Institute of Marine Research (IMR), Bergen, Norway

Department of Biological Sciences, University of Bergen, Bergen, Norway

T. S. Kristiansen et al. (eds.), *The Welfare of Fish*, Animal Welfare 20,
https://doi.org/10.1007/978-3-030-41675-1_17

17.1 Introduction

Introducing fish welfare, or more generally "catch welfare", to commercial wild-capture fisheries faces several challenges. Firstly, most target species are "non-charismatic" and their treatment during capture has, up until recently, held little interest in the general public (Bennett et al. 2015; Lundberg et al. 2019). Furthermore, fishing is a conservative industry, and many fishers find it counterintuitive to consider the welfare of animals that are about to be slaughtered (Breen, pers. comm.). Also, commercial wild-capture fisheries remain an essential component of global food security, particularly for developing nations, so the banning or restriction of unsuitable practices is likely to be politically unpalatable. More than 120 million people in the world depend directly on fisheries-related activities (fishing, processing, trading), the majority of them living in developing or emergent countries and 90% working in small-scale fisheries (HLPE 2014). Even given the moral and political will to introduce welfare-conscious practices, it could be argued that there are still issues of scale and practicality that could make them technically and economically prohibitive. However, in this chapter we discuss how taking a science-based approach to understanding catch welfare in commercial fisheries could lead to practical solutions to improving welfare that will not only have ethical benefits, but may also have tangible benefits for the fishery, including improved sustainability, product quality and shelf life, and hence profitability.

There are a broad range of views on what "fish welfare" is about (Huntingford et al. 2006; Diggles et al. 2011; Torgersen et al. 2011; Browman et al. 2018). While some consider welfare as the fish's own experience of its life quality, "what is good and what is bad" (feelings-based definitions), others consider good welfare as appropriate body functioning, regardless of how it is experienced by the fish concerned (function-based definitions) or that the animal should "lead a natural life, expressing naturally occurring behaviours" (nature-based definitions) (Diggles et al. 2011). Fisheries will generally only contribute negatively to welfare on the fish caught, and improved welfare is then, regardless of which definition that is used, about reducing negative impact to minimize stress and injury, and the number of individuals affected. So, while ethical discussion on the meaning and application of nature-based and feelings-based definitions of welfare has its place, if welfare-conscious practices are to be introduced to commercial fisheries, they will need to be pragmatic and utilitarian; and so take a functional approach to welfare. To promote support from the industry, establishing an empirical link between good welfare and product quality, and hence profit, would provide an important motivator. This functional approach to catch welfare will also need to recognise that it is not only the retained catch that experiences the capture process, but also the animals that are caught but later either escape or are actively released by fishers because they are "unwanted".

The FAO International Standard Statistical Classification of Fishing Gear, based on construction and mode of operation, recognises 58 different fishing methods, in 11 broad categories (FAO 1990). The vast majority of fishing methods are not benign, as fish are exposed to a range of stressors, which act upon the fish to elicit a stress response. The stress response is the naturally occurring sum of all physiological and behavioural adaptations made in the face of a stressor (see Chap. 11). Furthermore, commercial fishing operations affect many more animals than just those in the catch landed at the fish market. Since Holt's work in the late nineteenth century (Holt 1895), it has been recognised that fishing has the potential to cause unintentional or collateral mortality in the exploited populations, in addition to that accounted for by the total reported catch (Chopin et al. 1996; ICES 2005; Gilman et al. 2013). These potential "collateral mortalities" come from several sources, including illegal, unreported and unregulated fishing (IUU) (Agnew et al. 2009; Seafish 2013); discard mortality (Breen and Catchpole 2020; ICES 2014); escape mortality (Ingolfsson et al. 2007; Suuronen 2005); ghost fishing (Brown and Macfadyen 2007; Macfadyen et al. 2009); and the alteration of benthic habitats and communities (Eigaard et al. 2015; Buhl-Mortensen et al. 2013; Shephard et al. 2014).

Huntingford and Kadri (2009) suggested that, by using a "top-down" approach and addressing environmental degradation and incidental damage to non-target species in wild-capture fisheries, the welfare of individual fishes is likely to benefit, quoting "Commitment to better management of natural resources [is] a better ethical framework for fisheries (P. Hart, pers. comm.) than a concern for fish welfare". However, we believe a "bottom-up" approach is more likely to engage the fisher, and other stakeholders, in welfare-focused capture methods and practices, and will address several fundamental problems in fisheries management, including bycatch, environmental degradation, unaccounted fishing mortality (and associated uncertainties in stock assessment), reduced product quality and poor public perception. That is, "commitment to better fish/catch welfare is a more ethical [and engaging] framework for the management of natural resources".

Acute stress immediately prior to slaughter has been demonstrated to reduce flesh quality in several fish species, including Atlantic salmon (*Salmo salar*), sea trout (*Salmo trutta*), Atlantic cod (*Gadus morhua*), haddock (*Melanogrammus aeglefinus*), turbot (*Scophthalmus maximus*), sea bream (*Sparus aurata*) and sea bass (*Dicentrarchus labrax*) (Bagni et al. 2007; Bjørnevik and Solbakken 2010; Digre et al. 2010; Karlsson-Drangsholt et al. 2017; Kristoffersen et al. 2006; Matos et al. 2010; Morzel et al. 2003; Olsen et al. 2008; Sigholt et al. 1997; Stien et al. 2005). Thus "Catch Welfare" is more than just an ethical issue; it can potentially ensure good product quality from the catch and promote the survival of any unwanted catch, contributing to reduced total fishing mortality, which is essential for a sustainable fishery sector. These and the ethical concept of "good welfare" will promote confidence in the consumer with respect to both the quality of the product and sustainability of the fishery, which will give added value to the final product.

17.1.1 A Functional Definition of Good Catch Welfare Practices

To this end, this chapter will define good catch welfare practices as: "capture and handling methods that minimise the physical damage to, and allostatic load on, any retained fish until after they are either slaughtered or released", and thus promote the likelihood for post-release survival and/or good product quality". We believe that this is a pragmatic definition that can be accepted by advocates of both feelings-based and function-based views of welfare and that it makes catch welfare relatively straightforward to measure and quantify. We propose that by fostering good catch welfare and thus minimising the effects of stressors on the catch, welfare-conscious practices will promote the survival of any unwanted and released catch, hence the sustainability of the exploited stock and ecosystem, whilst also promoting the quality and hence profitability of the retained catch.

17.1.2 Chapter Objective

There has been little research to date that specifically addresses catch welfare in commercial fisheries (Kaiser and Huntingford 2009; Metcalfe 2009). Thus, this chapter will not confine itself solely to the welfare of just fish during capture, because many of the related stressors, and possible mitigations, are likely to be similar across taxa. Also, there is a substantial and growing body of literature on the fate and vitality of released animals from commercial fisheries; most recently catalysed by the introduction of the Landing Obligation (discard ban) in the European Union Common Fisheries Policy (Rihan et al. 2019; Breen and Catchpole 2020). Furthermore, there is much to be learned from the aquaculture industry with regard to good welfare practices and product quality, particularly regarding catch handling and slaughter. This chapter will utilise this available knowledge to develop a risk assessment-based framework for identifying capture-related stressors and suggest ways of mitigating their impact on the welfare of the catch, as well as on product quality. This framework will be developed in context with four contrasting case study capture methods: trawl, purse seine, gill/trammel nets and pots. Finally, it will conclude with a summary of the current research priorities and significant strategic challenges for developing welfare-conscious practices in commercial fisheries.

17.2 Developing a Framework for Assessing Welfare in the Capture Process

Before considering potential solutions to improving catch welfare during the capture process, it is necessary to define what stressors an animal is likely to experience therein. Capture can be broadly categorised as a four-step process: (1) Capture, (2) Retrieval, (3) Handling (and sorting) and (4) Endpoint, where the endpoint can be either (a) Release or (b) Slaughter (Fig. 17.1). Animals may also escape during capture or retrieval and will then not be exposed to all steps. Building on an approach from Davis (2002), the path taken by an animal from the point of capture to its subsequent release or slaughter can be conceptualised and the relevant stressors, and likely influential variables, identified (see also Broadhurst et al. 2006; Breen and Catchpole 2020; Fig. 17.1). Using this conceptual risk assessment approach, we identified 11 major capture-related stressors: hypoxia, fatigue/exhaustion,

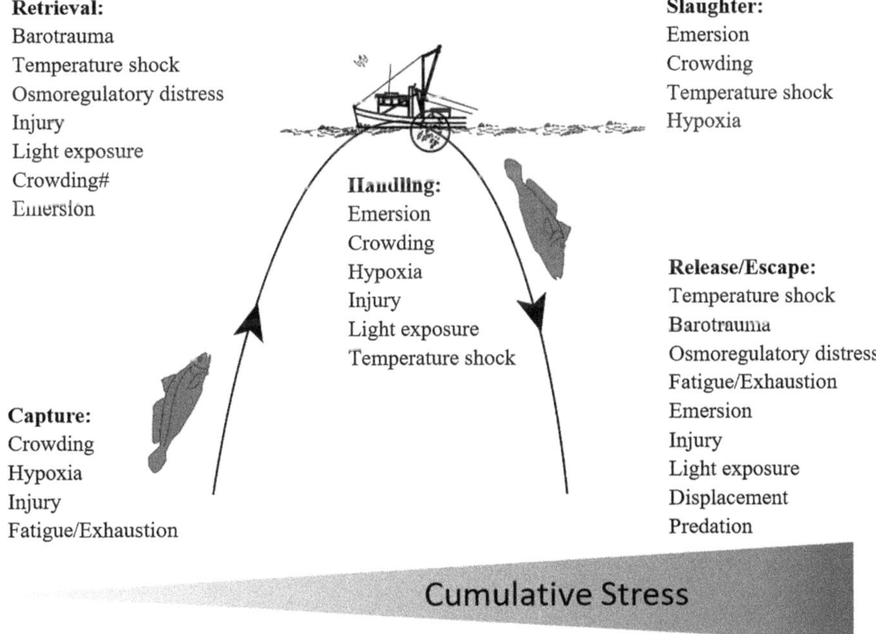

Fig. 17.1 Conceptual pathway analysis of the stressors encountered by an animal during the capture process: from the point of capture to its subsequent release or slaughter. "Retrieval" refers to the stage of the capture process when the fishing gear and its catch are retrieved from the fishing depth and brought aboard the vessel. "Handling" refers to the process of removal of the catch from the gear, containment and sorting operations on board the vessel. Adapted from Davis (2002), Broadhurst et al. (2006), Breen and Catchpole (2020)

barotrauma, temperature shock, osmoregulatory distress, crowding, injury, light exposure, emersion, displacement and predation (see Sect. 17.3 for details).

The relative importance of each of these stressors will be dependent upon the biology of the affected animal, the design and operation of the fishing method and the environment in which the animal is caught. Furthermore, these stressors are likely to have a cumulative effect on the affected animals over time, where the overall impact of the stressor will be the product of the severity of the stressor and the time exposed to it. In addition, there are likely to be compound/synergistic effects between stressors (Bonga 1997), as well as alternative stressors, for particular species and/or fisheries, not described here. To properly characterise the potential effects of capture-related stressors on a particular species/taxon, in a specific fishery/ environment, requires the development of suitable ways of assessing welfare, and this is discussed in Sect. 17.4.

Finally, in Sect. 17.5, this conceptual welfare assessment is applied to four example fishing methods to demonstrate the approach, in a generalised way, and identify potential ways of improving catch welfare, which are explored in more detail in Sect. 17.6.

17.3 Stressors Encountered During the Capture Process

The major capture-related stressors are discussed below, with respect to their causes and likely influential variables (for a more general and comprehensive introduction to stress and stress responses, see Chap. 11).

It could be argued that many of the stressors discussed below have the potential to cause pain in captive animals. Indeed, there is growing evidence that fish, and possibly crustacea, have the capacity of nociception, i.e. neuro-physiology and anatomy to sense damaging stimuli (e.g. Sneddon et al. 2014; Chap. 10). However, pain is the subjective experience of nociception, and it is currently debated whether fish are capable of perceiving subjective experiences, which requires some level of cognition (e.g. Rose et al. 2014; Brown 2015; Key 2016; Merker 2016; Browman et al. 2018). From the perspective of our functional definition of "good welfare practices", regardless of whether a stressor is experienced as pain, or only sensed as nociception, it will contribute negatively to the welfare of the animal.

17.3.1 Hypoxia

Insufficient oxygen, referred to as "hypoxia", is an important and potentially fatal stressor for all aerobic organisms, including fish (Hughes 1975; Domenici et al. 2012). In the context of fish capture, it can be caused by: low oxygen (hypoxic) concentrations in the water around the animal; a reduced functionality in the animal's ability to breath (asphyxia) due to injury or constriction; and/or complete emersion from the animal's breathing medium. The tolerance of hypoxia will vary between species and will be modulated by various environmental parameters, in particular temperature (Domenici et al. 2012; Rogers et al. 2016). Oxygen requirements are also strongly dependent on the activity level of the fish (Claireaux et al. 2000; Hvas et al. 2017), and stressed and struggling animals are thus more vulnerable to hypoxia than calm animals. However, prolonged hypoxia will eventually lead to cell death and the failure of critical biological systems (Hughes 1975).

17.3.2 Fatigue and Exhaustion

As part of the capture process (e.g. in trawls), while struggling to free themselves from a fishing gear (e.g. hooks and gillnets), or while on the deck of the fishing boat, captive animals may become fatigued or exhausted due to excessively strenuous activity. Exhaustive swimming can cause fatalities in some fish (Beamish 1966; Black 1958; Breen et al. 2004), due to physiological disruption from a build-up of lactate and other metabolic acids in the blood (Wood et al. 1983). Even if not fatal, the oxygen debt associated with exhaustion can limit an animal's metabolic capacity and compromise their ability to cope with other stressors. The effects of exhaustion will exacerbate, and in turn be exacerbated by, hypoxia (see Sect. 17.3.1).

17.3.3 Decompression and Barotrauma

A rapid ascent from depth will cause a reduction in hydrostatic pressure that can cause injury and stress in some aquatic animals via two related mechanisms: physical and physiological barotrauma (see also Chaps. 18 and 19). Physical barotrauma is due to a rapid and uncontrolled expansion in closed air spaces within the animal's body (e.g. physoclistous swim bladders in fish) in accordance with Boyle's Law (Humborstad and Mangor-Jensen 2013; Midling et al. 2012; Brown et al. 2012). This rapid expansion can cause the swim bladder wall to rupture, releasing gas into the abdomen, which with further expansion can cause internal organs to be everted through the mouth or anus (Feathers and Knable 1983; Rogers et al. 1986). Effects may be acute and readily observable at capture (e.g. Rummer and Bennett 2005) or chronic, developing over time post-capture (e.g. exophthalmia; Dehadrai 1966; Humborstad et al. 2016a). Effects tend to increase at depths greater than 40 m, where reduction in ambient hydrostatic pressure can cause greater than five-fold expansions in swim bladder volumes (Brown et al. 2012). Although effects are species and situation dependent, capture from shallower depths may also be damaging and deadlier than capture from greater depths (see Chap. 18; Midling et al. 2012). Physiological barotrauma is due to dissolved gases coming out of solution in the animal's blood and tissue fluids, in accordance with Henry's Law. This causes bubbles to form in the blood and tissues, which can impede blood supply and disrupt the functionality of some critical systems, especially the visual and neurological system (Humborstad et al. 2016a; Brown et al. 2012). In addition to the direct injury from barotrauma, the increased buoyancy resulting from gas expansion may prevent the descent of released individuals or cause individuals escaping during retrieval to float up to the surface. Floating makes fish more vulnerable to surface-related stressors (see Sects. 17.3.8, 17.3.9 and 17.3.11). Submergence and re-pressurisation (diving) may reduce mortality; however, the sublethal and long-term effects of barotrauma remain to be studied, for example reproductive success (Peregrin et al. 2015), and these may have more relevance to welfare than acute and rapid morality (Ferter et al. 2015).

17.3.4 Temperature Shock

Fish, and many other aquatic organisms, are poikilotherms; that is, their body temperature varies with the ambient environmental temperature. Rapid changes in the ambient temperature can cause disruption to the animal's metabolism called "temperature shock", which can sometimes result in death (Davis and Olla 2001, 2002; Davis et al. 2001; Donaldson et al. 2008; Gale et al. 2013). Animals captured in fishing gears are most likely to experience stressful temperature changes during the retrieval of the gear, when they can ascend through thermoclines, and in particular when exposed to ambient air temperatures, which could be far in excess of fatal temperature thresholds (e.g. less than −10 °C in the Barents Sea during winter; or greater than 30 °C in the Mediterranean in the summer).

17.3.5 Osmoregulatory Distress

Many animals need to maintain, "osmoregulate", an optimal concentration of dissolved salts in their blood and other tissue fluids for their metabolism to function effectively. This can be particularly challenging for water-breathing aquatic animals, which continually exchange water and dissolved ions across their respiratory surfaces (Greenwell et al. 2003). This challenge is heightened when the animal is stressed, because one of the key stress responses is to increase blood circulation to the respiratory surfaces in order to increase the potential uptake of dissolved oxygen from the water. This consequently increases water/ion exchange, leading to haemoconcentration (dehydration) in marine species and haemodilution in freshwater species (e.g. Wedemeyer et al. 1990; Barton 2002; Cooke et al. 2013). A substantial change in the salinity of the ambient water, i.e. by passing through a halocline during the retrieval of the gear, has been observed to cause osmoregulatory distress in some species (e.g. Harris and Ulmestrand 2004). Furthermore, injuries that result in blood loss and/or an increased exchange of water and dissolved ions could disrupt osmoregulation and consequentially reduce metabolic capacity to cope with other stressors (Smith 1993; Greenwell et al. 2003).

17.3.6 Crowding

Confinement and crowding can induce stress, injury and mortality in captive aquatic animals (Portz et al. 2006; Huse and Vold 2010; Tenningen et al. 2012). The stress experienced during crowding may lead to increased activity, such as burst swimming. Together with the restricted volume, this will further increase the likelihood of injury due to physical contact with the catch and fishing gear, as well as hypoxia, if the biomass is sufficiently large and/or dissolved oxygen supply is restricted. As such, crowding can be considered as a collective stressor (see also Sect. 17.3.9).

17.3.7 Physical Trauma and Injury

The potential for physical trauma during the capture, handling and release of animals is considerable and likely to result in a range of injuries, including skin abrasion, lacerations, puncture wounds, blunt force trauma and crushing. Moreover, many aquatic animals, particularly pelagic species, have evolved relatively delicate integuments (skins) and are not well adapted to physical contact with hard and abrasive surfaces (Elliott 2011a, b; Kitsios 2016). In addition to loss of functionality in the target organ, wounds from physical trauma may lead to the loss of blood and other tissue fluids, further reducing the metabolic capacity to cope with other stressors and increasing the likelihood of infections.

17.3.8 Light Exposure

Natural light levels are greatly reduced by attenuation at typical fishing depths in comparison to the surface (Johnsen 2012). At the surface, light intensity during the day will be many orders of magnitude higher than the animals are adapted to in their normal habitat. This is likely to cause disorientation and bleaching of sensory pigments in the eye (Pascoe 1990), leading to short-term or permanent blindness in some species (Frank and Widder 1994; Chapman et al. 2000). In addition to visible wavelengths of light, aquatic animals brought to the surface will be exposed to potentially damaging UV light, which has the capacity to burn exposed tissues and induce melanoma (Sweet et al. 2012).

17.3.9 Emersion

Removal from the water, "emersion", will likely present an aquatic animal with a potentially stressful array of novel stimuli. For example, lack of the supportive, hydrostatic properties of the water will mean that many animals will experience, for the first time, their own weight-in-air, to which they are not evolutionarily adapted. Furthermore, these novel stimuli will be compounded with an array of other stressors: hypoxia/asphyxiation (Sect. 17.3.1), barotrauma (Sect. 17.3.3), desiccation (Sect. 17.3.5), temperature shock (Sect. 17.3.4) and light exposure (Sect. 17.3.8). As such, it is common for "emersion" to be used as a collective term for the combined effect of all of these stressors.

17.3.10 Displacement

The location where the animal is released may not be close to the area or depth where it was originally caught, particularly if caught using a towed gear (e.g. trawl); this is referred to as "displacement". The new location may provide an inappropriate habitat for the released animal, with regard to environmental parameters (e.g. depth, water currents, temperature, salinity), as well as the provision of shelter and food. This is likely to further reduce the survivability of an already compromised animal. For physoclistous species, capture at depth and subsequent release at surface may induce barotrauma (see Sect. 17.3.3), and subsequent recovery behaviour can induce a type of vertical displacement, where the fish are forced to descend slowly to the original capture depth (Nichol and Chilton 2006).

17.3.11 Predation

The stressed and weakened state of released animals is likely to lead to behavioural impairment, which may decrease the likelihood of them being able to avoid predators (Olla et al. 1997; Ryer 2002, 2004; Ryer et al. 2004; Marçalo et al. 2013; Raby et al. 2014). Seabirds are likely to be one of the most prevalent predators of discarded animals because they forage at or just below the sea surface, are known to follow fishing boats to specifically prey on discards, and therefore present one of the first predatory threats to discarded animals at the point when most will be at their most vulnerable (e.g. Garthe et al. 1996; Depestele et al. 2016).

17.4 Developing Functional Welfare Metrics

The assessment of welfare in farmed fish is discussed extensively in Chap. 13. Here we will briefly discuss the key metrics most relevant to assessing the welfare of aquatic animals encountering commercial fishing gears and their associated stressors. Two broad approaches can be followed. Firstly, observation in laboratory settings, to establish physiological, behavioural and quality baselines, as well as describe responses to typical capture-related stressors under controlled conditions. Secondly, observation and monitoring of animals in situ, during commercial fishing operations, provides representative descriptions of the compound effects of these stressors on welfare.

17.4.1 Environmental

Describing the environmental changes that the animal experiences, as it is taken from the point of capture to its subsequent release or slaughter, is essential for identifying many of the most prevalent capture-related stressors. Simply measuring changes in depth, temperature, salinity, water movement, oxygen concentration and light intensity over time will identify the likelihood, and possibly magnitude, of decompression injury, temperature shock, osmoregulatory distress, exhaustion, hypoxia, light exposure and emersion, respectively (e.g. Breen et al. 2007; Breen and Catchpole 2020). Relatively inexpensive technologies are now available to measure most of these metrics, often in the simple form of a tag that can be easily attached to the fishing gear (e.g. Star-Oddi 2017).

17.4.2 Physiological

A broad array of physiological metrics exist that can describe the primary and secondary responses to capture-related stresses (see Chap. 11; Mckenzie et al. 2016), as well as underlying mechanisms determining subsequent effects on meat quality (e.g. Karlsson-Drangsholt et al. 2017) and post-release survival (e.g. Olsen et al. 2012). Of particular relevance to capture-related stressors is an understanding of the metabolic scope of affected species, because this can inform us on the capacity of individuals to cope with acute and potentially fatal stressors such as exhaustion, hypoxia and emersion (Killen et al. 2011). When assessing the effects of slaughter methods, using electro-encephalograms (EEG) to monitor brain activity is informative for determining the animal's state of consciousness (e.g. Van de Vis et al. 2003).

17.4.3 Quality

Using quality-related metrics has the potential to link welfare empirically to product characteristics that stakeholders can directly relate to. Although good quality and the relevant indicators are likely to vary between species and consumer preferences, using such tools as a catch-damage index (CDI) (Esaiassen et al. 2013) allows a sensory evaluation of product quality by examining factors such as bruising, abrasions and damages from gear by assigning a score. The appearance of the fillet, blood spotting, consistency and the degree of gaping (where the connective tissue between the fillet flakes breaks) are also useful qualitative metrics (Digre et al. 2016). More quantitative measures of quality include examining texture using a puncture test (Borderías et al. 1983), by examining the pH of the fillet (Roth et al. 2013) or by using reflectance spectroscopy to measure residual blood remaining in the meat (Olsen et al. 2008).

17.4.4 Behaviour and Vitality

Behavioural responses to a potential threat (stressor) are generally among the first reactions to be observed and form an integrated expression of the welfare status of the whole animal (Schreck et al. 1997; Davis 2010; Chap. 4). Also, as a welfare metric, behavioural change is easily observable, non-invasive to record and potentially non-intrusive (Dawkins 2004). Recent advances in underwater video technology now provide relatively low-cost solutions for making behavioural observations in the often-challenging conditions presented by commercial fishing operations (e.g. Anders et al. 2017; Bayse et al. 2016), as well as creating a permanent record for a thorough and repeatable analysis (Jury et al. 2001).

Mortality, in response to a stressor, is arguably the ultimate expression of "bad welfare" for an affected animal, and may indicate near-lethal conditions for the surviving individuals. There is a substantial and growing body of literature describing the mortality rates of animals released from commercial fishing operations (for reviews see: Broadhurst et al. 2006; Suuronen 2005; Veldhuizen 2017; Veldhuizen et al. 2018; Breen and Catchpole 2020), which provide an important insight into the welfare status of animals captured using different fishing methods under a wide range of conditions. Furthermore, assessing the impairment of reflexes/behaviours, as well as the occurrence of injuries, has been demonstrated to be a good predictor of mortality (e.g. Davis 2010; Benoît et al. 2010; Humborstad et al. 2016b) and is now being used to systematically assess the vitality of released animals (Breen and Catchpole 2020). Key observations from this work include: smaller animals are typically more likely to die; and mortality generally increases with increasing exposure to the capture method (i.e. haul duration, soak times, etc.) and emersion (i.e. sorting times and air exposure) (Veldhuizen 2017; Veldhuizen et al. 2018).

17.5 Assessing Welfare in the Capture Process

The nature of any capture-related stressors will be species- and capture method-specific, and will be modified by the environment and time-frame in which they take place (Breen and Catchpole 2020). With regard to the impact that a capture method can have upon the target and incidental catches, it can be informative to consider both their mode of operation and how they interact with the behaviour of the target animals. To this end, we will consider four different gear types, which represent contrastingly different modes of operation: Gill and trammel nets (passive entangling); purse seine (active surrounding); demersal trawls (active herding); and pots (passive attracting). The stressors associated with other fishing methods are described in detail in other chapters in this book: demersal seine (Chap. 18) and hook and line (Chap. 19). To avoid repetition, the processes of retrieval, handling and slaughter are discussed generally for all capture methods, while providing some method-specific example where relevant.

17.5.1 Capture

Gill and Trammel Nets (Passive Entangling, Fig. 17.2) Gill and trammel nets rely on the natural migratory and/or foraging behaviour of fish to capture them, as they attempt to swim through the low-visibility netting. Once a fish is entangled, it will inevitably struggle to escape, which will eventually fatigue/exhaust it. As fish are soft-bodied animals, attempts to struggle free can lead to abrasive injury. In addition, the restricted movement and constriction, particularly if gilled, can suffocate some animals. The survival of animals released from gillnets and trammel nets is highly variable depending on various factors including tolerance to hypoxia, taxa, animal

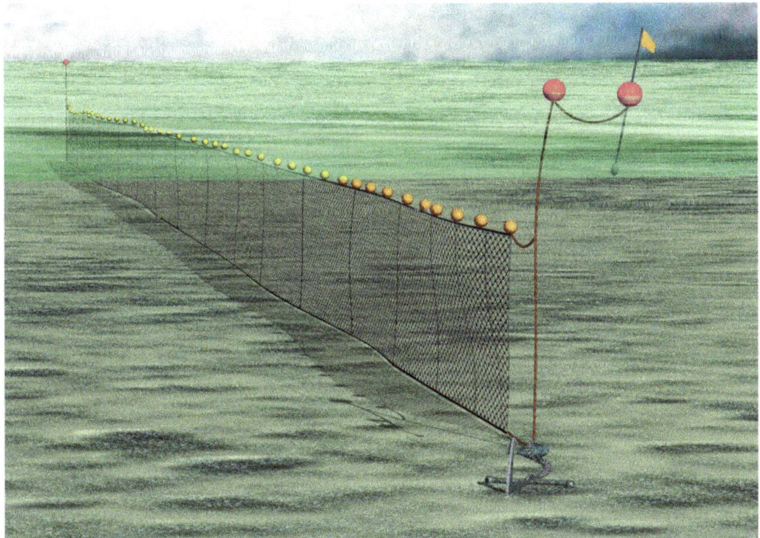

Fig. 17.2 ***Gillnets and trammel nets*** consist of panels of fine, low-visibility netting, suspended vertically in the water between a floatline and a weighted ground-rope, which entangle animals that encounter them. Gill nets have single netting sheets in each panel, while trammel nets have several netting layers which typically consist of a small-mesh inner layer, with larger-mesh outer layers. Source: Galbraith et al. (2004) [Crown Copyright]

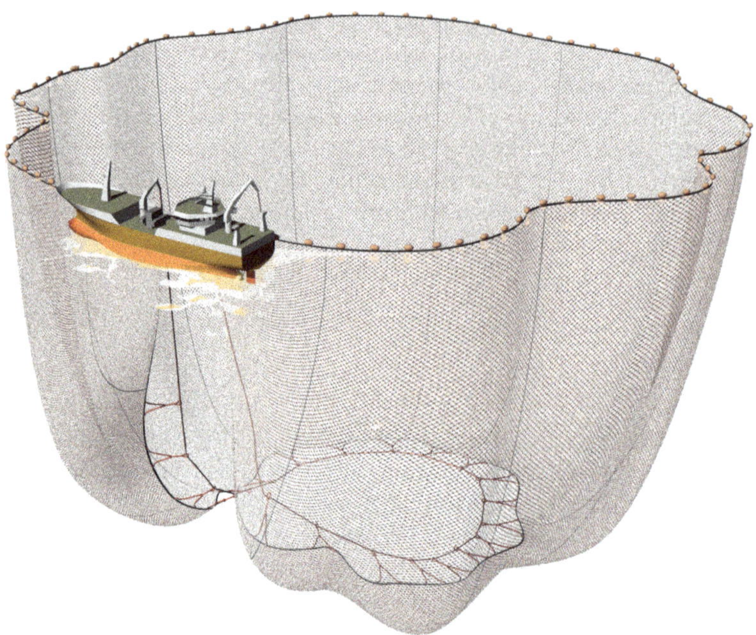

Fig. 17.3 ***Purse seines*** are surrounding nets that are deployed, or set, around natural aggregations of pelagic fish. They are essentially a wall of netting, up to 1000 m long and 200 m deep, that is deployed from a moving vessel to encircle the target school of fish. Once the school is encircled, a purse-wire on the bottom of the net is hauled in, closing the bottom of the net. Source: Galbraith et al. (2004) [Crown Copyright]

size and soak time (see Uhlmann and Broadhurst 2015, for review). Even if lost or abandoned, gill and trammel nets can continue catching animals, or "ghost-fishing" (FAO 2016), which will greatly increase the number of animals exposed to these stressors, and their exposure period.

Purse Seine (Active Surrounding, Fig. 17.3) As the net is hauled, the net volume gradually reduces, concentrating the catch close to the fishing vessel, where they can be taken on board, using either landing nets or pumping systems. The capture process starts to become stressful and injurious for the majority of the catch once the volume of the net has been reduced to the point at which the movement of the school becomes restricted and they become crowded in densities that they would not naturally encounter. At this point, there is an increased risk of physical contact with either the netting or other fish, resulting in abrasive injury. Crowding can become so dense that the catch is pressed against the netting with animals being exposed to compression and crushing injuries, as well as constricted movement preventing effective ventilation of the gills, which could lead to asphyxiation. In larger catches, this can be exacerbated because the large biomass of fish can quickly deplete the surrounding water of oxygen. Furthermore, crowding can also induce escape responses in some species, which will rapidly beat their tails to attempt to swim

Fig. 17.4 ***Pots*** are traps that use an attractor, typically bait, to entice target fish inside (Thomsen et al. 2010). Source: Galbraith et al. (2004) [Crown Copyright]

Table 17.1 Survival of animals caught and released from pots

Species	Survival (%)	References
Norwegian lobster (*Nephrops norvegicus*)	>96	Wileman et al. (1999)
Blue crab (*Callinectes sapidus*)	81	Darnell et al. (2010)
Deepwater red crab (*Chaceon quinquedens*)	95	Tallack (2007)
Cod (*Gadus morhua*)	79–96	Humborstad et al. (2016b)

away from the threatening stimuli. This can quickly lead to exhaustion and further exacerbates the risk of hypoxia and asphyxiation. Experiments have demonstrated a strong correlation between crowding density and mortality in fish released from purse seines (Huse and Vold 2010; Marçalo et al. 2010; Tenningen et al. 2012).

Pots (Passive Attracting, Fig. 17.4) Fundamental to the capture process in pots is that the animal enters voluntarily. Thus, at least until the animals has entered the pot, the capture process can be assumed to be non-stressful. Once caught, the animal risks injury from contact with netting and frame of the pot, as well as physical interaction with other captive animals. There may also be stress responses due to confinement. The behavioural observations of fish caught in pots show that after some initial escape attempts, the fish typically either swims around in the available space inside the pot or settles down and rests within it (Anders et al. 2017; Meintzer et al. 2017). Large animals, including fish (i.e. conger eel), octopus, crabs and lobsters, are also known to prey on smaller animals in the pots (Cole et al. 2001; Thomsen et al. 2010). However, high survival of animals caught and released from pots (Table 17.1) suggests that pots generally do not severely affect welfare; however, if the fish cannot submerge, mortality can be high (Humborstad et al. 2016b). It is generally believed that the most stressful and injurious part of the capture process in pots

Fig. 17.5 ***Demersal trawls*** are a complex gear, consisting of a conical net that is towed behind a fishing vessel, using long wire warps. The trawl net is held open vertically by a combination of a weighted ground gear and a buoyant headline, and laterally by trawl "doors" that are spread apart by hydrodynamic forces as the net is towed through the water. Source: Galbraith et al. (2004) [Crown Copyright]

begins at the point of retrieval (Humborstad et al. 2016b). As with gill and trammel nets, pots also have the capacity to continue "ghost-fishing" after they have been lost or abandoned (Macfadyen et al. 2009).

Demersal Trawls (Active Herding, Fig. 17.5) For non-mobile species, the trawl simply acts as a sieve collecting animals in the water column and on the seabed in its path. However, for more mobile and active species the behavioural interactions with a trawl can be complex and potentially injurious. Once in the mouth of the trawl, many fish will turn and try to swim ahead of the net. Eventually the fish become fatigued and begin to turn and fall back into the trawl. This may be repeated further into the net, when fish may attempt to swim in the direction of the towed net again, only to tire and drop back further into the net. Eventually they fall into the codend at the end of the trawl where, if unable to escape through the codend meshes or other selective devices, they ultimately become part of the catch amassing there.

As the fish progress in this way into the net, they become increasingly fatigued and, as the net narrows, their risk of injurious contact with the netting increases (Breen 2004). Injurious contact can also be made by animals in the mouth of the net, as they strike some component of the gear or pass under the ground gear (Ingolfsson et al. 2007). Fish collecting in the codend are exposed to further risk of injurious contact with other animals in the catch; and, as they become compressed within the catch, they may become constricted and, as a result, asphyxiated. The likelihood of fatigue and injury in a trawl is likely to be influenced by an individual's ability to swim, which can be dependent on many factors including: size, swimming mode, physical condition, temperature, towing speed and light levels (Suuronen 2005).

17.5.2 Retrieval

During this phase, the fishing gear and its catch are retrieved from the fishing depth and brought aboard the vessel. From the perspective of potential stressors for the captive animals, there are two key steps, which will compound additional stressors with those already experienced as part of the capture process (Breen and Morales Nin 2017):

1. *Ascent to the surface.* As the fishing gear is hauled to the surface the captive fish will experience a rapid reduction in hydrostatic pressure, which can cause barotrauma (see Sect. 17.3.3). In addition, this journey through the water column can cause other stressful changes in environmental conditions including water temperature (see Sect. 17.3.4), salinity (see Sect. 17.3.5) and light intensity (see Sect. 17.3.8).
2. *Retrieval from the water.* As the catch becomes emersed, it no longer has access to the dissolved oxygen they require for respiration (see Sect. 17.3.1). Also, temperature changes and exposure to light (particularly damaging ultra-violet) are likely to be exacerbated. Furthermore, movement of the gear (particularly in poor weather conditions) during this transition is likely to increase the risk of physical injuries as the fish contact the gear, the boat and other components of the catch. The lack of the supportive, hydrostatic properties of the water means that many animals will experience their own weight-in-air, for the first time. Furthermore, in large catches, this will be compounded with the weight of the catch around them and may lead to crushing injuries.

17.5.3 Handling

Once aboard the fishing vessel the catch must be handled directly to prepare it for processing and storage (Breen and Morales Nin 2017):

1. *Removal from gear.* The stressors experienced at this stage will depend upon the type of gear, how the catch is retained in that gear and the body-form and size of the animal. For example, trawl codends can generally be opened quickly and the catch efficiently emptied out. Gears using hooks will require some means of removing the catch from the hook; this is generally by means of a mechanical device, but dehooking by hand may also be used (e.g. Farrington et al. 2008). In gill- and trammel nets it can be challenging to remove the entangled catch quickly

and without injury, particularly crustaceans. In purse seines the catch is usually removed from the net directly using pump systems or landing nets while the net remains in the water concentrating the catch in a limited volume.

2. *Containment.* Once removed from the gear, if the catch is mixed and requires sorting, it is generally deposited either on the deck or into a container of some form. The nature of this containment will have important implications for the stressors experienced by the animals waiting to be sorted. If simply lying on deck, the fish is at risk of additional physical injury, including being stood on by the crew, as well as exposure to air and the desiccating effects of sunlight and wind. When in a catch bin or "hopper", these risks are reduced for the animals lying within the body of the catch, but in large catches there may be increased risk of injury from crushing and contact with other animals.
3. *Sorting.* This is usually done by hand by the crew, but mechanical size-sorting systems (typically grids) are used in some fisheries, as well as conveyor belts to extract the catch from the hopper. All may induce physical injury and trauma to varying degrees. How long the animals are exposed to the compounded stressors of emersion, temperature shock and sunlight will depend on the size of the catch and the efficiency of the crew. Crews are generally very efficient at sorting through a catch quickly, but they are less likely to be invested in handling the catch carefully to avoid further physical trauma and injury; particularly in the unwanted catch.

17.5.4 Release and Escape

Depending on the fishery, a substantial proportion of the catch may be unwanted by the fisher, because: it has little or no commercial value; it is undersize; or there is no available quota (Hall et al. 2000). These animals will experience all the stressors experienced by the retained catch, except for slaughter. In addition, the unwanted catch is generally discarded by simply being thrown overboard with associated exposure to emersion and risk of injury.

Fishing gears are not entirely effective at capturing their target animals, so a substantial proportion of animals can encounter a fishing gear and then escape. Furthermore, in an effort to reduce unwanted catches, there has been substantial investment in developing fishing gears that selectively catch target animals, while allowing the unwanted catch to escape (e.g. Hall et al. 2000). The stressors experienced by these escaping animals will be dependent upon the escape route and where in the capture process they escape (Breen 2004; Suuronen 2005). Although it has been shown that some escaping animals do die (e.g. Suuronen 2005; Breen et al. 2007; Ingolfsson et al. 2007; Broadhurst et al. 2006), it is clear they are likely to encounter fewer and less extreme stressors than released animals, and thus are more likely to survive an encounter with fishing gear (e.g. Wileman et al. 1999). The survival of trawl caught animals, both discarded and escaping, has been observed to be significantly correlated with individual size in several species (e.g. Sangster et al.

1996; Suuronen 2005) supporting the hypothesis that swimming ability is an important survival trait. Other factors linked with survival are: towing duration, catch size and composition, codend construction and mesh size.

17.5.5 Slaughter

Optimal slaughter methods should render animals unconscious, without avoidable distress, prior to killing (e.g. mechanical stunning and bleeding followed by chilling in ice water) (Poli et al. 2005; Robb 2008). Current slaughter practices on commercial fishing vessels vary, depending upon the morphology of the animal, fishing method, vessel size/age, as well as catch size and composition; but generally, they do not achieve this optimal welfare standard. For example, catches from trawl codends are typically deposited on deck or in dedicated bins, from where the fish are individually sorted, gutted and possibly bled, before being washed and transferred to a chilled/iced hold. Therefore, death will likely result from exsanguination while conscious, for those processed soon after the catch is brought on board, or asphyxiation/emersion on deck for the remainder. For very large catches, e.g. from purse seines, fish are typically transferred/pumped directly from the net into iced water, where they likely die from hypoxia, which has been demonstrated to be stressful, particularly for temperate and cold-water species (Skjervold et al. 2001). Where animals are removed individually from the gear, e.g. pots and longlines, there is potential to use welfare sympathetic slaughter methods; but generally animals will not be stunned before being processed (gutted/bled) and stored. However, where there is value to be gained from preserving the quality of a catch by avoiding a stressful death, fishers have invested in using optimal slaughter methods, for example tuna and the *iki jime* method (Poli et al. 2005; Lines and Spence 2014).

17.5.6 Summary

From this section it could be inferred that some capture methods are inherently more welfare-sympathetic than others; for example pots and purse seine, which minimise capture related stressors until late in the capture process. However, this conceptual risk assessment has been based on currently limited available knowledge about welfare in these example capture methods, and there has been no empirical testing of these hypotheses. Irrespective of capture method, this risk assessment has highlighted that the highest risk of stress is likely to be during the retrieval and handling phases (Fig. 17.6). While some exposure to these stressors may be unavoidable for the retained catch, it suggests that the welfare of any unwanted catch could be significantly improved by promoting their release at an early stage in the capture process, and certainly before the gear is retrieved.

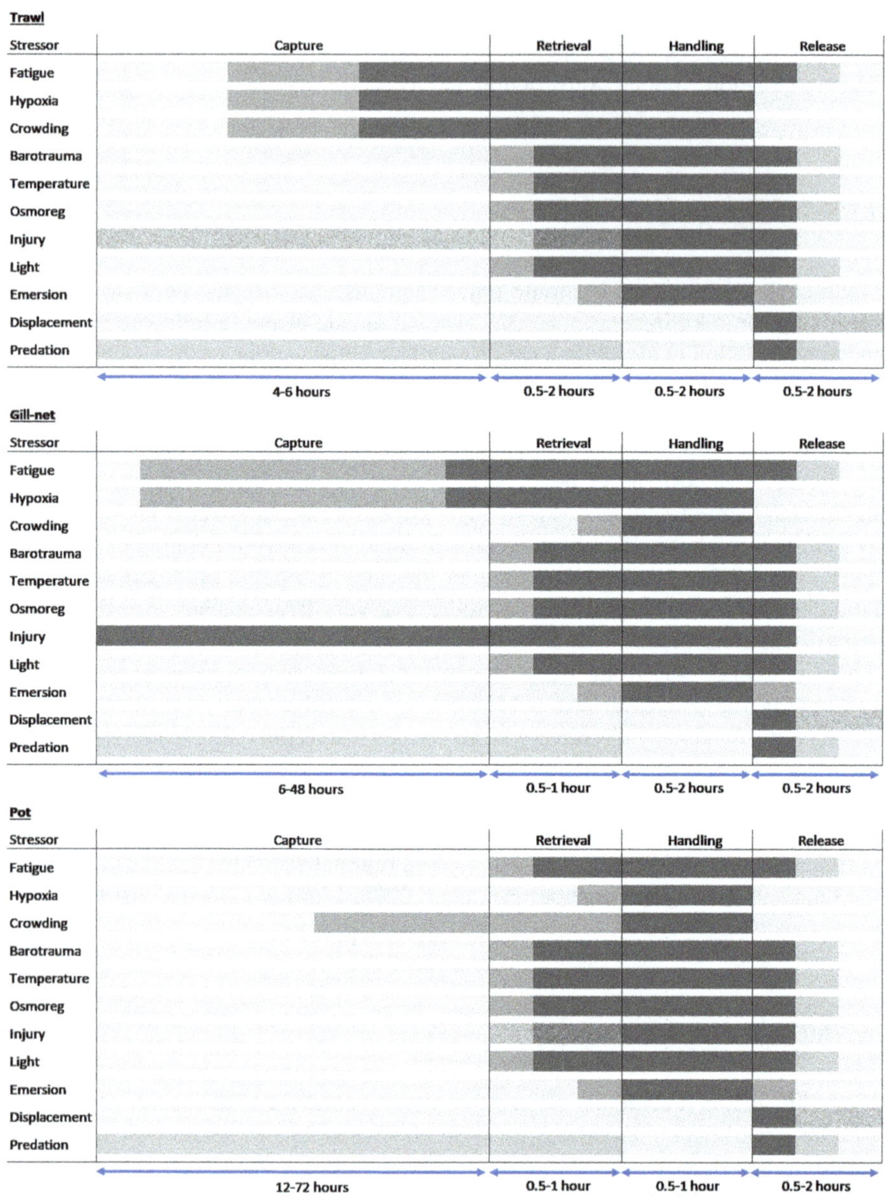

Fig. 17.6 A conceptualised example of a welfare risk assessment. Based on direct observations, or other empirical evidence, the likely risk of a stressor affecting a captive animal at each stage of the capture operation can be estimated. This example shows the conceptualised risks for a gadoid fish caught in and released from a trawl, gillnet and pot. It shows that the risk of encountering stressors may vary between capture method, although the greatest potential for stress is likely to be in the retrieval and handling phases. Key: low risk (light grey), moderate risk (grey) and high risk (dark grey)

17.6 Promoting Good Catch Welfare Practices

Knowledge of the severity and timing of particular stressors, as well as compound effects with other stressors and cumulative effects over time (Sect. 17.4), can be used to identify how and when in the capture process mitigating measures to improve the welfare of the catch could be applied most effectively. It may be necessary to prioritise the welfare of discarded or escaped individuals, over the retained catch, because if injured these animals may live with reduced welfare much longer than fish that are slaughtered.

The remainder of this section refers to output from discussions at the MINOUW workshop on Promoting Discard Survival (Breen and Morales Nin 2017).

General Principles

In some fisheries, the welfare of a substantial proportion of the catches could be greatly improved by avoiding catching unwanted animals. This can be achieved by modifying fishing practices to either:

1. *avoid unwanted catches* by not fishing in areas or times associated with substantial quantities of unwanted animals (e.g. Gullestad et al. 2014), or by,
2. *improving gear selectivity* to promote the escape of any unwanted catch before the gear is retrieved (e.g. Wileman et al. 1996; Holst et al. 1998).

Even where unwanted catches cannot be avoided, simple modification of fishing practices can still improve the welfare of any animals in the retained catch, for example:

- *Limiting the duration of fishing operations* (i.e. towing duration, soak times) will limit the exposure each animal has to capture-related stressors so that they are likely to have more capacity (i.e. better vitality) to cope with the stressors experienced during retrieval, handling and release.
- *Smaller catch volumes* will also be of benefit to the unwanted catch. The risk of injury and/or asphyxiation is likely to be reduced in smaller catches in trawl codends, for example. Also, smaller catches will mean the fishing crew will process the catch more quickly, thus reducing exposure to emersion-related stressors during the handling phase.

17.6.1 *Capture*

Gill and Trammel Nets

The survival of animals released from trammel nets is likely to be very low, due to the rough handling practices for removing the animals from the nets, as well as the relatively long soaking times that effectively kill all vertebrates in set nets (Breen and Morales Nin 2017). Therefore, to promote survival of unwanted animals the primary option will be to avoid capturing them.

Selectivity in gill and trammel nets can be modified by changing a range of gear-related parameters, including netting material, colour and thickness, mesh size, hanging ratio, etc. (Hamley 1975; Holst et al. 1998; Uhlmann and Broadhurst 2015). Furthermore, acoustic pingers have been successfully used to reduce by-catch of marine mammals (e.g. Carretta et al. 2008; Dawson and Slooten 2005). The addition of a panel, a "greca", to the bottom of the net to prevent entanglement can successfully reduce catches of unwanted crustaceans and other benthic invertebrates, as well as lessen predation on commercial catches by crabs and predatory snails (Catanese et al. 2018). In some fisheries the capture of by-catch over extended soak times (>24 h) is an important part of the capture process, to generate bait to attract the target animals, e.g. spiny lobsters (Catanese et al. 2018). In such cases, the use of bait sourced from previous catches and other fisheries and place in bait bags, as well as shorter soak times, is suggested as a possible alternative.

Ghost fishing is when lost or discarding fishing gears, particularly static gears like gill nets and pots, continue fishing without the fisher being able to utilise the catch. It can be addressed through a series of mitigation measures, including: modifying gears and operating practices to minimise gear loss; including biodegradable components, to reduce ghost fishing capacity, if lost; facilitating onshore gear disposal, to avoid abandonment at sea; and running programmes for the reporting and recovery of lost gears (Macfadyen et al. 2009; FAO 2016).

Purse Seine

Pre-catch characterisation: Purse seine is generally considered a non-selective fishing gear. However, the capture process can be highly selective, when the fishers have sufficient information about the catch to decide whether to take it or not in the early part of the fishing process (Breen et al. 2012; Marçalo et al. 2019). For example, very low by-catches have been reported for a purse seine fishery targeting sardine and anchovy in the north Aegean Sea. This fishery uses multiple floating lights to attract the sardine and anchovy; and during the collection of these lights by a rowing boat, if it can be seen that the proportion of by-catch species is too high, the catch is abandoned before the net is set (M. Costantini, pers. com.). Methods of pre-catch characterisation are also being developed in Norwegian purse seine, using hydro-acoustics to estimate the biomass and species of a target school and small cannon deployed trawls for taking samples, during the early capture phases, to determine size distribution and quality (Breen et al. 2012; Marçalo et al. 2019).

Slipping practices: if the early catch characterisation shows the catch is unsuitable or too large, practices are being developed to safely allow catches to be released from the net whilst still in the water (a process called "slipping"), with minimal risk of mortality (Vold et al. 2017; Anders et al. 2019a). For example, Marçalo et al. (2018) demonstrated that the survival of sardine can be significantly improved if the school is allowed to escape through a purposely formed opening in the net, rather than slipping them over the floatline.

Pots

Selectivity in pots can be achieved through modifications to the mesh size (to release small fish), through the introduction of escape gaps (to release small benthic animals) and by limiting entrance dimensions and shapes to prevent large predators (e.g. seals) from entering (Thomsen et al. 2010; Uhlmann and Broadhurst 2015). Pots can also be floated to prevent unwanted benthic animals, e.g. crustacea, from entering the pots (Furevik et al. 2008), the presence of which may increase handling times as well as injuring any fish catch.

Construction: some pots are designed to be collapsible on deck, for easy storage (e.g. Furevik et al. 2008). However, these pots also "collapse" when they are hauled to the surface, increasing the amount of and degree of fish to net contact. A suitable mechanism to prevent this could significantly improve catch welfare during retrieval.

Materials: As with trawl codends, the risk of injury to captive animals can be reduced by using non-abrasive netting materials, as well as ensuring there are no sharp edges, etc.

Pot Size: Having a large enough size of pot to accommodate the catch will reduce crowding effects. Moreover, there is evidence to suggest that it can also increase catch rates (Meintzer et al. 2017).

Ghost fishing: See gill and trammel nets.

Trawl

Trawl selectivity can be improved by simply altering codend mesh sizes (e.g. Hunt et al. 2014) and by including selective devices like square mesh panels (e.g. Graham et al. 2003) and selection grids (e.g. Larsen and Isaksen 1993). Species selection can also be improved by modifying the trawl entrance and main body, for example separator panels (e.g. Ferro et al. 2007) and topless trawls (e.g. Krag et al. 2015). To promote the survival of escaping animals from trawls, it is best to employ the selective device as early in the capture process as possible (Breen 2004; Suuronen 2005; Breen et al. 2007). Although, the introduction of selective technical measures into a fishery may be resisted by the industry, if they anticipate the loss of marketable catch.

Limiting *towing duration* is a practical option for reducing the exposure to capture-related stressors and has been demonstrated to improve discard survival (e.g. Uhlmann and Broadhurst 2007). However, this is unlikely to be acceptable to fishers if catch sizes are already small, although on the other hand shorter tow durations can produce better-quality fish in trawls (e.g. Wagner 1978; Digre et al. 2010).

Controlling catch size in trawl operations has been made more practical with a range of recent innovations which limit the volume of the catch held in the codend, and then automatically release any excess (see Grimaldo et al. 2014; ICES 2015c for examples).

Codend construction and materials can be modified to avoid injuries to captive animals by using non-abrasive materials, e.g. knotless netting (e.g. Barthel et al. 2003) and alternative mesh configurations, e.g. "T90" (Digre et al. 2010).

Lined codends have been used to reduce water flow in codends and thus reduce exhaustion and injury, to collect specimens for experiments (e.g. Breen et al. 2007). This has recently been developed further in commercial trawls to protect the catch and promote selectivity (e.g. Adams 2013; Precision Seafood Harvesting 2014).

17.6.2 Retrieval

Controlled decompression: Scientific protocols have been developed for decompressing fish with closed swim bladders at rates that avoid both physical barotrauma and to some extent temperature shock (e.g. Breen et al. 2007). However, it would be unrealistic to expect fishermen to haul their gear at these slow rates, because to haul from just 100 m would take many hours. So, in reality, the stressors associated with the environmental change during retrieval are practically unavoidable. The only practical solutions at present are to address the consequences and symptoms of these stressors (see Handling and Sorting, and Release).

Retrieval from Water

With urgency and care: The catch should be retrieved from the water as quickly and carefully as possible, to avoid risk of injury and predation in the surface waters, while avoiding striking the vessel and/or deck equipment.

Splitting catches: In trawl codends, particularly on smaller vessels, it is sometimes necessary to split the catch to safely lift the codend and its contents aboard the vessel and limit crushing effects inside the catch. However, the remainder of the catch stays in the water, inside the trawl extension, where it is at continued risk of abrasive injury and seabird predation. The benefits to the welfare of the catch will be dependent on several factors, including the size and composition of the catch, the surface conditions and the skill and efficiency of the crew in conducting the procedure.

Retain/retrieve in water: Retaining the catch in a small volume of water is frequently used to obtain viable, uninjured samples for scientific investigation, e.g. tagging, by avoiding emersion-related stressors. It can be achieved by partially lining the codend with an impermeable liner, to retain some water with the catch, as it is lifted aboard.

17.6.3 Handling and Sorting

Avoid emersion/hold in water: Where practical, the catch should be transferred from the fishing gear directly into water, where it can be held during the sorting process (e.g. Broadhurst et al. 2009). The first advantage of this is that direct transfer into water will help avoid injuries due to contact with hard surfaces or other components of the catch. Furthermore, provided the conditions within the water tank are suitable, this treatment will alleviate the effects of emersion. The water container should have a continuous supply of water with a temperature, salinity and dissolved oxygen content that will minimise effects of temperature shock, osmoregulatory stress and hypoxia. Ideally, the water should be supplied from the bottom of the container, via a network of small holes, to ensure that hypoxic zones do not develop in the tank where animals are most likely to aggregate (see also Chap. 18).

Where it is impractical, or even unsafe, to install a water-filled holding tank on the vessel, the crew should attempt to minimise the effects of physical/abrasive contact and try to spray the catch with a constant water supply. Even a water-soaked cloth covering the catch on the deck will help alleviate some effects of emersion.

Avoid direct sunlight: Wherever the catch is being held during sorting, it should be kept out of direct sunlight to avoid potential stress and injury from bright light.

With urgency and care: The catch should be sorted and handled as quickly and carefully as possible, to avoid risk of injury and further stress. With an efficient and conscientious crew, sorting by hand has some benefits over mechanical sorting with respect to this. However, with large catches this may need to be traded off with reducing sorting time and thus minimising emersion exposure, where mechanised sorting (e.g. grids) may be more beneficial.

Prioritisation: The sorting and release of unwanted catch should be prioritised over the processing of the landed catch to minimise emersion times. Protected and vulnerable animals should be dealt with first.

Vitality assessment: It may be advantageous to assess the vitality of the discarded components of the catch to ensure that any released animals do indeed have the potential to survive. These assessments need not be complex or time consuming and may simply be based upon an informative categorical scale; for more details see Breen and Catchpole (2020). This process will be facilitated by holding the catch in water, because it will be easier to identify active, more vital, animals as well as those with barotrauma that are unable to leave the surface of the water.

17.6.4 Release of Unwanted Catch

With urgency and care: As soon as an animal has been selected for release, it should be released into the water quickly and with care, via a route that promotes its escape from the surface, and minimises likelihood of further injury and encounters with predators.

Appropriate release location: Where practical and safe, they should be released in a location where they will quickly find a suitable habitat for shelter and food (i.e. ideally close to where they were first caught).

Assisted recompression: Animals suffering from physical barotrauma (i.e. excessively swollen or ruptured swim bladder), which would be unable to swim away from the surface, could be assisted using a device that enables recompression. Effective solutions to this problem have been developed by recreational fishers, and such devices could be developed for releasing animals from commercial fisheries; particularly if they are protected or endangered species. This device could be a simple cage for the compromised animals which can be lowered to a sufficient depth to relieve their symptoms. For example, lowering an affected animal to 50 m will reduce an equilibrated gas volume to one sixth of its volume at the surface. For examples of recompression devices, see Bellquist et al. (2019), DFO (2018) and NOAA (2020).

Discharge pipe: To give released animals some protection from predatory birds, they could be released via a pipe, with a constant flow of water through it, to a depth of around 5 m. While some diving birds can swim to this depth, most gulls cannot (Breen and Catchpole 2020). Moreover, the small increase in hydrostatic pressure will assist fish with barotrauma (reducing the gas volume to 2/3).

17.6.5 Slaughter

Implementing welfare-sympathetic slaughter practices is one of the most significant challenges to introducing good welfare practices in commercial fisheries. It could be argued that appropriate methods have already been developed by the aquaculture industry, e.g. automated stunning and bleeding technology (Lines and Spence 2014; Chap. 14). However, the effectiveness of some of these technologies is questionable; e.g. Robb et al. (2000) found that 50% of the shots in an automated spiking machine were inaccurate. Furthermore, even where fishers may be motivated to develop and use such methods, there are challenges with regard to effective operation and scale that could render these solutions impractical and/or cost-prohibitive in some fisheries. For example, in pelagic fisheries a single catch can be several hundred tonnes and contain hundreds of thousands of individual fish. Also, conditions on board offshore fishing vessels are substantially more unstable than at aquaculture harvesting facilities, which could further reduce the effectiveness of any automated technology. Research is currently ongoing to develop alternative humane slaughter methods, e.g. using anaesthesia or electrical stunning (e.g. Anders et al. 2019b). One interesting solution, which is explored in detail in Chap. 18, is the development of live capture methods, where fish are caught using low-stress methods and then held in welfare-sympathetic holding facilities, until they are slaughtered. Thus, the animals are able to recover from the initial capture, and harvesting is managed at an optimal rate for the available slaughter method.

17.6.6 Summary

There is potential to improve the welfare of animals encountering each of the capture methods considered here. Furthermore, there may be capacity for replacing one method with a more welfare-sympathetic method (e.g. pots; Utne-Palm et al. 2018; Humborstad et al. 2018). This is, however, likely to have consequences for catching efficiency, and therefore economic viability, because in many cases the preferred capture method has been developed and optimised for catching a specific target species/assemblage in a specific environment. In general, the greatest potential for experiencing capture-related stressors is during the retrieval and handling phases. However, the capacity to mitigate these stressors is limited, particularly those related to environmental change. Therefore, facilitating the release of the unwanted components of the catch early in the capture process, and certainly before retrieval, should be a priority.

17.7 Conclusion: Moving Forward

Good catch welfare makes sense. The introduction of good welfare practices to commercial capture fisheries is not just ethically sound, it has real potential to make fisheries more sustainable by reducing unwanted catches and collateral mortality, and could improve meat quality and product shelf life. This in itself may provide economic incentives for fishers to adopt good catch welfare practices, by opening up premium markets for high-quality, sustainable and ethically harvested food products. Furthermore, ethical harvesting practices could be incentivised and demand an associated price premium, through product certification schemes, e.g. Marine Stewardship Certification and Freedom Foods Scheme.

However, economic incentive alone may not be sufficient, or even relevant, for all fisheries. It may require regulatory incentives to better align fishing practices with shifts in societal ethics. One recent example of such a regulatory incentive is the introduction of a "Landing Obligation" (discard ban) in the European Union's Common Fisheries Policy (Salomon et al. 2014; Borges and Lado 2019; Karp et al. 2019). This was introduced in response to public condemnation of wasteful discarding practices that were not only allowed but were in fact driven by European fishing regulations (Borges and Lado 2019). Included in the Landing Obligation was a "High Survival Exemption" (Rihan et al. 2019), which was a pragmatic measure to avoid unnecessarily increasing fishing mortality by inadvertently forcing fishers to retain unwanted catch that would otherwise have survived. This exemption has motivated several fisheries to begin research programmes investigating the survival potential of various components of the unwanted catch (e.g. ICES 2014, 2015a, b, 2016a, b). This has increased fishers' and researchers' understanding of the stressors associated with the capture process and in turn motivated efforts to mitigate these stressors to promote high discard survival and catch welfare in general.

This chapter has identified several strategic challenges to introducing good welfare practices to commercial fishing that can prioritise future research in this area:

- Reducing unwanted catches—continued research in this area will significantly improve the welfare of the unwanted components of the catch in some fisheries, which may also have synergistic benefits for the retained catch (e.g. reduced catch size).
- Develop monitoring/assessment methods—species- and fishery-specific methods will be required to properly characterise stressors, and thus prioritise areas for improving catch welfare.
- Demonstrating good welfare promotes quality—empirical evidence of the link between stress and quality metrics could provide a strong incentive for stakeholders to adopt good welfare practices.

- Improving catch handling and slaughter methods—it is one of the most significant challenges which could negate any improvements in catch welfare and quality elsewhere in the capture process.

In conclusion, wild capture commercial fisheries are not only essential to global food security, they have the potential to produce premium "wild-caught/free-range" meat products: healthy, sustainable and harvested using humane, welfare-conscious practices from animals that have lived their entire lives in the wild. However, to achieve this idealistic goal we should not begin by condemning fishers' current harvesting practices, which have evolved over millennia to enable us to harvest a valuable food resource in a difficult and dangerous environment. Instead, we must work together, constructively, by learning more about capture-related stressors and how they can be mitigated, and then channelling this knowledge into practical and meaningful guidelines to inform fishers on best practices for good catch welfare.

Acknowledgements This chapter has been partly based on output and discussions from a workshop on Promoting the Survival of Discarded Fish, which is reported in Breen, M. & Morales Nin, B. (2017). We would like to thank our colleagues at the meeting for their input and lively discussion: Beatriz Morales Nin, Francesc Maynou, Miquel Palmer, Montse Demestre, Pilar Sánchez, Alfredo de Vinesa Gutiérrez, Mariona Segura, Claudio Viva, Alessandro Ligas, Catarina Adão, Ada Campos, Ana Marçalo and Hugues Benoit. The workshop was conducted as part of the EU funded project MINOUW (H2020-SFS-2014-2) [http://minouw-project.eu/] (European Commission contract n° 634495). Thanks also to Helen McGregor and Keith Mutch for providing the original images from Galbraith et al. (2004) [Crown Copyright], and to Sebastian Uhlmann for providing the original image for Fig. 17.1. Finally, we would especially like to thank the many anonymous fishers who have supported, and continue to support our research by allowing us to work alongside them, enabling us to better understand the fishing process and its effects.

References

Adams (2013) The future of commercial fishing. New Zealand Herald. http://www.nzherald.co.nz/business/news/article.cfm?c_id=3&objectid=11132876

Agnew DJ, Pearce J, Pramod G, Peatman T, Watson R et al (2009) Estimating the worldwide extent of illegal fishing. PLoS One 4(2):e4570. https://doi.org/10.1371/journal.pone.0004570

Anders N, Fernö A, Humborstad OB, Løkkeborg S, Utne-Palm AC (2017) Species specific behaviour and catchability of gadoid fish to floated and bottom set pots. ICES J Mar Sci 74 (3):769–779. https://doi.org/10.1093/icesjms/fsw200

Anders N, Breen M, Saltskår J, Totland B, Øvredal JT et al (2019a) Behavioural and welfare implications of a new slipping methodology for purse seine fisheries in Norwegian waters. PLoS One 14(3):e0213031. https://doi.org/10.1371/journal.pone.0213031

Anders N, Roth B, Grimsbø E, Breen M (2019b) Assessing the effectiveness of an; electrical stunning and chilling protocol for the slaughter of Atlantic mackerel (Scomber scombrus). PLoS One 14(9):e0222122. https://doi.org/10.1371/journal.pone.0222122

Bagni M, Civitareale C, Priori A, Ballerini A, Finoia M, Brambilla G, Marino G (2007) Pre-slaughter crowding stress and killing procedures affecting quality and welfare in sea bass (*Dicentrarchus labrax*) and sea bream (*Sparus aurata*). Aquaculture 263:52–60

Barthel BL, Cooke SJ, Suski CD, Philipp DP (2003) Effects of landing net mesh type on injury and mortality in a freshwater recreational fishery. Fish Res 63(2):275–282. ISSN: 0165-7836. https://doi.org/10.1016/S0165-7836(03)00059-6

Barton BA (2002) Stress in fishes: a diversity of responses with particular reference to changes in circulating corticosteroids. Integr Comp Biol 42:517–525
Bayse SM, Pol MV, He P (2016) Fish and squid behaviour at the mouth of a drop-chain trawl: factors contributing to capture or escape. ICES J Mar Sci 73(6):1545–1556. https://doi.org/10.1093/icesjms/fsw007
Beamish FWH (1966) Muscular fatigue and mortality in haddock (*Melanogrammus aeglifinus*) caught by otter trawl. J Fish Res Bd Can 23(10):1507–1521
Bellquist L, Beyer S, Arrington M, Maeding J, Siddall A, Fischer P, Hyde J, Wegner NC (2019) Effectiveness of descending devices to mitigate the effects of barotrauma among rockfishes (Sebastes spp.) in California recreational fisheries. Fish Res 215(2019):44–52. ISSN 0165-7836. https://doi.org/10.1016/j.fishres.2019.03.003
Benoît HP, Hurlbut T, Chassé J (2010) Assessing the factors influencing discard mortality of demersal fishes in four fisheries using a semi-quantitative indicator of survival potential. Fish Res 106:436–447
Bennett JR, Maloney R, Possingham HP (2015) Biodiversity gains from efficient use of private sponsorship for flagship species conservation. Proc B 282:1–7
Bjørnevik M, Solbakken V (2010) Pre-slaughter stress and subsequent effect on flesh quality in farmed cod. Aquac Res 41:467–474. https://doi.org/10.1111/j.1365-2109.2010.02498.x
Black EC (1958) Hyperactivity as a lethal factor in fish. J Fish Res Bd Can 15:573–586
Bonga SEW (1997) The stress response in fish. Physiol Rev 77(3):591–625
Borderías AJ, Lamua M, Tejada M (1983) Texture analysis of fish fillets and minced fish by both sensory and instrumental methods. Int J Food Sci Tech 18(1):85–95
Borges L, Lado EP (2019) Discards in the common fisheries policy. In: Uhlmann SS, Ulrich C, Kennelly S (eds) The evolution of the policy: The European discard policy–educing unwanted catches in complex multi-species and multi-jurisdictional fisheries. Springer. https://link.springer.com/chapter/10.1007/978-3-030-03308-8_2
Breen M (2004) Investigating the mortality of fish escaping from towed fishing gears – a critical analysis. PhD Thesis, University of Aberdeen. 313 pp
Breen M, Catchpole T (eds) (2020) ICES WKMEDS guidance on method for estimating discard survival. ICES Cooperative Research Report
Breen M, Morales Nin B (eds) (2017) Deliverable report 2.16: data on the survival of unwanted catch. Science, technology, and society initiative to minimize unwanted catches in European fisheries: project MINOUW. SFS-09-2014. http://minouw-project.eu/wp-content/uploads/2018/07/D2-16-Data-on-the-survival-of-unwanted-catch.pdf
Breen M, Dyson J, O'Neill FG, Jones E, Haigh M (2004) Swimming endurance of haddock (*Melanogrammus aeglefinus* L.) at prolonged and sustained swimming speeds, and its role in their capture by towed fishing gears. ICES J Mar Sci 61:1071–1079
Breen M, Huse I, Ingolfsson OA, Madsen N, Soldal AV (2007) Survival: an assessment of mortality in fish escaping from trawl codends and its use in fisheries management. Final report on EU contract Q5RS-2002-01603, 300 pp
Breen M, Isaksen B, Ona E, Pedersen AO, Pedersen G, Saltskår J, Svardal B, Tenningen M, Thomas PJ, Totland B, Øvredal JT, Vold A (2012) A review of possible mitigation measures for reducing mortality caused by slipping from purse-seine fisheries. ICES CM 2012/C:12
Broadhurst MK, Suuronen P, Hulme A (2006) Estimating collateral mortality from towed fishing gear. Fish Fish 7:180–218
Broadhurst MK, Millar RB, Brand CP, Uhlmann SS (2009) Modified sorting technique to mitigate the collateral mortality of trawled school prawns (*Metapenaeus macleayi*). Fish Bull 107:286–297
Browman HI, Cooke SJ, Cowx IG, Derbyshire SWG, Kasumyan A, Key B, Rose JD, Schwab A, Skiftesvik AB, Stevens ED, Watson CA, Arlinghaus R (2018) Welfare of aquatic animals: where things are, where they are going, and what it means for research, aquaculture, recreational angling, and commercial fishing. ICES J Mar Sci 76:82. https://doi.org/10.1093/icesjms/fsy067
Brown C (2015) Fish intelligence, sentience and ethics. Anim Cogn 18:1–17

Brown J, Macfadyen G (2007) Ghost fishing in European waters: impacts and management responses. Mar Pol 31:488–504

Brown RS, Pflugrath BD, Colotelo AH, Brauner CJ, Carlson TJ, Deng ZD, Seaburg AG (2012) Pathways of barotrauma in juvenile salmonids exposed to simulated hydroturbine passage: Boyle's law vs. Henry's law. Fish Res 121–122:43–50. ISSN: 0165-7836. https://doi.org/10.1016/j.fishres.2012.01.006

Buhl-Mortensen L, Aglen A, Breen M, Buhl-Mortensen P, Ervik A, Husa V, Løkkeborg S, Røttingen I, Stockhausen HH (2013) Impacts of fisheries and aquaculture on sediments and benthic fauna: suggestions for new management approaches. Fisken og Havet 3, 69 pp

Carretta J, Barlow J, Enriquez L (2008) Acoustic Pingers eliminate beaked whale bycatch in a gill net fishery. Publications, Agencies and Staff of the U.S. Department of Commerce. Paper 47. http://digitalcommons.unl.edu/usdeptcommercepub/47

Catanese G, Hinz H, del Mar Gil M, Palmer M, Breen M, Mira A, Pastor E, Grau A, Campos-Candela A, Koleva E, Grau AM, Beatriz Morales-Nin B (2018) Comparing the catch composition, profitability and discard survival from different trammel net designs targeting common spiny lobster (Palinurus elephas) in a Mediterranean fishery. Peer J 6:e4707. https://doi.org/10.7717/peerj.4707

Chapman CJ, Shelton PMJ, Shanks AM, Gaten E (2000) Survival and growth of the Norway lobster *Nephrops norvegicus* in relation to light-induced eye damage. Mar Biol 136:233–241

Chopin F, Inoue Y, Arimoto A (1996) Development of a catch mortality model. Fish Res 25:377–382

Claireaux G, Webber DM, Lagardere JP, Kerr SR (2000) Influence of water temperature and oxygenation on the aerobic metabolic scope of Atlantic cod (*Gadus morhua*). J Sea Res 44:257–265

Cole RG, Tindale DS, Blackwell RG (2001) A comparison of diver and pot sampling for blue cod (*Parapercis colias: Pinguipedidae*). Fish Res 52(3):191–201

Cooke SJ, Donaldson MR, O'Connor CM et al (2013) The physiological consequences of catch-and-release angling: perspectives on experimental design, interpretation, extrapolation and relevance to stakeholders. Fish Man Ecol 20:268–287

Darnell MZ, Darnell KM, McDowell RE, Rittschof D (2010) Postcapture survival and future reproductive potential of ovigerous blue crabs *Callinectes sapidus* caught in the central North Carolina pot fishery. Trans Am Fish Soc 139(6):1677–1687. https://doi.org/10.1577/T10-034.1

Davis MW (2002) Key principles for understanding fish bycatch discard mortality. Can J Fish Aq Sci 59:1834–1843

Davis MW (2010) Fish stress and mortality can be predicted using reflex impairment. Fish Fish 11:1467–2979

Davis MW, Olla BL (2001) Stress and delayed mortality induced in Pacific halibut *Hippoglossus stenolepis* by exposure to hooking, net towing, elevated seawater temperature and air: implications for management of bycatch. N Am J Fish Manag 21:725–732

Davis MW, Olla BL (2002) Mortality of lingcod towed in a net is related to fish length, seawater temperature and air exposure: a laboratory bycatch study. N Am J Fish Manag 22:395–404

Davis MW, Olla BL, Schreck CB (2001) Stress induced by hooking, net towing, elevated seawater temperature and air in sablefish: lack of concordance between mortality and physiological measures of stress. J Fish Biol 58:1–15

Dawkins MS (2004) Using behaviour to assess animal welfare. Anim Welf 13:S3–S7

Dawson SM, Slooten E (2005) Management of gillnet bycatch of cetaceans in New Zealand. J Cetacean Res Manag 7(1):59–64

Dehadrai PV (1966) Mechanism of gaseous exophthalmia in the Atlantic cod, *Gadus morhua* L. JFish Res Bd Can 23:909–914

Depestele J, Rochet M-J, Dorémus G, Laffargue P, Stienen EWM (2016) Favorites and leftovers on the menu of scavenging seabirds: modelling spatiotemporal variation in discard consumption. Can JFish Aquat Sci 73:1446–1459

DFO (2018) A review of the use of recompression devices as a tool for reducing the effects of barotrauma on rockfishes in British Columbia Canadian Science Advisory Secretariat (Pacific Region) Science Response 2018/043. https://waves-vagues.dfo-mpo.gc.ca/Library/40716120.pdf

Diggles BK, Cooke SJ, Rose JD, Sawynok W (2011) Ecology and welfare of aquatic animals in wild capture fisheries. Rev Fish Biol Fish 21:739–765

Digre H, Hansen UJ, Erikson U (2010) Effect of trawling with traditional and 'T90' trawl codends on fish size and on different quality parameters of cod *Gadus morhua* and haddock *Melanogrammus aeglefinus*. Fish Sci 76:549. https://doi.org/10.1007/s12562-010-0254-2

Digre H, Tveit GM, Solvang-Garten T, Eilertsen A, Aursand IG (2016) Pumping of mackerel (*Scomber scombrus*) onboard purse seiners, the effect on mortality, catch damage and fillet quality. Fish Res 176:65–75

Domenici P, Herbert NA, LeFrançois C, Steffensen JF, McKenzie DJ (2012) The effect of hypoxia on fish swimming performance and behaviour. In: Palstra AP, Planas JV (eds) Swimming physiology of fish. Springer, Berlin, pp 129–161

Donaldson MR, Cooke SJ, Patterson DA, Macdonald JS (2008) Cold shock and fish. J Fish Biol 73:1491–1530. https://doi.org/10.1111/j.1095-8649.2008.02061.x

Eigaard OR, Bastardie F, Breen M, Dinesen GE, Hintzen NT, Laffargue P, Mortensen LO, Nielsen JR, Nilsson HC, O'Neill FG, Polet H, Reid DG, Sala A, Skőld M, Smith C, Sørensen TK, Tully O, Zengin M, Rijnsdorp AD (2015) Estimating seabed pressure from demersal trawls, seines, and dredges based on gear design and dimensions. ICES J Mar Sci 73:i27. https://doi.org/10.1093/icesjms/fsv099

Eigaard OR, Bastardie F, Breen M, Dinesen GE, Hintzen NT, Laffargue P, Mortensen LO, Rasmus Nielsen J, Nilsson H, O'Neill FG, Polet H, Reid DG, Sala A, Sköld M, Smith C, Sørensen TK, Tully O, Zengin M, Rijnsdorp AD (2016) A correction to "Estimating seabed pressure from demersal trawls, seines and dredges based on gear design and dimensions". ICES J Mar Sci 73:2420. https://doi.org/10.1093/icesjms/fsw116

Elliott DG (2011a) The skin | Functional morphology of the integumentary system in fishes. In: Farrell AP (ed) Encyclopedia of fish physiology. Academic, San Diego

Elliott DG (2011b) The skin | The many functions of fish integument. In: Farrell AP (ed) Encyclopedia of fish physiology. Academic, San Diego

Esaiassen M, Akse L, Joensen S (2013) Development of a catch-damage-index to assess the quality of cod at landing. Food Control 29:231–235

FAO (1990) FAO international standard statistical classification of fishing gear (ISSCFG). http://www.fao.org/docrep/008/t0367t/t0367t00.htm

FAO (2016) Abandoned, lost or otherwise discarded gillnets and trammel nets: methods to estimate ghost fishing mortality, and the status of regional monitoring and management, by Eric Gilman, Francis Chopin, Petri Suuronen and Blaise Kuemlangan. FAO fisheries and aquaculture technical paper no. 600, Rome

Farrington M, Carr A, Pol M, Szymanski M (2008) Selectivity and survival of Atlantic cod (*Gadus morhua*) [and haddock (*Melangrammus aeglefinus*)] in the Northwest Atlantic longline fishery. Final report. NOAA/NMFS Saltonstall-Kennedy program. Grant number: NA86FD0108. http://archives.lib.state.ma.us/bitstream/handle/2452/429957/ocn960945695.pdf?sequence=1&isAllowed=y

Feathers MG, Knable AE (1983) Effects of decompression upon largemouth bass. N Am J Fish Manag 3:86–90

Ferro RST, Jones EG, Kynoch RJ, Fryer RJ, Buckett B-E (2007) Separating species using a horizontal panel in the Scottish North Sea whitefish fishery. ICES J Mar Sci 64:1543–1550

Ferter K, Weltersbach MS, Humborstad O-B, Fjelldal PG, Sambraus F, Strehlow HV, Vølstad JH (2015) Dive to survive: effects of capture depth on barotrauma and post-release survival of Atlantic cod (*Gadus morhua*) in recreational fisheries. ICES J Mar Sci 72:2467–2481

Frank TM, Widder EA (1994) Comparative study of behavioral sensitivity thresholds to near-UV and blue-green light in deep-sea crustaceans. Mar Biol 121:229–235

Furevik DM, Humborstad OB, Jørgensen T, Løkkeborg S (2008) Floated fish pot eliminates bycatch of red king crab and maintains target catch of cod. Fish Res 92(1):23–27

Galbraith RD, Rice A, Strange E (2004) An Introduction to commercial fishing gear and methods used in Scotland. Scottish Fisheries Information Pamphlet No. 25 2004. ISSN: 0309 9105, 44 pp

Gale MK, Hinch SG, Donaldson MR (2013) The role of temperature in the capture and release of fish. Fish Fish 14:1–33

Garthe S, Camphuysen K, Furness RW (1996) Amounts of discards by commercial fisheries and their significance as food for seabirds in the North Sea. MEPS 136:1–11

Gilman E, Suuronen P, Hall M, Kennelly S (2013) Causes and methods to estimate cryptic sources of fishing mortality. J Fish Biol 83:766–803. https://doi.org/10.1111/jfb.12148

Graham N, Kynoch RJ, Fryer RJ (2003) Square mesh panels in demersal trawls: further data relating haddock and whiting selectivity to panel position. Fish Res 62(3):361–375. ISSN: 0165-7836. https://doi.org/10.1016/S0165-7836(02)00279-5

Greenwell MG, Sherrill J, Clayton LA (2003) Osmoregulation in fish. Mechanisms and clinical implications. Vet Clin North Am Exot Anim Pract 6(1):169–189. vii

Grimaldo E, Sistiaga M, Larsen RB (2014) Development of catch control devices in the Barents Sea cod fishery. Fish Res 155:122–126

Gullestad P, Aglen A, Bjordal Å, Blom G, Johansen S, Krog J, Misund OA, Røttingen I (2014) Changing attitudes 1970–2012: evolution of the Norwegian management framework to prevent overfishing and to secure long-term sustainability. ICES J Mar Sci 71:173–182

Hall MA, Alverson DL, Metuzals KI (2000) By-catch: problems and solutions. Mar Pollut Bull 41 (1):210

Hamley JM (1975) Review of gillnet selectivity. J Fish Res Bd Can 32(11):1943–1969

Harris RR, Ulmestrand M (2004) Discarding Norway lobster (*Nephrops norvegicus* L.) through low salinity layers – mortality and damage seen in simulation experiments. ICES J. Mar Sci 61:127–139

HLPE (2014) Sustainable fisheries and aquaculture for food security and nutrition. A report by the High Level Panel of Experts on Food Security and Nutrition of the Committee on World Food Security, Rome 2014. http://www.fao.org/3/a-i3844e.pdf

Holst R, Madsen N, Moth-Poulsen T, Fonseca P, Campos A (1998) Manual for gillnet selectivity. European Commission. http://constat.dk/Papers/Gillman.pdf

Holt EWL (1895) An examination of the present state of the Grimsby trawl fishery: with especial reference to the destruction of immature fish. Revision of tables. J Mar Biol Assoc 3:337–448

Hughes GM (1975) Respiratory responses to hypoxia in fish. Am Zool 13:475–489

Humborstad O-B, Mangor-Jensen A (2013) Buoyancy adjustment after swimbladder puncture in cod *Gadus morhua*: an experimental study on the effect of rapid decompression in capture-based aquaculture. Mar Biol Res 9:383–393

Humborstad O-B, Utne-Palm AC, Breen M, Løkkeborg S (2018) Artificial light in baited pots substantially increases the catch of cod (Gadus morhua) by attracting active bait, krill (Thysanoessa inermis). ICES J Mar Sci. https://doi.org/10.1093/icesjms/fsy099

Humborstad OB, Ferter K, Kryvi H, Fjelldal P (2016a) Exophthalmia in wild-caught cod (*Gadus morhua* L.): development of a secondary barotrauma effect in captivity. J Fish Dis 40:41–49

Humborstad O-B, Breen M, Davis MW, Løkkeborg S, Mangor-Jensen A, Midling KØ, Olsen RE (2016b) Survival and recovery of longline- and pot -caught cod (*Gadus morhua*) for use in capture-based aquaculture (CBA). Fish Res 174:103–108

Hunt DE, Maynard DL, Gaston TF (2014) Tailoring codend mesh size to improve the size selectivity of undifferentiated trawl species. Fish Manag Ecol 21:503–508. https://doi.org/10.1111/fme.12099

Huntingford FA, Kadri S (2009) Taking account of fish welfare: lessons from aquaculture. J Fish Biol 75:2862–2867. https://doi.org/10.1111/j.1095-8649.2009.02465.x

Huntingford FA, Adams C, Braithwaite VA, Kadri S, Pottinger TG, Sandøe P, Turnbull JF (2006) Current issues in fish welfare. J Fish Biol 68:332–372

Huse I, Vold A (2010) Mortality of mackerel (*Scomber scombrus* L.) after pursing and slipping from a purse-seine. Fish Res 106(1):54–59
Hvas M, Folkedal O, Imsland A, Oppedal F (2017) The effect of thermal acclimation on aerobic scope and critical swimming speed in Atlantic salmon, *Salmo salar*. J Exp Biol 220:2757–2764
ICES (2005) Joint report of the study group on unaccounted fishing mortality (SGUFM) and the workshop on unaccounted fishing mortality (WKUFM). ICES CM 2005/B:08
ICES (2014) Report of the workshop on methods for estimating discard survival (WKMEDS), 17–21 February 2014, ICES HQ, Copenhagen, Denmark. ICES conference and meeting (CM) 2014/ACOM: 51, 114 pp
ICES (2015a) Report of the workshop on methods for estimating discard survival 2, 24–-28 November 2014, ICES HQ. ICES CM 2014\ACOM:66, 35 pp
ICES (2015b) Report of the workshop on methods for estimating discard survival 3 (WKMEDS 3), 20–24 April 2015, London. ICES CM 2015\ACOM:39, 47 pp
ICES (2015c) Final report of TOR innovative dynamic catch control devices in fishing. ICES-FAO Working Group on Fisheries Technology and Fish Behavior (WGFTFB). http://ices.dk/sites/pub/Publication%20Reports/Expert%20Group%20Report/SSGIEOM/2015/2015%20WGFTFB%20Annex%204%20Final%20report%20of%20TOR%20Innovative%20dynamic%20catch%20control%20devices%20in%20fishing.pdf
ICES (2016a) Report of the workshop on methods for estimating discard survival 4 (WKMEDS4), 30 November–4 December 2015, Ghent. ICES CM 2015\ACOM:39, 57 pp
ICES (2016b) Report of the workshop on methods for estimating discard survival 5 (WKMEDS 5), 23–27 May 2016, Lorient, France. ICES CM 2016/ACOM:56, 51 pp
Ingolfsson OA, Soldal AV, Huse I, Breen M (2007) Escape mortality of cod, saithe and haddock in a Barents Sea trawl fishery. ICES J Mar Sci 64:1836–1844
Johnsen S (2012) The optics of life: a biologist's guide to light in nature. Princeton University Press, Princeton. ISBN: 978-0-691-13990 6 (hbk); 978-0-691-13991-3 (pbk)
Jury SH, Howell H, O'Grady DF, Watson WH III (2001) Lobster trap video: in situ video surveillance of the behaviour of *Homarus americanus* in and around traps. Mar Freshw Res 52(8):1125–1132
Kaiser MJ, Huntingford FA (2009) Introduction to papers on fish welfare in commercial fisheries. J. Fish Biol 75:2852–2854
Karlsson-Drangsholt A, Svalheim RA, Aas-Hansen Ø, Olsen SH, Midling K, Breen M, Grimsbø E, Johnsen HK (2017) Recovery from exhaustive swimming and its effect on fillet quality in haddock (*Melanogrammus aeglefinus*). Fish Res 97:96–104. https://doi.org/10.1016/j.fishres.2017.09.006
Karp WA, Breen M, Borges L, Fitzpatrick M, Kennelly SJ, Kolding J, Nielsen KN, Viðarsson JR, Cocas L, Leadbitter D (2019) Strategies used throughout the world to manage fisheries discards – lessons for implementation of the eu landing obligation. In: Uhlmann SS, Ulrich C, Kennelly S (eds) The European discard policy-reducing unwanted catches in complex multi-species and multi-jurisdictional fisheries. Springer. https://link.springer.com/chapter/10.1007/978-3-030-03308-8_1
Key B (2016) Why fish do not feel pain. Animal Sentience 2016.003
Killen SS, Marras S, McKenzie DJ (2011) Fuel, fasting, fear: routine metabolic rate and food deprivation exert synergistic effects on risk-taking in individual juvenile European sea bass. J Anim Ecol 80:1024–1033. https://doi.org/10.1111/j.1365-2656.2011.01844.x
Kitsios E (2016) The nature and degree of skin damage in mackerel (*Scomber scombrus*) following mechanical stress: can skin damage lead to mortality following crowding in a purse seine? MSc Thesis, University of Bergen, 64 pp
Krag LA Herrmann B, Karlsen JD, Mieske B (2015) Species selectivity in different sized topless trawl designs: does size matter? Fish Res 172:243–249. ISSN: 0165-7836. https://doi.org/10.1016/j.fishres.2015.07.010

Kristoffersen S, Tobiassen T, Steinsund V, Olsen RL (2006) Slaughter stress, post-mortem muscle pH and rigor development in farmed Atlantic cod (*Gadus morhua* L.). Int J Food Sci Tech 41:861–864

Larsen RB, Isaksen B (1993) Size selectivity of rigid sorting grid in bottom trawls for Atlantic cod (*Gadus morhua*) and haddock (*Melanogrammus aeglefinus*). ICES Mar Sci Symp 196:178–182

Lines JA, Spence J (2014) Humane harvesting and slaughter of farmed fish. Rev Sci Tech 33 (1):255–264

Lundberg P, Vainio A, MacMillan DC, Smith RJ, Veríssimo D, Arponen A (2019) The effect of knowledge, species aesthetic appeal, familiarity and conservation need on willingness to donate. Anim Conserv 22:432–443. https://doi.org/10.1111/acv.12477

Macfadyen G, Huntington T, Cappell R (2009) Abandoned, lost or otherwise discarded fishing gear. UNEP regional seas reports and studies, no. 185. FAO fisheries and aquaculture technical paper, no. 523. Rome, UNEP/FAO, 115 p

Marçalo A, Marques T, Araújo J, Pousão-Ferreira P, Erzini K, Stratoudakis Y (2010) Fishing simulation experiments for predicting effects of purse seine capture on sardines (*Sardina pilchardus*). ICES J Mar Sci 67:334–344

Marçalo A, Araújo J, Pousão-Ferreira P, Pierce GJ, Stratoudakis Y, Erzini K (2013) Behavioural responses of sardines *Sardina pilchardus* to simulated purse-seine capture and slipping. J Fish Biol 83(3):480–500

Marçalo A, Guerreiro PM, Bentes L, Rangel M, Monteiro P, Oliveira F, Afonso CML, Pousão-Ferreira P, Benoit HP, Breen M, Erzini K, Gonçalves JMS (2018) Effects of different slipping methods on the mortality of sardine, *Sardina pilchardus*, after purse-seine capture off the Portuguese Southern coast (Algarve). PLoS One 13(5):e0195433. https://doi.org/10.1371/journal.pone.0195433

Marçalo A, Breen M, Tenningen M, Onandia I, Arregi L, Gonçalves JMS (2019) Chapter 11 – Mitigating slipping related mortality from purse seine fisheries for small pelagic fish: Case studies from European Atlantic waters. In: Uhlmann SS, Ulrich C, Kennelly S (eds) The European discard policy-reducing unwanted catches in complex multi-species and multi-jurisdictional fisheries. Springer. https://link.springer.com/chapter/10.1007/978-3-030-03308-8_15

Matos E, Gonçalves A, Nunes ML, Dinis MT, Dias J (2010) Effect of harvesting stress and slaughter conditions on selected flesh quality criteria of gilthead seabream (*Sparus aurata*). Aquaculture 305:66–72. https://doi.org/10.1016/j.aquaculture.2010.04.020

McKenzie DJ, Axelsson M, Chabot D, Claireaux G, Cooke SJ, Corner RA, De Boeck G, Domenici P, Guerreiro PM, Hamer B, Jørgensen C, Killen SS, Lefevre S, Marras S, Michaelidis B, Nilsson GE, Peck MA, Perez-Ruzafa A, Rijnsdorp AD, Shiels HA, Steffensen JF, Svendsen JC, Svendsen MBS, Teal LR, van der Meer J, Wang T, Wilson JM, Wilson RW, Metcalfe JD (2016) Conservation physiology of marine fishes: state of the art and prospects for policy. Conserv Physiol 4(1):cow046. https://doi.org/10.1093/conphys/cow046

Meintzer P, Walsh P, Favaro B (2017) Will you swim into my parlour? In situ observations of Atlantic cod (*Gadus morhua*) interactions with baited pots, with implications for gear design. PeerJ 5:e2953. https://doi.org/10.7717/peerj.2953

Merker B (2016) Drawing the line on pain. Animal Sentience 2016.030

Metcalfe JD (2009) Welfare in wild capture marine fisheries. J Fish Biol 75:2855–2861

Midling KØ, Koren C, Humborstad O-B, Sæther B-S (2012) Swimbladderhealing in Atlantic cod (*Gadus morhua*), after decompression and rupture incapture-based aquaculture. Mar Biol Res 8:373–379

Morzel M, Sohier D, Hans Van de Vis H (2003) Evaluation of slaughtering methods for turbot with respect to animal welfare and flesh quality. J Sci Food Agric 82:19–28. https://doi.org/10.1002/jsfa.1253

Nichol D, Chilton E (2006) Recuperation and behaviour of Pacific cod after barotrauma. ICES J Mar Sci 63:83–94

NOAA Fisheries Website (2020) Recompression devices: helping anglers fish smarter. https://videos.fisheries.noaa.gov/detail/videos/recreational-fishing/video/3619674964001/recompression-devices:-helping-anglers-fish-smarter?autoStart=true
Olla BL, Davis MW, Schreck CB (1997) Effects of simulated trawling on sablefish and walleye pollock: the role of light intensity, net velocity and towing duration. J Fish Biol 50:1181e1194
Olsen SH, Sørensen NK, Larsen R, Elvevoll EO, Nilsen H (2008) Impact of pre-slaughter stress on residual blood in fillet portions of farmed Atlantic cod (*Gadus morhua*) measured chemically and by visible and near-infrared spectroscopy. Aquaculture 284(1):90–97
Olsen RE, Oppedal F, Tenningen M, Vold A (2012) Physiological response and mortality caused by scale loss in Atlantic herring. Fish Res 129–130:21–27
Pascoe PL (1990) Light and capture of marine animals. In: Herring PJ, Campbell AK, Whitfield M, Maddock L (eds) Light and life in the sea. Cambridge University Press, Cambridge, 357 pp
Peregrin LS, Butcher PA, Broadhurst MK, Millar RB (2015) Angling-induced barotrauma in snapper *Chrysophrys auratus*: are there consequences for reproduction? PLoS One 10:1371
Poli BM, Parisi G, Scappini F, Zampacavallo G (2005) Fish welfare and quality as affected by preslaughter and slaughter management. Aquacult Int 13:29–49
Portz DE, Woodley CM, Cech JJ (2006) Stress-associated impacts of short-term holding on fish. Rev Fish Biol Fish 16:125–170
Precision Seafood Harvesting (2014). http://www.precisionseafoodharvesting.co.nz/
Raby GD, Packer JR, Danylchuk AJ, Cooke SJ (2014) The underappreciated and understudied role of predators in the mortality of animals released from fishing gears. Fish Fish 15:489–505
Rihan D, Uhlmann SS, Ulrich C, Breen M, Catchpole T (2019) Chapter 4 – Requirements for documentation, data collection and scientific evaluations. In: Uhlmann SS, Ulrich C, Kennelly S (eds) The European discard policy-reducing unwanted catches in complex multi-species and multi-jurisdictional fisheries. Springer. https://link.springer.com/chapter/10.1007/978-3-030-03308-8_3
Robb DHF (2008) Welfare of fish at harvest. In: Branson EJ (ed) Fish welfare. Blackwell, Oxford, 300 pp. ISBN-13: 978-1-4051-4629-6
Robb DHF, Wotton SB, McKinstry JL, Sørensen NK, Kestin SC (2000) Commercial slaughter methods used on Atlantic salmon: determination of the onset of brain failure by electroencephalography. Vet Rec 147:298–303
Rogers SG, Langston HT, Targett TE (1986) Anatomical trauma to sponge-coral reef fishes captured by trawling and angling. Fish Bull 84(3):697–704
Rogers NJ, Urbina MA, Reardon EE, McKenzie DJ, Wilson RW (2016) A new analysis of hypoxia tolerance in fishes using a database of critical oxygen level (Pcrit). Conserv Physiol 4(1): cow012. https://doi.org/10.1093/conphys/cow012
Rose JD, Arlinghaus R, Cooke SJ, Diggles BK, Sawynok W, Stevens ED, Wynne CDL (2014) Can fish really feel pain? Fish Fish 15:97–133. https://doi.org/10.1111/faf.12010
Roth B, Heia K, Skåra T, Sone I, Birkeland S, Jakobsen RA, Akse L (2013) Kvalitetsavvik sildefilet. Sluttrapport (In Norwegian). https://nofimaas.sharepoint.com/sites/public/_layouts/15/guestaccess.aspx?guestaccesstoken=csN3daighAgM5EZdXXyfXCtQPu%2Fq9%2BXqRQ2SXZa4lco%3D&docid=09f5f587768644956a1be1589813fc074
Rummer JL, Bennett WA (2005) Physiological effects of swim bladder overexpansion and catastrophic decompression on Red Snapper. Trans Am Fish Soc 134:1457–1470
Ryer CH (2002) Trawl stress and escapee vulnerability to predation in juvenile walleye pollock: is there an unobserved bycatch of behaviorally impaired escapees? Mar Ecol Prog Ser 232:269e279
Ryer CH (2004) Laboratory evidence for behavioural impairment of fish escaping trawls: a review. ICES J Mar Sci 61:1157–1164
Ryer CH, Ottmar ML, Sturm EA (2004) Behavioral impairment after escape from trawl codends may not be limited to fragile fish species. Fish Res 66:261e269

Salomon M, Markus T, Dross M (2014) Masterstroke or paper tiger – the reform of the EU's common fisheries policy. Mar Policy 47:76–84. ISSN: 0308-597X. https://doi.org/10.1016/j.marpol.2014.02.001
Sangster GI, Lehmann K, Breen M (1996) Commercial fishing experiments to assess the survival of haddock and whiting after escape from four sizes of diamond mesh codends. Fish Res 25:323–345
Schreck CB, Olla BL, Davis MW (1997) Behavioral responses to stress. Fish Stress Health Aquac 62:145–170
Seafish (2013) The seafish guide to illegal, unreported and unregulated fishing (IUU), September, 2013. http://www.seafish.org/media/publications/SeafishGuidetoIUU_201309.pdf
Shephard S, Minto C, Zölck M, Jennings S, Brophy D, Reid D (2014) Scavenging on trawled seabeds can modify trophic size structure of bottom-dwelling fish. ICES J Mar Sci 71:398–405
Sigholt T, Erikson U, Rustad T, Johansen S, Nordtvedt TS, Seland A (1997) Handling stress and storage temperature affect meat quality of farm-raised Atlantic salmon (*Salmo salar*). J Food Sci 62:898–905
Skjervold PO, Fjæra SO, Østby PB, Einen O (2001) Live-chilling and crowding stress before slaughter of Atlantic salmon (*Salmo salar*). Aquaculture 192:265–280
Smith LS (1993) Trying to explain scale loss mortality: a continuing puzzle. Rev Fish Sci 1 (4):337–355
Sneddon LU, Elwood RW, Adamo SA, Leach MC (2014) Defining and assessing animal pain. Anim Behav 97:201–212. https://doi.org/10.1016/j.anbehav.2014.09.007
Star-Oddi (2017). https://www.star-oddi.com/products/aquatic-animals
Stien L, Hirmas E, Bjørnevik M, Karlsen Ø, Nortvedt R, Rørå AMB, Sunde J, Kiessling A (2005) The effects of stress and storage temperature on the colour and texture of pre-rigor ¢lleted farmed cod (*Gadus morhua* L.). Aquac Res 36:1197
Suuronen P (2005) Mortality of fish escaping trawl gears. FAO fisheries technical paper, 478, 72 p
Sweet M, Kirkham N, Bendall M, Currey L, Bythell J et al (2012) Evidence of melanoma in wild marine fish populations. PLoS One 7(8):e41989. https://doi.org/10.1371/journal.pone.0041989
Tallack SML (2007) Escape ring selectivity, bycatch, and discard survivability in the New England fishery for deep-water red crab. ICES J Mar Sci 64:1579–1586
Tenningen M, Vold A, Olsen RE (2012) The response of herring to high crowding densities in purse-seines: survival and stress reaction. ICES J Mar Sci 69:1523–1531
Thomsen B, Humborstad OB, Furevik DM (2010) Fish pots: fish behaviour, capture processes, and conservation issues. In: He P (ed) Behaviour of marine fishes: capture processes and conservation challenges. Wiley, Iowa, pp 143–158
Torgersen T, Bracke MBM, Kristiansen TS (2011) Reply to Diggles et al. (2011): Ecology and welfare of aquatic animals in wild capture fisheries. Rev Fish Biol Fish 21:767–769
Uhlmann SS, Broadhurst MK (2007) Damage and partitioned mortality of teleosts discarded from two Australian penaeid fishing gears. Dis Aquat Org 76:173–186. https://doi.org/10.3354/dao076173
Uhlmann SS, Broadhurst MK (2015) Mitigating unaccounted fishing mortality from gillnets and traps. Fish Fish 16:183–229. https://doi.org/10.1111/faf.12049
Utne-Palm AC, Breen M, Løkkeborg S, Humborstad O-B (2018) Behavioural responses of krill and cod to artificial light in laboratory experiments. PLoS One 13(1):e0190918. https://doi.org/10.1371/journal.pone.0190918
Van de Vis H, Kestin S, Robb D, Oehlenschläger J, Lambooij B, Münkner W, Kuhlmann H, Kloosterboer K, Tejada M, Huidobro A, Otterå H, Roth B, Sørensen NK, Akse L, Hazel BH, Nesvadba P (2003) Is humane slaughter of fish possible for industry? Aquac Res 34:211–220
Veldhuizen LJL (2017) Understanding social sustainability of capture fisheries. PhD thesis, Wageningen University, Wageningen, 160 p. ISBN: 978-94-6257-964-4. https://doi.org/10.18174/392826

Veldhuizen LJL, Berentsen PBM, de Boer IJM, van de Vis JW, Bokkers EAM (2018) Fish welfare in capture fisheries: a review of injuries and mortality. Fish Res 204:41–48. ISSN: 0165-7836. https://doi.org/10.1016/j.fishres.2018.02.001

Vold A, Anders N, Breen M, Saltskår J, Totland B og Øvredal JT (2017) Best practices in slipping from purse seines. (Beste praksis for slipping fra not. Utvikling av standard slippemetode for makrell og sild i fiske med not. Faglig sluttrapport for FHF-prosjekt 900999.) Rapport fra Havforskningen no 6-2017. ISSN: 1893-4536 (online) (In Norwegian)

Wagner H (1978) Einfluss der Schleppzeiten und Steertfüllung auf die Qualität des Fisches (in German). Seewirtschaft 10:399–400

Wedemeyer GA, Barton BA, McLeay DJ (1990) Stress and acclimation. In: Schreck CB, Moyle PB (eds) Methods for fish biology. American Fisheries Society, Bethesda, pp 451–489

Wileman DA, Ferro RST, Fonteyne R, Millar RB (1996) Manual of the methods of measuring the selectivity of towed fishing gears. ICES coop. res. rep. no. 215. Copenhagen, 126 p

Wileman DA, Sangster GI, Breen M, Ulmestrand M, Soldal AV, Harris RR (1999) Roundfish and Nephrops survival after escape from commercial fishing gear. EU contract final report. EC contract no: FAIR-CT95-0753

Wood CM, Turner JD, Graham MS (1983) Why do fish die after severe exercise? J Fish Biol 22:189–201

Chapter 18
Fish Welfare in Capture-Based Aquaculture (CBA)

Odd-Børre Humborstad, Chris Noble, Bjørn-Steinar Sæther, Kjell Øivind Midling, and Mike Breen

Abstract Capture-based aquaculture (CBA) combines aquaculture practices with capture fisheries to keep the catch alive for either short or long periods of time, for feeding or for live storage. CBA enables us to market numerous species ranging from molluscs, scallops and crustaceans to fish such as tuna, cod, eel and groupers. In CBA, handling and adaptation to new environments have an additional influence upon the stressors to which fish are exposed during capture, and the duration of this impact increases dramatically from minutes and hours in traditional fishing to days and months in CBA. We show how a strong focus on welfare is already present in cod CBA fisheries and the rationale behind this focus. We present a case study on CBA of Atlantic cod (*Gadus morhua*) as a robust example and model species for detecting welfare risks and mitigating against them. We discuss the main welfare issues in relation to the three broad phases of capture, transport and live storage, and identify common current fish welfare challenges in CBA. We highlight the advantages of pursuing this approach using lessons learnt from an industry in which fisheries and aquaculture meet and where an existing and successful knowledge transfer process between fisheries and aquaculture is already under way.

Keywords Capture-Based Aquaculture · Fishing gear · Demersal seine · Live capture · Live transport · Live storage

O.-B. Humborstad (✉) · M. Breen
Institute of Marine Research, Bergen, Norway
e-mail: odd-boerre.humborstad@hi.no

C. Noble · B.-S. Sæther · K. Ø. Midling
Nofima, Tromsø, Norway

T. S. Kristiansen et al. (eds.), *The Welfare of Fish*, Animal Welfare 20,
https://doi.org/10.1007/978-3-030-41675-1_18

18.1 Capture-Based Aquaculture and Fish Welfare Rationale

Capture-based aquaculture (CBA) was first defined and distinguished from hatchery-based aquaculture (HBA) by Ottolenghi et al. (2004) as: *the practice of collecting "seed" material—from early life stages to adults—from the wild, and the subsequent on-growing of such material in captivity to marketable size by the use of aquaculture techniques*. CBA combines aquaculture practices with capture fisheries to keep the catch alive for either short or long periods of time, for feeding or for live storage; thus, on-growing is not a definitive prerequisite for CBA. CBA enables us to produce numerous species ranging from molluscs, scallops and crustaceans to fish species such as tuna, cod, eel, mullets and groupers. It is estimated that CBA is responsible for around 20% of marine aquaculture production, and it has a wide range of environmental, biodiversity and socioeconomic effects (Lovatelli 2011). Wild fish are captured for live storage for many different purposes including, but not limited to, food production, display in public aquaria, ornamental purposes (Chap. 15) and experimentation (Chap. 16) and thus share some welfare challenges with CBA. In this chapter, we focus on the welfare of fish in live-capture fisheries where (1) there is a commercial interest with regard to consumption and (2) live fish are held in a net cage.

The techniques used in CBA differ widely, depending on the phylum and species involved; however, for finfish the usual CBA process involves a broad three-phase procedure that incorporates (1) capture, (2) transport and (3) live storage. In CBA, potential handling and adaptation to new environments during the second and third phases can have an additional influence upon the stressors to which the fish are exposed during capture (Chap. 17), and the duration of this impact increases dramatically from minutes and hours in traditional fishing to days and months in CBA (Humborstad et al. 2009). As CBA changes the status of the fish from that of free-living animals subject to fisheries legislation (quotas, selectivity and environmental issues) to farmed animals subject to quite different regulations as regards handling, slaughter and welfare (Dreyer et al. 2008), a knowledge-based, interdisciplinary approach is a prerequisite for successful management of the industry.

A central tenet of CBA capture and transport has therefore been to minimise the exposure of the fish to capture and handling stressors as much as possible, in order to ensure that the process supplies consistently robust and resilient animals with a high survival potential for live storage during the aquaculture phase. Although this approach is primarily driven by quality and economic motives (Poli et al. 2005), fish welfare is still intrinsically linked with, and central to the entire process, making welfare issues more prominent in CBA than in traditional marine fisheries.

In a case study on Atlantic cod CBA, we show how this highly specialized fishery focuses on welfare, and we highlight the advantages of pursuing this approach in an industry where fisheries and aquaculture meet and where a successful transfer of best practice procedures between aquaculture and fisheries is already under way. After the treatment of CBA in cod, a broader discussion offers examples from other important CBA species.

18.2 CBA of Atlantic Cod: A Case Study

In Norway, CBA of Atlantic cod involves catching adult (above minimum landing size, MLS) fish with demersal seine-nets along the coast of the counties of Troms and Finnmark from March to June and holding them in marine sea cages to produce and supply fresh, high-quality fish throughout the year. Immature cod (3–5 years) that follow capelin (*Mallotus villosus*) on their spawning migrations to the coast dominate the catch, though spawning and spent cod are also common. In the course of the past three decades, cod CBA has reflected the available quotas for coastal vessels, with low quotas leading to high activity and vice versa (Dreyer et al. 2008). In Norway, annual CBA landings over the past 10 years have ranged from 1000 to 6000 tonnes. Cod CBA regulations were developed in 2006 with post-capture welfare as their primary focus. Here, we review the process with regard to the three different phases, the main stressors to which the fish are exposed and how they are mitigated against. For a detailed description of the fishery, see Dreyer et al. (2008).

18.2.1 Capture Phase

The main stressors associated with the capture phase (Fig. 18.1) of demersal seine for cod include physical contact, prolonged forced swimming and decompression/thermal stress during ascent from depth. These stressors can lead to mechanical injuries/wounds, exhaustion and barotraumas that reduce the welfare and survival potential of the fish.

18.2.1.1 Physical Impact

Several studies have observed gear-induced (mostly trawl) injuries in gadoids (e.g. Sangster and Lehmann 1993; Suuronen et al. 1996, 2005). Contact with the gear can be caused by direct impacts (Ingólfsson and Jørgensen 2006), abrasion as the fish rubs along the codend and inside the codend if forced to the netting wall by other captured fish, or while attempting to escape through the net mesh (Digre et al. 2010). Appropriate species-specific mesh sizes are required to ensure and minimise the risks of injuring juvenile fish as they pass through the seine (Ingólfsson et al. 2007; Soldal and Engås 1997; Soldal et al. 1991). Capture-related damage is frequently observed in cod taken by towed fishing gear, and such damage can have multiple effects on quality (Digre et al. 2010; Esaiassen et al. 2013;

Fig. 18.1 Demersal seine fishing involves locating, encircling and herding aggregations of cod into a seine net attached to weighted ropes (Eigaard et al. 2016; Sainsbury 1997). Encircling is carried out by deploying one rope (1000–2000 m, depending on depth and the size of the area to be covered) attached to a surface buoy, the seine and then the second rope in a triangular configuration. The buoy is then picked up and the vessel slowly moves forward, bringing the ropes closer together and herding the fish into the seine. When the ropes are parallel, fish are caught in the seine. The ropes are hauled in and the seine is raised from the bottom. The typical duration of the whole process from shooting the net until the fish are brought alongside the ship is about 1–1.5 h (Image courtesy of Seafish: www.seafish.org)

Margeirsson et al. 2007; Olsen et al. 2013; Rotabakk et al. 2011). It is well known that bigger hauls can lead to higher mortality (Olsen et al. 2013; Suuronen et al. 2005), probably caused by increased pressure on individual fish and an increased risk of asphyxiation in the catch.

To reduce the physical impact, the use of knotless netting materials is mandatory in the cod CBA seine fishery, and it is recommended that catches be kept low and that fishing is carried out only in good weather conditions, as such measures reduce stress, injuries and improve the survival rate and quality of the catch (Margeirsson et al. 2007). In addition to avoiding areas with high fish availability, catch control systems to reduce the catch size (Grimaldo et al. 2014) and thus lower the proportion of the fish that suffer pressure damage (Digre et al. 2010) are emerging.

18.2.1.2 Exhaustive Swimming

It has been observed that both cod (Beamish 1979) and haddock (Breen et al. 2004) can die during swimming endurance experiments. Reduced swimming performance and post-exhaustion stresses may thus explain the differences in mortality rates observed in CBA. It has been shown that fish that escape from trawl codends often display impaired swimming and behavioural deficits that leave them vulnerable to elevated predation risk and reduced feeding success (Ryer 2004). Such fish are likely to have poorer swimming capacity, as respiratory substances in white muscle tissue are depleted (Breen et al. 2004; Wood et al. 1983). Olsen et al. (2013) found that the longest hauls with the highest catches also had the highest mortality, and the fish had the lowest initial blood pH and elevated blood lactate level.

The basic precautionary measure to reduce exhaustive exercise is to reduce tow duration, which can be done either by using shorter ropes, faster towing speeds or setting narrower seines. However, some of these measures can negatively affect catch efficiency, and unless availability is high, fishers are likely to do the opposite to maximise their catch. Thus, a certain level of physiological impairment is unavoidable at present.

18.2.1.3 Barotrauma

During voluntary ascent, physoclistous fish slowly resorb the excess gas inside the swim bladder through a particularly highly vascularised area called "the oval" (Fänge 1953; Steen 1963). However, in capture situations, the speed of ascent usually exceeds the capacity of the compensatory mechanisms and the volume of the gas contained by the tissue increases, leading to rupture of the swim bladder (Tytler and Blaxter 1973). This phenomenon can be clearly observed in the seconds before the Danish seine net surfaces, as cascades of bubbles break the surface. When the swim bladder is distended, but does not rupture, the fish lose control of their buoyancy and are trapped (floating) at the surface (Midling et al. 2012), leaving them at risk of avian predation and thermal shock.

Intuitively, one would expect that swim bladder rupture and its effects would cause too much harm to allow for live storage. The legislation also states that capture depth should be "adjusted in a way that minimizes long term barotrauma injuries". To avoid puncture would imply capturing fish either from very shallow depths or at very slow rates of ascent, close to their natural rates. However, resorption is a slow process and cod need about 4 h to reduce swim bladder pressure by 50% (Arnold and Walker 1992; Tytler and Blaxter 1973), which would rule out Danish seine as capture gear. During the sorting process (see below), fish with visible signs of barotrauma such as floating, bloated eyes, everted stomachs (Rummer and Bennett 2005) are removed and slaughtered; however, most if not all cod captured by Danish seine that are suitable for CBA (i.e. post-sorting) have punctured but still inflated swim bladders (Midling et al. 2012). The reason behind this counterintuitive outcome is that cod possess a mechanism for gas release and rapid repair (Humborstad and Mangor-Jensen 2013; Midling et al. 2012), which allows cod to rid themselves of surplus gas.

The mitigation measure to avoid detrimental barotrauma is to fish deep and to reduce haulback speed during the last phase of the haul as this allows surplus gas to trickle out before surfacing. Cod for CBA are therefore normally fished at depths of

150–250 m, which ensures that the swim bladder will be punctured. The significance of the rate of ascent, however, is not fully understood, and there is reason to believe that the proportion of floaters is also dependent on the degree of crowding and the catch size (the authors' unpublished data).

18.2.2 Transport Phase

18.2.2.1 Loading

The transport phase starts with a loading process which involves either pumping fish from the codend or using a codend lift (Fig. 18.2). Both operations involve moving fish back and forth inside the extension of the codend along the shipside, and abrasion and other injuries such as to the epidermis, mouth, operculum, or eye damage or fin splitting may occur. Fish may suffer pressure damage inside the codend from other fish, especially when being lifted aboard the vessel, while during pumping direct impacts inside the vacuum/pressure chamber may occur, as well as pressure damage from valve closure.

In order to reduce welfare risks, a mandatory codend canvas bag that serves three purposes is used as it (1) reduces contact abrasion during landing, (2) reduces weight and pressure damage in the air and (3) reduces air exposure, especially when landings take place in low air temperatures. Further catch control is important as, for example, smaller hauls reduce the number of unloading sequences and reduce washing. In vacuum pumps, the main issue is to avoid sharp corners and reduce the

Fig. 18.2 A mandatory canvas bag is required when loading cod for CBA

suction speed through the hoses to a minimum in order to minimise deceleration damage when the fish enter the vacuum pump chamber.

18.2.2.2 Sorting and Auditing the Catch

Sorting is probably the most important stage of the CBA process in terms of reducing the risk of poor welfare during the later stages of CBA. It is the first opportunity that the fishers have to audit their catch and assess the suitability of the fish for storage during the aquaculture phase. However, the sorting process itself may be a risk factor for welfare. Physical impact from unloading from canvas bags or pumps, air exposure (Humborstad et al. 2009) and hypoxia are the most prominent threats. Moreover, prolonged exposure to solar radiation can damage the skin and compromise health (Kaweewat and Hofer 1997), while low temperature causes the skin mucus to freeze, a condition that may lead to frost-initiated wounds.

It is mandatory to sort out and slaughter all fish with positive buoyancy, visible injuries or reduced vitality. Many fishers normally pick unfit fish from a water-filled (or partially water filled) bin (Fig. 18.3). The reasons for doing so are primarily to (1) reduce mechanical damage (deceleration and abrasion), (2) reduce the risk of hypoxia and thermal injuries and (3) aid the identification of floaters, which are highly visible with their white bellies up in water.

Removing fish that show visual signs of damage has greatly reduced instantaneous and post-sorting mortality and thus the number of moribund fish with poor welfare. Some delayed mortality still occurs in apparently unharmed fish because the internal status of the animal cannot easily be evaluated by visual inspection. Depending on their size and condition, and the time since the fish entered the haul (swimming duration), they will normally display a wide range of physiological statuses. The level of exhaustion can be estimated by measuring physiological indicators such as lactate, glucose, pH, etc. (Olsen et al. 2013). However, these indicators can be labour-intensive and are therefore not suitable for routine use in commercial fisheries. More importantly, although they are useful for determining

Fig. 18.3 Left: Fishers sorting catch from a water-filled bin. Right: Positively buoyant cod, not fit for live storage

sub-lethal stress levels, they do not correlate well with mortality outcomes (Davis et al. 2001; Davis and Schreck 2005). There is also some concern that cod may develop secondary exophthalmia in captivity, an effect of barotrauma that does not manifest itself at the time of sorting (Humborstad et al. 2016b), but this effect has so far only been demonstrated under laboratory conditions.

Behaviour, in particular activity, in the sorting bin is one of the indicators most often used as a proxy for exhaustion and vitality. In a typical fishing scenario, the fish brought on board towards the end of the catch are the most exhausted, and the cessation of movement over a period can be highly visible to the trained eye. Fishers will also normally check reflexes when in doubt. The most common reflex is the tail reflex (the fish jerks when the tail is stimulated), followed by eye movement when the fish is rotated along its long axis (vestibular ocular response), the head complex (alternating breathing movement) and also fin erection. Cod RAMP (Reflex Action Mortality Predictor) curves indicate that fish with a reflex impairment of less than 50% are likely to recover from capture stress (Humborstad et al. 2009). In the CBA fishery, the precautionary principle is normally applied, which means that fish that lack more than one reflex response will be slaughtered. While experienced fishers can easily utilise these cues during sorting, new best practice guidelines that include reflex testing also help new CBA fishers to improve their sorting skills and obtain a better understanding of the technique.

18.2.2.3 Tank

When cod are released into the transport tank, the majority are negatively buoyant and exhausted, and swim to the bottom. During the initial transport period (usually 1–24 h), there is a high risk of hypoxia as fish can pile up on each other, depleting local oxygen resources (Figs. 18.4 and 18.5). After this period, the fish become physiologically restored, refill the swim bladder and lift off the bottom and start swimming.

It has long been known that hypoxia is the main lethal agent during the transport of cod (Sundnes 1957). At the start of cod CBA in the early 1990s, single-pipe water inlets were originally positioned above the bottom and replacement seawater would pass over the fish, leading to hypoxic conditions inside the densely packed fish layer at the bottom. This often resulted in extreme mortalities, exceeding 50%. Modifications introduced in the mid-1990s included installing a double bottom, perforated with one 8 mm diameter hole per 100 cm^2, located 10 cm above the original bottom. Replacement seawater pumped through the new bottom created an evenly distributed upwelling of water over the whole area of the tank, ensuring an adequate supply of oxygen-rich water to the resting fish. This innovation alone reduced mortality in the tanks to below 10% and has been mandatory since 2005.

Increasing the resting surface area of the tank can also increase survival. By mounting a horizontal separator floor dividing the fish tank in two equal volumes, the resting area is doubled and the transport capacity could be increased by at least 50% in comparison to single-floor tanks (Humborstad et al. 2010, Fig. 18.4).

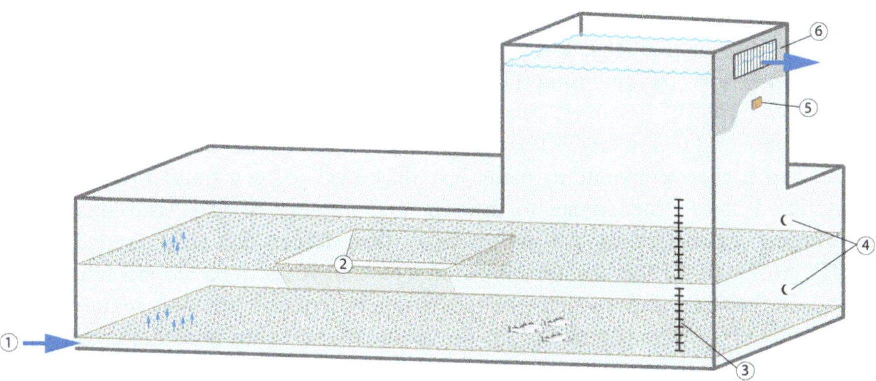

Fig. 18.4 A typical transport tank design for Atlantic cod CBA. Water supply is optimised by using flat, perforated double bottoms to ensure that each individual fish is flushed with replacement seawater. The available resting area is increased by mounting an additional separator floor (2). Scales (3) are used to estimate the level of crowding. Video monitoring (4) inside the tanks means behavioural indicators such as crowding, lethargy, increased ventilation rate, buoyancy problems or panic behaviour can be monitored. In such cases, the fisher will begin slaughter to avoid poor welfare, mortality and poor product quality. Oxygen monitoring (5) on outflow water is an essential means of identifying water supply problems or increasing the water supply during the periods of high oxygen consumption. This prevents hypoxia and promotes the removal of contaminants (6)

Fig. 18.5 During the first hours of recovery in the transport tank, crowding can be a problem, and a consistent supply of water from the tank bottom is a prerequisite for survival

Cod are currently transported at a density of around 150 kg m^{-3}, and although this figure varies greatly between vessels, it is far lower than the densities that have been used in previous experimental studies (>500 kg m^{-3}, Staurnes et al. 1994), which found that high/density transport can affect physiological parameters, raising plasma cortisol and glucose levels, but without leading to significant mortality. Cod can tolerate low oxygen saturation levels (Plante et al. 1998; Schurmann and Steffensen 1997); however, many factors affect hypoxia tolerance in fish (e.g. Chabot and Claireaux 2008). Oxygen solubility in water changes with temperature, and higher temperatures mean less dissolved oxygen (mg/L). Oxygen consumption is also proportional to the size and number of fish, with smaller fish consuming more oxygen per unit weight than larger fish (Colt and Tomasso 2001). Moreover, if the fish are subject to stress, have an impaired gill function, or if the oxygen-carrying capacity of the blood is reduced, the fish require higher dissolved oxygen concentrations in the water (Schurmann and Steffensen 1997; Sundnes 1957).

Oxygen monitoring is not mandatory during transport, partly because there is considerable uncertainty regarding suitable guidelines. However, many vessels do monitor oxygen, and a useful rule of thumb is to maintain the outflow water at 80% saturation or more, which so far seems to be a robust precautionary safe limit. Another broad safe limit is the required 0.5 l replacement rate kg^{-1} min^{-1}, which in addition to ensuring sufficient oxygen reduces the build-up of metabolic waste products. Further knowledge is required, especially regarding oxygen requirements over the range of stressful situation and environmental conditions encountered during cod CBA.

18.2.3 Live Storage Phase

18.2.3.1 Recovery

The final phase of the CBA process begins with pumping or netting the fish into recovery cages. In the 1980s and 1990s, the fishers used standard, flexible aquaculture nets for live holding and on-growing. This posed a problem, as a large proportion of the catch might still be exhausted or be suffering from barotrauma issues, so the fish needed to lie on the bottom of the cage to rest. Recovery from capture can take 24 h (Dreyer et al. 2008), and if they are not given a sufficient resting surface area, any fish that still require recovery time may suffer pressure damage, abrasive injuries and/or asphyxiation by conspecifics, as the net base deforms and fish pile up on top of each other.

This challenge led to the development of rigid flat-bottomed cages (Fig. 18.6) that provided the fish with a stable platform to recover and thus reduced the risk of suffocation and mortality (Dreyer et al. 2008). Fish can restore buoyancy control within 1 or 2 days after transfer and resume swimming. At this stage, they are ready for transfer to on-growing cages. The regulations require fish to be inspected daily during the recovery phase, and the holding density must not exceed 50 kg m^{-2} of the bottom panel area (Dreyer et al. 2008). Once the fish have recovered their buoyancy

Fig. 18.6 Cod resting inside a flat-bottomed recovery cage

and overcome their lethargy, they are transferred to holding cages for either short-term holding without feeding or long-term holding with feeding.

For the first 4 weeks, fish can be held without feed, after which they must be fed daily. After 12 weeks of live holding, the fish are governed by the Aquaculture Act and the CBA practitioner must hold the relevant licences for holding the fish and adhere to national slaughter procedures (Dreyer et al. 2008).

18.2.3.2 Storage Without Feeding

During the 4-week fasting period, the cod face several potential behavioural and physiological challenges, such as starvation/anorexia and cannibalism. Live storage without feeding does not subject the cod to atypical conditions, as they can regularly undergo long periods without food in the wild, especially during the winter months (Volkoff et al. 2009). To survive these conditions, the fish must mobilise nutrients present in their tissues (Beaulieu and Guderley 1998; Black and Love 1986). Tolerance of feed deprivation is therefore dependent upon the condition of the fish in terms of somatic energy reserves and will differ in relation to size, life stage, spawning condition of the cod or the time of year. If the fish are already in poor condition, with low energy reserves at the time of capture, or are at a life stage that cannot cope with long periods without feeding, both welfare and the economics of the operation may be compromised. Previous studies have reported that cod lose condition and cease to grow after 56 (Jobling et al. 1994) or 84 (Guderley et al. 2003) days of feed deprivation and lose condition when held for either 107 or 154 days without feed (Black and Love 1986). A long period of feed deprivation can also be detrimental to the hepatosomatic index, as some studies have found (Black and Love 1986) but not others (Jobling et al. 1994). This can also be detrimental to the physiological condition of the fish (Black and Love 1986) and may reduce stress tolerance (Olsen et al. 2008). Feed deprivation may also potentionally lead to cannibalism (Akse and Midling 1997) if size differences at landing are large, though very few studies are availiable for cod. Low energy reserves at transfer and during holding may also leave the cod prone to injuries or outbreaks of disease (see below).

Welfare risks regarding feed deprivation are therefore dependent upon a wide range of factors primarily associated with energetic status on their arrival in the cages (Sæther et al. 2016). Regular monitoring during the holding phase is therefore essential to ensure that the fish can tolerate the feed deprivation period (Dreyer et al. 2008). Robust size grading (which is mandatory) at cage stocking will reduce the risk of cannibalistic behaviour (Hermansen and Eide 2013).

18.2.3.3 Storage with Feeding

When the decision has been made to store the cod for a prolonged period with feed, the first challenge facing the CBA fisher is how to wean the fish onto artificial feed or bait fish. This weaning differs from the classical cod aquaculture challenge of weaning larvae from a live to an artificial diet during the hatchery phase (e.g. Brown et al. 2003; Fletcher et al. 2007), as the fishers have to wean wild-caught adult fish more than 3 years old (Dreyer et al. 2008) onto a possibly novel diet. This challenge differs from other CBA endeavours, such as the yellowtail *Seriola quinqueradiata* CBA fishery, which captures the fish as juveniles and then successfully weans them onto artificial granulated diets over a short period of time, while the fish are being transported to holding cages (Nakada 2008). The cod CBA industry is more comparable to the tuna CBA fishery that also uses purse seine fishing techniques and captures adult fish for on-growing (Dreyer et al. 2008; Ottolenghi 2008; Ottolenghi et al. 2004) and then has to wean them onto feed.

Weaning success can be affected by both the type of feed and how the feed is made available to the fish. These decisions cover the classic nutrition and feed management challenges that conventional aquaculturists regularly face. It has long been known that feeding in insufficient amounts (Hatlen et al. 2006) or using inappropriate feeding strategies (Noble et al. 2007) can be detrimental to fish welfare. Cod CBA practitioners generally feed their fish a diet of capelin or herring in the form of 25 kg frozen blocks that are dropped into the cages. These blocks float in the cages and are consumed by the fish as they defrost (Hermansen and Eide 2013). This feeding method and diet is generally preferred to artificial pelleted diets, as cod are generally difficult to wean onto dry pellets (Sæther et al. 2012). However, even with the current regime of feeding frozen blocks containing natural prey items, weaning success is generally only up to 70% (Sæther et al. 2012). The fish that do not begin feeding may lose weight and undergo a period of anorexia, depending on the length of the storage period.

Identification of these non-feeding (Misimi et al. 2014) fish at an early stage will allow the farmer to grade out individuals that will display signs of anorexia and its associated welfare challenges (Lovatelli 2011). In terms of operational techniques to eliminate or mitigate this problem, it has proven difficult to operationally identify these non-feeding fish on the basis of their size or condition factor at an early stage (Sæther et al. 2012). However, preliminary approaches using a grading method based on the individual fish's motivation to feed have shown some promise in both identifying feeders and non-feeders and sorting the landed stock (Sæther et al. 2012). This allows the fisher to harvest those fish that will ultimately adapt poorly to captivity and allow them to focus on producing and maintaining viable individuals that are both resilient and can thrive in holding cages.

18.2.3.4 Disease

The disease pattern in cod CBA is dominated by wounds and injuries inflicted during capture and handling (Sæther et al. 2012), which in turn may give rise to infections, blood loss and reduced control of osmolality. Weakened fish may also be more prone to suffer from blooms of common gill parasites like *Trichodina*, which can cause mass mortality in cultured cod (Khan 2004). CBA cod are likely to be susceptible to viral and bacterial infections also known in farmed cod (Samuelsen et al. 2006), although no statistics for CBA cod are currently available. Both atypical furunculosiss (*Aeromonas salmonicida*) and vibriosis (*Vibrio anguillarum*) have caused problems for farmed cod. *Vibrio salmonicida* has also been reported in wild-caught juvenile cod (Jørgensen et al. 1989). Most of the current concerns regarding disease in cod CBA are related to non-formulated wet feed (Dreyer et al. 2008) and disease interactions and transfer between wild and farmed populations (e.g. Johansen et al. 2011). The concern is due to both the latency of viruses in the feed fish (herring and capelin) and in the wild cod itself. This is especially the case with Viral Haemorrhagic Septicaemia Virus (VHSV), which frequently occurs in healthy fish too (Skall et al. 2005); stocking at high densities under suboptimal conditions may lead to stress and increase the probability of outbreaks.

Minimizing injuries caused by physical contact during capture and handling and a clear focus on sorting are required to avoid disease outbreaks. However, the current cod CBA practice where most captured cod are robust adults that are kept in cooler waters for relatively short periods of their lives reduces the probability of disease outbreaks. The current 10 km barrier between Atlantic salmon farming and cod CBA is regarded as a precautionary approach to prevent two-way spread of disease. Fresh or frozen feed fish are only regarded as a threat when feed fish and farmed fish are not from the same geographical area (Lyngstad et al. 2008), and for cod fed with capelin and herring, the risk is therefore regarded as low. In Norway, routine inspections focusing on the risks of infection, disease development in the production system, and spread to other sites, have been implemented for fish farming sites since the mid-1980s (Lyngstad et al. 2016). However, Norwegian legislation differs between the short- and long-term CBA of cod. If the cod CBA is for live storage and capture and is under 12 weeks, regulations only require veterinary inspection in case of increased mortality or if there is reason to suspect disease. If the cod CBA is for over 12 weeks, CBA farms must undergo routine inspections as with regular aquaculture sites.

18.3 Discussion

18.3.1 Cod as a Model Species for CBA and Indirect Welfare Research

Fish welfare studies in capture fisheries are still in their infancy (Metcalfe 2009), and few studies have been performed with the aim of improving specific welfare challenges for a specific gear or fishery (see Chap. 17), though the focus on this aspect is increasing (Mood 2010). Given the wide differences among commercial fisheries worldwide, the most logical way to introduce welfare concepts is to focus on the codes of best practice tailored for each individual fishery (Diggles et al. 2011). Lovatelli (2011) provides technical guidelines identifying CBA principles and guidelines for good practice for a wide range of species and fisheries.

Welfare is an intrinsic feature of CBA, where fisheries meet the welfare-focused aquaculture industry (Chap. 14). Although cod CBA is a relatively small industry, cod is probably the most intensively studied CBA species in terms of fish welfare (Lovatelli 2011; Lovatelli and Holthus 2008) and one of the most studied fish species by any standard in the Northern hemisphere. Fisheries legislation for cod CBA has been welfare oriented since its onset, and fishers need to document that they possess a sufficient level of knowledge and competence regarding welfare and that their vessels are equipped to fulfil welfare legislation requirements. For the cod CBA industry, this has led to a rise in publications dealing with welfare in CBA (e.g. Humborstad et al. 2009, 2016a, b; Midling et al. 2012; Misimi et al. 2014; Olsen et al. 2013), several of which utilise and refer to studies that could have an associative or indirect impact upon the welfare of captured fish. A comprehensive collection of publications and reviews deals with the mortality of escapees and discards from commercial fishing gears (Broadhurst et al. 2006; Chopin and Arimoto 1995; Suuronen 2005; Suuronen and Erickson 2010). Many of these (e.g. Broadhurst et al. 2006) also outline measures to reduce mortality by modifying fishing gears and improving operational and post-capture handling techniques. This approach is synonymous with CBA practices in terms of optimising the survival of the target fish species during live storage. It has long been recognized in farmed fish that high quality and extended shelf life can be achieved by reducing exposure to stressors (Poli et al. 2005). Also in traditional fisheries, a prerequisite for producing high-quality products is to keep stress and injuries low before processing (Borderias and Sanchez-Alonso 2011), and it is well documented that the capture technique and processing factors influence quality both between (Esaiassen et al. 2004; Rotabakk et al. 2011) and within (Olsen et al. 2013) fishing gears and seasons (Botta et al. 1987a, b). Cod is thus an excellent CBA model species, for which sufficient data already exist to explore fish welfare issues and serve as an example for potential risks and their mitigation in other species.

18.4 CBA Capture Methods Improve Welfare

Failures during the capture phase in any CBA fishery are decisive for the CBA outcome and the welfare of the fish. Apart from the vessel, fishing gear, and methodological and human requirements, several factors must be taken into consideration before attempting live capture. Several of these factors are related to the later phases of CBA, as the welfare issues only become apparent during transport and storage. For example, the size and species composition on a given fishing ground affects sorting time and oxygen consumption and the condition and feeding status of the fish affect stress tolerance and water contamination, while potential thermoclines and high surface temperatures affect metabolism and the stress response. Synthesising knowledge over temporal and spatial scales, especially with regard to conditions under which fishers should not carry out live capture, should be the focus of future work. The aspects of this work should focus on improving logbook data to include potential welfare risks and interviews with skippers and farmers for a joint understanding of the optimal fishing and rearing conditions for promoting good fish welfare. With specific regard to cod CBA, the standardisation of haul size by dynamic catch control (DCC) systems (Grimaldo et al. 2014; Ingolfsson et al., unpublished data) and its effects on vitality would seem to be a natural next step on the capture side.

In a broader context, we argue that if we wish to improve fish welfare, we should "fish as if you were going to store the fish alive". This principle can be applied to new and existing CBA fisheries, as a starting point for the identification of risk factors and to mitigate them but also to fisheries where live capture is not necessarily the goal (e.g. to achieve better quality, Chap. 17). These principles are even being applied to the development and building of new fishing boats in Norway. For example, high survival and improved quality can be obtained on board conventional bottom trawlers if CBA principles are applied (Olsen et al. 2013), although this has been questioned in a recent study (Digre et al. 2017). We suspect that the difference between these two studies lies in inadequate sorting. Further short-term storage, which on one side does prolong the duration of potential welfare impacts, allows for more welfare-friendly slaughtering methods such as electrical (Lambooij et al. 2012) or percussive stunning (Diggles 2015; Olsen et al. 2014; Salman et al. 2009) rather than asphyxiation or death by pressure.

There is a long tradition of live capture and short-term holding for pelagic species taken by purse seine fishing, which falls under the definition of CBA. In these fisheries, fishing methods have been improved through a process of trial and error led by the industry itself. If CBA capture methods are becoming increasingly more benign and more fish from the catch are able to survive holding, the adoption of these methods in non-CBA fisheries would arguably be advantageous in the contexts of both resource utilisation and fish welfare. Other potential welfare bridges between CBA and conventional fisheries can also be built, and these bridges are not limited to the welfare of the target species alone. For example, the release of non-target bycatch through gear selectivity devices is only rational if the fish survive, and CBA techniques may shed light on the consequences of each gear or technique, as regards the welfare of bycatch escapes and releases (Diggles et al. 2011; Wilson et al. 2014). In the first attempts to estimate mortality from net bursts in herring purse seine fisheries, CBA methods were used as a control (Misund and Beltestad 1995). In later experiments to reduce the detrimental effects of slipping (Tenningen et al. 2012), one of the main strengths of the study was that the methods employed for the capture, transfer and storage of fish were the same as those that are used for live capture of pelagic species in which survival is high. Ongoing projects also focus specifically on live capture methods for higher survival rates and improved welfare in purse seine fisheries.

18.4.1 Transport: Knowledge Transfer from Aquaculture Becomes Evident

There is a considerable potential for knowledge transfer between aquaculture and CBA and fisheries (Huntingford and Kadri 2009), and this becomes evident in the transport phase. In addition to generic underlying concepts of fish welfare stemming from aquaculture research (e.g. Diggles et al. 2011), more specific welfare issues such as transport (Southgate 2008), water quality (Portz et al. 2006), crowding (Brown et al. 2010; Damsgård et al. 2011; Fig. 18.5), handling (Demers and Bayne 1997; Sanz et al. 2012) and tank design (Portz et al. 2006; Fig. 18.4) are also directly applicable to CBA. Furthermore, CBA can take advantage of the knowledge base of widely known risks from aquaculture, such as injuries, crowding, handling and starvation, and of what actions to take when welfare is compromised (Brown et al. 2010; Noble et al. 2012b). Techniques for assessing and quantifying welfare threats using life-stage and species-specific welfare indicators (WIs) are a further area of significant research interest in aquaculture (e.g. Huntingford et al. 2006), and this is a key area in which knowledge transfer between aquaculture and fisheries is significant. To improve and evaluate fish welfare in both fisheries and aquaculture, robust methods for quantifying welfare must be developed and utilised. Numerous indicators stemming from aquaculture are available for assessing fish welfare, but many are unsuitable in commercial settings on board boats (or in towed net pens for tuna) where rapid, repeatable, robust and user-friendly operational welfare indicators (OWIs) are needed (Noble et al. 2012a). The suite of aquacultural OWIs available to the CBA practitioner include direct behavioural indicators such as swimming behaviour, ventilator activity, aggression (Huntingford et al. 2006; Martins et al. 2012; Noble et al. 2007), morphological indicators such as epidermal or fin damage or other external injuries (Noble et al. 2012b), physiological indicators such as glucose or lactate levels (Poli et al. 2005), health indicators (Segner et al. 2012), mortality (Ellis et al. 2012) and indirect indicators such as water quality (Person-Le Ruyet et al. 2008; Santos et al. 2010). In addition, reflex impairment (Davis 2010; Humborstad et al. 2009) and categorical vitality assessment using conditional reasoning (Benoît et al. 2012) are emerging, not just as measures of stress, but also as predictors of mortality. However, there is concern that many of these methods have not been assessed and validated for a wide range of species (e.g. tuna; Salman et al. 2009).

The stress-related effects of short-term holding are influenced by water quality, confinement density, holding container design and agonistic and predation-associated behaviours (Portz et al. 2006). As in cod CBA, the knowledge transfer in these areas is already considerable. However, water quality is an understudied area in CBA and the potential for knowledge transfer has not been fully exploited. Temperature, dissolved oxygen, ammonia, nitrite, nitrate, salinity, pH, carbon dioxide, alkalinity and water hardness are the most common water quality parameters that affect physiological stress (Portz et al. 2006), but only oxygen and temperature are routinely controlled in experiments, and seldom in CBA fisheries. Confinement density is often pushed to the limits of the vessel's carrying capacity, to increase the value of the catch, and this may lead to increased mortality due to stressful events (elevated temperatures, wave motion, poor water quality, etc.). Although sorting routines are improving with optimised tank designs, and increased monitoring (especially video for behavioural indices of stress) becoming more common in CBA, mass mortality can be a problem during transportation in a range of species, not just in cod CBA (e.g. bluefin tuna in towing cages, Ottolenghi 2008). The reasons underlying this mortality can be unclear. The key to understanding and preventing welfare-threatening events lies in combining the knowledge of fish behaviour, physiology, transport conditions and detailed records of potential adverse events in CBA. To achieve this, more attention needs to be paid to understanding the fish's status (size, condition, feeding status etc.), in addition to better monitoring of its holding environment and behaviour.

18.4.2 Live Storage: Adaptation to a Life in Captivity

CBA species are highly diverse (Lovatelli and Holthus 2008; Ottolenghi et al. 2004) ranging from juvenile glass eels (<1 g) to the stocking of large adult tunas (up to ca. 600 kg), and the welfare challenges during the live storage phase are therefore diverse. The main areas of focus on welfare during the live storage phase are related to the confinement of non-domesticated animals that are subject to either (1) short-term storage without feeding and thus have potential welfare problems associated with, for example, anorexia or cannibalism or (2) long-term storage with feeding, where the primarily welfare problems are associated with weaning and disease.

The welfare of the fish during the live storage phase depends on the species' ability to adapt to handling and new environments (allostasis, see Chap. 11). Every species has its own requirements with regard to the holding environment, and the knowledge of biological needs is crucial. Indeed, certain wild fish species may not even be suitable for live storage in CBA. For those that are, the selection of a suitable farm site is essential to avoid conditions under which the welfare needs of the fish are not met; for example turbid waters, currents that are beyond the swimming capacity of the fish and temperatures outwith their range of tolerance. Any of these, either alone or in combination, can lead to welfare problems and potentially to mortality (Ottolenghi 2008). Cod is a robust CBA species, and a great deal can be learned from using it as a case study for the welfare of other potential CBA species.

Optimising the diet and weaning the fish onto feed are critical aspects shared by many CBA species, and failure to wean may lead to starvation, poor health and poor welfare (Lovatelli 2011). Identifying and removing non-feeders at an early stage is therefore crucial to welfare. The reliance on using wild-caught food as a source of nutrition in CBA live storage is often regarded as problematic, as such feed items could otherwise be used for human consumption and they can also have a negative impact upon biosecurity, as they can be a vector for pathogen transfer or lead to inadequate nutrition (Nakada 2008). The development of formulated feed is therefore to be preferred, but this is a very challenging task and many species are still dependent on fresh or processed wild-caught aquatic resources as feed. Some species, e.g. juvenile yellowtail and groupers (*Epinephelus* spp.), are sensitive to food deprivation, and cannibalism may occur, particularly if the fish are kept in the holding tanks for long periods (Nakada 2008; Tupper and Sheriff 2008). This risk is particularly high if the fish have not undergone proper size grading. The onset of post-capture feeding may also be critical; for example, if the young fish are not fed for more than 3 days, they will usually fail to adapt to the artificial feeds (Nakada 2008), and similar findings have been made with wild-caught char (*Salvelinus alpinus*) in Norway (the authors' observations).

While disease has so far not been a major issue in cod CBA, other species can be prone to severe disease outbreaks (e.g. eels). Ottolenghi et al. (2004) have produced an in-depth overview of the current health status and disease pathogens in the four major CBA species groups of eels, yellowtails, groupers and tunas. CBA specimens may become more susceptible to disease because of the stress of capture, handling or sub-optimal live-holding conditions (Portz et al. 2006). Disease may subsequently spread rapidly among fish held at high densities. The susceptibility of each current and potential CBA species to disease under culture conditions therefore needs to be investigated (Ottolenghi et al. 2004). The transfer of disease (Lovatelli 2011) between regions and species is also of concern to many CBA fisheries, including but not limited to cod, groupers (Tupper and Sheriff 2008) and wrasses (Murray 2016).

18.5 Concluding Remarks

As opposed to the acute stress involved in conventional fisheries, the potential impacts of capture and handling upon the successful holding and rearing of fish in the later stages of CBA has made welfare a key consideration in the CBA of cod in Norway. By adopting this industry as a case study, we have shown how welfare issues during the three phases of capture, transport and live storage can be dealt with.

This may serve as a bridge to introduce welfare issues into other CBA fisheries as well as in fisheries in general. As we have pointed out above, if we wish to improve welfare in capture fisheries, we should "fish as if you were going to store the fish alive". Though research into welfare in fisheries is limited, cross-disciplinary bridges from studies on CBA, as well as on the survival and fate of escapees and discards, can be established. The potential of knowledge transfer from aquaculture becomes evident during transport in CBA, and this first opportunity to audit the catch and sort the fish makes it a decisive stage in determining how successful the CBA outcome will be. A strong focus on Operational Welfare Indicators (OWIs) is required and specialised equipment and procedures are needed. The welfare of the fish during the live storage phase depends on the ability of the individual species to adapt to handling and a new environment. The primary welfare factors to consider during storage are related to starvation, successful weaning and disease. Current challenges in cod CBA are related to when and where to perform live capture, plus the refinement of operational indicators and monitoring systems to identify potential welfare problems during transport and live storage at an early stage. Synthesising knowledge over a range of temporal and spatial scales, especially with regard to conditions where fishers ought not to perform live capture, should be the focus of future work. Even closer collaboration with aquaculture and their routines would further be advantageous in the transport and live storage phases, to ensure adequate exchange of knowledge and take advantage of existing best practices to detect and react to welfare challenges.

References

Akse L, Midling K (1997) Live capture and starvation of capelin cod (*Gadus morhua* L.) in order to improve the quality. Dev Food Sci 38:47–58

Arnold G, Walker MG (1992) Vertical movements of cod (*Gadus morhua* L.) in the open sea and the hydrostatic function of the swimbladder. ICES J Mar Sci 49:357–372

Beamish F (1979) Swimming capacity. Fish Physiol 7:101–187

Beaulieu M-A, Guderley H (1998) Changes in qualitative composition of white muscle with nutritional status of Atlantic cod, *Gadus morhua*. Comp Biochem Physiol A Mol Integr Physiol 121:135–141

Benoît HP, Hurlbut T, Chassé J, Jonsen ID (2012) Estimating fishery-scale rates of discard mortality using conditional reasoning. Fish Res 125:318–330

Black D, Love RM (1986) The sequential mobilisation and restoration of energy reserves in tissues of Atlantic cod during starvation and refeeding. J Comp Physiol B Biochem Syst Environ Physiol 156:469–479

Borderias AJ, Sanchez-Alonso I (2011) First processing steps and the quality of wild and farmed fish. J Food Sci 76:R1–R5. https://doi.org/10.1111/j.1750-3841.2010.01900.x

Botta J, Bonnell G, Squires B (1987a) Effect of method of catching and time of season on sensory quality of fresh raw Atlantic cod (*Gadus morhua*). J Food Sci 52:928–931

Botta J, Kennedy K, Squires B (1987b) Effect of method of catching and time of season on the composition of Atlantic cod (*Gadus morhua*). J Food Sci 52:922–924

Breen M, Dyson J, Oneill F, Jones E, Haigh M (2004) Swimming endurance of haddock (L.) at prolonged and sustained swimming speeds, and its role in their capture by towed fishing gears. ICES J Mar Sci 61:1071–1079. https://doi.org/10.1016/j.icesjms.2004.06.014

Broadhurst MK, Suuronen P, Hulme A (2006) Estimating collateral mortality from towed fishing gear. Fish Fish 7:180–218

Brown JA, Minkoff G, Puvanendran V (2003) Larviculture of Atlantic cod (*Gadus morhua*): progress, protocols and problems. Aquaculture 227:357–372

Brown JA, Watson J, Bourhill A, Wall T (2010) Physiological welfare of commercially reared cod and effects of crowding for harvesting. Aquaculture 298:315–324. https://doi.org/10.1016/j.aquaculture.2009.10.028

Chabot D, Claireaux G (2008) Environmental hypoxia as a metabolic constraint on fish: the case of Atlantic cod, *Gadus morhua*. Mar Pollut Bull 57:287–294. https://doi.org/10.1016/j.marpolbul.2008.04.001

Chopin F, Arimoto T (1995) The condition of fish escaping from fishing gears—a review. Fish Res 21:315–327

Colt JE, Tomasso JR (2001) Hatchery water supply and treatment. In: Wedemeyer GA (ed) Fish hatchery management, 2nd edn. American Fisheries Society, Bethesda, pp 91–186

Damsgård B, Bjørklund F, Johnsen HK, Toften H (2011) Short-and long-term effects of fish density and specific water flow on the welfare of Atlantic cod, *Gadus morhua*. Aquaculture 322:184–190

Davis MW (2010) Fish stress and mortality can be predicted using reflex impairment. Fish Fish 11:1–11. https://doi.org/10.1111/j.1467-2979.2009.00331.x

Davis MW, Schreck CB (2005) Responses by Pacific halibut to air exposure: lack of correspondence among plasma constituents and mortality. Trans Am Fish Soc 134:991–998

Davis M, Olla B, Schreck C (2001) Stress induced by hooking, net towing, elevated sea water temperature and air in sablefish: lack of concordance between mortality and physiological measures of stress. J Fish Biol 58:1–15

Demers NE, Bayne CJ (1997) The immediate effects of stress on hormones and plasma lysozyme in rainbow trout. Dev Comp Immunol 21:363–373

Diggles B (2015) Development of resources to promote best practice in the humane dispatch of finfish caught by recreational fishers. Fish Manag Ecol 23:200–207

Diggles BK, Cooke SJ, Rose JD, Sawynok W (2011) Ecology and welfare of aquatic animals in wild capture fisheries. Rev Fish Biol Fish 21:739–765. https://doi.org/10.1007/s11160-011-9206-x

Digre H, Hansen UJ, Erikson U (2010) Effect of trawling with traditional and 'T90' trawl codends on fish size and on different quality parameters of cod Gadus morhua and haddock Melanogrammus aeglefinus. Fish Sci 76:549–559. https://doi.org/10.1007/s12562-010-0254-2

Digre H, Rosten C, Erikson U, Mathiassen JR, Aursand IG (2017) The on-board live storage of Atlantic cod (*Gadus morhua*) and haddock (*Melanogrammus aeglefinus*) caught by trawl: fish behaviour, stress and fillet quality. Fish Res 189:42–54

Dreyer BM, Nøstvold BH, Midling KØ, Hermansen Ø (2008) Capture-based aquaculture of cod. In: Lovatelli A, Holthus PF (eds) Capture-based aquaculture global overview. FAO fisheries technical paper no. 508, Rome, pp 183–198

Eigaard OR, Bastardie F, Breen M, Dinesen GE, Hintzen NT, Laffargue P, Mortensen LO, Nielsen JR, Nilsson HC, O'Neill FG (2016) Estimating seabed pressure from demersal trawls, seines, and dredges based on gear design and dimensions. ICES J Mar Sci 73:i27–i43

Ellis T, Berrill I, Lines J, Turnbull JF, Knowles TG (2012) Mortality and fish welfare. Fish Physiol Biochem 38:189–199. https://doi.org/10.1007/s10695-011-9547-3

Esaiassen M, Nilsen H, Joensen S, Skjerdal T, Carlehög M, Eilertsen G, Gundersen B, Elvevoll E (2004) Effects of catching methods on quality changes during storage of cod (*Gadus morhua*). LWT Food Sci Technol 37:643–648. https://doi.org/10.1016/j.lwt.2004.02.002

Esaiassen M, Akse L, Joensen S (2013) Development of a catch-damage index to assess the quality of cod at landing. Food Control 29:231–235. https://doi.org/10.1016/j.foodcont.2012.05.065

Fänge R (1953) The mechanisms of gas transport in the euphysoclist swimbladder. Acta Physiol Scand Suppl 30:1–133

Fletcher R, Roy W, Davie A, Taylor J, Robertson D, Migaud H (2007) Evaluation of new microparticulate diets for early weaning of Atlantic cod (*Gadus morhua*): implications on larval performances and tank hygiene. Aquaculture 263:35–51

Grimaldo E, Sistiaga M, Larsen RB (2014) Development of catch control devices in the Barents Sea cod fishery. Fish Res 155:122–126

Guderley H, Lapointe D, Bédard M, Dutil J-D (2003) Metabolic priorities during starvation: enzyme sparing in liver and white muscle of Atlantic cod, *Gadus morhua* L. Comp Biochem Physiol A Mol Integr Physiol 135:347–356

Hatlen B, Grisdale-Helland B, Helland SJ (2006) Growth variation and fin damage in Atlantic cod (*Gadus morhua* L.) fed at graded levels of feed restriction. Aquaculture 261:1212–1221

Hermansen Ø, Eide A (2013) Bioeconomics of capture-based aquaculture of cod (*Gadus morhua*). Aquac Econ Manag 17:31–50

Humborstad O-B, Mangor-Jensen A (2013) Buoyancy adjustment after swimbladder puncture in cod *Gadus morhua*: an experimental study on the effect of rapid decompression in capture-based aquaculture. Mar Biol Res 9:383–393. https://doi.org/10.1080/17451000.2012.742546

Humborstad O-B, Davis MW, Løkkeborg S (2009) Reflex impairment as a measure of vitality and survival potential of Atlantic cod (*Gadus morhua*). Fish Bull 107:395–402

Humborstad O-B, Isaksen B, Midling K, Saltskår J, Totland B, Øvredal JT (2010) Optimal føringskapasitet og velferd for levende, villfanget torsk. Del 2: Praktiske forsøk-uttesting av etasjeskiller for økt hvileareal

Humborstad O-B, Breen M, Davis MW, Løkkeborg S, Mangor-Jensen A, Midling KØ, Olsen RE (2016a) Survival and recovery of longline- and pot-caught cod (*Gadus morhua*) for use in capture-based aquaculture (CBA). Fish Res 174:103–108

Humborstad OB, Ferter K, Kryvi H, Fjelldal P (2016b) Exophthalmia in wild-caught cod (*Gadus morhua* L.): development of a secondary barotrauma effect in captivity. J Fish Dis 40:41–49

Huntingford F, Kadri S (2009) Taking account of fish welfare: lessons from aquaculture. J Fish Biol 75:2862–2867

Huntingford FA, Adams C, Braithwaite V, Kadri S, Pottinger T, Sandøe P, Turnbull J (2006) Current issues in fish welfare. J Fish Biol 68:332–372

Ingólfsson ÓA, Jørgensen T (2006) Escapement of gadoid fish beneath a commercial bottom trawl: relevance to the overall trawl selectivity. Fish Res 79:303–312

Ingólfsson ÓA, Soldal AV, Huse I, Breen M (2007) Escape mortality of cod, saithe, and haddock in a Barents Sea trawl fishery. ICES J Mar Sci 64:1836–1844

Jobling M, Meløy O, Santos JD, Christiansen B (1994) The compensatory growth response of the Atlantic cod: effects of nutritional history. Aquac Int 2:75–90

Johansen L-H, Jensen I, Mikkelsen H, Bjørn P-A, Jansen P, Bergh Ø (2011) Disease interaction and pathogens exchange between wild and farmed fish populations, with special reference to Norway. Aquaculture 315:167–186

Jørgensen T, Midling K, Espelid S, Nilsen R, Stensvåg K (1989) Vibrio salmonicida, a pathogen in salmonids, also causes mortality in net-pen captured cod (*Gadus morhua*). Bull Eur Assoc Fish Pathol 9:42–44

Kaweewat K, Hofer R (1997) Effect of UV-B radiation on goblet cells in the skin of different fish species. J Photochem Photobiol B Biol 41:222–226

Khan R (2004) Disease outbreaks and mass mortality in cultured Atlantic cod, *Gadus morhua* L., associated with Trichodina murmanica (Ciliophora). J Fish Dis 27:181–184

Lambooij E, Digre H, Reimert H, Aursand I, Grimsmo L, Van de Vis J (2012) Effects of on-board storage and electrical stunning of wild cod (*Gadus morhua*) and haddock (*Melanogrammus aeglefinus*) on brain and heart activity. Fish Res 127:1–8

Lovatelli A (2011) Use of wild fishery resources for capture-based aquaculture. FAO technical guidelines for responsible fisheries. FAO, Rome, p 81

Lovatelli A, Holthus PF (2008) Capture-based aquaculture: global overview, vol 508. FAO Fisheries Technical Paper, Rome

Lyngstad TM, Høgåsen HR, Ørpetveit I, Hellberg H, Dale OB, Lillehaug A (2008) Faglig vurdering i forbindelse med bekjempelse av viral hemoragisk septikemi (VHS) i Storfjorden. Norwegian Veterinary Institute, Report 3:1–20

Lyngstad TM, Hellberg H, Viljugrein H, Jensen BB, Brun E, Sergeant E, Tavornpanich S (2016) Routine clinical inspections in Norwegian marine salmonid sites: a key role in surveillance for freedom from pathogenic viral haemorrhagic septicaemia (VHS). Prev Vet Med 124:85–95

Margeirsson S, Jonsson GR, Arason S, Thorkelsson G (2007) Influencing factors on yield, gaping, bruises and nematodes in cod (*Gadus morhua*) fillets. J Food Eng 80:503–508

Martins CI, Galhardo L, Noble C, Damsgård B, Spedicato MT, Zupa W, Beauchaud M, Kulczykowska E, Massabuau J-C, Carter T (2012) Behavioural indicators of welfare in farmed fish. Fish Physiol Biochem 38:17–41

Metcalfe J (2009) Welfare in wild-capture marine fisheries. J Fish Biol 75:2855–2861

Midling KØ, Koren C, Humborstad O-B, Sæther B-S (2012) Swimbladder healing in Atlantic cod (*Gadus morhua*), after decompression and rupture in capture-based aquaculture. Mar Biol Res 8:373–379

Misimi E, Martinsen S, Mathiassen JR, Erikson U (2014) Discrimination between weaned and unweaned Atlantic cod (*Gadus morhua*) in capture-based aquaculture (CBA) by X-ray imaging and radio-frequency metal detector. PLoS One 9:e95363

Misund OA, Beltestad AK (1995) Survival of herring after simulated net bursts and conventional storage in net pens. Fish Res 22:293–297

Mood A (2010) Worse things happen at sea: the welfare of wild-caught fish. Summary report. fishcount.org.uk

Murray A (2016) A modelling framework for assessing the risk of emerging diseases associated with the use of cleaner fish to control parasitic sea lice on salmon farms. Transbound Emerg Dis 63:270–277

Nakada M (2008) Capture-based aquaculture of yellowtails. In: Lovatelli A, Holthus PF (eds) Capture-based aquaculture global overview. FAO fisheries technical paper no. 508, Rome, pp 199–215

Noble C, Kadri S, Mitchell DF, Huntingford FA (2007) Influence of feeding regime on intraspecific competition, fin damage and growth in 1+ Atlantic salmon parr (*Salmo salar* L.) held in freshwater production cages. Aquac Res 38:1137–1143

Noble C, Berrill IK, Waller B, Kankainen M, Setälä J, Honkanen P, Mejdell CM, Turnbull JF, Damsgård B, Schneider O (2012a) A multi-disciplinary framework for bio-economic modeling in aquaculture: a welfare case study. Aquac Econ Manag 16:297–314

Noble C, Jones HAC, Damsgård B, Flood MJ, Midling KØ, Roque A, Sæther B-S, Cottee SY (2012b) Injuries and deformities in fish: their potential impacts upon aquacultural production and welfare. Fish Physiol Biochem 38:61–83

Olsen RE, Sundell K, Ringø E, Myklebust R, Hemre G-I, Hansen T, Karlsen Ø (2008) The acute stress response in fed and food-deprived Atlantic cod, *Gadus morhua* L. Aquaculture 280:232–241. https://doi.org/10.1016/j.aquaculture.2008.05.006

Olsen SH, Tobiassen T, Akse L, Evensen TH, Midling KØ (2013) Capture-induced stress and live storage of Atlantic cod (*Gadus morhua*) caught by trawl: consequences for the flesh quality. Fish Res 147:446–453. https://doi.org/10.1016/j.fishres.2013.03.009

Olsen SH, Digre H, Grimsmo L, Toldnes B, Eilertsen A, Evensen TH, Midling KØ (2014) Implementering av teknologi for optimal kvalitet i fremtidens prosesslinje på trålere "OPTIPRO"–Fase 1

Ottolenghi F (2008) Capture-based aquaculture of bluefin tuna. Capture-based aquaculture Global Overview. FAO fisheries technical paper 508, pp 169–182

Ottolenghi F, Silvestri C, Giordano P, Lovatelli A, New MB (2004) Capture-based aquaculture: the fattening of eels, groupers, tunas and yellowtails. FAO

Person-Le Ruyet J, Labbé L, Le Bayon N, Sévère A, Le Roux A, Le Delliou H, Quéméner L (2008) Combined effects of water quality and stocking density on welfare and growth of rainbow trout (*Oncorhynchus mykiss*). Aquat Living Resour 21:185–195

Plante S, Chabot D, Dutil JD (1998) Hypoxia tolerance in Atlantic cod. J Fish Biol 53:1342–1356

Poli B, Parisi G, Scappini F, Zampacavallo G (2005) Fish welfare and quality as affected by pre-slaughter and slaughter management. Aquac Int 13:29–49

Portz DE, Woodley CM, Cech JJ (2006) Stress-associated impacts of short-term holding on fishes. Rev Fish Biol Fish 16:125–170

Rotabakk BT, Skipnes D, Akse L, Birkeland S (2011) Quality assessment of Atlantic cod (*Gadus morhua*) caught by longlining and trawling at the same time and location. Fish Res 112:44–51. https://doi.org/10.1016/j.fishres.2011.08.009

Rummer JL, Bennett WA (2005) Physiological effects of swim bladder overexpansion and catastrophic decompression on red snapper. Trans Am Fish Soc 134:1457–1470. https://doi.org/10.1577/t04-235.1

Ryer C (2004) Laboratory evidence for behavioural impairment of fish escaping trawls: a review. ICES J Mar Sci 61:1157–1164. https://doi.org/10.1016/j.icesjms.2004.06.004

Sæther BS, Noble C, Humborstad O, Martinsen S, Veliyulin E, Misimi E, Midling KØ (2012) Fangstbasert akvakultur. Mellomlagring, oppfôring og foredling av villfanget fisk. Nofima Rep 14:50

Sæther B-S, Noble C, Midling KØ, Tobiassen T, Akse L, Koren C, Humborstad OB (2016) Velferd hos villfanget torsk i merd – Hovedvekt på hold uten fôring ut over 12 uker. Nofima Rep 16:32

Sainsbury J (1997) Commercial fishing methods: an introduction to vessels and gears. Oceanogr Lit Rev 11:1345

Salman J, Vannier P, Wierup M (2009) Species-specific welfare aspects of the main systems of stunning and killing of farmed tuna. Scientific opinion of the panel on animal health and welfare. ESFA J 1072:1–53

Samuelsen OB, Nerland AH, Jørgensen T, Schrøder MB, Svåsand T, Bergh Ø (2006) Viral and bacterial diseases of Atlantic cod *Gadus morhua*, their prophylaxis and treatment: a review. Dis Aquat Org 71:239–254

Sangster G, Lehmann K (1993) Assessment of the survival of fish escaping from commercial fishing gears. ICES CM, 6–7

Santos G, Schrama J, Mamauag R, Rombout J, Verreth J (2010) Chronic stress impairs performance, energy metabolism and welfare indicators in European seabass (*Dicentrarchus labrax*): the combined effects of fish crowding and water quality deterioration. Aquaculture 299:73–80

Sanz A, Furné M, Trenzado CE, de Haro C, Sánchez-Muros M (2012) Study of the oxidative state, as a marker of welfare, on Gilthead Sea bream, *Sparus aurata*, subjected to handling stress. J World Aquacult Soc 43:707–715

Schurmann H, Steffensen J (1997) Effects of temperature, hypoxia and activity on the metabolism of juvenile Atlantic cod. J Fish Biol 50:1166–1180

Segner H, Sundh H, Buchmann K, Douxfils J, Sundell KS, Mathieu C, Ruane N, Jutfelt F, Toften H, Vaughan L (2012) Health of farmed fish: its relation to fish welfare and its utility as welfare indicator. Fish Physiol Biochem 38:85–105

Skall HF, Olesen NJ, Mellergaard S (2005) Viral haemorrhagic septicaemia virus in marine fish and its implications for fish farming – a review. J Fish Dis 28:509–529

Soldal AV, Engås A (1997) Survival of young gadoids excluded from a shrimp trawl by a rigid deflecting grid. ICES J Mar Sci 54:117–124

Soldal AV, Isaksen B, Marteinsson JE, Engås A (1991) Scale damage and survival of cod and haddock escaping from a demersal trawl. ICES CM Documents B 44

Southgate PJ (2008) Welfare of fish during transport. In: Branson EJ (ed) Fish welfare. Blackwell, Oxford, pp 185–194

Staurnes M, Sigholt T, Pedersen HP, Rustad T (1994) Physiological effects of simulated high-density transport of Atlantic cod (*Gadus morhua*). Aquaculture 119:381–391

Steen J (1963) The physiology of the swimbladder in the eel *Anguilla vulgaris*. Acta Physiol 59:221–241
Sundnes G (1957) On the transport of live cod and coalfish. J Conseil 22:191–196
Suuronen P (2005) Mortality of fish escaping trawl gears. No 478. Food & Agriculture Org
Suuronen P, Erickson DL (2010) Mortality of animals that escape fishing gears or are discarded after capture: approaches to reduce mortality. In: He P (ed) Behavior of marine fishes: capture processes and conservation challenges. Wiley, Oxford, pp 265–293
Suuronen P, Lehtonen E, Tschernij V, Larsson P (1996) Skin injury and mortality of Baltic cod escaping from trawl codends equipped with exit windows. Arch Fish Mar Res 44:165–178
Suuronen P, Lehtonen E, Jounela P (2005) Escape mortality of trawl caught Baltic cod (*Gadus morhua*)—the effect of water temperature, fish size and codend catch. Fish Res 71:151–163. https://doi.org/10.1016/j.fishres.2004.08.022
Tenningen M, Vold A, Olsen RE (2012) The response of herring to high crowding densities in purse-seines: survival and stress reaction. ICES J Mar Sci 69:1523–1531
Tupper M, Sheriff N (2008) Capture-based aquaculture of groupers. In: Lovatelli A, Holthus PF (eds) Capture-based aquaculture global overview. FAO fisheries technical paper no. 508, Rome, pp 217–253
Tytler P, Blaxter J (1973) Adaptation by cod and saithe to pressure changes. Neth J Sea Res 7:31–45
Volkoff H, Xu M, MacDonald E, Hoskins L (2009) Aspects of the hormonal regulation of appetite in fish, with emphasis on goldfish, Atlantic cod and winter flounder: notes on actions and responses to nutritional, environmental and reproductive changes. Comp Biochem Physiol A Mol Integr Physiol 153:8–12
Wilson SM, Raby GD, Burnett NJ, Hinch SG, Cooke SJ (2014) Looking beyond the mortality of bycatch: sublethal effects of incidental capture on marine animals. Biol Conserv 171:61–72
Wood C, Turner J, Graham M (1983) Why do fish die after severe exercise? J Fish Biol 22:189–201

Chapter 19
Fish Welfare in Recreational Fishing

Keno Ferter, Steven J. Cooke, Odd-Børre Humborstad, Jonatan Nilsson, and Robert Arlinghaus

Abstract Recreational fishing is a popular activity around the globe, and fish welfare issues related to the activity have received increasing attention in some countries, particularly in central and northern Europe and Australia. This chapter offers an introduction to recreational fishing, reviews literature on fish welfare in relation to recreational fishing and provides an overview of potential biological impacts and ways to reduce such impacts. We first focus on the question on how to reduce impacts on the welfare of the fish during recreational fishing. Second, we describe two case studies highlighting that practical implications of the fish welfare discourse may be disjointed from the scientific information base and be rather about fundamental moral questions about the ethical acceptability of the activity per se. We end by providing an outlook on the future of recreational fishing in the light of the current fish welfare discourse.

Keywords Best practice guidelines · Catch-and-release · Fish welfare · Function-based approach · Recreational fishing · Sublethal impacts

Recreational fishing is a popular activity around the globe (Arlinghaus et al. 2015, 2019). Apart from the recognized biological and socio-economic importance of recreational fishing, fish welfare issues related to the activity have received increasing attention in some countries and in the academic literature (Huntingford et al. 2006; Arlinghaus et al. 2007a, b, 2012b; Cooke and Sneddon 2007; Volpato 2009; Arlinghaus and Schwab 2011). This chapter offers an introduction to recreational

K. Ferter (✉) · O.-B. Humborstad · J. Nilsson
Institute of Marine Research, Bergen, Norway
e-mail: Keno@hi.no

S. J. Cooke
Carleton University, Ottawa, ON, Canada

R. Arlinghaus
Leibniz-Institute of Freshwater Ecology and Inland Fisheries, Germany and Humboldt-Universität zu Berlin, Berlin, Germany

T. S. Kristiansen et al. (eds.), *The Welfare of Fish*, Animal Welfare 20,
https://doi.org/10.1007/978-3-030-41675-1_19

fishing and reviews literature on fish welfare in relation to recreational fishing. We will focus on the question of how to reduce impacts on the welfare of the fish during recreational fishing and provide an outlook on the future of recreational fishing in the light of the fish welfare discourse. We will not discuss, let alone answer, the question whether recreational fishing in general or angling are ethically acceptable, as this question has no objective solution and strongly depends on personal or cultural values. For example, one can ethically question the practice of catch-and-release as harming fish for no good reason (Volpato 2009), and one can take the stance that it is preferable to kill a recreationally captured fish than to release it (Webster 2005). Alternatively, one can conclude that it may be ethically impermissible to kill fish for human consumption (Bovenkerk and Braithwaite 2016). These extreme examples show how different the moral judgement of the very same practice—to catch and release a fish—can be in recreational fisheries. It is not our role as scientists to say what is right or wrong. Instead, we focus our chapter on the description of how to minimize negative impacts on the welfare of fish. That said, we will also highlight two case studies that show that in the "real world" the fish welfare discourse may be used to fundamentally question the ethical appropriateness of the entire activity. This is done to show that solving biological questions using robust, replicable science in terms of what impacts the well-being of fishes in recreational fishing may be of limited importance in practical fish welfare discourses (Riepe and Arlinghaus 2014).

19.1 A Short Overview of Recreational Fishing

19.1.1 Definitions

Recreational fishing is a diverse activity, and it is therefore challenging to clearly define. Although the separation from commercial fishing seems to be clear-cut as the main motivation for commercial fishing is economic revenue to the individual participant, there are recreational fisheries in which catches can also be sold (e.g. marine recreational fishing in Norway), or where recreational fishing happens in a commercial context (e.g. charter boat fishing). The separation from subsistence fisheries is even more difficult as many recreational fishers have subsistence-like motivations when harvesting fish (Macinko and Schumann 2007; Cooke et al. 2018). Accordingly, definitions of recreational fishing vary (Pawson et al. 2008; FAO 2012; ICES 2013). For example, FAO (2012) defines recreational fishing as "fishing of aquatic animals (mainly fish) that do not constitute the individual's primary resource to meet basic nutritional needs and are not generally sold or otherwise traded on export, domestic or black markets", while ICES (2013) considers recreational fishing as "the capture or attempted capture of living aquatic resources mainly for leisure and/or personal consumption". Pitcher and Hollingworth (2002) acknowledge that recreational fishing is separated from other forms of fishing by being mainly about fun rather than subsistence or economic revenue. However, a necessary component of the "fun" aspect is catching and possibly keeping fish for personal consumption, which contributes essential nutrients to the recreational harvester (Cooke et al. 2018). Despite being mainly about leisure, recreational fishing is responsible for a large economic activity, but the difference among commercial, subsistence and recreational fishing is that the main actor pursuing the recreational activity (the recreational fisher) is generally not interested in acquiring essential resources for their own survival and thus does not have economic interests himself or herself. Recreational fishing is the prime fishing activity in freshwater of all industrialized nations and grows in importance rapidly with economic development of societies (Arlinghaus et al. 2002; FAO 2012). It is also a common form of fishing in coastal areas, particularly in wealthy countries (Arlinghaus et al. 2019).

Although a range of passive fishing gears is employed by recreational fishers (e.g. gillnets, traps, hook-and-line), the main method used is angling using a rod and a reel (Arlinghaus et al. 2007a). We will therefore mainly focus on angling in this chapter but emphasize that other capture methods common to commercial fisheries (e.g. see Chap. 17) also have fish welfare implications in recreational fisheries. Apart from harvesting their catch, anglers often release a certain proportion of their catch due to regulations or personal motivations. This practice is called catch-and-release (C&R), defined as "the process of capturing fish by using hook and line, mostly assisted by rods and reels, and then releasing live fish back to the waters where they were captured, presumably to survive unharmed" (Arlinghaus et al. 2007a). C&R due to regulations (e.g. minimum landing sizes or bag limits) is referred to as regulatory C&R, while C&R of legally harvestable fish is referred to as voluntary C&R. If all captured fish are released, the term "total C&R" is used. Total C&R is rare in most recreational fisheries except some highly specialized fisheries, while some form of C&R probably occurs in all recreational fisheries worldwide (Arlinghaus et al. 2007a).

19.1.2 Relevance

Recreational fishing is a popular outdoor activity in inland and marine waters around the globe (Arlinghaus et al. 2019). Arlinghaus et al. (2015) estimated that around every tenth member of society engages in recreational fishing; they estimated that there are around 118 million recreational fishers in North America, Europe and Oceania alone. Data on participation rates in other parts of the world are insufficient or lacking entirely, but one can expect that global fishing participation is at least 220 million people (World Bank 2012). A recent study suggests that there might be 220 million anglers in China alone (China Society of Fisheries 2018). Locally high recreational fishing pressures can impact fish stocks (Post et al. 2002; Coleman et al. 2004; Cooke and Cowx 2004; Lewin et al. 2006). Moreover, several studies have shown that marine recreational fishing can account for a significant proportion of the total catch of some species (e.g. Post et al. 2002; Coleman et al. 2004; Strehlow et al. 2012; Herfaut et al. 2013; Brownscombe et al. 2014a; Kleiven et al. 2016; Hyder et al. 2018; Radford et al. 2018). Apart from its biological impacts, recreational fishing also provides socio-economic benefits, both to the individual fisher (e.g. food, nature experience, education and other personal rewards) and to society (e.g. jobs, social capital, management capital and economic benefits) (Weithman 1999; Arlinghaus and Cooke 2009; Parkkila et al. 2010; Tufts et al. 2015; Lynch et al. 2016; Griffiths et al. 2017), whilst remaining a source of concern for animal liberation and rights advocates (Arlinghaus et al. 2007a, 2012b).

19.1.3 Management Issues in Relation to Fish Welfare

Regulations associated with recreational fishing are highly variable across jurisdictions. In some parts of the world (e.g. most low- and middle-income countries), there is little if any active recreational fisheries management or associated regulatory frameworks (Potts et al. 2020). Some jurisdictions have state/provincial or federal licensing schemes (e.g. all states and provinces in North America and Australia) and science-based regulatory frameworks focused on the largely public waters. In much of Europe, property-right schemes are such that governments typically play small roles in recreational-fisheries management in inland waters but are responsible for marine recreational fisheries. Management of recreational fisheries is diverse; effort, harvest regulations, habitat management and stocking are employed to secure sustainable exploitation (Arlinghaus et al. 2016). As fishing segments can differ substantially in motives and expectations within the same recreational fishery, finding effective tools to manage the fishery as a whole in pursuit of ecological or social goals can be challenging (Radomski et al. 2001; Johnston et al. 2010; Beardmore et al. 2015). Common effort (input) regulations are limiting access through license systems, gear regulations and closed seasons. Output (e.g. harvest) is often regulated through bag limits and minimum-size limits, or other harvest regulations (Lewin et al. 2006). In recent years, total C&R has been used as a management tool to reduce fishing mortality of overexploited fish stocks in selected fisheries while maintaining angling opportunities. However, total C&R as a management tool has led to several fish welfare debates based on moral grounds in selected European countries (Aas et al. 2002; Arlinghaus et al. 2007a). The perspectives range from the capacity of C&R to maintain fish welfare by avoiding excessive mortality (Arlinghaus et al. 2007a, b) to total C&R being fishing for no good (culinary) reason at high welfare costs to the fish (Volpato et al. 2007).

Some non-binding technical guidelines for governance and management of recreational fisheries have been developed (EIFAC 2008; FAO 2012), along with various regional, national and international codes of practice (Arlinghaus et al. 2010), but it is unclear to what extent these have been adopted in the context of welfare (Arlinghaus et al. 2012a). Codes of practice generally promote sustainable recreational and responsible fishing practices that both maintain the population and minimize welfare impacts on angled fish. They can be consulted by individual anglers or used by angling clubs or management agencies to help develop their own outreach materials and internal policies and practices in line with the local and regional cultural contexts (Arlinghaus et al. 2012a). In general, regulations specific to fish welfare are uncommon across the globe, as it is difficult to regulate very specific angler behaviours (e.g. banning air exposure during C&R). However, in some jurisdictions, such as Germany, there are very specific rules and regulations that constrain practices that are considered unnecessarily harmful to fish welfare. For example, in Germany there are bans on the use of live bait fish, and there is the regulation that one has to kill a harvestable fish immediately after landing by percussive stunning and subsequent debleeding, which kills the fish, to minimize the harm on recreationally caught fish (Arlinghaus et al. 2007a). Moreover, natural resource management agencies have a mandate for education and outreach so there is an opportunity to improve fish welfare independent of any formal regulations (Cooke et al. 2013b).

19.2 Fish Welfare in the Context of Recreational Fishing

Considering fish welfare issues in relation to aquaculture has a long tradition (see Chaps. 1 and 11). During the last decade, increased attention has also been directed towards wild-capture fisheries, and recreational fisheries in particular (e.g. Davie and Kopf 2006; Huntingford et al. 2006; Cooke and Sneddon 2007; Arlinghaus 2008; Metcalfe 2009). In fact, the discussion on welfare of fishes in the context of capture fisheries seems to have been initiated in the context of recreational fisheries, with commercial capture fisheries (Chaps. 17 and 18) following somewhat later and more recently. It is impossible to catch a fish with rod and reel without causing some level of injury (particularly tissue damage from the hook) and physiological disturbance (particularly exercise during the escape response and fight). In addition, the fish are handled, possibly air exposed and either killed or released. In all of this, there is a potential for affecting fish welfare. Apart from the actual capture process (i.e. hooking, fighting and handling), a particular focus has been placed on C&R practices and killing (Davie and Kopf 2006; Huntingford et al. 2006).

There are different approaches to assessing fish welfare (see Chap. 13 for details). Two approaches that are commonly used in the context of recreational fishing are the feelings-based approach and the function-based approach; while the feelings-based approach focuses on the pain and suffering a fish may experience during the capture and release process in angling (Huntingford et al. 2006), the function-based approach focuses on the appropriate functioning of an individual (e.g. physiology, behaviour, health and fitness) (Huntingford et al. 2006; Arlinghaus et al. 2007b). The function-based approach has also been referred to as the pragmatic approach to fish welfare by Arlinghaus et al. (2009b), who suggested its use to objectively measure welfare indicators to assess fish welfare. In fisheries, welfare is about reducing or avoiding negative impacts on the fish. Irrespective of whether the impact is described in terms of feelings or function, most welfare approaches have the same goal, which is to avoid or minimize damage and stress and maintain the well-being of the individual fish as far as possible (Cooke and Sneddon 2007).

Most of the focus on the discussion of fish welfare has been directed towards the negative impacts of recreational fishing. These impacts can be categorized into sublethal and lethal impacts (Arlinghaus et al. 2007a, b). The sublethal impacts can be further portioned into those leading to primary (e.g. hormonal responses), secondary (e.g. mobilization of glucose) and tertiary (e.g. behavioural impairment) stress responses as well as injury and health impacts, with possible consequences for fitness surrogates (e.g. growth) or fitness (e.g. reduction of reproductive output and survival) (reviewed in Arlinghaus et al. 2007a, b; Cooke et al. 2013a). For example, catch-and-release can be a significant stressor and lead to injury (Muoneke and Childress 1994), behavioural changes (Ferter et al. 2015a), impaired feeding performance (Thompson et al. 2018), reduced growth (Klefoth et al. 2011), reduced reproductive output (Richard et al. 2013) and post-release mortality (Hühn and Arlinghaus 2011). However, there are several examples where recreational fishing-related activities have had or have positive impacts on welfare, e.g. by engaging in fisheries management that increases and conserves fish populations or by actions that reduce stressors on individual fishes (e.g. dams) in the wild. For example, anglers have been directly or indirectly involved in the rehabilitation of natural habitats (Granek et al. 2008), which has improved spawning grounds and the general health of the ecosystem (Nilsson et al. 2014). Moreover, the removal of dams enables fish to perform their natural spawning migrations, thereby contributing to improved welfare at the level of individual fishes. In general, the fact that many anglers engage or support fisheries management and improvement of natural habitat (Granek et al. 2008; Cooke et al. 2019) can be considered positive from a fish welfare perspective, despite the concept of fish welfare being an individual concept and not one focused on populations.

19.3 Ways to Promote Welfare

In the following, we will highlight the various areas where recreational fishing induces injury and stress to individual fish, thereby negatively affecting fish welfare, and we will also briefly mention ways by which such impacts could be minimized or avoided altogether. We will only deal with those issues that have received some level of scientific attention as evidenced by published scientific work. We will present the fish welfare impacts starting from the capture process, followed by the handling, and ultimately release or kill components.

19.3.1 Capture

The capture process has varying impacts on fish welfare depending on the capture method (Davis 2002). To be captured by angling, the fish must be hooked, which is bound to cause physical injury. In general, the fish is attracted by a bait or lure fitted

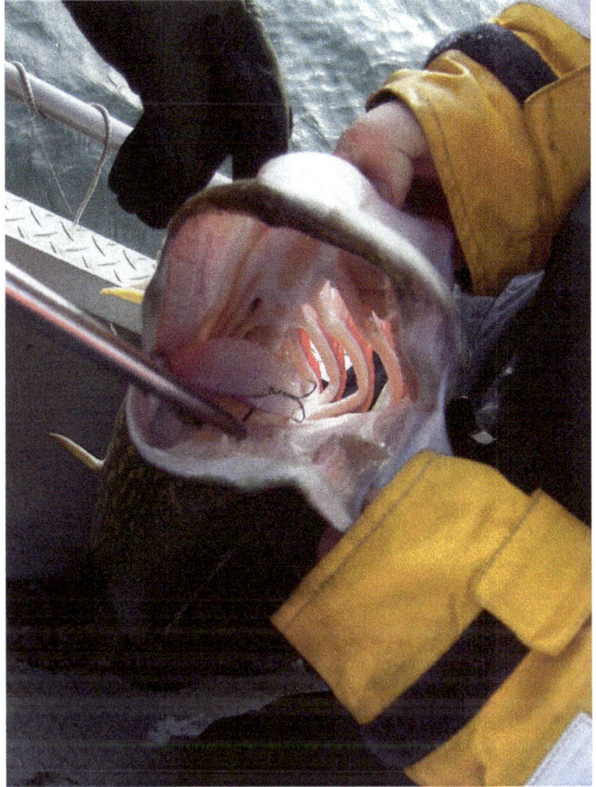

Fig. 20.1 Deep hooking can have lethal and sublethal impacts on the fish. When deep hooking is a problem, changing to larger artificial lures may reduce hooking in critical locations, as shown in Arlinghaus et al. (2008b) (photographer: Adaptfish IGB)

with a hook, which is ingested (Løkkeborg et al. 2014). Thus, the fish gets hooked in the lips, mouth, gills, esophagus or stomach (Alós et al. 2009; Weltersbach and Strehlow 2013). In some cases, the fish are involuntarily or voluntarily hooked on the outside, which is referred to as foul-hooking or snagging. Depending on the anatomical hooking location, the hook causes varying degrees of injury. When a fish is hooked in the jaws, for example, the injury is less than a fish that is hooked in the gills, the gullet, or other vital tissues (Eckroth et al. 2014; Stålhammar et al. 2014). The anatomical hooking location and the severity of hooking injury depend on several factors including, but not limited to, the hook size and type, lure or bait, fishing method (e.g. passive versus active angling) and size of the fish mouth relative to bait size (e.g. Grixti et al. 2007; Arlinghaus et al. 2008b; Alós et al. 2009). There is a large difference between hook types when it comes to anatomical hooking location. For example, when the hook is swallowed by a fish, traditional J-hooks are more likely to deep-hook a fish compared to circle hooks (Aalbers et al. 2004; Cooke and Suski 2004). The anatomical hooking location also depends on the bait or lure type used and its size (Fig. 20.1). Natural baits (e.g. worms) are more likely to be swallowed than an artificial lure (e.g. metal spoon) (Arlinghaus et al. 2008b), but this is also dependent on the fishing method (Payer et al. 1989; Rapp et al. 2008), and

Fig. 20.2 Once the fish is hooked, it has to be retrieved. Appropriate fishing tackle has to be used to ensure successful retrieval and minimize fighting time (photographer: Keno Ferter)

ultimately the size of the bait in combination with the size of the hooks (Wilde et al. 2003). Passive bait presentation gives the fish time to swallow the hook, while the hook is often set instantly when the bait or lure is actively fished (i.e., retrieved quickly as in fishing with artificial lures) (Schisler and Bergersen 1996; Sullivan et al. 2013). However, the likelihood of foul-hooking a fish might increase when the lure or bait is fished actively, and with particular lure types (e.g. crankbaits with two treble hooks; Arlinghaus et al. 2008b).

Once the fish is hooked, it has to be retrieved (Fig. 20.2). It is important to choose appropriate fishing tackle to ensure successful retrieval and avoid line breakage, which might leave the hook inside the fish (Arlinghaus et al. 2008a; Henry et al. 2009; Pullen et al. 2017). Retrieval or fighting time affects physical exhaustion and depends on factors like rod and line class, fish size and environmental conditions (e.g. water current) (Meka 2004; Meka and McCormick 2005; French et al. 2015). The longer it takes to retrieve a fish, the more exhausted the fish gets, which can have a negative impact on the welfare status (Meka 2004). Fighting time is generally positively correlated with the accumulation of lactate in the blood and muscle, and the plasma concentration of the stress hormone cortisol (Tracey et al. 2016). Higher water temperatures often exacerbate negative impacts on the fish (Gale et al. 2013). The lighter the rod or line class, the longer it will take to retrieve the fish, particularly when the fish is large.

Another factor which has an impact on fish welfare during retrieval is the capture depth. Capture depth does not only affect retrieval time (i.e. the deeper the longer it takes), but can also lead to barotrauma signs in some fish species (Ferter et al. 2015b). Barotrauma is caused when the swim bladder gas expands due to ambient pressure reduction during the forced ascent to the surface, and causes signs like swim

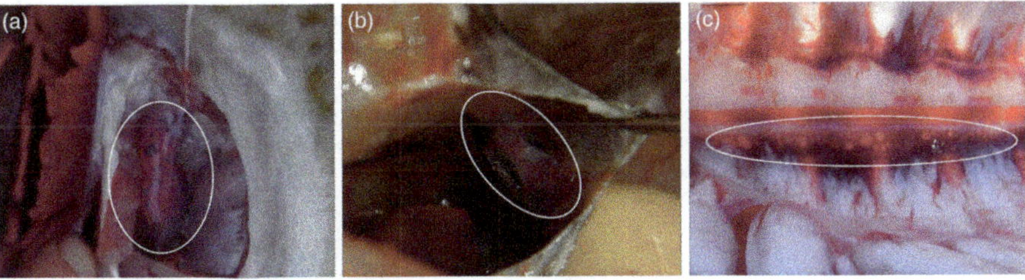

Fig. 20.3 Pictures of gas embolisms in the (**a**) *Vena cardinalis communis*, (**b**) *Vena hepatica* and (**c**) *Vena cardinalis caudalis* of Atlantic cod with barotrauma after rapid decompression. The affected veins are encircled with white rings (taken from Ferter et al. 2015b, licensed by CC BY 4.0)

bladder rupture, gas bubble formation in the blood (Fig. 20.3), exophthalmia and skin bubbles (Hannah et al. 2008; Brown et al. 2012; Ferter et al. 2015b). These signs have species-specific short- to long-lasting impacts on the welfare of the individual fish.

19.3.2 Handling

When the fish is retrieved, it must be landed. There are many different landing techniques that have different welfare implications. Some landing techniques (e.g. the use of a gaff, which consist of a pole with a large hook) are only suitable when the fish is killed immediately after landing. The most common landing techniques are the use of a landing net, hand landing, gaffing or stranding. When the fish is not supposed to be harvested, the best is to de-hook it inside the water to minimize handling and physical contact between the angler and the fish. Landing nets can cause damage to the mucus membrane of the fish, but this damage can be minimized by using knotless, fine-meshed netting (Colotelo and Cooke 2011). Landing the fish by hand can be gentle, but has to be done correctly to avoid injury (Barthel et al. 2003). While handling with towels may lead to mucus injuries during de-hooking, in some fish species, handling with towels has not been found to lead to significant sublethal impacts or death compared to handling with hands only (Schwabe et al. 2014). Nevertheless, the use of towels is more problematic than de-hooking with wet hands underwater. Common hand landing techniques include grabbing the fish in the mouth, under the gill cover or by the tail. When lifting the fish, it is important to support it with both hands to distribute the body weight and avoid damage to the spinal cord (Gould and Grace 2009). Some tools (e.g. lip grips) exist to assist hand landing, but these can cause significant damage on some fish species (Danylchuk et al. 2008). Gaffing the fish can lead to substantial injuries, because the gaff often penetrates vital organs, which is usually lethal. Gaffed fish should thus not be released and immediately killed. However, a fine gaff on a large fish which is pulled through the lips may cause injuries which are not more significant than the hooking injury itself. Stranding the fish (pulling the fish on shore) can lead to significant mucus membrane injury. Moreover, other parts of the

Fig. 20.4 When the fish is supposed to be released again, photographing should be done by lifting the fish shortly out of the water to minimize air exposure. Supporting the fish with both hands is recommended to distribute body weight and avoid damage to the spinal cord (photographer: Will Twardeck)

fish (e.g. the eyes or gills) may be damaged when they come into contact with soil, sand, rocks or vegetation, and thus this method is more damaging than hand landing from a fish welfare perspective.

Once the fish is landed, it is usually de-hooked. If the fish is supposed to be harvested, killing it before de-hooking minimizes impacts on the welfare of the fish (Diggles et al. 2011; FAO 2012) and preserves flesh quality. When the fish is supposed to be released, minimizing or avoiding air exposure is positive for fish welfare as it is generally a stressor for fish and leads to increased physiological disturbance (Arends et al. 1999; Suski et al. 2007; Rapp et al. 2012, 2014), especially when it occurs after exhausted exercise (Ferguson and Tufts 1992). Air exposure may even lead to death in some sensitive species (Arlinghaus and Hallermann 2007). To avoid skin or mucus damage, the captured fish may be placed on a de-hooking mat with a soft rubber surface during de-hooking (Arlinghaus 2007; Stålhammar et al. 2014). De-hooking devices can assist with deep-hooked fish. For some species, it may be beneficial to cut the line of deep-hooked fish close to the mouth instead of trying to de-hook the fish. While some species manage to eject the hook by themselves (Pullen et al. 2019), others can have difficulties (Hall et al. 2009; Weltersbach et al. 2016). Photographing the catch may either be done after the fish has been killed or, when the fish is supposed to be released again, by lifting the fish shortly out of the water to minimize air exposure (Fig. 20.4). Some anglers keep their catch in so-called keep nets or bags prior to killing or release, which may have negative impacts on the fish (Rapp et al. 2012, 2014). Keeping fish in live-wells is also common practice in live-release angling tournaments, and both retaining the fish and the extra handling during weighing can have negative impacts on the fish (Suski et al. 2003b).

19.3.3 Release

If the fish is to be released, doing this as quickly as possible after de-hooking is recommendable to avoid additional handling stress (Fig. 20.5). Only fish without substantial hooking injuries or without other injuries (e.g. due to barotrauma or landing) can be released unharmed, because fish with substantial injuries could suffer from lethal or sublethal short- and long-term impacts. In some regulations, however, the release of some fish species may be mandated independent of the fish condition (i.e. it would be illegal to kill or keep an undersized fish even if mortally wounded). Post-release mortality and sublethal impacts are species-specific and depend on many factors, including, but not limited to, hooking injury, fighting time, capture depth and water temperature (Bartholomew and Bohnsack 2005; Hühn and Arlinghaus 2011). Released fish can show behavioural alterations to varying degrees (Klefoth et al. 2008; Arlinghaus et al. 2009a; Ferter et al. 2015a), which can have implications for welfare. During the breeding season, for example, nesting behaviour could be impaired in some species, which could lead to reproductive failure (Suski et al. 2003a). When predator abundance is high, released fish with abnormal behaviour may be more prone to predation than usual (Brownscombe et al. 2014b). In such situations, it may be beneficial to use recovery bags or boxes in which fish can be held for a certain period of time prior to release (Brownscombe et al. 2013), although some of these may actually produce stress in some fish and lower activity post-release despite physical rehabilitation (Rapp et al. 2012, 2014). Severely exhausted fish may have difficulties recovering and drift downstream when released into strong currents. In these cases, assisted ventilation (i.e. moving the fish back and forth to increase water flow through the gills) is often recommended, but its benefits for the fish are questionable (Robinson et al. 2015). Fish suffering from barotrauma may have difficulties to submerge and are thus easy prey for avian predators. Venting can be used to release excess air from the swim bladder or coelomic cavity, but while this method can have benefits for some species

Fig. 20.5 Releasing the fish carefully shortly after de-hooking reduces sublethal impacts. Atlantic cod without substantial hook injuries or barotrauma return to normal behaviour shortly (<15 h) after the release when handled carefully, as shown in Ferter et al. (2015a) (photographer: Martin Wiech)

(e.g. black sea bass *Centropristis striata*; Collins et al. 1999), it has been shown to have negative or no effects on others (Wilde 2009). Moreover, venting has to be done correctly to avoid injury of vital organs. Another method to assist submergence is the use of release weights. These weights are fastened to the lips of the fish, and lowered to capture depth. Once the fish is recompressed, the weight is released from the fish and retrieved to the surface (Roach et al. 2011).

19.3.4 Killing

Rapidly killing the fish after landing is advisable from a welfare perspective (see Chap. 17 on slaughter). There are different killing methods used by recreational anglers, some of which can have negative implications for welfare (Davie and Kopf 2006). The ideal killing method leads to immediate unconsciousness and fast death. Air asphyxiation by leaving the fish outside the water or putting the fish alive into ice-chilled water is thus not ideal as it can take a long time until the fish dies. More welfare-friendly methods, which are used by recreational anglers and are legally mandated in some countries, are percussive stunning, pithing or ikejime (Davie and Kopf 2006; Diggles 2015) and sometimes shooting the fish (for large species). Guidelines on how to kill fish to promote both welfare and flesh quality have been made available online (e.g. DigsFish Services Pty Ltd 2019). Special tools to kill the fish with a blow to the head, so-called priests, are commercially available and often used to stun the fish prior to bleeding. Cutting the throat without prior stunning can lead to negative welfare as it can take the fish several minutes to die (Jensvoll 2007).

19.3.5 Stocking

An issue related to fish welfare that has as yet not seen a lot of discussion is stocking (Huntingford et al. 2006). While stocking can enhance and preserve fish populations (Lorenzen et al. 2012), the stocked fishes are usually brought into the wild from hatcheries and they are faced not only with the initial rearing phase in artificial environments but also with transport and release stress. As a consequence, stocked fish usually experience greater natural mortality than wild fishes (Lorenzen 2006; Lorenzen et al. 2012), suggesting there may be welfare issues associated with the practice. Throughout the whole process of rearing, fish welfare issues can emerge related to holding and handling-induced stress, which can lead to behavioural impairments and fin damage (Huntingford et al. 2006; Salvanes and Braithwaite 2006). Moreover, transport will affect the welfare of fishes (Barton et al. 1980). After release, when maladapted or when forced in competition with wild fishes, stocked fishes might suffer from large post-release mortality (Hühn et al. 2014), particularly in young fishes (Lorenzen 2005), rendering the stocking of less numerous, but larger, more robust fishes advisable from a fish welfare perspective. The benefit of stocking for recreational fishing is generally to maintain fish populations and fisheries, and failed stocking events are thus problematic as they are economically wasteful and impact the welfare of those fish that die. The conditions that increase the likelihood of a successful stocking event are equivalent to the conditions that minimize fish welfare impacts: only well-adapted fishes stocked at the right size with as little handling and transport stress as possible will do well after release. Thus, improving stocking for fisheries and for fish welfare goes hand in hand.

19.4 Case Studies on Fish Welfare Debates Related to Recreational Fishing

The following two case studies show how fish welfare has been dealt with in two countries. These case studies show that the ultimate treatment of recreational fishing under a fish welfare perspective often has surprisingly little to do with the purely biological question of how to minimize fish welfare as discussed earlier. The reason is that humans typically judge the moral acceptability of an action in relation to animals based on the motivation of the actor (Olson 2003), and less so by the degree to which a human action affects the welfare of the animal (Riepe and Arlinghaus 2014). Riepe and Arlinghaus (2014) showed that negative evaluations of specific angling practices such as C&R as well as the moral judgement of recreational fishing as a whole were predominantly explained by underlying values and animal rights-related attitudes, and less so or not at all by the degree to which people perceived animals to be able to experience human-like psychological states such as suffering. The following two case studies demonstrate that moral questions directed at the motives of the actor interacting with fishes, rather than scientific questions of how severe the interaction is to the affected fishes, have dictated which practices are considered good from a welfare perspective. This can lead to outcomes that ironically may even harm welfare (Browman et al. 2019).

19.4.1 Animal Welfare Law and Recreational Angling in Germany

Following German animal welfare law, one is only allowed to harm fish, e.g. during recreational fishing, if one has a so-called reasonable, or good-enough, reason. While the specific reasons justifying recreational fishing are not specified in the animal welfare law, a common argument substantiated by a range of court decisions is that a good-enough reason to harm fishes, and thereby affect their welfare negatively, is the actor's motivation to catch fish for dinner (Arlinghaus 2007). In turn, any actions and practices in recreational fishing that are not about getting food for personal consumption may be considered problematic and have typically been banned or are under normative reprehension by the public and by fellow anglers. Examples are the stocking of legal-sized fish for immediate recapture (because one could directly consume the fish prior to release), voluntary C&R of legally sized fish (which is considered to indicate the lack of a harvest motive and to be mainly about playing with food, Aas et al. 2002) or competitive fishing with associated C&R of the catch. However, if one has a reasonable reason prior to casting the line out into the water, the angler is legally allowed to affect the welfare of fishes negatively, by catching it, by releasing it if undersized or a non-targeted bycatch, and by killing it.

These examples show that judging whether a certain recreational fishing practice is considered permissible or reprehensible from a fish welfare perspective may not be as easily judged from a natural science perspective in terms of what happens to the fish during the process as elaborated in previous sections. For example, the very same practice, e.g. C&R, may be considered legally and morally acceptable if it happens to an undersized fish, say a pike of 49 cm at a minimum-length limit of 50 cm. While the same release event of a pike of 50.5 cm may bring the angler to court if it is done voluntarily in the absence of a general harvest motive. Yet, the fish welfare impacts of the two C&R events to the fish are identical. More importantly, because fishing without a harvest motive is considered unethical in Germany, this essentially means that legally speaking all fish that are caught and legal-sized would have to be taken home for dinner. This in turn means that killing of fish is considered ethically superior to voluntarily releasing part or all catch. However, it is not immediately obvious whether killing or releasing is more problematic from the perspective of the individual focal fish. In fact, one can argue that a fish that is released quickly and in good state may quickly recover from capture- and handling-induced physiological stress, and return into reproductive mode and survive. Likely, from the perspective of the individual fish whose primary interest of life is to stay alive and contribute genes to the next generation (defined as biological fitness), the best situation would be to not be captured at all, followed by catch-and-release without major impacts as the second-best, and being killed as the third-best.

19.4.2 Discussion of Catch-and-Release Practice of Marine Fishes in Norway

Both voluntary and regulatory C&R are common in Norwegian marine recreational fisheries (Ferter et al. 2013, 2015a). Although Norway has implemented a general discard ban, the release of viable fish is allowed according to Norwegian fisheries regulations. This rule applies for both voluntary and regulatory C&R for the most of Norway, but in the Skagerrak voluntary C&R of handheld hook-and-line caught fish has recently been prohibited (Forskrift om utøvelse av fisket i sjøen 2013). Voluntary C&R practice has led to several public debates due to potential welfare issues. Particularly C&R of Atlantic halibut (*Hippoglossus hippoglossus*) has recently become a hot topic. This species is also one of the most popular voluntary C&R species for marine angling tourists and resident anglers due to its size and powerful endurance during the fight (K. Ferter, personal observation). Thus, this species supports a significant fishing tourism industry in Norway (Borch et al. 2011) and is also important for domestic recreational fishers, particularly in the northern part of Norway. In recent years, voluntary C&R practice for halibut has received substantial media attention and caused several public debates as it is seen as animal abuse for no good reason (e.g. NRK 2016), similar to the case in Germany. The Norwegian Food Safety Authority recently evaluated voluntary C&R of halibut on inquiry of the Ministry of Trade, Industry and Fisheries (Mattilsynet 2015). They concluded that voluntary C&R is problematic from an animal welfare point of view and should be forbidden for all marine species, because the only acceptable motivation for fishing is to fish with a harvest aim. Recently, this statement has been revised by specifying that this ban applies if the only intention of the angler was to experience joy and excitement by practicing voluntary C&R (Mattilsynet 2019). Like in the German case, this judgement is much less about what happens to the fish, but about the moral judgments of the motivations and intentions of the actors (Riepe and Arlinghaus 2014). Given the complexity of the debate, the Norwegian Food Safety Authority further recommended implementing a ban on voluntary C&R as a guideline rather than an actual fisheries regulation, although they may consider an actual change in fisheries regulations if this ban is not followed. Recently, a maximum landing size of 200 cm has been implemented for halibut requiring the release of very large individuals. However, even with this maximum landing size in place, such a C&R ban can be problematic as it can lead to unforeseen ecological consequences. Large specimens which are below the maximum landing size are often released voluntarily because they are important spawners and are not suitable as food due to the coarse meat texture and higher concentrations of environmental contaminants. Thus, if such large spawners now would have to be landed, and fishing pressure remains high, this could ultimately have negative consequences for the fish stocks and the people eating the fish. It may therefore be wise to carefully evaluate the necessity and consequences of voluntary C&R regulations before their implementation because policies driven by individual-level welfare considerations may bring about important population-level conservation problems related to overfishing due to overharvest.

19.5 The Future of Recreational Fishing in the Light of Fish Welfare Concerns

Angling and other recreational fishing practices inevitably have some negative impacts on fish welfare, e.g. by causing stress and injury to the individual fish, and we have shown that many of these issues can be reduced practically by altering fishers' behaviour and practices. Yet, as our case studies have shown, ultimately, the question of whether recreational fishing practices are morally acceptable from a welfare perspective may depend on the intention of the actor (Olson 2003), and human values and standpoints (Riepe and Arlinghaus 2014), which cannot be answered by natural sciences. However, natural sciences can contribute to promote fish welfare by studying welfare indicators, which can serve as a basis to improve fish welfare in recreational fisheries. Moreover, such studies can serve as a basis to evaluate the welfare impacts of angling and angling practices such as C&R and weigh those impacts against other human interests (Arlinghaus et al. 2007b).

That said, as our case studies have shown, the issue of fish welfare quickly extends into the moral domain by relating to ethically acceptable or unacceptable intentions of recreational fishers. In this context, the most visible welfare debate centres around voluntary C&R, which is considered ethically problematic in some countries if the actor fishes without the basic intention to harvest, while the very same practice is seen as the desired solution to minimize fishing impacts in other cultures (e.g. USA, Canada, UK). In recent years, the moral debate in recreational fishing, and fishing in general, has extended to question whether fishing for food is acceptable (Bovenkerk and Braithwaite 2016). Such development could, in turn, lead to voluntary C&R becoming a morally superior practice to catch-and-kill, again based on ethical welfare arguments. Alternatively, morally questioning whether killing of fish is acceptable could be equated with limitations on recreational fishing, if the only purpose of recreational fishing considered appropriate is to fish for personal consumption. As these examples are showing, dealing with fish welfare becomes complicated the moment one moves beyond the simple applied biological question of how to minimize fish welfare impacts.

Dawkins (2017) advised that fish welfare can be improved without referring to contentious topics such as consciousness or suffering. Similar arguments have been suggested in the context of recreational fishing (Arlinghaus et al. 2007b; Browman et al. 2019). We have shown that much effort has been made to minimize negative welfare impacts on the individual fish by developing best practice guidelines (EIFAC 2008; FAO 2012), but a lot of further improvement can be achieved at the species-specific level (Cooke and Suski 2005). However, if one considers the moral judgement of a recreational fishing practice solely in terms of the intention of the actor (e.g. whether you fish for harvest or not), the ethical judgement of, and recommendations for, the activity may be divorced from the biological underpinning of fish welfare. In other words, merely judging the ethical appropriateness of recreational fishing in terms of morally acceptable intentions of the angler may unintentionally undermine fish welfare, as less attention is given to what actually happens to the fish. We suggest avoiding grand moral reasoning and instead using science to promote changes practices and fishers' behaviours in a way that minimizes or avoids fish welfare impacts in recreational fisheries. Our work presents concrete steps into that direction.

References

Aalbers SA, Stutzer GM, Drawbridge MA (2004) The effects of catch-and-release angling on the growth and survival of juvenile white seabass captured on offset circle and J-type hooks. N Am J Fish Manag 24:793–800

Aas Ø, Thailing CE, Ditton RB (2002) Controversy over catch-and-release recreational fishing in Europe. In: Pitcher TJ, Hollingworth CE (eds) Recreational fisheries: ecological, economic and social evaluation. Blackwell Science, Oxford, pp 95–106

Alós J, Arlinghaus R, Palmer M, March D, Álvarez I (2009) The influence of type of natural bait on fish catches and hooking location in a mixed-species marine recreational fishery, with implications for management. Fish Res 97:270–277

Arends R, Mancera J, Munoz J, Bonga SW, Flik G (1999) The stress response of the gilthead sea bream (*Sparus aurata L.*) to air exposure and confinement. J Endocrinol 163:149–157

Arlinghaus R (2007) Voluntary catch and release can generate conflict within the recreational angling community: a qualitative case study of specialised carp, *Cyprinus carpio*, angling in Germany. Fish Manag Ecol 14:161–171

Arlinghaus R (2008) The challenge of ethical angling: the case of C&R and its relation to fish welfare. In: AAS Ø (ed) Global challenges in recreational fisheries. Blackwell, Oxford, pp 223–236

Arlinghaus R, Cooke SJ (2009) Recreational fisheries: socioeconomic importance, conservation issues and management challenges. In: Dickson B, Hutton J, Adams WM (eds) Recreational hunting, conservation and rural livelihoods: science and practice. Oxford, Wiley-Blackwell, pp 39–58

Arlinghaus R, Hallermann J (2007) Effects of air exposure on mortality and growth of undersized pikeperch, *Sander lucioperca*, at low water temperatures with implications for catch-and-release fishing. Fish Manag Ecol 14:155–160

Arlinghaus R, Schwab A (2011) Five ethical challenges to recreational fishing: what they are and what they mean. In: American fisheries society symposium, pp 219–234

Arlinghaus R, Mehner T, Cowx I (2002) Reconciling traditional inland fisheries management and sustainability in industrialized countries, with emphasis on Europe. Fish Fish 3:261–316

Arlinghaus R, Cooke SJ, Lyman J, Policansky D, Schwab A, Suski C, Sutton SG, Thorstad EB (2007a) Understanding the complexity of catch-and-release in recreational fishing: an integrative synthesis of global knowledge from historical, ethical, social, and biological perspectives. Rev Fish Sci 15:75–167

Arlinghaus R, Cooke SJ, Schwab A, Cowx IG (2007b) Fish welfare: a challenge to the feelings-based approach, with implications for recreational fishing. Fish Fish 8:57–71

Arlinghaus R, Klefoth T, Gingerich A, Donaldson M, Hanson K, Cooke S (2008a) Behaviour and survival of pike, *Esox lucius*, with a retained lure in the lower jaw. Fish Manag Ecol 15:459–466

Arlinghaus R, Klefoth T, Kobler A, Cooke SJ (2008b) Size selectivity, injury, handling time, and determinants of initial hooking mortality in recreational angling for northern pike: the influence of type and size of bait. N Am J Fish Manag 28:123–134

Arlinghaus R, Klefoth T, Cooke SJ, Gingerich A, Suski C (2009a) Physiological and behavioural consequences of catch-and-release angling on northern pike (*Esox lucius* L.). Fish Res 97:223–233

Arlinghaus R, Schwab A, Cooke S, Cowx I (2009b) Contrasting pragmatic and suffering-centred approaches to fish welfare in recreational angling. J Fish Biol 75:2448–2463

Arlinghaus R, Cooke SJ, Cowx IG (2010) Providing context to the global code of practice for recreational fisheries. Fish Manag Ecol 17:146–156

Arlinghaus R, Beard TD Jr, Cooke SJ, Cowx IG (2012a) Benefits and risks of adopting the global code of practice for recreational fisheries. Fisheries 37:165–172

Arlinghaus R, Schwab A, Riepe C, Teel T (2012b) A primer on anti-angling philosophy and its relevance for recreational fisheries in urbanized societies. Fisheries 37:153–164

Arlinghaus R, Tillner R, Bork M (2015) Explaining participation rates in recreational fishing across industrialised countries. Fish Manag Ecol 22:45–55

Arlinghaus R, Lorenzen K, Johnson BM, Cooke SJ, Cowx IG (2016) Management of freshwater fisheries: addressing habitat, people and fishes. In: Craig JF (ed) Freshwater fisheries ecology. Wiley, Chichester, pp 557–579

Arlinghaus R, Abbott JK, Fenichel EP, Carpenter SR, Hunt LM, Alós J, Klefoth T, Cooke SJ, Hilborn R, Jensen OP, Wilberg MJ, Post JR, Manfredo MJ (2019) Opinion: governing the recreational dimension of global fisheries. Proc Natl Acad Sci 116:5209–5213

Barthel B, Cooke S, Suski C, Philipp D (2003) Effects of landing net mesh type on injury and mortality in a freshwater recreational fishery. Fish Res 63:275–282

Bartholomew A, Bohnsack J (2005) A review of catch-and-release angling mortality with implications for no-take reserves. Rev Fish Biol Fish 15:129–154

Barton BA, Peter RE, Paulencu CR (1980) Plasma cortisol levels of fingerling rainbow trout (*Salmo gairdneri*) at rest, and subjected to handling, confinement, transport, and stocking. Can J Fish Aquat Sci 37:805–811

Beardmore B, Hunt LM, Haider W, Dorow M, Arlinghaus R (2015) Effectively managing angler satisfaction in recreational fisheries requires understanding the fish species and the anglers. Can J Fish Aquat Sci 72:500–513

Borch T, Moilanen M, Olsen F (2011) Marine fishing tourism in Norway: structure and economic effects. Økonomisk Fiskeriforskning 21:1–17

Bovenkerk B, Braithwaite V (2016) Beneath the surface: killing of fish as a moral problem. In: Meijboom FLB, Stassen EN (eds) The end of animal life: a start for ethical debate: ethical and societal considerations on killing animals. Wageningen Academic, Wageningen, pp 225–250

Browman HI, Cooke SJ, Cowx IG, Derbyshire SW, Kasumyan A, Key B, Rose JD, Schwab A, Skiftesvik AB, Stevens ED, Watson CA, Arlinghaus R (2019) Welfare of aquatic animals: where things are, where they are going, and what it means for research, aquaculture, recreational angling, and commercial fishing. ICES J Mar Sci 76:82–92

Brown RS, Pflugrath BD, Colotelo AH, Brauner CJ, Carlson TJ, Deng ZD, Seaburg AG (2012) Pathways of barotrauma in juvenile salmonids exposed to simulated hydroturbine passage: Boyle's law vs. Henry's law. Fish Res 121–122:43–50

Brownscombe JW, Thiem JD, Hatry C, Cull F, Haak CR, Danylchuk AJ, Cooke SJ (2013) Recovery bags reduce post-release impairments in locomotory activity and behavior of bonefish (*Albula* spp.) following exposure to angling-related stressors. J Exp Mar Biol Ecol 440:207–215

Brownscombe JW, Bower SD, Bowden W, Nowell L, Midwood JD, Johnson N, Cooke SJ (2014a) Canadian recreational fisheries: 35 years of social, biological, and economic dynamics from a national survey. Fisheries 39:251–260

Brownscombe JW, Nowell L, Samson E, Danylchuk AJ, Cooke SJ (2014b) Fishing-related stressors inhibit refuge-seeking behavior in released subadult Great Barracuda. Trans Am Fish Soc 143:613–617

China Society of Fisheries (2018) The development report of China's recreational fishery. China Fish 12:20–30

Coleman FC, Figueira WF, Ueland JS, Crowder LB (2004) The impact of United States recreational fisheries on marine fish populations. Science 305:1958–1960

Collins MR, McGovern JC, Sedberry GR, Meister HS, Pardieck R (1999) Swim bladder deflation in Black Sea bass and vermilion snapper: potential for increasing postrelease survival. N Am J Fish Manag 19:828–832

Colotelo AH, Cooke SJ (2011) Evaluation of common angling-induced sources of epithelial damage for popular freshwater sport fish using fluorescein. Fish Res 109:217–224

Cooke S, Cowx I (2004) The role of recreational fishing in global fish crises. Bioscience 54:857–859

Cooke SJ, Sneddon LU (2007) Animal welfare perspectives on recreational angling. Appl Anim Behav Sci 104:176–198

Cooke S, Suski C (2004) Are circle hooks an effective tool for conserving marine and freshwater recreational catch-and-release fisheries? Aquat Conserv Mar Freshwat Ecosyst 14:299–326
Cooke S, Suski C (2005) Do we need species-specific guidelines for catch-and-release recreational angling to effectively conserve diverse fishery resources? Biodivers Conserv 14:1195–1209
Cooke SJ, Donaldson MR, O'Connor CM, Raby GD, Arlinghaus R, Danylchuk AJ, Hanson KC, Hinch SG, Clark TD, Patterson DA, Suski CD (2013a) The physiological consequences of catch-and-release angling: perspectives on experimental design, interpretation, extrapolation and relevance to stakeholders. Fish Manag Ecol 20:268–287
Cooke SJ, Suski CD, Arlinghaus R, Danylchuk AJ (2013b) Voluntary institutions and behaviours as alternatives to formal regulations in recreational fisheries management. Fish Fish 14:439–457
Cooke SJ, Twardek WM, Lennox RJ, Zolderdo AJ, Bower SD, Gutowsky LFG, Danylchuk AJ, Arlinghaus R, Beard D (2018) The nexus of fun and nutrition: recreational fishing is also about food. Fish Fish 19:201–224
Cooke SJ, Twardek WM, Reid AJ, Lennox RJ, Danylchuk SC, Brownscombe JW, Bower SD, Arlinghaus R, Hyder K, Danylchuk AJ (2019) Searching for responsible and sustainable recreational fisheries in the Anthropocene. J Fish Biol 94:845–856
Danylchuk AJ, Adams A, Cooke SJ, Suski CD (2008) An evaluation of the injury and short-term survival of bonefish (*Albula spp.*) as influenced by a mechanical lip-gripping device used by recreational anglers. Fish Res 93:248–252
Davie P, Kopf R (2006) Physiology, behaviour and welfare of fish during recreational fishing and after release. N Z Vet J 54:161–172
Davis MW (2002) Key principles for understanding fish bycatch discard mortality. Can J Fish Aquat Sci 59:1834–1843
Dawkins MS (2017) Animal welfare with and without consciousness. J Zool 301:1–10
Diggles B (2015) Development of resources to promote best practice in the humane dispatch of finfish caught by recreational fishers. Fish Manag Ecol 23:200–207
Diggles B, Cooke S, Rose J, Sawynok W (2011) Ecology and welfare of aquatic animals in wild capture fisheries. Rev Fish Biol Fish 21:739–765
DigsFish Services Pty Ltd (2019) Humane killing of fish [Online]. http://www.ikijime.com/fish/. Accessed 08 Mar 2019
Eckroth JR, Aas-Hansen Ø, Sneddon LU, Bichão H, Døving KB (2014) Physiological and behavioural responses to noxious stimuli in the Atlantic Cod (*Gadus morhua*). PLoS One 9: e100150
EIFAC (2008) EIFAC code of practice for recreational fisheries, EIFAC. Occasional Paper No. 42, Rome, 45
FAO (2012) Recreational fisheries. FAO technical guidelines for responsible fisheries. No. 13, Rome, FAO, 176
Ferguson R, Tufts B (1992) Physiological effects of brief air exposure in exhaustively exercised rainbow trout (*Oncorhynchus mykiss*): implications for "catch and release" fisheries. Can J Fish Aquat Sci 49:1157–1162
Ferter K, Borch T, Kolding J, Vølstad JH (2013) Angler behaviour and implications for management – catch-and-release among marine angling tourists in Norway. Fish Manag Ecol 20:137–147
Ferter K, Hartmann K, Kleiven AR, Moland E, Olsen EM (2015a) Catch-and-release of Atlantic cod (*Gadus morhua*): post-release behaviour of acoustically pretagged fish in a natural marine environment. Can J Fish Aquat Sci 72:252–261
Ferter K, Weltersbach MS, Humborstad O-B, Fjelldal PG, Sambraus F, Strehlow HV, Vølstad JH (2015b) Dive to survive: effects of capture depth on barotrauma and post-release survival of Atlantic cod (*Gadus morhua*) in recreational fisheries. ICES J Mar Sci 72:2467–2481
Forskrift om utøvelse av fisket i sjøen (2013) Kapittel X. Forbud mot utkast og oppmaling [Online]. https://lovdata.no/forskrift/2004-12-22-1878/§48. Accessed 28 Mar 2017

French RP, Lyle J, Tracey S, Currie S, Semmens JM (2015) High survivorship after catch-and-release fishing suggests physiological resilience in the endothermic shortfin mako shark (*Isurus oxyrinchus*). Conserv Physiol 3:cov044

Gale MK, Hinch SG, Donaldson MR (2013) The role of temperature in the capture and release of fish. Fish Fish 14:1–33

Gould A, Grace B (2009) Injuries to barramundi *Lates calcarifer* resulting from lip-gripping devices in the laboratory. N Am J Fish Manag 29:1418–1424

Granek EF, Madin EM, Brown M, Figueira W, Cameron DS, Hogan Z, Kristianson G, de Villiers P, Williams JE, Post J (2008) Engaging recreational fishers in management and conservation: global case studies. Conserv Biol 22:1125–1134

Griffiths SP, Bryant J, Raymond HF, Newcombe PA (2017) Quantifying subjective human dimensions of recreational fishing: does good health come to those who bait? Fish Fish 18:171–184

Grixti D, Conron SD, Jones PL (2007) The effect of hook/bait size and angling technique on the hooking location and the catch of recreationally caught black bream *Acanthopagrus butcheri*. Fish Res 84:338–344

Hall K, Broadhurst M, Butcher P, Rowland S (2009) Effects of angling on post-release mortality, gonadal development and somatic condition of Australian bass *Macquaria novemaculeata*. J Fish Biol 75:2737–2755

Hannah RW, Rankin PS, Penny AN, Parker SJ (2008) Physical model of the development of external signs of barotrauma in Pacific rockfish. Aquat Biol 3:291–296

Henry NA, Cooke SJ, Hanson KC (2009) Consequences of fishing lure retention on the behaviour and physiology of free-swimming smallmouth bass during the reproductive period. Fish Res 100:178–182

Herfaut J, Levrel H, Thébaud O, Véron G (2013) The nationwide assessment of marine recreational fishing: a French example. Ocean Coast Manag 78:121–131

Hühn D, Arlinghaus R (2011) Determinants of hooking mortality in freshwater recreational fisheries: a quantitative meta-analysis. Am Fish Soc Symp 75:141–170

Hühn D, Lübke K, Skov C, Arlinghaus R (2014) Natural recruitment, density-dependent juvenile survival, and the potential for additive effects of stock enhancement: an experimental evaluation of stocking northern pike (*Esox lucius*) fry. Can J Fish Aquat Sci 71:1508–1519

Huntingford FA, Adams C, Braithwaite V, Kadri S, Pottinger T, Sandøe P, Turnbull J (2006) Current issues in fish welfare. J Fish Biol 68:332–372

Hyder K, Weltersbach MS, Armstrong M, Ferter K, Townhill B, Ahvonen A, Arlinghaus R, Baikov A, Bellanger M, Birzaks J, Borch T, Cambie G, Graaf M, Diogo HMC, Dziemian Ł, Gordoa A, Grzebielec R, Hartill B, Kagervall A, Kapiris K, Karlsson M, Kleiven AR, Lejk AM, Levrel H, Lovell S, Lyle J, Moilanen P, Monkman G, Morales-Nin B, Mugerza E, Martinez R, O'Reilly P, Olesen HJ, Papadopoulos A, Pita P, Radford Z, Radtke K, Roche W, Rocklin D, Ruiz J, Scougal C, Silvestri R, Skov C, Steinback S, Sundelöf A, Svagzdys A, Turnbull D, Hammen T, Voorhees D, Winsen F, Verleye T, Veiga P, Vølstad JH, Zarauz L, Zolubas T, Strehlow HV (2018) Recreational sea fishing in Europe in a global context – participation rates, fishing effort, expenditure, and implications for monitoring and assessment. Fish 19:225–243

ICES (2013) Report of the ICES working group on recreational fisheries surveys (2013) (WGRFS), 22–26 April 2013, Esporles, Spain. ICES CM 2013/ACOM:23

Jensvoll J (2007) Avlivning og utblødning av torsk. Master thesis, University of Tromsø, 71

Johnston FDJFD, Arlinghaus RAR, Dieckmann UDU (2010) Diversity and complexity of angler behaviour drive socially optimal input and output regulations in a bioeconomic recreational-fisheries model. Can J Fish Aquat Sci 67:1507–1531

Klefoth T, Kobler A, Arlinghaus R (2008) The impact of catch-and-release angling on short-term behaviour and habitat choice of northern pike (*Esox lucius*). Hydrobiologia 601:99–110

Klefoth T, Kobler A, Arlinghaus R (2011) Behavioural and fitness consequences of direct and indirect non-lethal disturbances in a catch-and-release northern pike (*Esox lucius*) fishery. Knowl Manag Aquat Ecosyst 11:18

Kleiven AR, Fernandez-Chacon A, Nordahl J-H, Moland E, Espeland SH, Knutsen H, Olsen EM (2016) Harvest pressure on coastal Atlantic cod (*Gadus morhua*) from recreational fishing relative to commercial fishing assessed from tag-recovery data. PLoS One 11:e0149595

Lewin W-C, Arlinghaus R, Mehner T (2006) Documented and potential biological impacts of recreational fishing: insights for management and conservation. Rev Fish Sci 14:305–367

Løkkeborg S, Siikavuopio S, Humborstad O-B, Utne-Palm A, Ferter K (2014) Towards more efficient longline fisheries: fish feeding behaviour, bait characteristics and development of alternative baits. Rev Fish Biol Fish 24:985–1003

Lorenzen K (2005) Population dynamics and potential of fisheries stock enhancement: practical theory for assessment and policy analysis. Philos Trans R Soc Lond B Biol Sci 360:171–189

Lorenzen K (2006) Population management in fisheries enhancement: gaining key information from release experiments through use of a size-dependent mortality model. Fish Res 80:19–27

Lorenzen K, Beveridge M, Mangel M (2012) Cultured fish: integrative biology and management of domestication and interactions with wild fish. Biol Rev 87:639–660

Lynch AJ, Cooke SJ, Deines AM, Bower SD, Bunnell DB, Cowx IG, Nguyen VM, Nohner J, Phouthavong K, Riley B, Rogers MW, Taylor WW, Woelmer W, Youn S-J, Beard TD (2016) The social, economic, and environmental importance of inland fish and fisheries. Environ Rev 24:115–121

Macinko S, Schumann S (2007) Searching for subsistence: in the field in pursuit of an elusive concept in small-scale fisheries. Fisheries 32:592–600

Mattilsynet (2015) Fang og slipp av marin fisk. Letter sent to the Ministry of Trade. Industry and Fisheries, Oslo

Mattilsynet (2019) Fritidsfiske og dyrevelferdsloven [Online]. https://www.mattilsynet.no/fisk_og_akvakultur/fiskevelferd/fritidsfiske_og_dyrevelferdsloven.21109. Accessed 08 Mar 2018

Meka JM (2004) The influence of hook type, angler experience, and fish size on injury rates and the duration of capture in an Alaskan catch-and-release rainbow trout fishery. N Am J Fish Manag 24:1309–1321

Meka JM, McCormick SD (2005) Physiological response of wild rainbow trout to angling: impact of angling duration, fish size, body condition, and temperature. Fish Res 72:311–322

Metcalfe JD (2009) Welfare in wild-capture marine fisheries. J Fish Biol 75:2855–2861

Muoneke MI, Childress WM (1994) Hooking mortality: a review for recreational fisheries. Rev Fish Sci 2:123–156

Nilsson J, Engstedt O, Larsson P (2014) Wetlands for northern pike (*Esox lucius* L.) recruitment in the Baltic Sea. Hydrobiologia 721:145–154

NRK (2016) Her svømmer Daniel med en kveite på 2,5 meter [Online]. https://www.nrk.no/nordland/her-svommer-daniel-med-en-kveite-pa-2_5-meter-1.13068206. Accessed 28 Mar 2017

Olson L (2003) Contemplating the intentions of anglers: the ethicist's challenge. Environ Ethics 25:267–277

Parkkila K, Arlinghaus R, Artell J, Gentner B, Haider W, Aas Ø, Barton D, Roth E, Sipponen M (2010) European inland fisheries advisory commission methodologies for assessing socio-economic benefits of European inland recreational fisheries. EIFAC Occasional Paper, 112

Pawson MG, Glenn H, Padda G (2008) The definition of marine recreational fishing in Europe. Mar Policy 32:339–350

Payer RD, Pierce RB, Pereira DL (1989) Hooking mortality of walleyes caught on live and artificial baits. N Am J Fish Manag 9:188–192

Pitcher TJ, Hollingworth CE (2002) Fishing for fun: where's the catch? Blackwell, Oxford, pp 1–16

Post JR, Sullivan M, Cox S, Lester NP, Walters CJ, Parkinson EA, Paul AJ, Jackson L, Shuter BJ (2002) Canada's recreational fisheries: the invisible collapse? Fisheries 27:6–17

Potts WM, Downey-Breedt N, Obregon P, Hyder K, Bealey R, Sauer WHH (2020) What constitutes effective governance of recreational fisheries?—A global review. Fish Fish 21: 91–103

Pullen CE, Hayes K, O'Connor CM, Arlinghaus R, Suski CD, Midwood JD, Cooke SJ (2017) Consequences of oral lure retention on the physiology and behaviour of adult northern pike (*Esox lucius* L.). Fish Res 186(Part 3):601–611

Pullen CE, Arlinghaus R, Lennox RJ, Cooke SJ (2019) Telemetry reveals the movement, fate, and lure-shedding of northern pike (*Esox lucius*) that break the line and escape recreational fisheries capture. Fish Res 211:176–182

Radford Z, Hyder K, Zarauz L, Mugerza E, Ferter K, Prellezo R, Strehlow HV, Townhill B, Lewin W-C, Weltersbach MS (2018) The impact of marine recreational fishing on key fish stocks in European waters. PLoS One 13:e0201666

Radomski PJ, Grant GC, Jacobson PC, Cook MF (2001) Visions for recreational fishing regulations. Fisheries 26:7–18

Rapp T, Cooke SJ, Arlinghaus R (2008) Exploitation of specialised fisheries resources: the importance of hook size in recreational angling for large common carp (*Cyprinus carpio* L.). Fish Res 94:79–83

Rapp T, Hallermann J, Cooke SJ, Hetz SK, Wuertz S, Arlinghaus R (2012) Physiological and behavioural consequences of capture and retention in carp sacks on common carp (*Cyprinus carpio* L.), with implications for catch-and-release recreational fishing. Fish Res 125:57–68

Rapp T, Hallermann J, Cooke SJ, Hetz SK, Wuertz S, Arlinghaus R (2014) Consequences of air exposure on the physiology and behavior of caught-and-released common carp in the laboratory and under natural conditions. N Am J Fish Manag 34:232–246

Richard A, Dionne M, Wang J, Bernatchez L (2013) Does catch and release affect the mating system and individual reproductive success of wild Atlantic salmon (*Salmo salar* L.)? Mol Ecol 22:187–200

Riepe C, Arlinghaus R (2014) Explaining anti-angling sentiments in the general population of Germany: an application of the cognitive hierarchy model. Hum Dimens Wildl 19:371–390

Roach J, Hall K, Broadhurst M (2011) Effects of barotrauma and mitigation methods on released Australian bass *Macquaria novemaculeata*. J Fish Biol 79:1130–1145

Robinson KA, Hinch SG, Raby GD, Donaldson MR, Robichaud D, Patterson DA, Cooke SJ (2015) Influence of postcapture ventilation assistance on migration success of adult sockeye salmon following capture and release. Trans Am Fish Soc 144:693–704

Salvanes AGV, Braithwaite V (2006) The need to understand the behaviour of fish reared for mariculture or restocking. ICES J Mar Sci 63:345–354

Schisler GJ, Bergersen EP (1996) Postrelease hooking mortality of rainbow trout caught on scented artificial baits. N Am J Fish Manag 16:570–578

Schwabe M, Meinelt T, Phan TM, Cooke SJ, Arlinghaus R (2014) Absence of handling-induced saprolegnia infection in juvenile rainbow trout with implications for catch-and-release angling. N Am J Fish Manag 34:1221–1226

Stålhammar M, Fränstam T, Lindström J, Höjesjö J, Arlinghaus R, Nilsson PA (2014) Effects of lure type, fish size and water temperature on hooking location and bleeding in northern pike (*Esox lucius*) angled in the Baltic Sea. Fish Res 157:164–169

Strehlow HV, Schultz N, Zimmermann C, Hammer C (2012) Cod catches taken by the German recreational fishery in the western Baltic Sea, 2005–2010: implications for stock assessment and management. ICES J Mar Sci 69:1769–1780

Sullivan CL, Meyer KA, Schill DJ (2013) Deep hooking and angling success when passively and actively fishing for stream-dwelling trout with baited J and circle hooks. N Am J Fish Manag 33:1–6

Suski C, Svec J, Ludden J, Phelan F, Philipp D (2003a) The effect of catch-and-release angling on the parental care behavior of male smallmouth bass. Trans Am Fish Soc 132:210–218

Suski CD, Killen SS, Morrissey MB, Lund SG, Tufts BL (2003b) Physiological changes in largemouth bass caused by live-release angling tournaments in southeastern Ontario. N Am J Fish Manag 23:760–769

Suski CD, Cooke SJ, Danylchuk AJ, O'Connor CM, Gravel M-A, Redpath T, Hanson KC, Gingerich AJ, Murchie KJ, Danylchuk SE (2007) Physiological disturbance and recovery

dynamics of bonefish (*Albula vulpes*), a tropical marine fish, in response to variable exercise and exposure to air. Comp Biochem Physiol A Mol Integr Physiol 148:664–673
Thompson M, Van Wassenbergh S, Rogers SM, Seamone SG, Higham TE (2018) Angling-induced injuries have a negative impact on suction feeding performance and hydrodynamics in marine shiner perch, *Cymatogaster aggregata*. J Exp Biol 221:jeb180935
Tracey SR, Hartmann K, Leef M, McAllister J (2016) Capture-induced physiological stress and postrelease mortality for Southern bluefin tuna (*Thunnus maccoyii*) from a recreational fishery. Can J Fish Aquat Sci 73:1547–1556
Tufts B, Holden J, DeMille M (2015) Benefits arising from sustainable use of North America's fishery resources: economic and conservation impacts of recreational angling. Int J Environ Stud 72:850–868
Volpato GL (2009) Challenges in assessing fish welfare. ILAR J 50:329–337
Volpato GL, Gonçalves-de-Freitas E, Fernandes-de-Castilho M (2007) Insights into the concept of fish welfare. Dis Aquat Org 75:165–171
Webster J (2005) Animal welfare: limping towards Eden. Wiley-Blackwell, Oxford, p 279
Weithman AS (1999) Socioeconomic benefits of fisheries. In: Kohler CC, Hubert WA (eds) Inland fisheries management in North America, 2nd edn. Bethesda, American Fisheries Society, pp 193–213
Weltersbach MS, Strehlow HV (2013) Dead or alive – estimating post-release mortality of Atlantic cod in the recreational fishery. ICES J Mar Sci 70:864–872
Weltersbach MS, Ferter K, Sambraus F, Strehlow HV (2016) Hook shedding and post-release fate of deep-hooked European eel. Biol Conserv 199:16–24
Wilde GR (2009) Does venting promote survival of released fish? Fisheries 34:20–28
Wilde GR, Pope KL, Durham BW (2003) Lure-size restrictions in recreational fisheries. Fisheries 28:18–26
World Bank (2012) Hidden harvest: the global contribution of capture fisheries. Report No. 66469-GLB, Washington, International Bank for Reconstruction and Development, p 152

Chapter 20
Impacts of Human-Induced Pollution on Wild Fish Welfare

Kathryn Hassell, Luke Barrett, and Tim Dempster

Abstract The natural environment has been altered by anthropogenic actions for several centuries. For example, land clearing, water diversion and abstraction for agriculture have changed aquatic ecosystems, as have inputs from various diffuse and point-source pollution sources. The alteration of natural waterbodies leads to water quality and habitat changes that ultimately impact the welfare of resident fishes and may compromise their existence. In this chapter, we review different classes of pollutants and provide key examples of impacts observed in wild fish populations from freshwater and marine environments worldwide. This includes case studies on major pollution events and key pollution sources. Impacts ranging from direct toxicity and physiological perturbations to behavioural changes and alterations in species compositions have all been documented, highlighting the need for on-going management of anthropogenic inputs to aquatic environments.

Keywords Anthropogenic pollution · Health · Oxidative stress · Sublethal stress

20.1 Introduction

Anthropogenic disturbances are a major consideration for the welfare of wild fish. Disturbances include any external forces that alter ecosystem structure, such as toxic chemical pollutants that may cause direct and/or indirect mortality or habitat alteration that compromises living spaces and resource availability. Pulse disturbances

K. Hassell (✉)
Centre for Aquatic Pollution Identification and Management (CAPIM), School of BioSciences, University of Melbourne, Melbourne, VIC, Australia

Aquatic Environmental Stress (AQUEST) Research Group, School of Science, RMIT University, Bundoora, VIC, Australia
e-mail: khassell@unimelb.edu.au

L. Barrett · T. Dempster
Sustainable Aquaculture Laboratory – Tropical and Temperate (SALTT), School of BioSciences, University of Melbourne, Melbourne, VIC, Australia

T. S. Kristiansen et al. (eds.), *The Welfare of Fish*, Animal Welfare 20,
https://doi.org/10.1007/978-3-030-41675-1_20

are short-term perturbations (e.g. accidental chemical spills or flood events), while press disturbances are longer term perturbations that remain in an altered state after the initial disturbance (e.g. barriers caused by dams or elevated sediment contaminant concentrations). Ramp disturbance occurs when the disturbance increases over time with or without an upper boundary or asymptote (e.g. increasing sedimentation in a wetland) (Lake 2000). The biotic response to each of these disturbance types can equally be categorised as pulse (short-term with return to baseline), press (longer term with new baseline/altered steady state) or ramp (increasing/decreasing response over time); and responses to the same disturbance will be different for different types of organisms. Fish inhabit diverse environments and may be exposed to press, pulse and ramp disturbances throughout their lifetime. The ability to live and reproduce in such environments is based on differences in physiology and ecology which contribute to species' resilience and ability to recover from disturbance, such as that caused by pollution.

Some fishes are capable of living in extreme conditions and exhibit evolutionary adaptations to hostile environments. For example, annual killifish (*Austrofundulus limnaeus*) inhabit ephemeral tropical ponds in South America and produce embryos that are tolerant to UV radiation, salinity changes, anoxia and desiccation (Wagner et al. 2018). The embryos can depress their metabolism in a state of diapause and accordingly have become an important vertebrate model for investigating the genetic mechanisms that enable these unique biological adaptations (Wagner et al. 2018). Arctic charr (*Salvelinus alpinus*) are the most northerly distributed freshwater fish, which inhabit polar regions with thermal extremes and up to 24 h daylight during summer and 24 h darkness in winter. This species can alter activity rhythms based on photoperiod and forages in low light conditions at temperatures below 1 °C (Hawley et al. 2017).

The above examples illustrate the diversity in the physical and behavioural needs of fish and the likely consequences of disturbances to welfare across fishes. Furthermore, what is considered acceptable in some circumstances (e.g. fish production for aquaculture) may not be acceptable in others (e.g. laboratory research). Subsequently, defining 'good fish welfare' is complex and necessarily species specific. A good general definition should incorporate good health, meeting behavioural (and social) needs and being free from pain or fear (Huntingford and Kadri 2008).

The most effective way to determine fish health and overall welfare is through regular monitoring. While this is possible for captive fish, in wild fish it is much more difficult and requires both stock assessment (to determine estimates of abundance and size/age) and biomonitoring to establish health status. This makes it particularly difficult to link specific disturbances to effects in wild fish, which is reflected in the low number of published scientific studies (Henry 2015; Hamilton et al. 2016).

In this chapter, we review the effects of human-induced pollution in aquatic environments on the welfare of wild fish, first via summarising known effects for specific pollutant classes, and then through case studies that summarise known effects within an ecosystem through time.

20.2 Part 1: Types of Pollution

Point source pollution originates from a known location, often a deliberate site of waste removal, such as a sewage outfall. Conversely, non-point source or diffuse pollution may be an unintentional waste spill comprising diverse inputs that are difficult to identify and trace back to a source. Diffuse pollution is often associated with specific land uses, such as stormwater pollution with urban land use, or pesticide and nutrient pollution with agricultural use. Aquatic pollution, regardless of the source, is a major anthropogenic disturbance with negative impacts on fish welfare globally.

20.2.1 Nutrients

Human-induced increases in nutrient loadings have led to eutrophication of numerous areas around the globe, especially in developed countries. Eutrophication is now considered one of the greatest threats to surface waters worldwide (Xu et al. 2014). Eutrophication occurs when waterways experience excessive plant growth due to nitrogen (N) and phosphorus (P) inputs, derived from sources such as fertilisers, land clearing, animal production and discharge of human and animal wastes (Cloern 2001). Direct impacts of eutrophication include selective mortality, leading to changes in community structure (selected for nutrient tolerant species), while indirect effects include causing increased turbidity and low dissolved oxygen, leading to food web alterations and eventually changes in community structure (Smith and Schindler 2009; Budria 2017).

Reduced water clarity as a result of eutrophication can affect sexual selection in species that use visual-based selection cues, such as sticklebacks, sand gobies, pipefish and cichlids (reviewed by Alexander et al. 2017). Eutrophication also influences host–parasite interactions, resulting in increases in opportunistic parasite infections in fish (Budria 2017), while Warry et al. (2018) reported negative associations between demersal species richness in juvenile fish assemblages in estuaries with the degree of catchment fertilisation.

During the twentieth century, nutrient transport to surface waters has increased dramatically, especially from agriculture. Global inputs are estimated to be 67 million ton/year N and 9 million ton/year P and contribute >50% of total input rates (Beusen et al. 2016). This is predicted to be further exacerbated with future climate change, as N loadings into river catchments from agricultural soils increase during heavy rainfall events (Jeppesen et al. 2011).

Nitrogen, in the form of ammonia (NH_4), nitrate (NO_3) and nitrite (NO_2) are all very toxic to fish at high concentrations (Handy and Poxton 1993); however, such levels greatly exceed values that are normally observed in natural environments.

20.2.2 Hypoxia

A consequence of eutrophication is increased algal blooms and associated oxygen demand when the algae decomposes (Rabalais et al. 2010). Reduced dissolved oxygen concentrations in water lead to the development of hypoxia when levels reach <2.0 ml O_2/l (~35% S; 2.9 mg/l) and severe hypoxia when levels reach <0.5 ml O_2/l (Diaz and Rosenberg 2008). Based on mean lethal concentrations for a range of fish and invertebrate species, Vaquer-Sunyer and Duarte (2008) suggested a threshold of 4.6 mg/l (~62.5% S; 3.2 ml O_2/l) should be used to define hypoxia. Regardless of which threshold is used, it remains that severely hypoxic waters limit where species can live, and globally, more than 400 systems have been identified as adversely affected (Diaz and Rosenberg 2008). The number of coastal sites affected by hypoxia is rapidly increasing at a rate of 5.5% per year, based on nearly 100 years of data (Vaquer-Sunyer and Duarte 2008). Moreover, in temperate regions there are usually seasonal patterns of hypoxia (Gobler and Baumann 2016).

There are large variations in oxygen thresholds between species, and a range of behavioural and physiological responses occur in response to hypoxia (Wu 2002; Vaquer-Sunyer and Duarte 2008; Rabalais et al. 2010). For example, some benthic species actively avoid low dissolved oxygen areas, moving into shallower, more oxygenated waters to avoid hypoxia. Physiological responses in fish range from increased ventilation rates and oxygen binding capacity of haemoglobin to downregulation of several metabolic pathways to conserve energy, before eventually shifting to anaerobic respiration (Wu 2002). There are several welfare concerns associated with these responses, including increased risk of predation (by moving into shallower water), reduced growth and fitness, as well as the potential for transgenerational and epigenetic effects on offspring (Wu 2002; Vaquer-Sunyer and Duarte 2008; Wang et al. 2016).

20.2.3 Ocean Acidification

Predictions have been made that oceanic carbon dioxide (CO_2) concentration oscillations may increase up to 10-fold by 2100 if atmospheric CO_2 emissions continue to rise (Meinshausen et al. 2011; McNeil and Sasse 2016). This would result in CO_2 levels in surface waters increasing from current estimates of around 390 μatm to values >1000 μatm (Melzner et al. 2013; McNeil and Sasse 2016). Additionally, the same processes responsible for driving hypoxia in marine systems are also implicated in ocean acidification and are likely to exacerbate effects since low dissolved oxygen and associated low pH contribute to the formation of carbonic acid and resultant acidification of waters (Gobler and Baumann 2016).

Acidification has four main effects on fish: (1) altered otolith formation at high levels of atmospheric CO_2 (Checkley et al. 2009); (2) altered extra- and intracellular acid-base status, affecting physiological processes (Melzner et al. 2009); (3) altered acid-base status, which affects energy budgets (Melzner et al. 2013); and (4) altered behaviours (e.g. preferences for salinity and temperature; Pistevos et al. 2017). Fish can compensate by adjusting their extra-cellular acid-base equilibrium, which can lead to hyper-calcification and behavioural changes (Melzner et al. 2009; McNeil and Sasse 2016).

Increased CO_2 levels in water may interfere with oxygen transport and uptake (Hannan and Rummer 2018) as well as impairing physiological functions and causing the diversion of energy away from important fitness needs (i.e. locomotion, reproduction and avoidance of predation and environmental stressors) (Brewer and Peltzer 2009). The combination of low DO and low pH causes variable but generally adverse effects on growth and survival of the early life stages of estuarine forage fishes (*Menidia beryllina*, *M. menidia*, *Cyprinodon variegatus*). Both additive and synergistic negative effects on fitness and survival may occur (DePasquale et al. 2015). In orange clownfish (*Amphiprion perula*), increased pCO_2 in seawater affected the ability of larvae to detect adult cues, and the larvae were attracted to cues that they would normally avoid (Munday et al. 2009). Acidification caused impaired sensory ability in fish larvae, which negatively affected their ability to locate suitable settlement sites. In adult fish, chronic exposure to elevated CO_2 affected reproductive output in two Australian reef fishes (Welch and Munday 2016): in *Amphiprion perula*, an increase in the number of clutches and the number of eggs per clutch were observed, while in *Acanthochromis polyacanthus*, the opposite occurred, with a decrease in the number and size of the egg clutches produced.

Clearly ocean acidification has immense potential to negatively impact welfare in wild fish populations, and given that responses are species-specific, accurate predictions of likely impacts are difficult. Furthermore, even with improved management of all anthropogenic inputs that influence eutrophication, hypoxia and acidification, future climate change will intensify the amplitude of these stressors and therefore the potential for widespread impacts on fish welfare.

20.2.4 Heavy Metals

Metal pollution is a global problem resulting from mining and industry, as well as natural geological processes. Metals enter aquatic systems predominantly through point sources such as industrial effluents, as well as diffuse sources like stormwater and urban runoff. Some metals (and metalloids) are essential, in that organisms require trace concentrations of them for normal physiological function (e.g. iron in haemoglobin to facilitate oxygen binding), while the presence of non-essential metals is often associated with stress and detoxification responses. The toxicity of metals depends on bioavailability, which is in turn influenced by the organic and inorganic complexes formed by the metals in the natural environment (Wang and Rainbow 2008). The key metals that negatively affect fish welfare are lead, mercury, cadmium, chromium, nickel, copper, zinc, tin, aluminium and arsenic.

In the widespread freshwater fish, *Galaxias maculatus*, metal pollution causes a range of impacts, including increased oxidative stress and ionoregulatory disturbances (McRae et al. 2018), alterations in behavioural responses to alarm cues (Thomas et al. 2016) and delayed embryo development and poor quality larvae with reduced phototactic responses (Barbee et al. 2014). Physiological responses to metal pollution may also be increased by additional stressors such as salinity or handling stress (Harley and Glover 2014; Glover et al. 2016). In fish early life stages, metal toxicity can delay growth and cause developmental retardation, reduced survival and increased rates of deformities. Skeletal deformities are of particular concern, since they affect essential characteristics such as the ability to swim, with subsequent effects on other behaviours such as predator avoidance and migration (Kennedy 2011; Sfakianakis et al. 2015).

Metals can accumulate in specific fish tissues, which over time can affect metabolism and cause oxidative stress, lead to cellular alterations, chromosomal damage, and potentially cause multi-generational effects (Giguère et al. 2005; Mieiro et al. 2011; Mohmood et al. 2012; Pereira et al. 2016; Defo et al. 2018), Metals may also interfere with physiological processes that can affect dominance behaviours and social interactions in fishes (Sloman 2007).

To assess food web interactions and their influence on metal bioaccumulation, 26 different fish species were sampled from the upper Yangtze River in China and the highest concentrations of metals (As, Cr, Cd, Hg, Cu, Zn, Pb, Fe) were measured in large predatory fish and benthic feeding species (Yi et al. 2017). Similarly, in perch (*Perca fluviatilis*) and roach (*Rutilis rutilis*) sampled from a Polish lake, some species-specific differences in tissue accumulation were observed and the concentration of Cu in perch livers was positively correlated with the hepatosomatic index (HSI), whilst the concentration of Hg in perch gonads was negatively correlated with the gonadosomatic index (GSI) (Luczynska et al. 2018).

In sand flathead (*Platycephalus bassensis*) sampled from a metal contaminated estuary in Australia, Fu et al. (2017) reported significant upregulation of genes associated with metal homeostasis, detoxification and oxidative stress, relative to fish sampled from a reference site. They also noted a higher prevalence of certain gill histopathologies in the fish from polluted sites, yet, the prevalence of gill parasites was lower in fish from polluted sites than the reference site (Fu et al. 2017).

Chronic exposure to metals may lead to adaptation in local populations, resulting in higher metal tolerances and altered physiological responses and bioaccumulation kinetics (Durrant et al. 2011; Hamilton et al. 2016; Abril et al. 2018). For example, populations of brown trout (*Salmo trutta*) from a British river contaminated by metals from historic mining activities were found to be genetically distinct (forming separate populations) across a very short distance (1 km) due to a 'chemical barrier' formed by the heavily polluted section of the river (Durrant et al. 2011).

20.2.5 Persistent Organic Pollutants

Persistent organic pollutants (POPs) are derived from a range of industrial and agricultural processes and are a globally important pollution class. The most well-studied and environmentally important POPs include dioxins and furans, polychlorinated biphenyls (PCBs), polycyclic aromatic hydrocarbons (PAHs), polybrominated diphenyl ethers (PBDEs) and some per- and poly-fluoroalkyl substances (PFASs). POPs persist in the environment, often bioaccumulating in exposed organisms, and can function as reproductive toxicants. Many POPs have been banned internationally under the Stockholm Convention, and several other chemicals have tight restrictions on use or phase-out strategies (UNEP 2001). Toxic responses to POPs are mostly mediated through the aryl hydrocarbon receptor (AhR) pathway (Zhou et al. 2010), and there are several well-documented cases of POPs affecting fish welfare (Cook et al. 2003; Letcher et al. 2010; Henry 2015; Akortia et al. 2016).

The most potent dioxin, 2,3,7,8-Tetrachlorodibenzo-p-dioxin (TCDD) is extremely toxic to the early life stages of fish and causes a distinctive and predictable spectrum of toxic responses, including yolk sac and pericardial oedema, haemorrhaging and vascular damage, craniofacial malformations and hyperpigmentation (Zabel et al. 1995; Walker et al. 1996; Cook et al. 2003). Other AhR agonist chemicals can cause similar effects in fish larvae, and high loadings of POPs entering Lake Ontario during the twentieth century were associated with the collapse of a Lake trout (*Salvelinus namaycush*) population (Cook et al. 2003).

Some POPs are also endocrine disruptors. For example, Baldigo et al. (2006) reported that ratios of sex steroids (17β-estradiol and 11-ketotestosterone) and vitellogenin protein were correlated with lipid-based PCB residues in the tissue of male fish of four species (carp—*Cyprinus carpio*; bass—*Micropterus salmoides* and *Micropterus dolomieui*; and bullhead—*Ameiurus nebulosus*) from the Hudson River, New York.

20.2.6 Endocrine Disrupting Chemicals

Environmental pollutants that can interfere with the normal functioning of the endocrine system are known as endocrine disrupting chemicals (EDCs). This grouping includes many different types of chemicals, such as pharmaceuticals, some metals, POPs, pesticides and plasticisers. These chemicals are capable of binding or blocking hormone receptors, which leads to upregulation or downregulation of hormone production and cascading effects within specific endocrine pathways (Colborn et al. 1993). The most well-studied systems in fish are the hypothalamic-pituitary-gonadal (HPG) axis, which regulates all aspects of reproduction, and those involved with glucose regulation and thyroid metabolism pathways (Trudeau and Tyler 2007). Fish exposed to EDCs may exhibit behavioural changes that affect sexual selection and reproductive outcomes (copulation success), as well as physiological and morphological changes.

Exposure to EDCs can have transgenerational and epigenetic impacts, leading to poor reproductive outcomes, reduced survival of offspring and lower fecundity in subsequent generations (Guerrero-Bosagna et al. 2007). Several anthropogenic pollutants are classified as EDCs, including natural and synthetic estrogens from sewage discharges, as well as pesticides, pharmaceuticals and personal care products from point source and diffuse sources (Tijani et al. 2016). A review of chemicals known to elicit endocrine-mediated adverse effects on wildlife concluded that legacy compounds such as tri-organic tins and POPs caused greater impacts than most current use compounds, with the exception of 17α-ethinylestradiol (EE2) and other sewage-associated estrogenic compounds that have been widely studied and demonstrated to cause adverse effects on fish populations (Matthiessen et al. 2018). A whole-lake experiment conducted in the Canadian Experimental Lakes Area demonstrated adverse impacts in fish chronically exposed to the potent synthetic estrogen, 17α-ethinylestradiol (EE2) (Kidd et al. 2007; Palace et al. 2009): fathead minnows (*Pimephales promelas*) displayed endocrine disruption impacts, including feminisation of male gonads, increased production of the egg yolk precursor vitellogenin and altered oogenesis in females. Over multiple years, decreased catch rates were observed, and reproductive failure led to near-complete collapse of the minnow population within 2 years of the initial exposure (Kidd et al. 2007). However, other fish species in the lake were not affected as severely (Palace et al. 2009), indicating that life history characteristics (i.e. length of life cycle) are important considerations for predicting the impacts of endocrine disruptors on wild fish populations. Subsequent studies on the same populations have indicated that recovery (of fish abundance and distribution) occurred following the cessation of EE2 exposure (Blanchfield et al. 2015).

In the United Kingdom, several years of research has also demonstrated that environmental estrogens cause reproductive alterations in fish (Tyler and Routledge 1998; Sumpter and Jobling 2013). The development of intersex gonads, where male testis develop oocytes, has been linked with decreased fertility (Harris et al. 2011) and in severe cases can result in complete sex reversal and feminisation. However, despite a large body of scientific literature from both lab and field studies, establishing population-level impacts in English rivers due to endocrine disruption has been difficult and is still not completely resolved (Sumpter and Jobling 2013).

20.2.7 Pesticides

Pesticides are an essential pest management tool for intensive agriculture globally. Several different classes of pesticides, with different modes of action (and therefore toxicity), have been developed to target specific organisms, such as insecticides for invertebrate pests and herbicides for weed species. Effects on fish welfare (in laboratory settings) are well documented, especially for older pesticides, many of which are now banned due to their known persistent, bioaccumulative and toxic properties (Jorgenson 2001). General effects of pesticides on fish are direct toxicity, sublethal stress responses (i.e. upregulation of detoxification enzymes and protective proteins) and reproductive toxicity. Furthermore, several pesticides bioaccumulate (Lazartigues et al. 2013) and some are also classified as EDCs due to demonstrated interference with hormonal systems (McKinlay et al. 2008; Brander et al. 2016).

In fishes sampled from a river in Eastern Spain, Belenguer et al. (2014) observed a significant relationship between tissue concentrations of the organophosphate insecticide, diazinon, and Fulton's condition factor, suggesting growth may be affected by pesticide exposure. The synthetic pyrethroids are a class of insecticide used widely in agricultural, veterinarian and domestic/household uses. Pyrethroids bioaccumulate in wild fish (Corcellas et al. 2015), and following a chemical spill from an industrial area in North Eastern Italy, a large fish kill involving multiple species was attributed to pyrethroid pollution (Bille et al. 2017). Whilst in Brazil, common carp (*Cyprinus carpio*) reared in an irrigated rice-farming system that uses multiple pesticides were shown to bioaccumulate certain synthetic pyrethroids and fungicides and display altered enzymatic activity in brain, liver, gills and muscle, as well as increased lipid peroxidation and protein oxidation (Clasen et al. 2018).

20.2.8 Emerging Pollutants and Problems

As populations grow and new technologies are developed globally, emerging pollutants continue to increase in the environment. Emerging pollutants are natural or synthetic chemicals that can be detected in the environment, but are not yet subject to environmental regulations. Emerging pollutants are likely to bring welfare concerns for wild fish, but in many cases, there is insufficient data on their effects.

Plastic pollution is a direct product of anthropogenic development. It is a manmade product that has been widely used for all kinds of applications since the 1940s. Since plastic does not degrade, once it enters the environment, it remains, often breaking into smaller fragments (Li et al. 2016). Marine plastic pollution has been linked with welfare issues in fish, most notably mortality due to starvation and obstruction of the digestive system by ingested plastic, as well as exposure to contaminants sorbed to ingested plastic particles (Gall and Thompson 2015; Rummel et al. 2016; Wardrop et al. 2016). In a comparative study of anthropogenic debris in fish sold for human consumption, Rochman et al. (2015) reported that 55% of all species sampled from Indonesian fish markets, and 67% of all species sampled from US fish markets, contained debris (plastic or fibres) in their digestive tracts, which included fish from various trophic levels and habitat types, including small foraging species, to large predators.

Per- and poly-fluoroalkyl substances (PFASs) are another emerging group of pollutants that may have impacts on fish welfare. PFASs are widely used in industrial applications from firefighting foams and stain resistant fabrics to non-stick frypans and water repellent clothing. They persist in the environment and bioaccumulate. Two widely used PFASs, perfluorooctanoic acid (PFOA) and perfluorooctane sulfonate (PFOS), are harmful pollutants, with phase-outs and future bans in place under the Stockholm Convention. Detectable concentrations of PFOA and PFOS in blood samples from wild eels (*Anguilla anguilla*) from two Italian waterways were associated with liver macrophage aggregates and lipid vacuolation (Giari et al. 2015), while in carp (*Cyprinus carpio*) and eels (*A. anguilla*) sampled from multiple Belgian rivers, hepatic PFOS concentrations were significantly correlated with changes in serum alanine aminotransferase activity, protein content and electrolyte levels (Hoff et al. 2005).

Each year several new chemicals are registered for use in industrial and agricultural processes, and often the properties that make them useful (i.e. long-lasting and hard-wearing) contribute to their persistence once they enter aquatic ecosystems. Therefore, prevention or minimisation of any chemicals entering waterways is the most effective way of ensuring there will be no welfare impacts on wild fish.

20.3 Part 2: Case Studies—Pollutants in the Environment and Their Impacts on Wild Fish

20.3.1 Aquaculture Pollution and Its Effects on Wild Fish Populations

Aquaculture, especially sea cage fish farming, introduces a range of environmental pollutants that affect wild fish populations. Foremost, high stocking densities produce large quantities of waste in the form of faeces and spilled feed. This localised nutrient input can lead to altered benthic communities and low dissolved oxygen conditions where water exchange is insufficient and nutrient thresholds exceed those that receiving environments can biologically assimilate (Wu et al. 1994). However, this waste also provides an attractive trophic subsidy (Dempster et al. 2002, 2009, 2011; Sanchez-Jerez et al. 2011). In both tropical and temperate systems, fish farms thus act as 'hot spots' of wild fish aggregation, with substantial increase in abundance and diversity in the near vicinity of farms. By feeding preferentially at farms, wild fish ameliorate and disperse nutrient loading (Vita et al. 2004), but also undergo dietary changes, including a shift from marine-derived long-chain polyunsaturated fatty acids to terrestrial short-chain fatty acids (Fernandez-Jover et al. 2011; Arechavala-Lopez et al. 2015). Likely effects of this dietary shift are poorly understood in wild populations (Salze et al. 2005; Bogevik et al. 2012). In addition to nutritional changes, fish that are attracted to farms by waste feed face increased risk of disease transmission from farmed to wild fish (Zlotkin et al. 1998; Diamant et al. 2000; Colorni et al. 2002; Glover et al. 2013) and are vulnerable to increased fishing pressure and predation (Bagdonas et al. 2012; Callier et al. 2017).

Contaminants from feed, antibiotics, parasiticides and anti-fouling are also possible near aquaculture sites (Burridge et al. 2010; Taranger et al. 2015). Elevated levels of mercury (2.0–2.1×: DeBruyn et al. 2006; Bustnes et al. 2011) and organohalogens (Bustnes et al. 2010) have been reported in the tissues of farm-associated wild fish, although it is unclear whether this effect is driven by contaminated feed or biomagnification due to the elevated trophic level of farm-associated fish assemblages. The effects of these levels on wild fish remain untested.

Where antimicrobials are used at fish farms, residues can appear in the tissues of wild fish (oxytetracycline: Björklund et al. 1990; oxolinic acid: Samuelsen et al. 1992; flumequine: Ervik et al. 1994), leading to selection for resistant genes that over time can increase the rate of antimicrobial resistance. Selective microbial growth can also alter the biodiversity of normal skin and gut flora, which may compromise fish immunity and reduce resilience (Cabello et al. 2013). Antimicrobial resistance is a major threat to the aquaculture industry since disease outbreaks can rapidly spread in intensive holding conditions, and without effective treatment options, infection can cause widespread mortality that can easily decimate populations (Cabello et al. 2013; Watts et al. 2017).The development of vaccines has allowed fish farmers in some countries (e.g. Norway, Scotland) to largely cease antimicrobial use, but use remains high elsewhere, such as Chile (Watts et al. 2017). The aquaculture industry is developing a range of new techniques to minimise effects of pollution, including new feed delivery technologies that reduce spillage, moving farms offshore where waste disperses better, protecting farm-associated fish from fishing, and disease control measures. Successful improvements along these fronts will minimise negative welfare impacts on wild fish.

20.3.2 *Acid Rain*

Freshwater fish populations in many regions are vulnerable to acidic deposition or 'acid rain'. Industrial emissions of sulphur dioxide and nitrogen oxide react with atmospheric moisture to form sulphuric and nitric acid, resulting in atmospheric moisture with a pH range of 3.5–5.0. Acid is then deposited by precipitation (wet deposition) or contact between the atmosphere and the earth's surface (dry deposition), before accumulating in aquatic environments via surface runoff. Long-term exposure can erode the buffering capacity of water bodies, with pH in highly vulnerable lakes falling below 5.0.

Acid rain caused fish kills in thousands of Scandinavian lakes and rivers from the 1950s until the 1990s (Hesthagen et al. 1999; Leivestad and Muniz 1976), and large-scale fish population declines were also documented throughout western Europe and North America during that period (Menz and Seip 2004). Declines in fish populations may initially be an indirect result of acidification, as invertebrates are often the first affected by changes to pH and resulting declines in prey abundance can lead to starvation for predatory fishes before direct effects occur (Schindler 1988). Early fish life stages are also more vulnerable than adults, causing recruitment failure (Alabaster and Lloyd 1982). For example, pH $<5.1–5.9$ was sufficient to reduce egg and larval survival in several Canadian freshwater fishes (Holtze and Hutchinson 1989). Fatal pH levels for adult fish vary widely across species and are strongly dependent on interacting factors such as adaptation, exposure duration, water hardness, free carbon dioxide concentration and the presence of other pollutants such as aluminium, which is mobilised by low pH (Alabaster and Lloyd 1982; Schindler 1988). Adults in acid-adapted populations may survive at least temporary exposure to pH 3.7, but more often pH <5.0 is lethal (Alabaster and Lloyd 1982). The most productive fish populations are found at pH >6.3, with lower productivity in mildly acidic waterways perhaps reflecting low food availability and sublethal effects such as physiological stress, sensory disruption and behavioural changes that are likely to reduce reproductive output. For example, reproductive behaviours in salmonids are suppressed at pH <6.4 (Ikuta et al. 2003).

Fish populations can recover once acid deposition is reduced, but there may be a lag of several years or even decades depending on the rate of return to natural pH and recovery of lower trophic levels, especially where recolonisation or restocking is necessary after local extinction (Hesthagen et al. 2011; Menz and Seip 2004; Mills et al. 2000).

20.3.3 Oil Spill Impacts on Wild Fish

Marine oil spills have become an inevitable consequence of fossil fuel extraction and refinement. The various pollutants that enter the environment as a result cause a range of toxic effects on wild fish populations. Oils are complex mixtures of components including linear hydrocarbons, polycyclic aromatic hydrocarbons (PAHs), pentacyclic hopanes, and benzene, toluene, ethylbenzene and xylene (BTEX). BTEX tend to degrade quickly in seawater, whereas PAHs are most persistent and can accumulate in sediments (Murawski et al. 2016).

In 2010, the BP Deepwater Horizon oil rig was damaged due to fire, resulting in the release of >3 million barrels of crude oil into the Gulf of Mexico (Beyer et al. 2016). The estimated extent of the contamination of coastal and continental shelf areas with oils was 144,192 km^2 and 88% of these areas had concentrations of PAHs greater than those known to cause toxic impacts in marine life (Murawski et al. 2016). Hydrocarbon concentrations in the water column were 160-fold higher than levels measured prior to 2010 and reached mean (±95% CI) levels of 104 ± 17 ppb (total hydrocarbons) and 43 ± 17 ppb (PAHs) near the surface (Murawski et al. 2016).

Measuring PAH metabolites in fish bile is a sensitive indicator of exposure to hydrocarbons, and while very high concentrations of naphthalene equivalents were measured in some fish from the Gulf of Mexico in 2011 (470,000 ng/g bile), levels have been gradually decreasing since then (Beyer et al. 2016). Similarly, reductions in the activity of several hepatic biomarkers known to be upregulated by exposure to PAHs (ethoxyresorufin-*O*-deethylase (EROD), glutathione transferase (GST) and glutathione peroxidase (GPx)) were observed in red snapper (*Lutjanus campechanus*) and gray triggerfish (*Balistes capriscus*) over a 3-year period following the Deepwater Horizon spill (Smeltz et al. 2017). Contrary to expectation, fish abundances in the Gulf of Mexico increased after the event due to fisheries closures and reduced predation (due to mortality in seabirds and other large predators). Therefore, it has been difficult to accurately gauge the impacts of the oil spill on fish assemblages, but changes in species compositions and patterns of fish recruitment in the Gulf of Mexico are beginning to emerge (Schaefer et al. 2016). For example, exceptionally high recruitment has been observed in Gulf menhaden (*Brevoortia patronus*) for multiple years following the oil spill due to a loss of predators (Short et al. 2017).

Developmental and subsequent physiological impacts are also beginning to emerge, such as malformations of the hearts of large predatory species, with associated reductions in fitness and swimming capacity (Incardona et al. 2014). Such changes are likely to have profound effects on shaping the fish communities and overall ecosystem integrity of the Gulf of Mexico in the future.

20.3.4 Radionuclide Impacts on Wild Fish

Energy production using nuclear technologies is widely utilised, and accidental releases of radioactive materials into marine ecosystems can have serious impacts on wild fish populations. In 2011, an earthquake in Japan triggered a large tsunami that flooded the Fukushima Daiichi nuclear power plant, leading to a catastrophic explosion and subsequent release of large quantities of radioactive cesium (^{134}Cs, ^{137}Cs) and other isotopes into the Pacific Ocean. The two major sources of radionuclides to the environment following the Fukushima incident were atmospheric fallout and discharge of contaminated seawater from the power plant. Groundwater and river runoff contribute additional and ongoing sources of contamination (Buesseler et al. 2017).

Radiocesium isotopes have long half-lives (^{134}Cs—2.06 years; ^{137}Cs—30.2 years) and were detected widely in surface seawater and marine biota immediately following the accident (Wada et al. 2016; Buesseler et al. 2017). Other radionuclides that were released include ^{90}Sr, ^{239,240}Pu and ^{129}I, all of which carry health concerns due to their long (>1 year) half-lives. Radionuclides can damage cells and chromosomal DNA, causing a range of adverse welfare effects. Reduced fitness, due to changes in blood composition and immunosuppression, as well as reduced reproductive output due to gonad abnormalities, reduced fertility and increases in mortality and abnormalities in fish early life stages have all been observed following chronic radiation exposure (Sazykina and Kryshev 2003; Kong et al. 2016; Hurem et al. 2018). Radioactive pollution is measured in Becquerel (Bq) units, which represent the quantity of radioactive material per unit time. Large quantities are reported as penta Becquerels (PBq, 10^{15} Bq). Estimates of the total nuclear fallout from the Fukushima accident are in the range of 8.8–50 PBq (Buesseler et al. 2017).

Cesium uptake into fish occurs through water as well as food ingestion, and Cs has moderate bioconcentration (uptake from water) and biomagnification (increase with trophic level) factors (Madigan et al. 2017). In 2011, about half of all fish sampled in coastal areas surrounding Fukushima exceeded the Japanese regulatory limit for Cs of 100 Bq/kg, and the levels tended to be higher in demersal than pelagic fishes. Within 4 years, less than 1% of samples exceeded the regulatory limits (Buesseler et al. 2017). In heavily contaminated areas within Fukushima harbour, some fishes still exceed guideline values, and as such, netting barriers have been installed to prevent these fish leaving the contaminated harbour. Reductions in fish Cs levels are occurring more slowly in demersal than pelagic species, due to ongoing exposure through feeding on contaminated benthic infauna. Continued monitoring of wild fish populations in the Fukushima region is needed, because whilst to date there has been no evidence of adverse effects, chronic radiation exposure is known to reduce fitness and reproductive output and therefore could affect fish welfare in the future.

20.4 Conclusion

Pollution and other anthropogenic disturbances affect all aspects of animal welfare, from direct toxicity due to chemicals such as heavy metals and hydrocarbons to avoidance behaviours and species alterations due to eutrophication and associated hypoxia and ocean acidification. Quantifying stress and welfare is especially difficult in wild fish and requires a versatile range of endpoints that can be measured from individual fish to entire fish communities.

To ensure fish populations are sustainable and that adequate welfare requirements are achieved, consideration needs to be given not only to existing pollution levels and suitable abatement/reduction methods, but also to the probable effects that future climate change will have on fish, in particular the altered dynamics of processes such as the nitrogen cycle and acid chemistry in both freshwaters and oceanic ecosystems.

Animal ethics legislation already assists in managing captive fish and direct human interactions with fish (i.e. fishing regulations), but establishing and maintaining appropriate welfare standards in wild fish requires close, adaptive management of not only physical disturbances (i.e. habitat loss), but of chemical pollutants and wastes from agriculture and aquaculture. These are important issues that require a global response to protect and preserve our diverse global fish populations.

References

Abril SIM, Costa PG, Bianchini A (2018) Metal accumulation and expression of genes encoding for metallothionein and copper transporters in a chronically exposed wild population of the fish *Hyphessobrycon luetkenii*. Comp Biochem Physiol C-Toxicol Pharmacol 211:25–31

Akortia E, Okonkwo JO, Lupankwa M, Osae SD, Daso AP, Olukunle OI, Chaudhary A (2016) A review of sources, levels, and toxicity of polybrominated diphenyl ethers (PBDEs) and their transformation and transport in various environmental compartments. Environ Rev 24:253–273

Alabaster JS, Lloyd RS (1982) Water quality criteria for freshwater fish, 2nd edn. Butterworths, London

Alexander TJ, Vonlanthen P, Seehausen O (2017) Does eutrophication-driven evolution change aquatic ecosystems? Philos Trans R Soc B 372:20160041

Arechavala-Lopez P, Sæther BS, Marhuenda-Egea F, Sanchez-Jerez P, Uglem I (2015) Assessing the influence of salmon farming through total lipids, fatty acids, and trace elements in the liver and muscle of wild saithe *Pollachius virens*. Mar Coast Fish 7:59–67

Bagdonas K, Humborstad O-B, Løkkeborg S (2012) Capture of wild saithe (*Pollachius virens*) and cod (*Gadus morhua*) in the vicinity of salmon farms: three pot types compared. Fish Res 134–136:1–5

Baldigo BP, Sloan RJ, Smith SB, Denslow ND, Blazer VS, Gross TS (2006) Polychlorinated biphenyls, mercury, and potential endocrine disruption in fish from the Hudson River, New York, USA. Aquat Sci 68:206–228

Barbee NC, Ganio K, Swearer SE (2014) Integrating multiple bioassays to detect and assess impacts of sublethal exposure to metal mixtures in an estuarine fish. Aquat Toxicol 152:244–255

Belenguer V, Martinez-Capel F, Masiá A, Picó Y (2014) Patterns of presence and concentration of pesticides in fish and waters of the Júcar River (Eastern Spain). J Hazard Mater 265:271–279

Beusen AHW, Bouwman AF, Van Beek LPH, Mogollon JM, Middelburg JJ (2016) Global riverine N and P transport to ocean increased during the 20th century despite increased retention along the aquatic continuum. Biogeosciences 13:2441–2451

Beyer J, Trannum HC, Bakke T, Hodson PV, Collier TK (2016) Environmental effects of the Deepwater Horizon oil spill: a review. Mar Pollut Bull 110:28–51

Bille L, Binato G, Gabrieli C, Manfrin A, Pascoli F, Pretto T, Toffan A, Pozza MD, Angeletti R, Arcangeli G (2017) First report of a fish kill episode caused by pyrethroids in Italian freshwater. Forensic Sci Int 281:176–182

Björklund H, Bondestam J, Bylund G (1990) Residues of oxytetracycline in wild fish and sediments from fish farms. Aquaculture 86:359–367

Blanchfield PJ, Kidd KA, Docker MF, Palace VP, Park BJ, Postma LD (2015) Recovery of a wild fish population from whole-lake additions of a synthetic estrogen. Environ Sci Technol 49:3136–3144

Bogevik AS, Natário S, Karlsen Ø, Thorsen A, Hamre K, Rosenlund G, Norberg B (2012) The effect of dietary lipid content and stress on egg quality in farmed Atlantic cod *Gadus morhua*. J Fish Biol 81:1391–1405

Brander SM, Gabler MK, Fowler NL, Connon RE, Schlenk D (2016) Pyrethroid pesticides as endocrine disruptors: molecular mechanisms in vertebrates with a focus on fishes. Environ Sci Technol 50:8977–8992

Brewer PG, Peltzer ET (2009) Limits to marine life. Science 324:347–348

Budria A (2017) Beyond troubled waters: the influence of eutrophication on host-parasite interactions. Funct Ecol 31:1348–1358

Buesseler K, Dai MH, Aoyama M, Benitez-Nelson C, Charmasson S, Higley K, Maderich V, Masque P, Morris PJ, Oughton D, Smith JN, Annual R (2017) Fukushima Daiichi-derived radionuclides in the ocean: transport, fate, and impacts. Annu Rev Mar Sci 9:173–203

Burridge L, Weis JS, Cabello F, Pizarro J, Bostick K (2010) Chemical use in salmon aquaculture: a review of current practices and possible environmental effects. Aquaculture 306:7–23

Bustnes JO, Lie E, Herzke D, Dempster T, Bjørn PA, Nygård T, Uglem I (2010) Salmon farms as a source of organohalogenated contaminants in wild fish. Environ Sci Technol 44:8736–8743

Bustnes JO, Nygard T, Dempster T, Ciesielski T, Jenssen BM, Bjorn PA, Uglem I (2011) Do salmon farms increase the concentrations of mercury and other elements in wild fish? J Environ Monit 13:1687–1694

Cabello FC, Godfrey HP, Tomova A, Ivanova L, Dolz H, Millanao A, Buschmann AH (2013) Antimicrobial use in aquaculture re-examined: its relevance to antimicrobial resistance and to animal and human health. Environ Microbiol 15(7):1917–1942

Callier MD, Byron CJ, Bengtson DA, Cranford PJ, Cross SF, Focken U, Jansen HM, Kamermans P, Kiessling A, Landry T, O'Beirn F, Petersson E, Rheault RB, Strand Ø, Sundell K, Svåsand T, Wikfors GH, McKindsey CW (2017) Attraction and repulsion of mobile wild organisms to finfish and shellfish aquaculture: a review. Rev Aquac 10:924. https://doi.org/10.1111/raq.12208

Checkley DM, Ayon P, Baumgartner TR, Bernal M, Coetzee JC, Emmett R, Guevara-Carrasco R, Hutchings L, Ibaibarriaga L, Nakata H, Oozeki Y, Planque B, Schweigert J, Stratoudakis Y, van der Lingen CD (2009) Habitats. Cambridge Univ Press, Cambridge

Clasen B, Loro VL, Murussi CR, Tiecher TL, Moraes B, Zanella R (2018) Bioaccumulation and oxidative stress caused by pesticides in *Cyprinus carpio* reared in a rice-fish system. Sci Total Environ 626:737–743

Cloern JE (2001) Our evolving conceptual model of the coastal eutrophication problem. Mar Ecol Prog Ser 210:223–253

Colborn T, Saal FSV, Soto AM (1993) Developmental effects of endocrine-disrupting chemicals in wildlife and humans. Environ Health Perspect 101:378–384

Colorni A, Diamant A, Eldar A, Kvitt H, Zlotkin A (2002) *Streptococcus iniae* infections in Red Sea cage-cultured and wild fishes. Dis Aquat Org 49:165–170

Cook PM, Robbins JA, Endicott DD, Lodge KB, Guiney PD, Walker MK, Zabel EW, Peterson RE (2003) Effects of aryl hydrocarbon receptor-mediated early life stage toxicity on lake trout populations in Lake Ontario during the 20th century. Environ Sci Technol 37:3864–3877

Corcellas C, Eljarrat E, Barcelo D (2015) First report of pyrethroid bioaccumulation in wild river fish: a case study in Iberian river basins (Spain). Environ Int 75:110–116

DeBruyn AMH, Trudel M, Eyding N, Harding J, McNally H, Mountain R, Orr C, Urban D, Verenitch S, Mazumder A (2006) Ecosystemic effects of salmon farming increase mercury contamination in wild fish. Environ Sci Technol 40:3489–3493

Defo MA, Bernatchez L, Campbell PGC, Couture P (2018) Temporal variations in kidney metal concentrations and their implications for retinoid metabolism and oxidative stress response in wild yellow perch (*Perca flavescens*). Aquat Toxicol 202:26–35

Dempster T, Sanchez-Jerez P, Bayle-Sempere JT, Giminez-Casualdero F, Valle C (2002) Attraction of wild fish to sea-cage fish farms in the south-western Mediterranean Sea: spatial and short-term variability. Mar Ecol Prog Ser 242:237–252

Dempster T, Uglem I, Sanchez-Jerez P, Fernandez-Jover D, Bayle-Sempere J, Nilsen R, Bjørn P (2009) Coastal salmon farms attract large and persistent aggregations of wild fish: an ecosystem effect. Mar Ecol Prog Ser 385:1–14

Dempster T, Sanchez-Jerez P, Fernandez-Jover D, Bayle-Sempere JT, Nilsen R, Bjørn PA, Uglem I (2011) Proxy measures of fitness suggest coastal fish farms can act as population sources and not ecological traps for wild gadoid fish. PLoS One 6:e15646–e15646

DePasquale E, Baumann H, Gobler CJ (2015) Vulnerability of early life stage Northwest Atlantic forage fish to ocean acidification and low oxygen. Mar Ecol Prog Ser 523:145–156

Diamant A, Banet A, Ucko M, Colorni A, Knibb W, Kvitt H (2000) Mycobacteriosis in wild rabbitfish *Siganus rivulatus* associated with cage farming in the Gulf of Eilat, Red Sea. Dis Aquat Org 39:211–219

Diaz RJ, Rosenberg R (2008) Spreading dead zones and consequences for marine ecosystems. Science 321:926–929

Durrant CJ, Stevens JR, Hogstrand C, Bury NR (2011) The effect of metal pollution on the population genetic structure of brown trout (*Salmo trutta* L.) residing in the river Hayle, Cornwall, UK. Environ Pollut 159(12):3595–3603

Ervik A, Thorsen B, Eriksen V, Lunestad BT, Samuelsen OB (1994) Impact of administering antibacterial agents on wild fish and blue mussels *Mytilus edulis* in the vicinity of fish farms. Dis Aquat Org 18:45–51

Fernandez-Jover D, Martinez-Rubio L, Sanchez-Jerez P, Bayle-Sempere JT, Lopez Jimenez JA, Martínez Lopez FJ, Bjørn P-A, Uglem I, Dempster T (2011) Waste feed from coastal fish farms: a trophic subsidy with compositional side-effects for wild gadoids. Estuar Coast Shelf Sci 91:559–568

Fu D, Bridle A, Leef M, Norte Dos Santos C, Nowak B (2017) Hepatic expression of metal-related genes and gill histology in sand flathead (*Platycephalus bassensis*) from a metal contaminated estuary. Mar Environ Res 131:80–89

Gall SC, Thompson RC (2015) The impact of debris on marine life. Mar Pollut Bull 92:170–179

Giari L, Guerranti C, Perra G, Lanzoni M, Fano EA, Castaldelli G (2015) Occurrence of perfluorooctanesulfonate and perfluorooctanoic acid and histopathology in eels from north Italian waters. Chemosphere 118:117–123

Giguère A, Campbell PGC, Hare L, Cossu-Leguille C (2005) Metal bioaccumulation and oxidative stress in yellow perch (Perca flavescens) collected from eight lakes along a metal contamination gradient (Cd, Cu, Zn, Ni). Can J Fish Aquat Sci 62:563–577

Glover KA, Sørvik AGE, Karlsbakk E, Zhang Z, Skaala Ø (2013) Molecular genetic analysis of stomach contents reveals wild Atlantic cod feeding on piscine reovirus (PRV) infected Atlantic salmon originating from a commercial fish farm. PLoS One 8:e60924

Glover CN, Urbina MA, Harley RA, Lee JA (2016) Salinity-dependent mechanisms of copper toxicity in the galaxiid fish, *Galaxias maculatus*. Aquat Toxicol 174:199–207

Gobler CJ, Baumann H (2016) Hypoxia and acidification in ocean ecosystems: coupled dynamics and effects on marine life. Biol Lett 12:20150976

Guerrero-Bosagna C, Valladares L, Gore AC (2007) Endocrine disruptors, epigenetically induced changes, and transgenerational transmission of characters and epigenetic states. In: Gore AC (ed) Endocrine-disrupting chemicals: from basic research to clinical practice. Humana, Totowa, pp 175–189

Hamilton PB, Cowx IG, Oleksiak MF, Griffiths AM, Grahn M, Stevens JR, Carvalho GR, Nicol E, Tyler CR (2016) Population-level consequences for wild fish exposed to sublethal concentrations of chemicals – a critical review. Fish 17(3):545–566

Handy RD, Poxton MG (1993) Nitrogen pollution in mariculture – toxicity and excretion of nitrogenous compounds by marine fish. Rev Fish Biol Fish 3:205–241

Hannan KD, Rummer JL (2018) Aquatic acidification: a mechanism underpinning maintained oxygen transport and performance in fish experiencing elevated carbon dioxide conditions. J Exp Biol 221:jeb154559

Harley RA, Glover CN (2014) The impacts of stress on sodium metabolism and copper accumulation in a freshwater fish. Aquat Toxicol 147:41–47

Harris CA, Hamilton PB, Runnalls TJ, Vinciotti V, Henshaw A, Hodgson D, Coe TS, Jobling S, Tyler CR, Sumpter JP (2011) The consequences of feminization in breeding groups of wild fish. Environ Health Perspect 119:306–311

Hawley KL, Rosten CM, Haugen TO, Christensen G, Lucas MC (2017) Freezer on, lights off! Environmental effects on activity rhythms of fish in the Arctic. Biol Lett 13:20170575

Henry TB (2015) Ecotoxicology of polychlorinated biphenyls in fish – a critical review. Crit Rev Toxicol 45:643–661

Hesthagen T, Sevaldrud IH, Berger HM (1999) Assessment of damage to fish populations in Norwegian lakes due to acidification. Ambio 28:112–117

Hesthagen T, Fjellheim A, Schartau AK, Wright RF, Saksgård R, Rosseland BO (2011) Chemical and biological recovery of Lake Saudlandsvatn, a formerly highly acidified lake in southernmost Norway, in response to decreased acid deposition. Sci Total Environ 409:2908–2916

Hoff PT, Van Campenhout K, de Vijver K, Covaci A, Bervoets L, Moens L, Huyskens G, Goemans G, Belpaire C, Blust R, De Coen W (2005) Perfluorooctane sulfonic acid and organohalogen pollutants in liver of three freshwater fish species in Flanders (Belgium): relationships with biochemical and organismal effects. Environ Pollut 137:324–333

Holtze KE, Hutchinson NJ (1989) Lethality of low pH and Al to early life stages of six fish species inhabiting Precambrian shield waters in Ontario. Can J Fish Aquat Sci 46:1188–1202

Huntingford FA, Kadri S (2008) Welfare and fish. In: Branson EJ (ed) Fish welfare. Blackwell, Oxford, pp 19–31
Hurem S, Gomes T, Brede DA, Mayer I, Lobert VH, Mutoloki S, Gutzkow KB, Teien HC, Oughton D, Alestrom P, Lyche JL (2018) Gamma irradiation during gametogenesis in young adult zebrafish causes persistent genotoxicity and adverse reproductive effects. Ecotoxicol Environ Saf 154:19–26
Ikuta K, Suzuki Y, Kitamura S (2003) Effects of low pH on the reproductive behavior of salmonid fishes. Fish Physiol Biochem 28:407–410
Incardona JP, Gardner LD, Linbo TL, Brown TL, Esbaugh AJ, Mager EM, Stieglitz JD, French BL, Labenia JS, Laetz CA, Tagal M, Sloan CA, Elizur A, Benetti DD, Grosell M, Block BA, Scholz NL (2014) Deepwater Horizon crude oil impacts the developing hearts of large predatory pelagic fish. Proc Natl Acad Sci USA 111:E1510–E1518
Jeppesen E, Kronvang B, Olesen JE, Audet J, Sondergaard M, Hoffmann CC, Andersen HE, Lauridsen TL, Liboriussen L, Larsen SE, Beklioglu M, Meerhoff M, Ozen A, Ozkan K (2011) Climate change effects on nitrogen loading from cultivated catchments in Europe: implications for nitrogen retention, ecological state of lakes and adaptation. Hydrobiologia 663:1–21
Jorgenson JL (2001) Aldrin and dieldrin: a review of research on their production, environmental deposition and fate, bioaccumulation, toxicology and epidemiology in the United States. Environ Health Perspect 109:113–139
Kennedy CJ (2011) The toxicology of metals in fishes. In: Farrell AP (ed) Encyclopedia of fish physiology: from genome to environment, vol 3. Academic, San Diego, pp 2061–2068
Kidd KA, Blanchfield PJ, Mills KH, Palace VP, Evans RE, Lazorchak JM, Flick RW (2007) Collapse of a fish population after exposure to a synthetic estrogen. Proc Natl Acad Sci USA 104:8897–8901
Kong EY, Cheng SH, Yu KN (2016) Zebrafish as an in vivo model to assess epigenetic effects of ionizing radiation. Int J Mol Sci 17(12):2108
Lake PS (2000) Disturbance, patchiness, and diversity in streams. J N Am Benthol Soc 19:573–592
Lazartigues A, Thomas M, Banas D, Brun-Bellut J, Cren-Olive C, Feidt C (2013) Accumulation and half-lives of 13 pesticides in muscle tissue of freshwater fishes through food exposure. Chemosphere 91:530–535
Leivestad H, Muniz IP (1976) Fish kill at low pH in a Norwegian river. Nature 259:391–392
Letcher RJ, Bustnes JO, Dietz R, Jenssen BM, Jorgensen EH, Sonne C, Verreault J, Vijayan MM, Gabrielsen GW (2010) Exposure and effects assessment of persistent organohalogen contaminants in Arctic wildlife and fish. Sci Total Environ 408:2995–3043
Li WC, Tse HF, Fok L (2016) Plastic waste in the marine environment: a review of sources, occurrence and effects. Sci Total Environ 566:333–349
Luczynska J, Paszczyk B, Luczynski MJ (2018) Fish as a bioindicator of heavy metals pollution in aquatic ecosystem of Pluszne Lake, Poland, and risk assessment for consumer's health. Ecotoxicol Environ Saf 153:60–67
Madigan DJ, Baumann Z, Snodgrass OE, Dewar H, Berman-Kowalewski M, Weng KC, Nishikawa J, Dutton PH, Fisher NS (2017) Assessing Fukushima-derived radiocesium in migratory Pacific predators. Environ Sci Technol 51:8962–8971
Matthiessen P, Wheeler JR, Weltje L (2018) A review of the evidence for endocrine disrupting effects of current-use chemicals on wildlife populations. Crit Rev Toxicol 48:195–216
McKinlay R, Plant JA, Bell JNB, Voulvoulis N (2008) Endocrine disrupting pesticides: implications for risk assessment. Environ Int 34:168–183
McNeil BI, Sasse TP (2016) Future Ocean hypercapnia driven by anthropogenic amplification of the natural CO_2 cycle. Nature 529:383
McRae NK, Gaw S, Glover CN (2018) Effects of waterborne cadmium on metabolic rate, oxidative stress, and ion regulation in the freshwater fish, inanga (*Galaxias maculatus*). Aquat Toxicol 194:1–9

Meinshausen M, Smith SJ, Calvin K, Daniel JS, Kainuma MLT, Lamarque JF, Matsumoto K, Montzka SA, Raper SCB, Riahi K, Thomson A, Velders GJM, van Vuuren DPP (2011) The RCP greenhouse gas concentrations and their extensions from 1765 to 2300. Clim Chang 109:213–241

Melzner F, Gutowska MA, Langenbuch M, Dupont S, Lucassen M, Thorndyke MC, Bleich M, Portner HO (2009) Physiological basis for high CO_2 tolerance in marine ectothermic animals: pre-adaptation through lifestyle and ontogeny? Biogeosciences 6:2313–2331

Melzner F, Thomsen J, Koeve W, Oschlies A, Gutowska MA, Bange HW, Hansen HP, Kortzinger A (2013) Future Ocean acidification will be amplified by hypoxia in coastal habitats. Mar Biol 160:1875–1888

Menz FC, Seip HM (2004) Acid rain in Europe and the United States: an update. Environ Sci Pol 7:253–265

Mieiro CL, Pereira ME, Duarte AC, Pacheco M (2011) Brain as a critical target of mercury in environmentally exposed fish (Dicentrarchus labrax)—Bioaccumulation and oxidative stress profiles. Aquat Toxicol 103:233–240

Mills KH, Chalanchuk SM, Allan DJ (2000) Recovery of fish populations in Lake 223 from experimental acidification. Can J Fish Aquat Sci 57:192–204

Mohmood I, Mieiro CL, Coelho JP, Anjum NA, Ahmad I, Pereira E, Duarte AC, Pacheco M (2012) Mercury-induced chromosomal damage in wild fish (Dicentrarchus labrax L.) Reflecting aquatic contamination in contrasting seasons. Arch Environ Contam Toxicol 63:554–562

Munday PL, Donelson JM, Dixson DL, Endo GGK (2009) Effects of ocean acidification on the early life history of a tropical marine fish. Proc R Soc B 276:3275–3283

Murawski SA, Fleeger JW, Patterson WF, Hu CM, Daly K, Romero I, Toro-Farmer GA (2016) How did the Deepwater Horizon oil spill affect coastal and continental shelf ecosystems of the Gulf of Mexico? Oceanography 29:160–173

Palace VP, Evans RE, Wautier KG, Mills KH, Blanchfield PJ, Park BJ, Baron CL, Kidd KA (2009) Interspecies differences in biochemical, histopathological, and population responses in four wild fish species exposed to ethynylestradiol added to a whole lake. Can J Fish Aquat Sci 66:1920–1935

Pereira LS, Ribas JLC, Vicari T, Silva SB, Stival J, Baldan AP, Valdez Domingos FX, Grassi MT, Cestari MM, Silva de Assis HC (2016) Effects of ecologically relevant concentrations of cadmium in a freshwater fish. Ecotoxicol Environ Saf 130:29–36

Pistevos JCA, Nagelkerken I, Rossi T, Connell SD (2017) Ocean acidification alters temperature and salinity preferences in larval fish. Oecologia 183:545–553

Rabalais NN, Diaz RJ, Levin LA, Turner RE, Gilbert D, Zhang J (2010) Dynamics and distribution of natural and human-caused hypoxia. Biogeosciences 7:585–619

Rochman CM, Tahir A, Williams SL, Baxa DV, Lam R, Miller JT, Teh FC, Werorilangi S, Teh SJ (2015) Anthropogenic debris in seafood: plastic debris and fibers from textiles in fish and bivalves sold for human consumption. Sci Rep 5:14340

Rummel CD, Loder MGJ, Fricke NF, Lang T, Griebeler EM, Janke M, Gerdts G (2016) Plastic ingestion by pelagic and demersal fish from the North Sea and Baltic Sea. Mar Pollut Bull 102:134–141

Salze G, Tocher DR, Roy WJ, Robertson DA (2005) Egg quality determinants in cod (*Gadus morhua* L.): egg performance and lipids in eggs from farmed and wild broodstock. Aquac Res 36:1488–1499

Samuelsen OB, Lunestad BT, Husevag B, Holleland T, Ervik A (1992) Residues of oxolinic acid in wild fauna following medication in fish farms. Dis Aquat Org 12:111–119

Sanchez-Jerez P, Fernandez-Jover D, Uglem I, Arechavala-Lopez P, Dempster T, Bayle-Sempere JT, Valle Pérez C, Izquierdo D, Bjørn P-A, Nilsen R (2011) Coastal fish farms as fish aggregation devices (FADs). In: Bortone SA, Brandini FP, Fabi G, Otake S (eds) Artificial reefs in fishery management. CRC, Taylor & Francis Group, Boca Raton, pp 187–208

Sazykina TG, Kryshev AI (2003) EPIC database on the effects of chronic radiation in fish: Russian/ FSU data. J Environ Radioact 68:65–87

Schaefer J, Frazier N, Barr J (2016) Dynamics of near-coastal fish assemblages following the Deepwater Horizon oil spill in the northern Gulf of Mexico. Trans Am Fish Soc 145:108–119
Schindler DW (1988) Effects of acid rain on freshwater ecosystems. Science 239:149–157
Sfakianakis DG, Renieri E, Kentouri M, Tsatsakis AM (2015) Effect of heavy metals on fish larvae deformities: a review. Environ Res 137:246–255
Short JW, Geiger HJ, Haney JC, Voss CM, Vozzo ML, Guillory V, Peterson CH (2017) Anomalously high recruitment of the 2010 Gulf Menhaden (*Brevoortia patronus*) year class: evidence of indirect effects from the Deepwater Horizon blowout in the Gulf of Mexico. Arch Environ Contam Toxicol 73:76–92
Sloman KA (2007) Effects of trace metals on salmonid fish: the role of social hierarchies. Appl Anim Behav Sci 104:326–345
Smeltz M, Rowland-Faux L, Ghiran C, Patterson WF, Garner SB, Beers A, Mievre Q, Kane AS, James MO (2017) A multi-year study of hepatic biomarkers in coastal fishes from the Gulf of Mexico after the Deepwater Horizon oil spill. Mar Environ Res 129:57–67
Smith VH, Schindler DW (2009) Eutrophication science: where do we go from here? Trends Ecol Evol 24:201–207
Sumpter JP, Jobling S (2013) The occurrence, causes, and consequences of estrogens in the aquatic environment. Environ Toxicol Chem 32:249–251
Taranger GL, Karlsen Ø, Bannister RJ, Glover KA, Husa V, Karlsbakk E, Kvamme BO, Boxaspen KK, Bjørn PA, Finstad B et al (2015) Risk assessment of the environmental impact of Norwegian Atlantic salmon farming. ICES J Mar Sci 72:997–1021
Thomas ORB, Barbee NC, Hassell KL, Swearer SE (2016) Smell no evil: copper disrupts the alarm chemical response in a diadromous fish, *Galaxias maculatus*. Environ Toxicol Chem 35:2209–2214
Tijani JO, Fatoba OO, Babajide OO, Petrik LF (2016) Pharmaceuticals, endocrine disruptors, personal care products, nanomaterials and perfluorinated pollutants: a review. Environ Chem Lett 14:27–49
Trudeau V, Tyler C (2007) Endocrine disruption. Gen Comp Endocrinol 153:13–14
Tyler CR, Routledge EJ (1998) Oestrogenic effects in fish in English rivers with evidence of their causation. Pure Appl Chem 70:1795–1804
UNEP (2001) Stockholm convention on persistent organic pollutants. United Nations Environment Programme. http://chm.pops.int
Vaquer-Sunyer R, Duarte CM (2008) Thresholds of hypoxia for marine biodiversity. Proc Natl Acad Sci USA 105:15452–15457
Vita R, Marin A, Madrid JA, Jimenez-Brinquis B, Cesar A, Marin-Guirao L (2004) Effects of wild fishes on waste exportation from a Mediterranean fish farm. Mar Ecol Prog Ser 277:253–261
Wada T, Fujita T, Nemoto Y, Shimamura S, Mizuno T, Sohtome T, Kamiyama K, Narita K, Watanabe M, Hatta N, Ogata Y, Morita T, Igarashi S (2016) Effects of the nuclear disaster on marine products in Fukushima: an update after five years. J Environ Radioact 164:312–324
Wagner JT, Singh PP, Romney AL, Riggs CL, Minx P, Woll SC, Roush J, Warren WC, Brunet A, Podrabsky JE (2018) The genome of *Austrofundulus limnaeus* offers insights into extreme vertebrate stress tolerance and embryonic development. BMC Genomics 19:155
Walker MK, Cook PM, Butterworth BC, Zabel EW, Peterson RE (1996) Potency of a complex mixture of polychlorinated dibenzo-p-dioxin, dibenzofuran, and biphenyl congeners compared to 2,3,7,8-tetrachlorodibenzo-p-dioxin in causing fish early life stage mortality. Fundam Appl Toxicol 30:178–186
Wang WX, Rainbow PS (2008) Comparative approaches to understand metal bioaccumulation in aquatic animals. Comp Biochem Physiol C 148:315–323
Wang SY, Lau K, Lai KP, Zhang JW, Tse ACK, Li JW, Tong Y, Chan TF, Wong CKC, Chiu JMY, Au DWT, Wong AST, Kong RYC, Wu RSS (2016) Hypoxia causes transgenerational impairments in reproduction of fish. Nat Commun 7:12114

Wardrop P, Shimeta J, Nugegoda D, Morrison PD, Miranda A, Tang M, Clarke BO (2016) Chemical pollutants sorbed to ingested microbeads from personal care products accumulate in fish. Environ Sci Technol 50:4037–4044

Warry FY, Reich P, Cook PLM, Mac Nally R, Woodland RJ (2018) The role of catchment land use and tidal exchange in structuring estuarine fish assemblages. Hydrobiologia 811:173–191

Watts JEM, Schreier HJ, Lanska L, Hale MS (2017) The rising tide of antimicrobial resistance in aquaculture: sources, sinks and solutions. Mar Drugs 15:158–158

Welch MJ, Munday PL (2016) Contrasting effects of ocean acidification on reproduction in reef fishes. Coral Reefs 35:485–493

Wu RSS (2002) Hypoxia: from molecular responses to ecosystem responses. Mar Pollut Bull 45:35–45

Wu RSS, Lam KS, Mackay DW, Lau TC, Yam V (1994) Impact of marine fish farming on water-quality and bottom sediment – a case-study in the subtropical environment. Mar Environ Res 38:115–145

Xu YH, Peng H, Yang YQ, Zhang WS, Wang SL (2014) A cumulative eutrophication risk evaluation method based on a bioaccumulation model. Ecol Model 289:77–85

Yi YJ, Tang CH, Yi T, Yang ZF, Zhang SH (2017) Health risk assessment of heavy metals in fish and accumulation patterns in food web in the upper Yangtze River, China. Ecotoxicol Environ Saf 145:295–302

Zabel EW, Walker MK, Hornung MW, Clayton MK, Peterson RE (1995) Interactions of polychlorinated dibenzo-p-dioxin, dibenzofuran, and biphenyl congeners for producing rainbow-trout early-life stage mortality. Toxicol Appl Pharmacol 134:204–213

Zhou HL, Wu HF, Liao CY, Diao XP, Zhen JP, Chen LL, Xue QZ (2010) Toxicology mechanism of the persistent organic pollutants (POPs) in fish through AhR pathway. Toxicol Mech Methods 20:279–286

Zlotkin A, Hershko H, Eldar A (1998) Possible transmission of *Streptococcus iniae* from wild fish to cultured marine fish. Appl Environ Microbiol 64:4065–4067

Chapter 21
What Have We Learned?

Anders Fernö, Michail A. Pavlidis, Hans van de Vis, and Tore S. Kristiansen

Abstract So, what have we learned from The Welfare of Fish book? To understand and evaluate the welfare of fishes is certainly a challenge. However, given the available information, we have in many cases been able to draw quite firm conclusions. Although some of them can certainly be modified and perhaps even refuted by some researchers, we will here skip phrases and words such as "this suggests", "it cannot be excluded", "mostly", "presumably", and the like. This is the editors' current understanding of the cognitive capacity, consciousness, and welfare of fishes after we have read the book chapters . We also suggest three important areas for research.

Keywords Fish welfare · Aquaculture · Synthesis · Cognition · Predictions · Allostasis

So, what have we learned from this book? To understand and evaluate the welfare of fishes is certainly a challenge. However, given the available information, we have in many cases been able to draw quite firm conclusions. Although some of them can certainly be modified and perhaps even refuted by some researchers, we will here skip phrases and words such as "this suggests", "it cannot be excluded", "mostly", "presumably", and the like. This is the editors' current understanding of the cognitive capacity, consciousness, and welfare of fishes.

A. Fernö (✉)
Department of Biology, University of Bergen, Bergen, Norway

M. A. Pavlidis
Department of Biology, University of Crete, Heraklion, Greece

H. van de Vis
IMARES, Wageningen University & Research Centre, Yerseke, The Netherlands

T. S. Kristiansen
Institute of Marine Research, Bergen, Norway
e-mail: torek@hi.no

T. S. Kristiansen et al. (eds.), *The Welfare of Fish*, Animal Welfare 20,
https://doi.org/10.1007/978-3-030-41675-1_21

1. The cognitive abilities of many fish species match those of other vertebrates. Fish can easily learn to associate various stimuli with resources and dangers. How fast they learn is influenced by the costs of missing an actual relationship versus the costs of acting based on a relationship that does not exist. Fish can also do more complicated things, including cooperating with and manipulating other fish, problem-solving and using tools. We know that they *can* do this and often understand *why* they do it, but we don't know *how* they do it. The mechanisms involved may be quite different from what we initially assume.
   But although the fish of one species possess a given skill, this does not mean that all fish share it. Fish show the greatest diversity in the number of species and habitats, and greatest variation in brain anatomy and brain function of all vertebrates, as well as a remarkable plasticity in their response to environmental challenges. The possession of a high mental capacity is costly. A larger brain confers cognitive benefits and can improve feeding success, mate choice, and antipredator behaviour, but at the same time, a higher metabolic rate may lower fecundity and growth rate and lead to impaired immunity if it is not compensated for. Fish occupy virtually every aquatic habitat, from tropical reefs to abyssal depths, and the mental capacity of any given species depends on the environmental and social complexity it encounters and differs not only between species but also between populations, coping styles, life stages, and sexes. Fish species faced with cognitively demanding challenges possess larger brains. Hence, we need to maintain an ecological perspective on the cognitive and learning capacities of fish.
2. The mammalian brain is not reactive but predictive, and we assume that similar predictive processing also plays a key role in the fish brain. Based on experiences from similar situations, the brain continuously attempts to predict the sensory inputs and their most probable causes. The predictions are compared to the incoming sensory signals and continuously adjusted to reduce prediction errors. This turns the view of sensory perception upside down and means that animals are proactive "predictavores" trying to stay one-step ahead of the incoming waves of sensory stimulation. Therefore, experienced welfare is dependent on what they predict and on how well they can adapt to and adjust to new situations. Fish must also be able to make sense of their environment and predict which behaviour is most likely to be adaptive.
   Predictions can be based on past experiences from seconds to many years since, and genetically based (inherited) predictions should last their entire life. Predictions can also be of various degrees of complexity. Classical conditioning results in simple predictions, but predictions could also be made in more complex situations. Predictions based on experience can have long-lasting effects, and we want to offer farmed fish an environment that makes them look at the bright side of life. Predictions should help the fish to tune their state to the future situation, but they can also trap them in situations with chronically negative expectations that can result in anorectic "looser" fish. A monotonous life will make the fish vulnerable to stress and leave them with low regulatory capacity when exposed to new challenges.

3. The available evidence suggests that fish have some kind of consciousness and experience subjective feelings, but until now this has been controversial. In agreement with European laws and regulations, we take the precautionary approach and claim that fish deserve moral consideration. In particular, we have obligations towards fish in farms, aquaria, or experimental tanks because they are dependent on our care and we are in a specific relationship to them. It is thus meaningful to improve fish welfare.
   The on-going debate about whether fish experience conscious feelings and pain is often too polarised with scientists on both sides making faith-based interpretations and generalisations based on limited available evidence. The subjective experiences of different fish species should differ. For instance, several teleost species display significant physiological and behavioural changes in response to pain, but as the importance of pain would be expected to differ based on the range of options the fish have in nature, we would expect different species to perceive pain differently. We should join forces and seek some kind of synthesis and ask ourselves what kind of consciousness a fish species experience, and the kind and degree of pain it experiences.
4. Not all fish are suitable for traditional farming, capture-based aquaculture, or public aquaria. The species have different "personalities" and can be classified from adaptive/flexible/plastic species to non-flexible species. Even if we try to adapt the farming environment to the requirements of a particular species, there will be periods with suboptimal conditions. The ability to adapt to a suboptimal environment is influenced by the temporal and spatial environmental variability that a species experiences in its natural environment. Furthermore, fish evolved in systems with restricted possibilities of finding better conditions should experience less dissatisfaction with suboptimal farming and aquarium conditions ("stoic species"). In multi-species assemblages in public aquaria, it is also crucial that we choose species that fit well together; otherwise an aggressive species could stress and injure individuals of other species.
   Even within a species, there are different "personality" types. The actions that comprise the proactive coping style are based, to a greater degree, on the predictions of the environment than the reactive coping style, and the predictable conditions in intensive rearing systems favour risk-taking proactive individuals. Even so, proactive individuals are more aggressive and have a stronger tendency to follow routines that are a disadvantageous if the conditions change, and even if proactive individuals do better in mixed groups, this does not necessarily mean than they do better in single-species groups.
5. Farmed fish are not evolutionarily adapted to the confined environment and extremely high densities in tanks and sea cages. There has been little domestication of most aquaculture species, and cultured fish retain most of the natural behaviour of their wild counterparts. Some problems arise because fish in culture display natural responses in circumstances where this is inappropriate, and others because they fail to show natural responses when to do so would be more appropriate. To promote welfare in captive environments, it is essential that housing, handling, and routine care are tailored to the needs of each species,

and environmental enrichment and variability can mimic some of the features of the natural habitat. In the course of time, domestication will affect not only growth rate and disease resistance but also agonistic behaviour and social communication. Lessons learned from terrestrial farmed animals ought therefore to be taken into account when designing genetic selection breeding programmes for commercial fish species.

6. The environment that farmed fish experience changes certain aspects of the brain, physiology, and behaviour and influences which adult phenotypes that are expressed. Learning about man-made objects and the procedures employed in aquaculture is essential for welfare, and fish can even learn to react positively to initially aversive events. However, a farming environment that is poor in critical stimuli may impair the capacity to learn new things, and the high fish densities and frequent disturbances are cognitively demanding. Unpredictable chronic stress reduces learning capacity, and social behaviour can override strong incentives such as food. Environmental enrichment can improve the cognitive capacity.
7. The core task of the brain is not to sustain homeostasis but rather to regulate the body budget and functions efficiently by a balancing act known as allostasis. The brain constantly monitors large numbers of external and internal parameters to anticipate changing needs, evaluate priorities, and prepare the organism to satisfy them before they lead to errors. Physiological stress is a natural response to challenges and it enables fish to take effective countermeasures, but under intensive culture conditions, prolonged stress in response to real or perceived threats can have adverse effects on fish performance and welfare. Active responses such as avoidance and aggression are adaptive when stressors are mild, predictable, and of short duration and do not necessarily impair welfare. However, stress-induced long-term behavioural inhibition can result in pathological situations, depression-like states, and impaired welfare.
8. As we believe that fish experience emotions and can suffer, we claim their life should be made as pleasant as possible. But this does not mean that we must avoid all situations in which fish might experience brief periods of impaired welfare. In a conflict situation, we should within certain limits prioritise the needs of our species, trading-off costs and benefits for fish and humans via a utilitarian approach (Huntingford et al. 2006). Nevertheless, it is of the utmost importance that we maintain a certain standard with regard to the factors that influence the welfare of fish in farms, aquaria, and experimental research, as well as when we capture wild fish. We should make a prioritised list of the most and least important factors for each species and aquaculture and fishing practices and make serious efforts to solve one problem after another. In cases where the fish suffer from unacceptable standards of welfare, and it is not possible in practice to improve them, we should completely ban certain practices or uses of an area for fish farming.

   However, we should only try to remove stressors that impair welfare up to a certain point. In the prioritised list with the most important problem at the top, there will be diminishing returns as we work through the less serious problems.

Somewhere along this line, we will have to stop. Fish will not experience poor welfare if they are exposed to moderate challenges that they can cope with. As most species have, to a certain extent, evolved to work for benefits, individuals living in an environment with too little challenges could even experience worse welfare than fish that are challenged from time to time. "Stress is the salt of life" (Selye 1976). We could thus add another point to The Five Freedoms for ensuring animal welfare (UK Farm Animal Welfare Council 1995)—The Sixth Freedom: "Freedom from absolute freedom". As wild fish species have to compete in nature for their survival, we should not worry too much if farmed fish are occasionally challenged, provided that the situation soon returns to normal. Exposing fish to an environment to which they cannot adapt and that results in maldevelopment, chronic stress or injury is quite another matter.

9. We must continue our search for operational indicators of welfare and identify relevant and practical measures. The capacity of fish to learn opens new ways to assess welfare. Some indicators are more interesting from the scientific than the practical point of view. An underrated indicator is the crude welfare indicator mortality rate. Mortality is the ultimate expression of "bad welfare". Of course, we cannot claim that low mortality automatically means good welfare, but once we exclude short-term events like technical breakdowns and predatory events, high mortality is definite proof that the fish that die have been living in a situation with unacceptable welfare and also indicates near-lethal conditions for the surviving individuals. We therefore suggest that a maximum rate of overall mortality should be set. If this value is repeatedly exceeded in a fish farm, robust action must be taken.
10. Fish face many challenges in their natural environment too, and human impacts do not make their life easier. The way in which fish categorise objects is preadapted and tailored to their ecological niche, and wild fish that encounter fishing gear often categorise the novel object incorrectly and are caught. Anthropogenic environmental changes may bring fish close to their tolerance level and have negative effects on cognition and welfare.

Fish that are captured by fishing gear are only exposed to a critical situation for a short time. This is equally the case for fish that are used for human consumption and those that are caught and discarded or released. But the welfare can still be impaired during stressful events, and fish that are released or escape from gear may also suffer from negative physical and psychological long-term effects. Should we accept this from a welfare perspective? When we also consider other important issues such as food security, human nutrition, and economical aspects, the answer must be yes. But acceptance of this demands that the fish are exposed to as little stress and suffering as possible during the catch phase. Simple modifications of fishing practices such as smaller catch volumes can improve the welfare of fish in the retained catch. Acute stress prior to slaughter reduces flesh quality, and establishing an empirical link between good welfare and product quality would be an important motivation for improving welfare. Ethical harvesting practices could be encouraged through product certification schemes, for instance based on a maximum time on deck before the fish lose consciousness and die.

Catch-and-release is more complicated. Can only the pleasure of fishing justify exposing fish to harmful events? Whether catch-and-release is ethically acceptable depends on personal values. But once we take into account the positive aspects of recreational fishing-related activities, such as anglers fulfilling a role as watchdogs of the environment and engaging in the rehabilitation of natural habitats and conservation of fish populations, catch-and-release may be acceptable provided that fishing is performed in a responsible way that minimises physical injury and stress with negligible long-term negative effects. However, similar practice targeting birds and mammals would not be acceptable in the public opinion, so we may be "culturally blind" when we accept this for fish.

11. As is often claimed: "We need more research". But the research needs to focus on the bottlenecks that most severely limit the welfare of fish in the situations to which we expose them. We suggest three important areas of research:

    (i) Subjective experiences
    Although we will never have a complete understanding how fish experience their world, highly focused experiments on the kind and degree of consciousness that fish possess should help us to make more informed decisions on welfare issues. Furthermore, research on fish will improve our understanding of higher organisms such as the mammals we exploit and could even help us to better understand our own species.

    (ii) The role of predictability and controllability
    The actions of fish are driven by predictions, so unless we understand the role played by predictions, we will be unable to comprehend why a fish is doing what it does. Questions on how predictions lead to a selective choice of sensory inputs from the environment may be more concrete and testable than questions about consciousness. So perhaps we should stop asking to what degree fish are conscious and instead try to find out how smart they need to be to make the correct predictions and decisions? If a fish can predict what to do in complex situations and act adaptively taking into account its state and various other interrelated factors, this should tell us something about their state of consciousness. How deeply rooted experience-based predictions are and how prediction errors modify subsequent predictions should also be further studied, as well as the role of predictions in proactive versus reactive fish. Finally, we should try to establish the optimal degree of predictability in farming systems.

    (iii) Ranking different stressors
    If we are to remove the situations that impair welfare the most, it is essential to rank different stressors. Although it has been observed that fish will pay a cost to accessing analgesia and sacrifice access to a favourable area to obtain pain relief (Sneddon 2015), it has, to the best of our knowledge, not been specifically addressed whether fish experience

chronic stress as worse than acute, intensive stress. So far we have assumed that long-lasting stress is the worst, but it may be that they can get used to suboptimal situations. By enabling fish to choose between different situations, it would be possible to rank the severity of different stressors. Chronic stress of moderate level (e.g. reduced oxygen level) could be compared with acute high-level stress by training fish to escape from the suboptimal situation by passing a barrier where they are exposed to acute stress (e.g. electric shock). This would enable us to titrate different kinds of stressors against each other. Such knowledge would enable us to modify farming conditions and fishing techniques in order to improve fish welfare.

We need more fundamental knowledge about all species, especially the many species about which we know very little. However, given the enormous diversity of fish, it is unrealistic to characterise. The cognitive and adaptive capacities of all, at least within the foreseeable future. Besides, individuals of all species have moral value and integrity and ought to be treated with respect, and to be protected against unnecessary, harmful human impact.

References

Huntingford FA, Adams C, Braithwaite VA, Kadri S, Pottinger TG, Sandoe P, Turnbull JF (2006) Current issues in fish welfare. J Fish Biol 68:332–372

Selye H (1976) Forty years of stress research: principal remaining problems and misconceptions. Can Med Assoc J 115:53–56

Sneddon LU (2015) Pain in aquatic animals. J Exp Biol 218:967–976

UK Farm Animal Welfare Council (FAWC) (1995) Five freedoms of the farm animal welfare council. http://webarchive.nationalarchives.gov.uk/20121007104210/http:/www.fawc.org.uk/freedoms.htm. Accessed 20 February 2020

GPSR Compliance
The European Union's (EU) General Product Safety Regulation (GPSR) is a set of rules that requires consumer products to be safe and our obligations to ensure this.

If you have any concerns about our products, you can contact us on

ProductSafety@springernature.com

In case Publisher is established outside the EU, the EU authorized representative is:

Springer Nature Customer Service Center GmbH
Europaplatz 3
69115 Heidelberg, Germany

www.ingramcontent.com/pod-product-compliance
Ingram Content Group UK Ltd.
Pitfield, Milton Keynes, MK11 3LW, UK
UKHW021832270726
14058UKWH00001B/110
9783030416775